Applications and Properties of Magnetic Nanoparticles

Applications and Properties of Magnetic Nanoparticles

Editor

Paolo Arosio

MDPI • Basel • Beijing • Wuhan • Barcelona • Belgrade • Manchester • Tokyo • Cluj • Tianjin

Editor
Paolo Arosio
Physics Department
Università degli Studi
di Milano
Milano
Italy

Editorial Office
MDPI
St. Alban-Anlage 66
4052 Basel, Switzerland

This is a reprint of articles from the Special Issue published online in the open access journal *Nanomaterials* (ISSN 2079-4991) (available at: www.mdpi.com/journal/nanomaterials/special_issues/appl_magnetic).

For citation purposes, cite each article independently as indicated on the article page online and as indicated below:

LastName, A.A.; LastName, B.B.; LastName, C.C. Article Title. *Journal Name* **Year**, *Volume Number*, Page Range.

ISBN 978-3-0365-6208-7 (Hbk)
ISBN 978-3-0365-6207-0 (PDF)

© 2023 by the authors. Articles in this book are Open Access and distributed under the Creative Commons Attribution (CC BY) license, which allows users to download, copy and build upon published articles, as long as the author and publisher are properly credited, which ensures maximum dissemination and a wider impact of our publications.

The book as a whole is distributed by MDPI under the terms and conditions of the Creative Commons license CC BY-NC-ND.

Contents

About the Editor vii

Paolo Arosio
Applications and Properties of Magnetic Nanoparticles
Reprinted from: *Nanomaterials* 2021, *11*, 1297, doi:10.3390/nano11051297 1

Ulrich M. Engelmann, Ahmed Shalaby, Carolyn Shasha, Kannan M. Krishnan and
Hans-Joachim Krause
Comparative Modeling of Frequency Mixing Measurements of Magnetic Nanoparticles Using
Micromagnetic Simulations and Langevin Theory
Reprinted from: *Nanomaterials* 2021, *11*, 1257, doi:10.3390/nano11051257 5

Hong Jae Cheon, Quynh Huong Nguyen and Moon Il Kim
Highly Sensitive Fluorescent Detection of Acetylcholine Based on the Enhanced
Peroxidase-Like Activity of Histidine Coated Magnetic Nanoparticles
Reprinted from: *Nanomaterials* 2021, *11*, 1207, doi:10.3390/nano11051207 21

Alexander Omelyanchik, Valentina Antipova, Christina Gritsenko, Valeria Kolesnikova,
Dmitry Murzin and Yilin Han et al.
Boosting Magnetoelectric Effect in Polymer-Based Nanocomposites
Reprinted from: *Nanomaterials* 2021, *11*, 1154, doi:10.3390/nano11051154 33

Mohamed S. A. Darwish, Hohyeon Kim, Minh Phu Bui, Tuan-Anh Le, Hwangjae Lee and
Chiseon Ryu et al.
The Heating Efficiency and Imaging Performance of Magnesium Iron Oxide@tetramethyl
Ammonium Hydroxide Nanoparticles for Biomedical Applications
Reprinted from: *Nanomaterials* 2021, *11*, 1096, doi:10.3390/nano11051096 55

Oksana Petrichenko, Aiva Plotniece, Karlis Pajuste, Martins Rucins, Pavels Dimitrijevs and
Arkadij Sobolev et al.
Evaluation of Physicochemical Properties of Amphiphilic 1,4-Dihydropyridines and
Preparation of Magnetoliposomes
Reprinted from: *Nanomaterials* 2021, *11*, 593, doi:10.3390/nano11030593 69

Vitalij Novickij, Ramunė Stanevičienė, Rūta Gruškienė, Kazimieras Badokas, Juliana Lukša
and Jolanta Sereikaitė et al.
Inactivation of Bacteria Using Bioactive Nanoparticles and Alternating Magnetic Fields
Reprinted from: *Nanomaterials* 2021, *11*, 342, doi:10.3390/nano11020342 87

Kiran Reddy Kanubaddi, Pei-Yu Huang, Ya-Lin Chang, Cheng Hsin Wu, Wei Li and Ranjith
Kumar Kankala et al.
Deviation of Trypsin Activity Using Peptide Conformational Imprints
Reprinted from: *Nanomaterials* 2021, *11*, 334, doi:10.3390/nano11020334 101

Saeko Mizoguchi, Masamitsu Hayashida and Takeshi Ohgai
Determination of Cobalt Spin-Diffusion Length in Co/Cu Multilayered Heterojunction
Nanocylinders Based on Valet–Fert Model
Reprinted from: *Nanomaterials* 2021, *11*, 218, doi:10.3390/nano11010218 115

Marc Fuhrmann, Anna Musyanovych, Ronald Thoelen, Sibylle von Bomhard and Hildegard Möbius
Magnetic Imaging of Encapsulated Superparamagnetic Nanoparticles by Data Fusion of Magnetic Force Microscopy and Atomic Force Microscopy Signals for Correction of Topographic Crosstalk
Reprinted from: *Nanomaterials* **2020**, *10*, 2486, doi:10.3390/nano10122486 129

Francesca Brero, Martina Basini, Matteo Avolio, Francesco Orsini, Paolo Arosio and Claudio Sangregorio et al.
Coating Effect on the ^{1}H—NMR Relaxation Properties of Iron Oxide Magnetic Nanoparticles
Reprinted from: *Nanomaterials* **2020**, *10*, 1660, doi:10.3390/nano10091660 141

Nikolai A. Usov and Elizaveta M. Gubanova
Application of Magnetosomes in Magnetic Hyperthermia
Reprinted from: *Nanomaterials* **2020**, *10*, 1320, doi:10.3390/nano10071320 155

Ashish Chhaganlal Gandhi, Tai-Yue Li, B. Vijaya Kumar, P. Muralidhar Reddy, Jen-Chih Peng and Chun-Ming Wu et al.
Room Temperature Magnetic Memory Effect in Cluster-Glassy Fe-Doped NiO Nanoparticles
Reprinted from: *Nanomaterials* **2020**, *10*, 1318, doi:10.3390/nano10071318 169

Alexander Omelyanchik, María Salvador, Franco D'Orazio, Valentina Mameli, Carla Cannas and Dino Fiorani et al.
Magnetocrystalline and Surface Anisotropy in $CoFe_2O_4$ Nanoparticles
Reprinted from: *Nanomaterials* **2020**, *10*, 1288, doi:10.3390/nano10071288 183

Silvio Dutz, Norbert Buske, Joachim Landers, Christine Gräfe, Heiko Wende and Joachim H. Clement
Biocompatible Magnetic Fluids of Co-Doped Iron Oxide Nanoparticles with Tunable Magnetic Properties
Reprinted from: *Nanomaterials* **2020**, *10*, 1019, doi:10.3390/nano10061019 195

Thomas Vangijzegem, Dimitri Stanicki, Adriano Panepinto, Vlad Socoliuc, Ladislau Vekas and Robert N. Muller et al.
Influence of Experimental Parameters of a Continuous Flow Process on the Properties of Very Small Iron Oxide Nanoparticles (VSION) Designed for T_1-Weighted Magnetic Resonance Imaging (MRI)
Reprinted from: *Nanomaterials* **2020**, *10*, 757, doi:10.3390/nano10040757 215

Mohamed S. A. Darwish, Hohyeon Kim, Hwangjae Lee, Chiseon Ryu, Jae Young Lee and Jungwon Yoon
Synthesis of Magnetic Ferrite Nanoparticles with High Hyperthermia Performance via a Controlled Co-Precipitation Method
Reprinted from: *Nanomaterials* **2019**, *9*, 1176, doi:10.3390/nano9081176 233

Yaofei Chen, Weiting Sun, Yaxin Zhang, Guishi Liu, Yunhan Luo and Jiangli Dong et al.
Magnetic Nanoparticles Functionalized Few-Mode-Fiber-Based Plasmonic Vector Magnetometer
Reprinted from: *Nanomaterials* **2019**, *9*, 785, doi:10.3390/nano9050785 255

About the Editor

Paolo Arosio

Paolo Arosio is currently an associate professor at the Physics Department of Università degli Studi di Milano.

His research activity regards the magnetism of magnetic materials and morpho-dimensional studies on biological samples in the magnetism and nanometric systems group of the Physics Department, which collaborates with numerous Italian and European research groups. Since 2011, P.A. has been responsible for experimental activity on NMR and magnetic hyperthermia in his research group.

Editorial

Applications and Properties of Magnetic Nanoparticles

Paolo Arosio [1,2,3]

1 Dipartimento di Fisica and INFN, Università degli Studi di Milano, Via Celoria 16, 20133 Milan, Italy; paolo.arosio@unimi.it
2 Consorzio Interuniversitario Nazionale per la Scienza e Tecnologia dei Materiali–Milano Unit, Via Celoria 16, 20133 Milan, Italy
3 INFN, Istituto Nazionale di Fisica Nucleare–Milano Unit, Via Celoria 16, 20133 Milan, Italy

Citation: Arosio, P. Applications and Properties of Magnetic Nanoparticles. *Nanomaterials* 2021, *11*, 1297. https://doi.org/10.3390/nano11051297

Received: 6 May 2021
Accepted: 13 May 2021
Published: 14 May 2021

Publisher's Note: MDPI stays neutral with regard to jurisdictional claims in published maps and institutional affiliations.

Copyright: © 2021 by the author. Licensee MDPI, Basel, Switzerland. This article is an open access article distributed under the terms and conditions of the Creative Commons Attribution (CC BY) license (https://creativecommons.org/licenses/by/4.0/).

In the last few decades, magnetic nanoconstructs have attracted increasing attention due to, among others, their specific magnetic properties and huge number of applications in completely different fields. The purpose of this Special Issue is to cover the new developments in synthesis, characterization, and fundamental studies of magnetic nanoconstructs, ranging from conventional metal oxide nanoparticles, to novel molecule-based or hybrid multifunctional nano-objects. At the same time, this Special Issue is intended to focus on and explore the potential of these novel magnetic nanoconstructs in nanomedicine and biology, energy harvesting and storage applications, sensing applications, pollution remediation, data storage, and several other possible applications.

Without claiming to be exhaustive in such a broad research field, this Special Issue provides at least a partial snapshot of the state-of-the-art magnetic nanomaterial studies, showing several recent advances, new ideas, and open problems.

The collection includes 17 research articles by authors from 18 countries, all over the world, which shows once more the global scientific interest in the magnetism of nano-objects and its crucial role in many new technological areas.

The original articles of this Special Issue cover several scientific themes and arguments associated with synthesis, characterization, and the use of magnetic nanoconstructs. The topics covered by the collection are summarized here below.

Engelmann et al. studied a dual frequency magnetic excitation of magnetic nanoparticles for biosensing applications [1], comparing their experimental results with a thermal equilibrium model, based on Langevin function and a micromagnetic Monte-Carlo (MC)-simulation method that models the non-equilibrium dynamics of nanoparticles.

Cheon et al. [2] proposed new magnetic nanoparticles, functionalized with histidine as nano-enzymes for the detection of acetylcholine, with high specificity and sensitivity.

In Ref. [3] Omelyanchik and co-workers focused on the preparation and study of polymer-based nanocomposites that include magnetic nanoparticles with intriguing piezoelectric properties. The obtained results suggested a possible future application of these magnetoelectric composites as bioactive surfaces for guiding the differentiation of neuronal stem cells.

Hydrophilic magnesium iron oxide nanoparticles, presented by Darwish et al. exhibited high heating efficiency as hyperthermic system and a good Magnetic Particle Imaging (MPI) signal with adequate spatial resolution [4]. Bringing together these physical properties, the proposed magnetic nanomaterials could be interesting for various biomedical applications, such as image-guided cancer treatment.

Petrichenko and co-workers produced magnetoliposomes starting from 1,4-dihydropyridines with different substituents and demonstrated that the properties of liposomes did not change varying the length of alkyl chain or changing substituents at position 4 of 1,4-dihydropyridines [5].

Novickij et al. used nisin-loaded magnetic nanoparticles stimulated by an alternating magnetic field as a method for the treatment of drug-resistant foodborne pathogens and, in particular, they showed its efficacy to decrease the resistance to the drug of *Listeria innocua* bacteria [6].

The immobilization of porcine pancreatic alpha-trypsin (PPT) by means of molecularly imprinted polymers, fabricated on magnetic nanoparticles, was obtained by Kanubaddi et al. The authors demonstrated that immobilized enzyme maintains its catalytic activity also after several cycles of use, paving the way for a new method of point immobilization of enzymes [7].

Co/Cu multi-layered nanocylinders, electrodeposited into anodized aluminium oxide nanochannels, were synthesized and characterized by Mizoguchi and co-workers [8]. The nanocylinders showed high current-in-plane giant magnetoresistance performances and a cobalt spin–diffusion length that can be described by Valet–Fert equation.

Fuhrmann et al. proposed a novel approach for the magnetic imaging of superparamagnetic nanoparticles encapsulated in a polymer matrix, using data fusion of Magnetic Force and Atomic Force Microscopies [9]. The numerical method developed by authors allowed us to minimize topographic crosstalk and consequently, to characterize local magnetic properties of nanoparticles.

Meanwhile, Brero and co-workers discussed the effect of negative polyelectrolyte coatings on the magnetic resonance properties of maghemite nanoparticles with two core dimensions [10]. The authors stated that the chosen nanoparticles are promising superparamagnetic T_2 contrast agents for MRI, but the polyelectrolyte coatings poorly influenced their longitudinal and transverse relaxometric properties.

In Ref. [11], Usov and Gubanova used a theoretical approach based on the stochastic Landau–Lifshitz equation to investigate which are the optimal physical characteristics of magnetosome chains, used as heat mediator in magnetic hyperthermia.

Room-temperature magnetic memory effect from Fe-doped NiO nanoparticles, synthesized and characterized by Gandhi et al. was correlated to the formation of defect clusters that enhanced intraparticle interactions. Consequently, the magnetic anisotropy of the nanoparticles changed, causing a different behaviour of magnetic moments relaxation processes that could be intriguing for the design of future spintronic devices [12].

Omelyanchik et al. deeply studied the magnetic properties of nanocomposites consisting of Co–ferrite nanoparticles, dispersed in a SiO_2 matrix. The results presented by authors showed that the annealing temperature changed the magnetocrystalline anisotropy and the saturation magnetization of the nanocomposite, modifying the particle size and also influencing the magnetic disorder of their surface [13].

Furthermore, Dutz and co-workers presented Co-doped iron oxide nanoparticles with tunable magnetic properties. The most promising particle–magnetic field combination seemed to be suitable for application in magnetic fluid hyperthermia, given that cell viability analysis also did not show an increased toxicity for the Co-doped particles, compared to the pure iron oxide ones [14].

Vangijzegem et al. proposed a continuous flow system preparation of very small iron oxide nanoparticles by thermal decomposition. Studying the influence of experimental parameters (e.g., ligand concentration and nature, temperature, etc.) on the nanoparticles magneto-crystalline and relaxometric properties, the authors showed that they had incidence on relaxometric and magnetic properties, while they did not affect size properties of the nanoparticles [15].

Darwish et al. reported interesting results on magnetic ferrite nanoparticles, with high performances again for the magnetic hyperthermia "on fire" topic [16]. In particular, they compared cobalt–ferrite (8 ± 2 nm) and zinc–cobalt ferrite (25 ± 5 nm) nanoparticles coated with sodium citrate, evaluating their heat release, and testing their cytocompatibility.

In the last article, Chen et al. developed a plasmonic vector magnetometer, based on a side-polished few-mode-fiber, functionalized by MNPs [17]. The optical-fiber magnetic sensor exhibited high sensitivity to both the magnetic-field intensity and orientation. Therefore, considering its compact size and online detection scheme, it could be used in the future for the detection of weak magnetic-field vectors.

In conclusion, this Special Issue of *Nanomaterials* collects a series of original research articles of front-line researchers, with an interdisciplinary approach on exploring the use of magnetic nano-objects in a broad range of applications. I am confident that this Special Issue will provide the reader with an overall view of the latest prospects in this fast evolving and cross-disciplinary field.

Funding: P.A. acknowledges the financial support from Dipartimento di Fisica, Università degli Studi di Milano, Italy.

Acknowledgments: The Guest Editor thanks all the authors for submitting their work to this Special Issue and for contributing to its successful completion. A special thank you belongs to all the reviewers participating in the peer-review process of the submitted manuscripts for enhancing their quality and impact. I am also grateful to Cassie Zhang for her precious, extremely professional, efficient and helpful work during the months of the Special Issue. Lastly, I thank the editorial assistants who made the entire Special Issue creation a smooth and efficient process.

Conflicts of Interest: The authors declare no conflict of interest.

References

1. Engelmann, U.M.; Shalaby, A.; Shasha, C.; Krishnan, K.M.; Krause, H.J. Comparative modeling of frequency mixing measurements of magnetic nanoparticles using micromagnetic simulations and Langevin theory. *Nanomaterials* **2021**, *11*, 1257. [CrossRef]
2. Cheon, H.J.; Nguyen, Q.H.; Kim, M.I. Highly sensitive acetylcholine detection based on the enhanced peroxidase-like activity of histidine coated magnetic nanoparticles. *Nanomaterials* **2021**, *11*, 1207. [CrossRef]
3. Omelyanchik, A.; Antipova, V.; Gritsenko, C.; Kolesnikova, V.; Murzin, D.; Han, Y.; Turutin, A.V.; Kubasov, I.V.; Kislyuk, A.M.; Ilina, T.S.; et al. Boosting Magnetoelectric Effect in Polymer-Based Nanocomposites. *Nanomaterials* **2021**, *11*, 1154. [CrossRef] [PubMed]
4. Darwish, M.S.A.; Kim, H.; Bui, M.P.; Le, T.A.; Lee, H.; Ryu, C.; Lee, J.Y.; Yoon, J. The Heating Efficiency and Imaging Performance of Magnesium Iron Oxide@tetramethyl Ammonium Hydroxide Nanoparticles for Biomedical Applications. *Nanomaterials* **2021**, *11*, 1096. [CrossRef] [PubMed]
5. Petrichenko, O.; Plotniece, A.; Pajuste, K.; Rucins, M.; Dimitrijevs, P.; Sobolev, A.; Sprugis, E.; Cēbers, A. Evaluation of Physicochemical Properties of Amphiphilic 1,4-Dihydropyridines and Preparation of Magnetoliposomes. *Nanomaterials* **2021**, *11*, 593. [CrossRef] [PubMed]
6. Novickij, V.; Stanevičienė, R.; Gruškienė, R.; Badokas, K.; Lukša, J.; Sereikaitė, J.; Mažeika, K.; Višniakov, N.; Novickij, J.; Servienė, E. Inactivation of Bacteria Using Bioactive Nanoparticles and Alternating Magnetic Fields. *Nanomaterials* **2021**, *11*, 342. [CrossRef] [PubMed]
7. Kanubaddi, K.R.; Huang, P.Y.; Chang, Y.L.; Wu, C.H.; Li, W.; Kankala, R.K.; Tai, D.F.; Lee, C.H. Deviation of Trypsin Activity Using Peptide Conformational Imprints. *Nanomaterials* **2021**, *11*, 334. [CrossRef] [PubMed]
8. Mizoguchi, S.; Hayashida, M.; Ohgai, T. Determination of Cobalt Spin-Diffusion Length in Co/Cu Multilayered Heterojunction Nanocylinders Based on Valet–Fert Model. *Nanomaterials* **2021**, *11*, 218. [CrossRef] [PubMed]
9. Fuhrmann, M.; Musyanovych, A.; Thoelen, R.; von Bomhard, S.; Möbius, H. Magnetic Imaging of Encapsulated Superparamagnetic Nanoparticles by Data Fusion of Magnetic Force Microscopy and Atomic Force Microscopy Signals for Correction of Topographic Crosstalk. *Nanomaterials* **2020**, *10*, 2486. [CrossRef] [PubMed]
10. Brero, F.; Basini, M.; Avolio, M.; Orsini, F.; Arosio, P.; Sangregorio, C.; Innocenti, C.; Guerrini, A.; Boucard, J.; Ishow, E.; et al. Coating Effect on the ^{1}H-NMR Relaxation Properties of Iron Oxide Magnetic Nanoparticles. *Nanomaterials* **2020**, *10*, 1660. [CrossRef] [PubMed]
11. Usov, N.A.; Gubanova, E.M. Application of Magnetosomes in Magnetic Hyperthermia. *Nanomaterials* **2020**, *10*, 1320. [CrossRef] [PubMed]
12. Gandhi, A.C.; Li, T.Y.; Kumar, B.V.; Reddy, M.; Peng, J.C.; Wu, C.M.; Wu, S.Y. Room Temperature Magnetic Memory Effect in Cluster-Glassy Fe-Doped NiO Nanoparticles. *Nanomaterials* **2020**, *10*, 1318. [CrossRef] [PubMed]
13. Omelyanchik, A.; Salvador, M.; D'Orazio, F.; Mameli, V.; Cannas, C.; Fiorani, D.; Musinu, A.; Rivas, M.; Rodionova, V.; Varvaro, G.; et al. Magnetocrystalline and Surface Anisotropy in CoFe2O4 Nanoparticles. *Nanomaterials* **2020**, *10*, 1288. [CrossRef] [PubMed]
14. Dutz, S.; Buske, N.; Landers, J.; Gräfe, C.; Wende, H.; Clement, J.H. Biocompatible Magnetic Fluids of Co-Doped Iron Oxide Nanoparticles with Tunable Magnetic Properties. *Nanomaterials* **2020**, *10*, 1019. [CrossRef] [PubMed]
15. Vangijzegem, T.; Stanicki, D.; Panepinto, A.; Socoliuc, V.; Vekas, L.; Muller, R.N.; Laurent, S. Influence of Experimental Parameters of a Continuous Flow Process on the Properties of Very Small Iron Oxide Nanoparticles (VSION) Designed for T1-Weighted Magnetic Resonance Imaging (MRI). *Nanomaterials* **2020**, *10*, 757. [CrossRef] [PubMed]
16. Darwish, M.S.A.; Kim, H.; Lee, H.; Ryu, C.; Lee, J.Y.; Yoon, J. Synthesis of Magnetic Ferrite Nanoparticles with High Hyperthermia Performance via a Controlled Co-Precipitation Method. *Nanomaterials* **2020**, *9*, 1176. [CrossRef] [PubMed]
17. Chen, Y.; Sun, W.; Zhang, Y.; Liu, G.; Luo, Y.; Dong, J.; Zhong, Y.; Zhu, W.; Yu, J.; Chen, Z. Magnetic Nanoparticles Functionalized Few-Mode-Fiber-Based Plasmonic Vector Magnetometer. *Nanomaterials* **2019**, *9*, 785. [CrossRef] [PubMed]

Article

Comparative Modeling of Frequency Mixing Measurements of Magnetic Nanoparticles Using Micromagnetic Simulations and Langevin Theory

Ulrich M. Engelmann [1,*], Ahmed Shalaby [1], Carolyn Shasha [2], Kannan M. Krishnan [2,3] and Hans-Joachim Krause [1,4,*]

1. Department of Medical Engineering and Applied Mathematics, FH Aachen University of Applied Sciences, 52428 Jülich, Germany; ahmed.shalaby@alumni.fh-aachen.de
2. Department of Physics, University of Washington, Seattle, WA 98195, USA; cshasha@uw.edu (C.S.); kannanmk@uw.edu (K.M.K.)
3. Department of Materials Science and Engineering, University of Washington, Seattle, WA 98195, USA
4. Institute of Biological Information Processing—Bioelectronics (IBI-3), Forschungszentrum Jülich, 52425 Jülich, Germany
* Correspondence: engelmann@fh-aachen.de (U.M.E.); h.-j.krause@fz-juelich.de (H.-J.K.)

Abstract: Dual frequency magnetic excitation of magnetic nanoparticles (MNP) enables enhanced biosensing applications. This was studied from an experimental and theoretical perspective: non-linear sum-frequency components of MNP exposed to dual-frequency magnetic excitation were measured as a function of static magnetic offset field. The Langevin model in thermodynamic equilibrium was fitted to the experimental data to derive parameters of the lognormal core size distribution. These parameters were subsequently used as inputs for micromagnetic Monte-Carlo (MC)-simulations. From the hysteresis loops obtained from MC-simulations, sum-frequency components were numerically demodulated and compared with both experiment and Langevin model predictions. From the latter, we derived that approximately 90% of the frequency mixing magnetic response signal is generated by the largest 10% of MNP. We therefore suggest that small particles do not contribute to the frequency mixing signal, which is supported by MC-simulation results. Both theoretical approaches describe the experimental signal shapes well, but with notable differences between experiment and micromagnetic simulations. These deviations could result from Brownian relaxations which are, albeit experimentally inhibited, included in MC-simulation, or (yet unconsidered) cluster-effects of MNP, or inaccurately derived input for MC-simulations, because the largest particles dominate the experimental signal but concurrently do not fulfill the precondition of thermodynamic equilibrium required by Langevin theory.

Keywords: magnetic nanoparticles; frequency mixing magnetic detection; Langevin theory; micromagnetic simulation; nonequilibrium dynamics; magnetic relaxation

1. Introduction

Magnetic nanoparticles (MNPs) have a plethora of applications not only in biomedical diagnostics, mainly determined by sample preparation, but also in detection [1,2]. MNPs are used as markers for biomolecules in immunoassays; in addition to the well-established techniques of AC-Susceptometry [3] and Relaxometry [4], Frequency-Mixing Magnetic Detection (FMMD) [5] has been increasingly applied during the past decade because of its high selectivity. This technique relies on simultaneously applying a low-frequency magnetic driving field, which brings the particles close to saturation, and a high-frequency excitation field, which probes the particles' susceptibility. Due to the nonlinear magnetization of the MNP, harmonics of both the individual incident frequencies and the intermodulation products of both frequencies are generated. Their signal can usually be picked up using a receive coil; however, other magnetic detectors can also be employed [6]. Due to its high sensitivity, FMMD has been successfully applied to realize magnetic immunoassays for the detection of a multitude of different analytes, for instance, viruses [7], antibiotics [8], or bacterial toxins [9]. Furthermore, FMMD enables the distinction of different types of MNP based on their different frequency mixing response spectrum [10–12], which opens the potential for multi-analyte detection.

In this work, we present offset-field-dependent FMMD measurements of MNPs and compare them quantitatively with two different modeling approaches: a simple Langevin-function-based thermal equilibrium model based on a lognormal size distribution, and a micromagnetic Monte Carlo (MC)-simulation method [13]. Although particles are immobilized in the experiment and can thus only relax according to Néel relaxation, the MC-simulation also includes Brownian relaxation. With this, micromagnetic MC-simulations provide new insight into the frequency mixing excitation of MNPs by modeling their non-equilibrium dynamics. For the Langevin model, the nature of relaxation is irrelevant. The core size distribution parameters derived from the Langevin model were used as starting values for the MC-simulation. As a consistency check, the MC results were again compared against the Langevin model. Prospects and limitations of both models are discussed, and suggestions for future research are developed.

2. Materials and Methods

2.1. Experimental Setup for Frequency Mixing Magnetic Detection

A custom-built measurement setup was employed for simultaneous sample excitation and demodulation signal detection. This comprises two excitation and two pick-up coils in a measurement head unit, allowing digital frequency demodulation directly from the detected signal. Details on the setup can be found in earlier works [5,14,15]. The applied alternating magnetic excitation field (AMF) was:

$$H(t) = H_0 + H_1 \sin(2\pi f_1 t) + H_2 \sin(2\pi f_2 t) \quad (1)$$

where H_0 denotes the static magnetic offset field, $H_1 = 1.29$ mT/μ_0 is the magnetic field amplitude at high frequency $f_1 = 30,543$ Hz, and $H_2 = 16.4$ mT/μ_0 is the amplitude at low frequency $f_2 = 62.95$ Hz. With this low-frequency amplitude, the particles are driven well into the nonlinear regime of the magnetization curve. The almost 400-fold higher f_1 yields well-detectable voltages at the mixing frequencies.

ABICAP® columns from Senova GmbH (Weimar, Germany) containing polyethylene filters with pore sizes of approximately 50 μm, a height of 5 mm, and a diameter of 5 mm were primed with ethanol in a desiccator to remove air bubbles from the filter. After washing the column twice with 500 μL of distilled water, 450 μL of nanomag®-D SPIO (Prod.#: 79-00-201; Micromod Partikeltechnologie GmbH, Rostock, Germany) was flushed through the column, followed by another washing step to ensure homogeneous MNP distribution and to remove unbound MNPs. The sample was dried at ambient conditions. Then the columns with the immobilized MNPs were measured at varying static field strength $H_0 = (0, \ldots, 24)$ mT/μ_0 in steps of 0.48 mT/μ_0. Due to coil-current resistive heating, the temperature T in the measurement head was approximately 318 K. The first four nonlinear magnetic moment demodulation components from frequency mixing (both real and imaginary part) were stored: $f_1 + n \cdot f_2$ with $n = 1, 2, 3, 4$. Background subtraction was performed using data from reference measurements without a sample, and phase correction for frequency-dependent phase shift inside the induction coil and amplification chain was performed.

2.2. Thermodynamic Langevin Model of a Magnetic Nanoparticle Ensemble

In the classical thermodynamic model description, the MNP sample can be described as an ensemble of noninteracting particles. (The validity of this assumption for our sample is assessed in Appendix A by an estimation of the dipole–dipole energy.) Neglecting surface effects, the saturation magnetic moment of a spherical particle with a core diameter d_c is given by:

$$|\mathbf{m}_p| = m_p(d_c) = \frac{M_s \pi d_c^3}{6} \tag{2}$$

with M_s denoting the saturation magnetization of the MNP. For simplicity, all particles are assumed to be spheres.

The total magnetic moment of the particle ensemble in thermodynamic equilibrium is calculated by averaging over a Boltzmann distribution of the orientations of the individual particle moments, yielding a dependence on the amplitude of the applied magnetic field $|\mathbf{H}| = H$, which is governed by a Langevin function [16]:

$$\mathcal{L}(\xi) = \coth(\xi) - \frac{1}{\xi} \tag{3}$$

with the dimensionless magnetic field variable:

$$\xi = \frac{m_p \mu_0 H}{k_B T} \tag{4}$$

and with temperature T, Boltzmann's constant $k_B = 1.38 \times 10^{-23}$ J/K, and the permeability of vacuum $\mu_0 = 4\pi \times 10^{-7}$ Vs/Am [17].

In the average over the particle ensemble, it has to be considered that the saturation magnetic moment m_p of each particle depends on its diameter d_c. Usually, particle ensembles exhibit a lognormal size distribution with a probability density function PDF(d_c) given by [17]:

$$\text{PDF}(d_c, d_0, \sigma) = \frac{1}{\sqrt{2\pi} \cdot d_c \cdot \sigma} \cdot \exp\left(-\frac{\ln^2(d_c/d_0)}{2\sigma^2}\right), \tag{5}$$

with the median diameter d_0 and the standard deviation σ of the diameters' natural logarithm.

The total magnetic moment of the ensemble of N_p particles is then calculated by integrating over the lognormal distribution:

$$m_{tot} = N_p \int_0^\infty dd_c \cdot \text{PDF}(d_c) \cdot m_p(d_c) \cdot \mathcal{L}\left(\frac{M_s \pi d_c^3}{6 k_B T} \mu_0 H\right). \tag{6}$$

In our FMMD scheme [5], the particle ensemble is exposed to a two-frequency excitation with static offset magnetic field (see Equation (1)). Inserting Equations (1) and (2) into (6) yields:

$$m_{tot} = \frac{N_p M_s \pi}{6} \int_0^\infty dd_c \cdot \text{PDF}(d_c) \cdot d_c^3 \cdot$$
$$\cdot \mathcal{L}\left(\frac{M_s \pi d_c^3 \mu_0}{6 k_B T}[H_0 + H_1 \sin(2\pi f_1 t) + H_2 \sin(2\pi f_2 t)]\right) \tag{7}$$

As shown in [5], the nonlinearity of the magnetization curve gives rise to the emergence of intermodulation products $m \cdot f_1 \pm n \cdot f_2$ of the total magnetic moment of the particle ensemble (with m and n denoting integers). In particular, the frequency mixing components $f_1 + f_2, f_1 + 2 \cdot f_2, f_1 + 3 \cdot f_2$ and $f_1 + 4 \cdot f_2$ appear. In the limit of small excitation amplitudes H_1 and H_2, these frequency mixing responses can be calculated with a Taylor expansion of Equation (3), yielding offset field dependencies of the mixing components proportional to the higher order derivatives of the Langevin function [5]. In the case of larger excitation amplitudes H_1 and H_2, the Taylor approximation is no longer valid and has to be replaced by the respective Fourier components of Equation (7). For instance, the average nonlinear moment response $m_1(d_c)$ at frequency $f_1 + f_2$ of one particle with diameter d_c is given by:

$$m_1(d_C) = \frac{M_s \pi d_c^3}{6} \cdot \frac{2}{k} \sum_{i=0}^{k} \cos[2\pi(f_1 + f_2)t_i] \cdot \\ \cdot \mathcal{L}\left(\frac{M_s \pi d_c^3 \mu_0}{6 k_B T} [H_0 + H_1 \sin(2\pi f_1 t_i) + H_2 \sin(2\pi f_2 t_i)]\right) \quad (8)$$

The factor $2/k$ normalizes the sum and accounts for the full-period average over $\sin^2(..)$ which is $\frac{1}{2}$. The sampling time steps t_i should be chosen such that a sufficient number of samples is taken in one high frequency period. In our calculations, we took 10 steps in a period $1/f_1$, i.e., $\Delta t = t_i - t_{i-1} = 0.1/f_1$, yielding sufficient numerical precision. Although in our experiments, the high frequency f_1 was 485 times larger than the low frequency f_2, it was sufficient to select $f_1 = 20 \cdot f_2$ for our numerical calculations. Thus, $k = 200$ was used.

In a similar fashion, the response component $m_2(d_c)$ at frequency $f_1 + 2 \cdot f_2$ is obtained from:

$$m_2(d_c) = \frac{M_s \pi d_c^3}{6} \cdot \frac{2}{k} \sum_{i=0}^{k} \sin[2\pi(f_1 + 2f_2)t_i] \cdot \\ \cdot \mathcal{L}\left(\frac{M_s \pi d_c^3 \mu_0}{6 k_B T} [H_0 + H_1 \sin(2\pi f_1 t_i) + H_2 \sin(2\pi \cdot 2 f_2 \cdot t_i)]\right). \quad (9)$$

Note the $\cos[..]/\sin[..]$ alternation in the reference frequency term behind the sum symbol in Equations (8) and (9), which is due to the fact that with increasing order, the frequency mixing responses are alternately uneven (point-symmetric) and even (axisymmetric) functions. Components $m_3(d_c)$ and $m_4(d_c)$ are calculated similarly.

The total magnetic moment component $m_{n,tot}$ at the frequency mixing component $f_1 + n \cdot f_2$ is then obtained by integration over the lognormally weighted particle ensemble:

$$m_{n,tot}(d_0, \sigma, N_p) = N_p \int_0^\infty dd_c \cdot \text{PDF}(d_c, d_0, \sigma) \cdot m_n(d_c). \quad (10)$$

Equation (10) constitutes our forward model for calculating the frequency mixing signals. The model contains just three fitting parameters, the lognormal distribution characteristics median diameter d_0 and width σ, and the total number of particles N_p. The measured nonlinear magnetic moment components of nanoparticle samples at frequencies $f_1 + n \cdot f_2$, $n = 1, 2, 3, 4$, were fitted with this model using the Levenberg–Marquardt least-squares algorithm.

2.3. Micromagnetic Monte Carlo (MC-)Simulation

The nonlinear particle relaxation dynamics in nonequilibrium conditions under the influence of an applied AMF, **H**, can be described by combined Néel–Brownian relaxation [13]. The Néel relaxation of the direction of the magnetic moment of a single MNP, **m**$_p$, is governed by the Landau–Lifshitz–Gilbert equation (LLG) [18]:

$$\frac{d\mathbf{m}_p}{dt} = \frac{\mu_0 \gamma}{1 + \alpha^2} \cdot \left(\mathbf{H}_{\text{eff}} \times \mathbf{m}_p + \alpha \mathbf{m}_p \times (\mathbf{H}_{\text{eff}} \times \mathbf{m}_p)\right) \quad (11)$$

with the permeability of free space, μ_0, the electron gyromagnetic ratio, γ, the damping parameter, α, and the effective field $\mathbf{H}_{\text{eff}}$. The Brownian rotation of the MNP easy axis, **n**, is described via the generalized torque, Θ, as follows [19]:

$$\frac{d\mathbf{n}}{dt} = \frac{\Theta}{6 \eta V_H} \times \mathbf{n} \quad (12)$$

with the carrier matrix viscosity, η, and the MNP hydrodynamic volume, $V_h = \pi/6 \cdot d_h^3$, in which d_h is the hydrodynamic particle diameter. Néel and Brownian relaxation are coupled in the internal particle energy:

$$U = -\mu_0 \cdot m_p(\mathbf{m}_p \cdot \mathbf{H}) - K \cdot V_c (\mathbf{m}_p \cdot \mathbf{n})^2 + \varepsilon_{IA} \quad (13)$$

where $m_p = |\mathbf{m}_p| = V_c \cdot M_S$ (cf. Equation (2)) is the magnitude of the MNP magnetic moment, and $V_c = \pi/6 \cdot d_c^3$ is the MNP core volume. The first term describes the Zeeman energy including the applied field **H**. The second term incorporates the anisotropy energy via the effective anisotropy constant, K, and under the assumption of uniaxial anisotropy and spherically shaped particles. The third term in Equation (13) includes magnetic dipole–dipole interaction. However, thermal energy, $\varepsilon_{\text{therm}}$, dominates magnetic interaction energy by two orders of magnitude in our MNP samples, so that particle–particle interaction is negligible ($\varepsilon_{IA} \ll \varepsilon_{\text{therm}}$). Please refer to Appendix A for a detailed estimation of the effect of magnetic dipole–dipole interaction energy that corroborates our assumption. Thermal fluctuations are included by adding the terms $\mathbf{H}_{\text{th}}$ and $\mathbf{\Theta}_{\text{th}}$, which are implemented as Gaussian-distributed white noise with zero mean values ($\langle \mathbf{H}_{th}^i(t) \rangle = 0$ and $\langle \mathbf{\Theta}_{th}^i(t) \rangle = 0$) and variances $\langle \mathbf{H}_{th}^i(t) \mathbf{H}_{th}^j(t') \rangle = (2 k_B T \cdot (1+\alpha^2))/(\gamma m_p \alpha) \cdot \delta_{ij} \delta(t-t')$ and $\langle \mathbf{\Theta}_{th}^i(t) \mathbf{\Theta}_{th}^j(t') \rangle = 12 k_B T \eta V_H \cdot \delta_{ij} \delta(t-t')$, respectively. Here T is the global temperature in the system. With this, the effective field and generalized torque read:

$$\mathbf{H}_{\text{eff}} = -\frac{1}{m_p \cdot \mu_0} \cdot \frac{\partial U}{\partial \mathbf{m}} + \mathbf{H}_{\text{th}} = \mathbf{H} + \frac{2K \cdot V_c}{m_p \cdot \mu_0} \cdot (\mathbf{m}_p \cdot \mathbf{n}) \mathbf{n} + \mathbf{H}_{\text{th}} \qquad (14)$$

$$\mathbf{\Theta} = \frac{\partial U}{\partial \mathbf{n}} \times \mathbf{n} + \mathbf{\Theta}_{\text{th}} = -2K \cdot V_c (\mathbf{m}_p \cdot \mathbf{n})(\mathbf{m}_p \times \mathbf{n}) + \mathbf{\Theta}_{\text{th}} \qquad (15)$$

We apply the Stratonovic–Heun scheme to solve the system of coupled stochastic differential Equations (11) through (15) and implement a Monte Carlo method routine as described in our previous works [20–22]. The full source code is available as listed in the Data Availability Statement and its results are denoted as MC-simulations henceforth. Simulation parameters were chosen as listed in Table 1 with the MNP properties matching the experimentally determined values from Table 2. Furthermore, the damping parameter α was set to unity [23]. One thousand particles were simulated simultaneously and initialized with randomized directions of magnetization and easy axes for each MNP. The MNPs were then thermalized for one-fifth of the total number of time steps, N, before the AMF was applied. We used $N = 50,000$ and averaged the magnetization over five independent simulation runs to achieve a good compromise between accuracy and acceptable computation time. The time step sizes were then 10 ns. The simulations were performed with the open-access Python code referenced in the Data Availability Statement. Calculations were carried out on a PC cluster consisting of 2 × CPU Intel Xeon E5-2687W, 3.1/3.8 GHz, with 8 clusters each and RAM 64 GB each. The typical calculation time for one offset field value was approx. 53 h.

To approximate experimental data for comparison, the excitation field parameters were chosen as $H_1 = 1$ mT/μ_0, $f_1 = 40,000$ Hz, $H_2 = 16$ mT/μ_0, $f_2 = 2000$ Hz, and the static field was varied from $H_0 = (0, \ldots, 24)$ mT/μ_0 with a step size of 1 mT/μ_0.

Table 1. Simulation parameters applied in the micromagnetic simulation model.

Effective Anisotropy Constant	Saturation Magnetization [1] M_S	Mass Density of Magnetite	Viscosity of Surrounding (Water)	Temperature
11 kJ/m^3	476 kA/m	5.2 g/cm^3	8.9×10^{-4} Pa·s	300 K

[1] The literature value for bulk magnetite from [24] was used.

Table 2. Material properties of MNPs from fitting the experimental data with the Langevin model.

Core Diameter d_c	Log-Normal Distribution Width σ	Polydispersity Index (PDI)	Hydrodynamic Diameter [2] d_h	Concentration [1]
7.81 nm	0.346	0.127	20 nm	2.4 mg(Fe)/mL

[1] The concentration c is taken from the datasheet of the manufacturer. The concentration in the filter might be smaller due to unbound particles being washed out undetected. [2] The hydrodynamic diameter d_h is taken from the datasheet of the manufacturer.

3. Results

3.1. Experimental Results and Thermodynamic Langevin Model Fitting

The Langevin model (Equation (10), with inputs (8) and (9), see Section 2.2) was fitted to the measured real parts of the experimental data using the Levenberg–Marquardt least-square algorithm routine, whose results are plotted in Figure 1. Although the immobilized MNPs in the experimental setup are blocked in Brownian rotation, this step is justified because Langevin theory approximates the magnetization of the entire ensemble of MNP independently of the underlying mechanism of relaxation. The imaginary parts of the response signal were found to be two orders of magnitude weaker and were therefore disregarded. For all four demodulation components $f_1 + n \cdot f_2$ with $n = 1, 2, 3, 4$, a very good agreement between experimental and simulated results was observed, confirmed by a coefficient of determination of $R^2 > 0.98$. Only for component $f_1 + f_2$ at high offset field did the simulation predict slightly higher values than measurement, whereas for $f_1 + 2 \cdot f_2$ and $f_1 + 3 \cdot f_2$, the simulation slightly underestimated measurements.

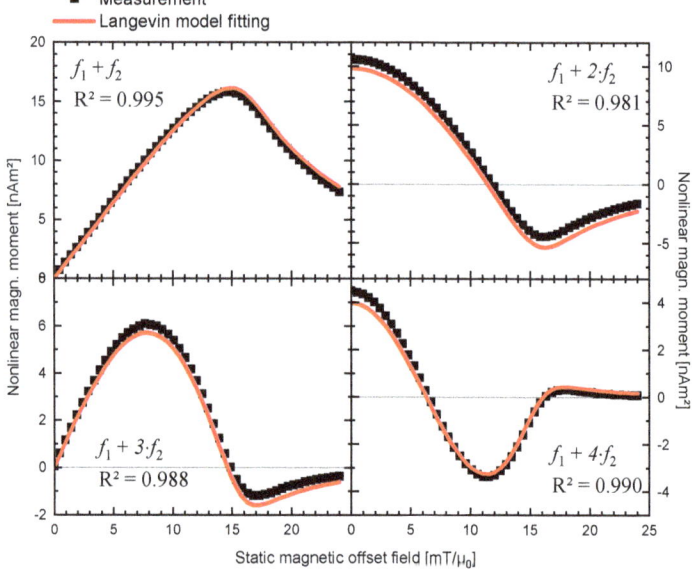

Figure 1. MNP nonlinear magnetic moment for dual frequency excitation at mixing frequencies $f_{LF} + n \cdot f_{HF}$ with $n = 1, 2, 3, 4$ from experimental measurement ($H_1 = 1.29$ mT/μ_0, $f_1 = 30{,}534$ Hz, $H_2 = 16.4$ mT/μ_0, $f_2 = 62.95$ Hz) and fitted with the Langevin model of Equation (10) with the same parameters.

Fitting yielded the MNP material properties of median core diameter, d_c, and its log-normal distribution width, σ, which are listed in Table 2. The hydrodynamic diameter, d_h, and the concentration of the MNP were taken from the datasheet of the manufacturer.

3.2. Micromagnetic MC-Simulation Results

We used the MNP material properties derived from fitting the Langevin model to the experimental data as described in the previous Section 3.1 (see Table 2) as input parameters for the micromagnetic MC-simulations. These simulations yield magnetization curves ($M(H)$-loops), which are shown for exemplary static offset fields in Figure 2.

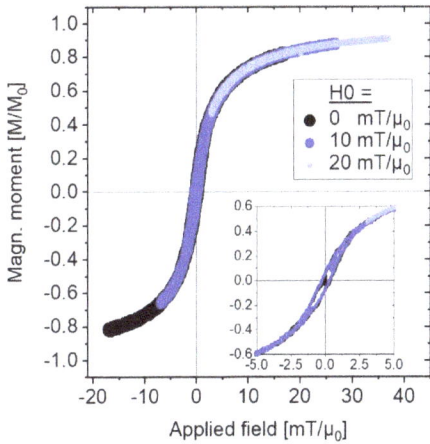

Figure 2. Exemplary magnetization curves ($M(H)$-loops) generated from micromagnetic MC-simulations for different static offset fields $H_0 = (0, 10, 20)$ mT/μ_0. Inset shows magnification of small applied fields, revealing a slight opening of the loops.

The magnetization curves are (almost) identically overlapping, but shift towards the magnetization saturation of MNP as the static offset field, H_0, increases. The $M(H)$-loops are very slightly opened, revealing minor hysteresis area (see Figure 2, inset). For values of $H_0 > (H_1 + H_2)$, e.g., for $H_0 = 20$ mT/μ_0 in Figure 2, the applied field is constantly keeping the MNPs in (almost) saturation and no hysteresis area is present.

From the $M(H)$-loops, the first four demodulation components $f_1 + n \cdot f_2$ were calculated following Equations (8) and (9). For comparison, the experimental measurement data and the MC-simulated data were normalized to their respective highest value, M_{max}, and plotted alongside each other in Figure 3. The agreement between experimental and simulated data is acceptable, as confirmed by a coefficient of determination of $R^2 > 0.76$ for all four demodulation components. We observe for component $f_1 + f_2$ that MC-simulation consistently underestimates the measurement results. However, the peak at $H_0 = 15$ mT/μ_0 coincides. For the mixing term $f_1 + 2 \cdot f_2$, MC-simulation underestimates the measurement results for $H_0 < 18$ mT/μ_0 and $H_0 > 20$ mT/μ_0 and shows a mismatch of over 50% for the peak value at $H_0 = 15$ mT/μ_0. MC-simulation and measurement of component $f_1 + 3 \cdot f_2$ coincide for values up to $H_0 = 10$ mT/μ_0 and again for $H_0 \geq 20$ mT/μ_0. However, around the peak at $H_0 = 16$ mT/μ_0, MC-simulations overestimate experimental data by approx. 70%. For mixing term $f_1 + 2 \cdot f_2$, MC-simulations match experimental data, except around the peaks at $H_0 = 11$ mT/μ_0 and $H_0 = 16$ mT/μ_0.

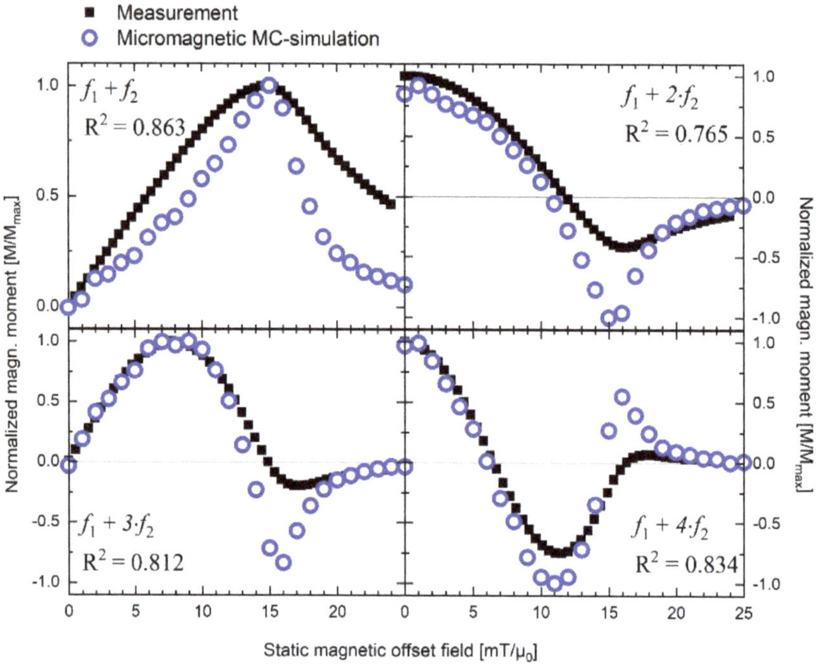

Figure 3. Normalized MNP nonlinear magnetic moment for dual frequency excitation at mixing frequencies $f_1 + n \cdot f_2$ with $n = 1, 2, 3, 4$ comparing experimental results ($H_1 = 1.29$ mT/μ_0, $f_1 = 30{,}534$ Hz, $H_2 = 16.4$ mT/μ_0, $f_2 = 62.95$ Hz) and predictions from micromagnetic MC-simulations ($H_1 = 1$ mT/μ0, $f_1 = 40{,}000$ Hz, $H_2 = 16$ mT/μ_0, $f_2 = 2000$ Hz).

3.3. Comparing Micromagnetic MC-Simulation Results and Thermodynamic Langevin Model Fitting

To test whether predictions from the Langevin model and MC-simulations show reproducible results, we fitted the Langevin model directly to the results from MC-simulation. We used the same input parameters for MC-simulations and Langevin model fitting of $H_1 = 1$ mT/μ_0, $f_1 = 40{,}000$ Hz, $H_2 = 16$ mT/μ_0, $f_2 = 2000$ Hz, and fixed the mean core diameter with $d_c = 7.81$ nm and variable distribution parameter σ for the fitting. The results are plotted in Figure 4, confirming overall good agreement with a coefficient of determination of $R^2 > 0.989$ for all four demodulation components. From the qualitative comparison, one sees that Langevin model fitting and MC-simulations coincide, except for the secondary peaks for $n = 3$ and $n = 4$ (cf. Figure 4). The fitting yields a distribution width of $\sigma = 1.466$. This is significantly different than the input parameters to MC-simulations ($\sigma = 0.346$), questioning our hypothesis that assumes identical outputs from the Langevin model and MC-simulation. As we discuss in detail in the next section, we suspect the reasons for this lie with MNP properties that are not (yet) accurately represented in the modeling.

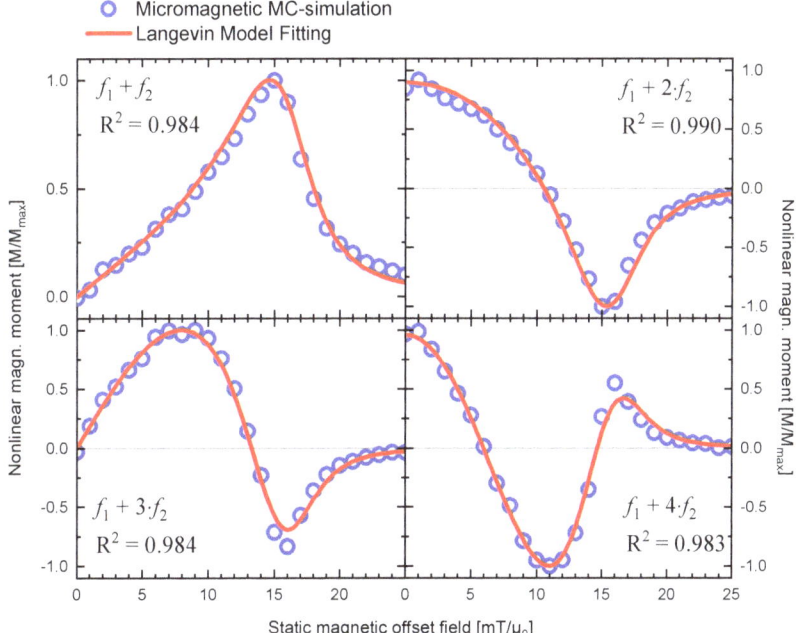

Figure 4. Normalized MNP nonlinear magnetic moment for dual frequency excitation at mixing frequencies $f_1 + n \cdot f_2$ with $n = 1, 2, 3, 4$ from micromagnetic MC-simulations fitted with the thermodynamic Langevin model with fixed core diameter $d_C = 7.813$ nm.

4. Discussion

The Langevin model has been extensively applied to describe MNP magnetization for various applications such as diagnostic biosensing [25], Magnetic Particle Imaging (MPI) [26], and therapeutic Magnetic Fluid Hyperthermia (MFH) [27]. Its application to frequency mixing excitation seems therefore naturally reasonable. This assumption is fostered by our results reporting very good agreement between experimental data and Langevin model fitting ($R^2 > 0.98$; cf. Figure 1). The resulting fitting parameters ($d_c \approx 7.8$ nm and $\sigma \approx 0.35$; cf. Table 2) are furthermore in good agreement with literature values reporting a core diameter between $d_c \approx 7$ nm [28] and $d_c \approx 11$ nm [29] for nanomag®-D SPIO. The broad size distribution (with width $\sigma > 0.3$) reflects a heterogeneous particle size with an effectively range of $d_c = (3 - 25)$ nm, cf. Figure 5.

Nevertheless, the Langevin model is only valid for noninteracting, single particle ensembles with uniaxial anisotropy in thermodynamic equilibrium and (quasi)static magnetic fields [16]. Therein lies its major limitation, because the Langevin model cannot model an opening of $M(H)$-loops (hysteresis), which are, however, expected for AMF excitations at applied frequencies that approximate the inverse MNP relaxation time, $f \sim \tau^{-1}$, as the magnetic moment of each MNP begins to lag behind the driving AMF [2]. In contrast, micromagnetic MC-simulations provide new insight in the frequency mixing excitation of MNP by modeling the non-equilibrium dynamics of MNP by including thermal fluctuations: The MC-simulations reveal a minor hysteresis in the $M(H)$-loop of the MNP (cf. Figure 2), which is inaccessible via equilibrium Langevin theory. We attribute this minor hysteresis to the small portion of large particles, whose magnetic relaxation is (thermally) blocked by the volume (and therefore size cubed) dependent anisotropy barrier: $K \cdot V_C$.

This can be assessed by assuming an AC-field-dependent Brownian relaxation time of [30]:

$$\tau_B(H_{AC}) = \frac{\tau_B}{\sqrt{1 + 0.21 \cdot \xi(H_{AC})^2}} \quad (16)$$

and an AC-field dependent Néel relaxation time of [16]:

$$\tau_N(H_{AC}) = \tau_0 \cdot \exp\left(\frac{K \cdot V_C}{k_B T} \cdot \left(1 - \frac{H_{AC}}{H_K}\right)^2\right) \quad (17)$$

where H_{AC} denotes the AMF amplitude, $\tau_B = 3\eta V_H/(k_B T)$ the field-independent Brownian relaxation time, $\xi(H_{AC}) = m H_{AC}/(k_B T)$ the reduced field parameter (cf. Equation (4)), $\tau_0 = 10^{-9}$ s the time constant, and $H_K = 2K/\mu_0 M_S$ the (uniaxial) anisotropy field strength. From Equations (16) and (17), the effective relaxation time follows with:

$$\tau = \frac{1}{\tau_B^{-1} + \tau_N^{-1}}. \quad (18)$$

Using the (mean) values from Tables 1 and 2, and the experimental setup parameters with $H_{AC} = H_1 + H_2 \sim 17$ mT/μ_0, one calculates $\tau \sim \tau_0 \sim 10^{-9}$ s so that for $f \sim (30-40)$ kHz the condition for the onset of hysteresis, $f \sim \tau^{-1}$, is not fulfilled. Because the measured imaginary parts of the mixing components are hundredfold weaker than the real parts, we can conclude that dissipation is indeed negligible. However, for the small portion of large particles with $d_C \geq 20$ nm and naturally increasing the effective hydrodynamic diameter $d_h > 20$ nm, and presumably decreasing effective anisotropy constant for these larger core particles, $K \sim 5$ kJ/m^3 [31,32], the effective relaxation time increases by several orders of magnitude. The relaxation time for these large particles is dominated by the Brownian relaxation mechanism, fulfilling the condition $f \sim \tau^{-1}$. Consequently, the minor opening of the $M(H)$-loop observed in Figure 2 from MC-simulations is a direct contribution from these large size particles relaxing with Brownian rotation. This assumption could, however, not be verified experimentally in our current study, because the MNPs were immobilized due to sample preparation, blocking Brownian rotation. Nevertheless, Figures 2 and 4 demonstrate the potential insight to be gained on the micromagnetic level in the underlying mechanisms in dual frequency excitation responses of MNP (see discussion regarding future experiments below).

Interestingly, these large size particles apparently also contribute most dominantly to the Langevin model fitting to MC-simulation data (cf. Figure 4): When we deliberately restrict the particle size range for Langevin model calculation to the largest portion of core sizes (i.e., limiting the minimum core size d_c), we find that 90% of the calculated signal is contributed from the particles with $d_c \geq 12.1$ nm. Similarly, 99% of the signal stems from $d_c \geq 9.2$ nm. With the reverse of the cumulative distribution function (CDF) of the lognormal distribution (see Equation (A3) in Appendix B), the corresponding size fractions (quantiles) of the distribution are calculated, beginning from the large-sized tail. This finding is visualized in Figure 5. The largest 10.3% of the particles contribute 90% of the FMMD signal. Furthermore, 99% of the signal is produced by the largest 31.8% of particles. In other words, this means that almost all of the FMMD signal originates from the large-sized tail of the size distribution.

This could also explain why the Langevin model can be fitted with very good agreement to MC-simulation data (cf. Figure 4; $R > 0.989$), even though it is physically unable to model nonequilibrium dynamic relaxation processes: The mathematical fitting routine can force the fitting parameters (d_0, σ, N_p) to values outside the model's range of validity (for further details see Appendix B). As the opening $M(H)$-loop and our relaxation time approximations (see above) suggest, thermodynamic equilibrium is not valid anymore for large particles $d_C \geq 20$ nm at frequencies $f \sim (30-40)$ kHz. Therefore, fitting the model to the experimental data could lead to unreliable distribution parameters.

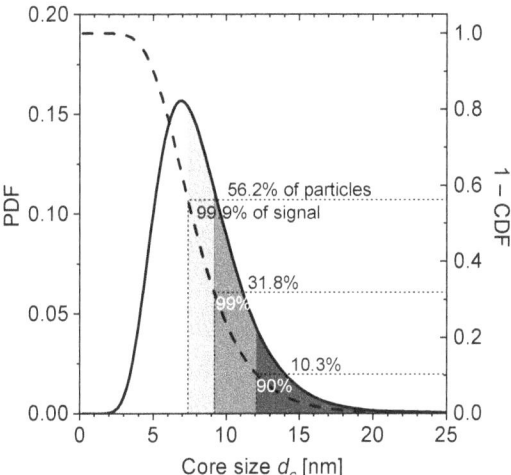

Figure 5. PDF of lognormal distribution (solid line) with $d_0 = 7.81$ nm and $\sigma = 0.346$ and its reverse CDF, counted from large sizes (dashed line). The quantiles which yield 90% (99%; 99.9%) of the FMMD signal are shaded in dark grey (grey; light gray), they consist of 10.3% (31.8%; 56.2%) of particles on the large-sized tail of the distribution.

For other frequency-dependent biomedical applications of iron-oxide MNPs, such as MPI [33,34] or MFH [32,35], an optimal core size of $d_c \sim 25$ nm has been suggested from experiment and MC-simulation, stressing the importance of open hysteresis loops for signal generation. Our results from Figures 4 and 5 could therefore indicate that an opening of the $M(H)$-loop is also favorable for generating strong signals in mixed frequency measurements. As mentioned above, our current sample preparation does not allow experimental verification of this assumption, because Brownian rotation is blocked for the immobilized MNP. Even more so, this assumption remains to be tested in future experiments using MNP with larger core sizes, ideally suspended and freely rotatable in solution. In future experiments, the question of whether the Brownian mechanism dominates the MNP relaxation mechanism could be further probed by suspending MNP in gelated matrices (e.g., agarose or poly-acrylamid gels), in which Brownian relaxation has been shown to be controllably blocked in MFH measurements with particles of the same size as used in the present study [36].

MC-simulations deviated quantitatively in intensity at the peaks by up to 70% from the experimental values (cf. Figure 3) when directly compared. The first reason for this could be that Brownian relaxation mechanism is blocked for immobilized MNPs in the experimental sample, while it adds to the signal in MC-simulations (see paragraph above). A second potential reason for deviations is that the parameters used for MC-simulation input slightly differ from the experimental parameters. Another reason could be that MNPs form clusters of a few particles in the experimental setting, which Dennis et al. consistently found for nanomag-D SPIO from small-angle neutron scattering (SANS) and transmission electron microscopy [37]. Clustering of MNPs has been suggested to lead to increased particle–particle interaction, changed effective anisotropy, and restriction of MNP rotatability [38–40]. Although the interplay of these clustering effects on MNP relaxation behavior is the subject of ongoing discussions [39–42], it is overall assumed to diminish the experimental signal intensity [43]. Both the rise of particle–particle interaction—although purposefully excluded from this study—and the influence of varying K could be investigated in future MC-simulations studies (e.g., by diffusion-limited colloidal aggregation [42]) to advance the understanding of its influence on mixed frequency excitation signal generation.

Finally, we are aware that our current study is limited by the extraction of input parameters for the MC-simulations from Langevin model fitting to experimental data (cf. Table 2). Future studies will complement these assumptions by experimental methods, e.g., measuring MNP core sizes via transmission electron microscopy (TEM) or the MNP magnetic core size distribution from magnetic measurement [44], in addition to determining the hydrodynamic diameters from dynamic light scattering (DLS) experiments [20], which will serve as input parameters to further validate our fitting and modelling.

5. Conclusions

With the current work, we present the first insights into dual-frequency magnetic excitation of MNP from interpreting experimental results in terms of both the thermodynamic Langevin model and nonequilibrium dynamic MC-simulations. In summary, we draw the following conclusions from our comparative study:

1. Both the Langevin model and MC-simulations reproduce the shape of the experimental signal satisfactorily (see Figures 1 and 3). However, punctual deviations between experimental data and MC-simulations are observed (see Figure 3).
2. MC-simulations allow the investigation of the dynamic hysteresis ($M(H)$)-loop during AMF excitation, revealing a minor opening (cf. Figure 2). This opening is attributed to the small portion of large, thermally blocked particles.
3. Langevin model fitting suggests that 90% of the experimentally detected FMMD signal intensity is generated by the largest 10% of the particles (cf. Figure 5).
4. For the large particles ($d_c > 20$ nm) which dominate the FMMD signal, relaxation cannot be neglected. However, this effect is not included in our Langevin model. We suspect this is the reason for observed deviations between the Langevin model and MC-simulations.

We are aware that these findings raise more questions and could serve as the starting point for further investigations. Based on these conclusions we propose the following respective future research directions:

- Complementary experimental methods should be used to derive MNP properties for more accurate input in the MC-simulations (to address and remedy point 1, above). From this, we also plan to further increase our predictive accuracy in the future by coordinating simulation and experimental parameters to be identical (both materials properties and field parameters).
- Experimental sample preparation should be expanded to allow—ideally gradually controllable—Brownian rotation of MNPs (e.g., in poly-acrylamide gels) in order to precisely analyze the role of the Brownian relaxation mechanism for signal generation (see also next point below).
- Furthermore, MC-simulations could be expanded by including magnetic particle–particle interaction effects and the inclusion of clustering-effects of MNPs. This should be investigated with special regard to point 3 (above), because magnetic interaction scales with increasing particle core size. Additionally, the influence of effective anisotropy could be studied systematically with MC-simulations. Finally, the dominant relaxation mechanism in dual frequency excitation could be further investigated by weighting Néel and Brownian relaxation mechanisms systematically in MC-simulations, and comparing the results to experimental findings.
- Point 1 (above) suggests the necessity for multi-theory approaches to interpret dual-frequency MNP excitation responses. Therefore, systematic parameter studies, e.g., varying MNP properties such as effective anisotropy, median core sizes and shell sizes, and AMF parameters (H_1 H_2, f_1, f_2), should be performed to generate a multi-parameter repository from MC-simulations. This could serve as a basis for a unified model to explain FMMD signal generation in the future. However, such a study must be well designed to ensure acceptable computational effort.

Author Contributions: Conceptualization, U.M.E. and H.-J.K.; methodology, U.M.E. and H.-J.K.; software, U.M.E., A.S. and C.S.; validation, C.S. and K.M.K.; investigation, U.M.E., A.S. and H.-J.K.; resources, U.M.E., K.M.K., and H.-J.K.; data curation, U.M.E., A.S. and H.-J.K.; writing—original draft preparation, U.M.E. and H.-J.K.; writing—review and editing, all authors; supervision, K.M.K.; project administration, U.M.E.; funding acquisition, U.M.E., K.M.K. and H.-J.K. All authors have read and agreed to the published version of the manuscript.

Funding: U.E. was funded by German Federal Ministry of Culture and Science of North Rhine-Westphalia, grant Karrierewege FH-Professur.

Data Availability Statement: Source code used for the nonlinear particle relaxation dynamics MC-simulations can be accessed at https://github.com/cshasha/nano-simulate.

Acknowledgments: We gratefully acknowledge experimental setup support from Stefan Achtsnicht, assistance with measurements from Mrinal Nambipareechee, and phase corrections from Ali Pourshahidi, all from Forschungszentrum Jülich. Simulations were performed with computing resources provided by FH Aachen University of Applied Sciences, Dept. of Medical Engineering and Applied Mathematics, under guidance of Martin Reißel.

Conflicts of Interest: The authors declare no conflict of interest. The funders had no role in the design of the study; in the collection, analyses, or interpretation of data; in the writing of the manuscript, or in the decision to publish the results.

Appendix A. Estimating the Effect of Magnetic Dipole–Dipole Interaction Energy

The magnetic dipole–dipole interaction energy exerted on a single magnetic nanoparticle by N neighboring particles with interparticle distance vector $\mathbf{d}_i$ can be described by [20]:

$$\varepsilon_{IA} = \sum_i \frac{\mu_0}{4\pi d_i^3} \cdot \left(\frac{3 \cdot (\mathbf{m}_0 \cdot \mathbf{d}_i) \cdot (\mathbf{m}_i \cdot \mathbf{d}_i)}{\mathbf{d}_i^2} - \mathbf{m}_0 \cdot \mathbf{m}_i \right) \tag{A1}$$

For two neighboring particles with magnetic moments $|\mathbf{m}_0| = |\mathbf{m}_i| = m_p = V_C \cdot M_S$ (cf. Equation (2)), Equation (A1) can be simplified to the extremal energies:

$$\varepsilon_{IA,min} = \frac{\mu_0 m_p^2}{4\pi d^3} \text{ and } \varepsilon_{IA,max} = 3\frac{\mu_0 m_p^2}{4\pi d^3} \tag{A2}$$

Using Equation (A2) and the MNP properties from Table 2, we calculated a corridor for ε_{IA} in dependence of the interparticle distance, shown in Figure A1. The thermal excitation energy of $\varepsilon_{therm} = k_B \cdot T = 25.8$ meV for $T = 300$ K is also marked in Figure A1.

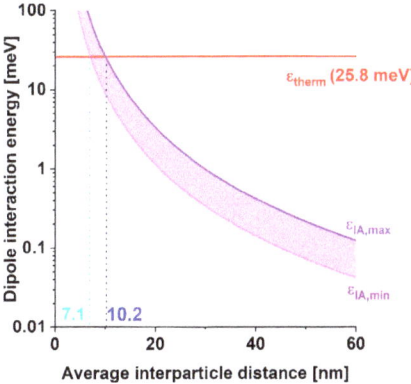

Figure A1. Estimated corridor for magnetic dipole–dipole interaction energy between neighboring MNPs (shaded) versus the average interparticle distance, estimated from Equation (A2). Thermal excitation energy is shown for reference.

From the intersection of ε_{therm} and the corridor, the range of interparticle distances in which magnetic dipole–dipole interaction energy is of the same order of magnitude as the thermal excitation energy can be estimated with $d = (7.1 - 10.2)$ nm (see mark in Figure A1). In other words, for average interparticle distances $d > 10.2$ nm, thermal excitation can be assumed to dominate magnetic interaction.

The average interparticle distance d_{avg} can be estimated from the number of particles in solution, z, by taking the inverse third root: $d_{avg} = z^{-1/3}$. With $z = 4.3 \cdot 10^{15} \frac{1}{cm^3}$ (taken from nanomag-D SPIO manufacturer's datasheet) we estimate an average interparticle distance of $d_{avg} \approx 61.5$ nm. From Figure A1 one can see that for such an average interparticle distance the thermal excitation energy is two orders of magnitude larger than magnetic dipole–dipole interaction energy, $\varepsilon_{therm} \gg \varepsilon_{IA}$, corroborating our assumption to neglect magnetic interaction between MNP for our calculations.

Appendix B. Log-Normal Distribution Probability Density Function (PDF)

The log-normal distribution probability density function (PDF) for a general particle size d is defined as shown in Equation (5) with the median diameter d_0 and the standard deviation σ of the diameters' natural logarithm. Integration over PDF from 0 to d_c yields the cumulative distribution function (CDF) [17]:

$$\text{CDF}(d_c, d_0, \sigma) = \int_0^{d_c} dd \cdot \text{PDF}(d, d_0, \sigma) = \frac{1}{2} + \left(1 + \text{erf}\left(\frac{\ln(d_c/d_0)}{\sqrt{2} \cdot \sigma}\right)\right) \quad (A3)$$

Standard CDF cumulates particles starting from the small-sized tail of the distribution. For FMMD, large particles contribute most of the signal. To quantify their contribution, it is advantageous to consider the reverse cumulative distribution starting from the large size tail, i.e., 1—CDF. The exemplary log-normal distribution PDF(d_c,d_0,σ) and its reverse CDF using the parameters from Table 2 are plotted in Figure 5.

By fitting the Langevin model with fixed $d_c = 7.81$ nm to the MC-simulation results, a significantly larger width parameter $\sigma = 1.466$ was obtained (see Section 3.3). The corresponding lognormal distribution and its reverse CDF are plotted in Figure A2. The FMMD signal calculated with this distribution is plotted in Figure 4. Analogously to the quantile analysis described in Section 4, the quantiles yielding 90%, 99%, and 99.9% of the FMMD signal were also determined for this wider distribution, as shown in Figure A2.

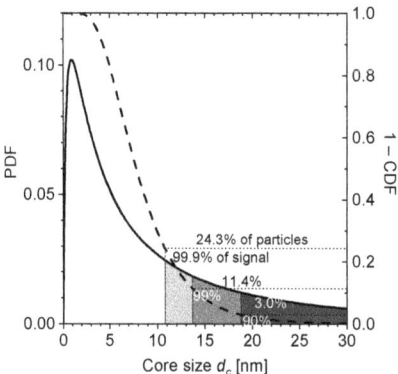

Figure A2. PDF of lognormal distribution (solid line) with $d_0 = 7.81$ nm and $\sigma = 1.466$ and its reverse CDF, counted from large sizes (dashed line). The quantiles which yield 90% (99%; 99.9%) of the FMMD signal are shaded in dark grey (grey; light gray), they consist of 3.0% (11.4%; 24.3%) of particles on the large-sized tail of the distribution.

For this wider distribution, it was found that the largest 3.0% of the particles (with $d_c > 18.8$ nm) contribute 90% of the FMMD signal. The largest 11.4% of particles (with $d_c > 13.7$ nm) contribute 99% of signal. Almost all of the signal (99.9%) is generated by the largest 24.3% of the particles above 10.8 nm core size.

References

1. Thanh, N.T.K. *Magnetic Nanoparticles: From Fabrication to Clinical Applications*; CRC Press: Boca Raton, FL, USA, 2012.
2. Krishnan, K.M. Biomedical nanomagnetics: A spin through possibilities in imaging, diagnostics, and therapy. *IEEE Trans. Magn.* **2010**, *46*, 2523–2558. [CrossRef] [PubMed]
3. Kriz, K.; Gehrke, J.; Kriz, D. Advancements toward magneto immunoassays. *Biosens. Bioelectron.* **1998**, *13*, 817–823. [CrossRef]
4. Lange, J.; Kötitz, R.; Haller, A.; Trahms, L.; Semmler, W.; Weitschies, W. Magnetorelaxometry—A new binding specific detection method based on magnetic nanoparticles. *J. Magn. Magn. Mater.* **2002**, *252*, 381–383. [CrossRef]
5. Krause, H.-J.; Wolters, N.; Zhang, Y.; Offenhäusser, A.; Miethe, P.; Meyer, M.H.F.; Hartmann, M.; Keusgen, M. Magnetic particle detection by frequency mixing for immunoassay applications. *J. Magn. Magn. Mater.* **2007**, *311*, 436–444. [CrossRef]
6. Cardoso, S.; Leitao, D.C.; Dias, T.M.; Valadeiro, J.; Silva, M.D.; Chicharo, A.; Silverio, V.; Gaspar, J.; Freitas, P.P. Challenges and trends in magnetic sensor integration with microfluidics for biomedical applications. *J. Phys. D Appl. Phys.* **2017**, *50*, 213001. [CrossRef]
7. Wu, K.; Saha, R.; Su, D.; Krishna, V.D.; Liu, J.; Cheeran, M.C.-J.; Wang, J.-P. Magnetic-nanosensor-based virus and pathogen detection strategies before and during covid-19. *ACS Appl. Nano Mater.* **2020**, *3*, 9560–9580. [CrossRef]
8. Pietschmann, J.; Dittmann, D.; Spiegel, H.; Krause, H.-J.; Schröper, F. A Novel Method for Antibiotic Detection in Milk Based on Competitive Magnetic Immunodetection. *Foods* **2020**, *9*, 1773. [CrossRef] [PubMed]
9. Achtsnicht, S.; Neuendorf, C.; Faßbender, T.; Nölke, G.; Offenhäusser, A.; Krause, H.-J.; Schröper, F. Sensitive and rapid detection of cholera toxin subunit B using magnetic frequency mixing detection. *PLoS ONE* **2019**, *14*, e0219356. [CrossRef]
10. Lenglet, L. Multiparametric magnetic immunoassays utilizing non-linear signatures of magnetic labels. *J. Magn. Magn. Mater.* **2009**, *321*, 1639–1643. [CrossRef]
11. Tu, L.; Wu, K.; Klein, T.; Wang, J.-P. Magnetic nanoparticles colourization by a mixing-frequency method. *J. Phys. D Appl. Phys.* **2014**, *47*, 155001. [CrossRef]
12. Achtsnicht, S.; Pourshahidi, A.M.; Offenhäusser, A.; Krause, H.-J. Multiplex Detection of Different Magnetic Beads Using Frequency Scanning in Magnetic Frequency Mixing Technique. *Sensors* **2019**, *19*, 2599. [CrossRef]
13. Shasha, C.; Krishnan, K.M. Nonequilibrium Dynamics of Magnetic Nanoparticles with Applications in Biomedicine. *Adv. Mater.* **2020**, e1904131. [CrossRef]
14. Hong, H.; Lim, E.-G.; Jeong, J.-C.; Chang, J.; Shin, S.-W.; Krause, H.-J. Frequency Mixing Magnetic Detection Scanner for Imaging Magnetic Particles in Planar Samples. *J. Vis. Exp.* **2016**. [CrossRef]
15. Achtsnicht, S. Multiplex Magnetic Detection of Superparamagnetic Beads for the Identification of Contaminations in Drinking Water. Ph.D. Thesis, RWTH Aachen University, Aachen, Germany, 2020.
16. Krishnan, K.M. *Fundamentals and Applications of Magnetic Materials*; Oxford University Press: Oxford, UK, 2016; ISBN 0199570442.
17. Gubin, S.P. *Magnetic Nanoparticles*; WILEY-VCH Verlag GmbH & Co.: Weinheim, Germany, 2009.
18. Gilbert, T.L. Classics in Magnetics A Phenomenological Theory of Damping in Ferromagnetic Materials. *IEEE Trans. Magn.* **2004**, *40*, 3443–3449. [CrossRef]
19. Usov, N.A.; Liubimov, B.Y. Dynamics of magnetic nanoparticle in a viscous liquid: Application to magnetic nanoparticle hyperthermia. *J. Appl. Phys.* **2012**, *112*, 23901. [CrossRef]
20. Engelmann, U.M. *Assessing Magnetic Fluid Hyperthermia: Magnetic Relaxation Simulation, Modeling of Nanoparticle Uptake inside Pancreatic Tumor Cells and In Vitro Efficacy*, 1st ed.; Infinite Science Publication: Lübeck, Germany, 2019; ISBN 978-3945954584.
21. Shasha, C. Nonequilibrium Nanoparticle Dynamics for the Development of Magnetic Particle Imaging. Ph.D. Thesis, University of Washington, Seattle, WA, USA, 2019.
22. Engelmann, U.M.; Shasha, C.; Slabu, I. Magnetic Nanoparticle Relaxation in Biomedical Application: Focus on Simulating Nanoparticle Heating. In *Magnetic Nanoparticles in Human Health and Medicine: Current Medical Applications and Alternative Therapy of Cancer*; in press; John Wiley & Sons, Inc.: Hoboken, NJ, USA, 2021; ISBN 978-1119754671.
23. Shah, S.A.; Reeves, D.B.; Ferguson, R.M.; Weaver, J.B.; Krishnan, K.M. Mixed Brownian alignment and Néel rotations in superparamagnetic iron oxide nanoparticle suspensions driven by an ac field. *Phys. Rev. B* **2015**, *92*, 94438. [CrossRef]
24. Aali, H.; Mollazadeh, S.; Khaki, J.V. Single-phase magnetite with high saturation magnetization synthesized via modified solution combustion synthesis procedure. *Ceram. Int.* **2018**, *44*, 20267–20274. [CrossRef]
25. Darton, N.J.; Ionescu, A.; Llandro, J. *Magnetic Nanoparticles in Biosensing and Medicine*; Cambridge University Press: Cambridge, UK, 2019; ISBN 1107031095.
26. Yu, E.Y.; Bishop, M.; Zheng, B.; Ferguson, R.M.; Khandhar, A.P.; Kemp, S.J.; Krishnan, K.M.; Goodwill, P.W.; Conolly, S.M. Magnetic particle imaging: A novel in vivo imaging platform for cancer detection. *Nano Lett.* **2017**, *17*, 1648–1654. [CrossRef]
27. Soto-Aquino, D.; Rinaldi, C. Nonlinear energy dissipation of magnetic nanoparticles in oscillating magnetic fields. *J. Magn. Magn. Mater.* **2015**, *393*, 46–55. [CrossRef]

28. Grüttner, C.; Teller, J.; Schütt, W.; Westphal, F.; Schümichen, C.; Paulke, B.R. Preparation and Characterization of Magnetic Nanospheres for in Vivo Application. In *Scientific and Clinical Applications of Magnetic Carriers*; Häfeli, U., Schütt, W., Teller, J., Zborowski, M., Eds.; Springer: Boston, MA, USA, 1997; ISBN 978-1-4757-6482-6_4.
29. Kallumadil, M.; Tada, M.; Nakagawa, T.; Abe, M.; Southern, P.; Pankhurst, Q.A. Suitability of commercial colloids for magnetic hyperthermia. *J. Magn. Magn. Mater.* **2009**, *321*, 1509–1513. [CrossRef]
30. Yoshida, T.; Enpuku, K. Simulation and Quantitative Clarification of AC Susceptibility of Magnetic Fluid in Nonlinear Brownian Relaxation Region. *Jpn. J. Appl. Phys.* **2009**, *48*, 127002. [CrossRef]
31. Ludwig, F.; Remmer, H.; Kuhlmann, C.; Wawrzik, T.; Arami, H.; Ferguson, R.M.; Krishnan, K.M. Self-consistent magnetic properties of magnetite tracers optimized for magnetic particle imaging measured by ac susceptometry, magnetorelaxometry and magnetic particle spectroscopy. *J. Magn. Magn. Mater.* **2014**, *360*, 169–173. [CrossRef] [PubMed]
32. Engelmann, U.M.; Shasha, C.; Teeman, E.; Slabu, I.; Krishnan, K.M. Predicting size-dependent heating efficiency of magnetic nanoparticles from experiment and stochastic Néel-Brown Langevin simulation. *J. Magn. Magn. Mater.* **2019**, *471*, 450–456. [CrossRef]
33. Tay, Z.W.; Hensley, D.W.; Vreeland, E.C.; Zheng, B.; Conolly, S.M. The relaxation wall: Experimental limits to improving MPI spatial resolution by increasing nanoparticle core size. *Biomed. Phys. Eng. Express* **2017**, *3*, 35003. [CrossRef] [PubMed]
34. Shasha, C.; Teeman, E.; Krishnan, K.M. Nanoparticle core size optimization for magnetic particle imaging. *Biomed. Phys. Eng. Express* **2019**, *5*, 55010. [CrossRef]
35. Tong, S.; Quinto, C.A.; Zhang, L.; Mohindra, P.; Bao, G. Size-dependent heating of magnetic iron oxide nanoparticles. *ACS Nano* **2017**, *11*, 6808–6816. [CrossRef]
36. Engelmann, U.M.; Seifert, J.; Mues, B.; Roitsch, S.; Ménager, C.; Schmidt, A.M.; Slabu, I. Heating efficiency of magnetic nanoparticles decreases with gradual immobilization in hydrogels. *J. Magn. Magn. Mater.* **2019**, *471*, 486–494. [CrossRef]
37. Dennis, C.L.; Krycka, K.L.; Borchers, J.A.; Desautels, R.D.; van Lierop, J.; Huls, N.F.; Jackson, A.J.; Gruettner, C.; Ivkov, R. Internal Magnetic Structure of Nanoparticles Dominates Time-Dependent Relaxation Processes in a Magnetic Field. *Adv. Funct. Mater.* **2015**, *25*, 4300–4311. [CrossRef]
38. Engelmann, U.M.; Buhl, E.M.; Draack, S.; Viereck, T.; Ludwig, F.; Schmitz-Rode, T.; Slabu, I. Magnetic relaxation of agglomerated and immobilized iron oxide nanoparticles for hyperthermia and imaging applications. *IEEE Magn. Lett.* **2018**, *9*, 1–5. [CrossRef]
39. Branquinho, L.C.; Carrião, M.S.; Costa, A.S.; Zufelato, N.; Sousa, M.H.; Miotto, R.; Ivkov, R.; Bakuzis, A.F. Effect of magnetic dipolar interactions on nanoparticle heating efficiency: Implications for cancer hyperthermia. *Sci. Rep.* **2013**, *3*, 2887. [CrossRef]
40. Ilg, P.; Kröger, M. Dynamics of interacting magnetic nanoparticles: Effective behavior from competition between Brownian and Néel relaxation. *Phys. Chem. Chem. Phys.* **2020**, *22*, 22244–22259. [CrossRef]
41. Landi, G.T. Role of dipolar interaction in magnetic hyperthermia. *Phys. Rev. B* **2014**, *89*, 014403. [CrossRef]
42. Ilg, P. Equilibrium magnetization and magnetization relaxation of multicore magnetic nanoparticles. *Phys. Rev. B* **2017**, *95*, 214427. [CrossRef]
43. Ficko, B.W.; NDong, C.; Giacometti, P.; Griswold, K.E.; Diamond, S.G. A Feasibility Study of Nonlinear Spectroscopic Measurement of Magnetic Nanoparticles Targeted to Cancer Cells. *IEEE Trans. Biomed. Eng.* **2017**, *64*, 972–979. [CrossRef] [PubMed]
44. Chantrell, R.; Popplewell, J.; Charles, S. Measurements of particle size distribution parameters in ferrofluids. *IEEE Trans. Magn.* **1978**, *14*, 975–977. [CrossRef]

Article

Highly Sensitive Fluorescent Detection of Acetylcholine Based on the Enhanced Peroxidase-Like Activity of Histidine Coated Magnetic Nanoparticles

Hong Jae Cheon [†], Quynh Huong Nguyen [†] and Moon Il Kim *

Department of BioNano Technology, Gachon University, 1342 Seongnamdae-ro, Sujeong-gu, Seongnam 13120, Gyeonggi, Korea; hjchun1201@naver.com (H.J.C.); pruedence122@gmail.com (Q.H.N.)
* Correspondence: moonil@gachon.ac.kr; Tel.: +82-31-750-8563
† These authors contributed equally to this work.

Abstract: Inspired by the active site structure of natural horseradish peroxidase having iron as a pivotal element with coordinated histidine residues, we have developed histidine coated magnetic nanoparticles (His@MNPs) with relatively uniform and small sizes (less than 10 nm) through one-pot heat treatment. In comparison to pristine MNPs and other amino acid coated MNPs, His@MNPs exhibited a considerably enhanced peroxidase-imitating activity, approaching 10-fold higher in catalytic reactions. With the high activity, His@MNPs then were exploited to detect the important neurotransmitter acetylcholine. By coupling choline oxidase and acetylcholine esterase with His@MNPs as peroxidase mimics, target choline and acetylcholine were successfully detected via fluorescent mode with high specificity and sensitivity with the limits of detection down to 200 and 100 nM, respectively. The diagnostic capability of the method is demonstrated by analyzing acetylcholine in human blood serum. This study thus demonstrates the potential of utilizing His@MNPs as peroxidase-mimicking nanozymes for detecting important biological and clinical targets with high sensitivity and reliability.

Keywords: histidine coated magnetic nanoparticles; peroxidase mimics; nanozyme; acetylcholine detection; Alzheimer's disease

Citation: Cheon, H.J.; Nguyen, Q.H.; Kim, M.I. Highly Sensitive Fluorescent Detection of Acetylcholine Based on the Enhanced Peroxidase-Like Activity of Histidine Coated Magnetic Nanoparticles. Nanomaterials 2021, 11, 1207. https://doi.org/10.3390/nano11051207

Academic Editor: Paolo Arosio

Received: 1 April 2021
Accepted: 28 April 2021
Published: 1 May 2021

Publisher's Note: MDPI stays neutral with regard to jurisdictional claims in published maps and institutional affiliations.

Copyright: © 2021 by the authors. Licensee MDPI, Basel, Switzerland. This article is an open access article distributed under the terms and conditions of the Creative Commons Attribution (CC BY) license (https://creativecommons.org/licenses/by/4.0/).

1. Introduction

Acetylcholine (ACh) is a vitally important neurotransmitter that is closely related to functions of both the peripheral and central nervous systems that control various human physiological processes and behaviors. Abnormal amounts of ACh, thus, are able to provoke several cognitive or neural disorders, consisting of progressing dementia, schizophrenia, and Parkinson's and Alzheimer's diseases [1–4]. Alzheimer's disease, in fact, belongs to the top 10 leading causes of death in the United State and is considered as one of the major healthcare and economic burdens across the globe [5,6]. In order to determine the level of ACh in clinical samples, several conventional methods such as mass spectrometry, gas chromatography, and high performance liquid chromatography have been exploited [2,7,8]. Though these approaches are sensitive and reliable, they are generally expensive and somewhat burdensome to perform, as they require sample pretreatment and sophisticated instrumentations as well as skilled technicians, which can only be found in centralized hospitals and laboratories [2,7,8]. The prolonged time for diagnosis is also a hurdle that prevents these methods from being broadly utilized in regular testing of ACh in clinical samples. Therefore, it is crucial to develop a simple, sensitive, but reliable approach for frequent ACh level examination that can not only help detect and diagnose the disease at premature stages but also have better clinical preventions and treatments [9].

Nanomaterials that can perform enzyme-like activities are defined as nanozymes and considered to be the rival for natural enzymes in aspects such as high stability, low cost, and the robustness to be stored for a relatively long time [10]. Since the first peroxidase-mimicking nanozymes, magnetic nanoparticles (MNPs), were discovered in 2007 [11], plenty of nanozymes have been developed and studied [10,12–16]. However, MNPs were still considered as fascinating materials in numerous biosensing fields [17–21] and biomedical applications such as drug delivery, magnetic hyperthermia, and biomedical imaging, owing to their inherent biocompatibility, low toxicity, as well as their exotic response to the magnetic field [22,23]. Though the peroxidase-mimicking MNPs are more stable in a broad range of pH and temperature values than natural horseradish peroxidase (HRP), their practical applications in the sensing field are somewhat limited due to their relatively low catalytic activity and specificity [18]. Thus, further modifications for size and morphology optimization, or surface coating to ameliorate enzymatic activity of MNPs with a simple synthetic procedure, are still an attractive topic in this research field. In fact, there have been many attempts to modify the surface of MNPs with other molecules, including SiO_2, 3-aminopropyl triethoxysilane (APTES), dextran, and poly (ethylene glycol) (PEG). Unfortunately, these alterations could lead to surface shield effects that prevent catalytic sites of MNPs from substrates, resulting in lower catalytic performances [11]. Other studies utilized active polymers such as chitosan to protect MNPs from oxidation as well as provide active amino groups for subsequent immobilizations or attachments instead of boosting catalytic activity [24,25]. Therefore, it is necessary to develop other efficient strategies to modify MNPs that perform better peroxidase activity.

Motivated by the necessity to enhance the catalytic activity of MNPs, we have developed histidine coated magnetic nanoparticles (His@MNPs), inspired by the natural architecture of the active site of HRP, which contains a heme group having iron as central to its catalytic molecules and two coordinated histidine residues. The as-synthesized His@MNPs possessed uniform spherical morphology with application-friendly sizes of less than 10 nm and exhibited much higher peroxidase activity compared to pristine or other amino acids coated MNPs. With the amplified catalytic activity, the newly developed His@MNPs were applied to detect ACh and choline levels in clinical samples. The simple synthesis and effortless detection procedure of this system shows a promise for robust and trustworthy detection of ACh with reduced costs and intricacies.

2. Materials and Methods

2.1. Materials

Iron (II) chloride tetrahydrate, L-alanine (Ala), L-arginine (Arg), L-histidine (His), L-lysine (Lys), L-methionine (Met), sodium hydroxide, choline chloride, acetylcholine chloride (ACh), choline oxidase (ChOx), acetylcholinesterase (AChE) from human erythrocytes, human serum, 3,3′,5,5′-tetramethylbenzidine (TMB), lactose, aspartic acid, glycine, glutamic acid, and glutathione were purchased from Sigma-Aldrich (Milwaukee, WI). Hydrogen peroxide (30%, H_2O_2) was purchased from Junsei Chemical Co. (Tokyo, Japan). Amplex UltraRed (AUR) reagent was purchased from Thermo Fisher Scientific (Waltham, MA, USA). Choline/Acetylcholine assay kit (ab6534) was purchased from Abcam company (Cambridge, MA, USA). All other chemicals were of analytical grade or higher, and used without further purification. All solutions were prepared with DI water purified by a Milli-Q Purification system (Millipore Sigma, Burlington, MA, USA).

2.2. One-Pot Synthesis of Histidine and other Amino Acids Coated Magnetic Nanoparticles

His@MNPs was synthesized as follows. Briefly, 0.3 g of His was added into 50 mL of 75.84 mM aqueous $FeCl_2 \cdot 4H_2O$ solution, followed by vigorous stirring for 12 h at 25 °C. Subsequently, the above mixture was added dropwise to 50 mL of 200 mM sodium hydroxide solution with continuous stirring and incubated during the course of 30 min for reaction. Afterwards, the black precipitates were collected by the magnet, followed by several washing steps with ethanol and distilled water. Finally, the resulting His@MNPs were dried in a vacuum oven at 60 °C. As controls, other amino acids coated MNPs, including Met, Arg, Ala, and Lys coated MNPs, were also prepared as aforementioned procedures except the kinds of employed amino acid.

2.3. Characterization of His@MNPs

The as-prepared His@MNPs were characterized by using scanning transmission electron microscopy (STEM). To examine the element compositions, the energy-dispersive spectrometer (EDS) (Bruker, Billerica, MA, USA) was employed. Fourier transform infrared (FT-IR) spectra of His@MNPs were obtained using FT-IR spectrophotometer (FT-IR-4600, JASCO, Easton, MD, USA). The crystal structure of nanomaterials was determined by utilizing X-ray diffraction (Rigaku Corporation, Tokyo, Japan).

2.4. Investigation for the Peroxidase-Like Activity of His@MNPs and Controls

The peroxidase-mimicking activity of His@MNPs was demonstrated via the oxidation of TMB in the presence of H_2O_2. Generally, 100 µg/mL of His@MNPs was added into sodium acetate buffer (0.1 M, pH 4.0) containing 100 mM H_2O_2 and 5 mM TMB, followed by 10 min incubation at 40 °C. After the reaction, the color change from colorless to blue of oxidized TMB can be observed by the naked eye, and data were recorded in scanning mode by employing a microplate reader (Synergy H1, BioTek, Winooski, VT, USA). The oxidation of controls, including pristine MNPs as well as other amino acids coated MNPs (Met@MNPs, Arg@MNPs, Ala@MNPs and Lys@MNPs), were also carried out and recorded for comparisons.

The peroxidase-imitating activities of His@MNPs and pristine MNPs were further elucidated via steady-state kinetic analysis deploying TMB and H_2O_2 as substrates. The investigations were designed by varying the concentration of one substrate concentration while fixing another substrate at a saturated concentration. For the kinetic assay of TMB, 1 M of H_2O_2 was added to 1 mL of reaction buffer with varying concentrations of TMB from 3 µM to 800 µM. For H_2O_2-dependent kinetic assay, 800 µM of TMB was utilized with varying concentrations of H_2O_2 from 0.03 M to 1 M. Kinetic parameters including the apparent Michaelis-Menten constant (K_m) and maximum initial velocity (V_{max}) were computed using the Lineweaver-Burk plots and Michaelis-Menten equation as follows: $v = V_{max} \times 1/(K_m + [S])$, where v is the initial velocity and $[S]$ is the concentration of substrate [3,11]. The catalytic constant k_{cat} was stemmed from $k_{cat} = V_{max}/[E]$, where $[E]$ is the iron concentration quantified by inductively coupled plasma atomic emission spectroscopy (ICP-AES, Polyscan 60E, Thermo Jarrell Ash, Franklin, MA, USA) method.

2.5. Detection of H_2O_2, Choline, and ACh Using His@MNPs

Fluorescent detection of H_2O_2, choline, and ACh was conducted by utilizing AUR as a substrate in a black 96-well plate, and the data were recorded by using a microplate reader. For the detection of choline and ACh, Tris-acetate buffer at pH 7.0 was utilized due to the pH-dependent nature of ChOx and AChE, which show their activities at around neutral pH [26,27]. Regarding H_2O_2 detection, a total 200 µL solution consisting of 100 µg/mL His@MNPs, 10 µM AUR, and H_2O_2 at various concentrations (0–250 mM) was incubated at 40 °C for 15 min. The resulting fluorescence signals of oxidized AUR were recorded by a microplate reader with the excitation and emission wavelength at 490 nm and 590 nm, respectively. In terms of choline determination, the experiment was conducted via the following procedure. A total 200 µL of mixture reaction containing His@MNP (100 µg/mL), ChOx (0.6 U/mL), AUR (10 µM), and different concentrations of choline (0–200 µM) in Tris-acetate buffer (10 mM, pH 7.0) was incubated at 40 °C for 15 min. The measurement was performed by following the same aforementioned procedures for detection of H_2O_2. ACh detection by His@MNPs was carried out by incubating His@MNP (100 µg/mL), AChE (5 U/mL), ChOx (0.6 U/mL), AUR (10 µM), and ACh with different concentrations (0–250 µM) in the Tris-acetate buffer for 15 min at 40 °C. The signal results were recorded by the same procedure described above.

For the selectivity assay toward ACh, several potentially interfering biological compounds, such as carbohydrates (glucose, lactose), amino acids (aspartic acid, glycine, and glutamic acid), and biothiol (glutathione), were employed with 10-times higher concentration (100 µM) in comparison with target ACh (10 µM).

The concentrations of ACh in spiked human blood serum samples were determined through the standard addition method. Prior to addition, the human serum was diluted to 10% and predetermined amounts of ACh (3, 5, and 8 µM) were added to make spiked samples. Accordingly, the recovery rate [recovery (%) = measured value/added value × 100] and the coefficient of variation [CV (%) = SD/average × 100] were computed as the described equations to assess the precision and reproducibility of the assays.

3. Results and Discussion

3.1. Synthesis and Characterization of His@MNPs

A simple but efficient one-pot heat-treatment method was developed to synthesize His@MNPs that exhibit significantly amplified peroxidase-like activity and were utilized to detect ACh. The detection regime was based on the consecutive catalytic reactions triggered from ACh, with the involvement of multiple enzymes including AChE and ChOx, to create H_2O_2 as a final, direct product for peroxidase reaction. The peroxidase reaction was performed by His@MNPs, which oxidize the AUR substrate into a highly fluorescent product (AURox) (Scheme 1). The prepared materials then were well characterized by different techniques. Firstly, the morphology and size of His@MNPs and pristine MNPs were examined via TEM analysis, as shown in Figure 1a,b. The TEM images revealed that both pristine MNPs and His@MNPs were well prepared with uniform, spherical shapes and sizes of less than 10 nm, which are desirable for catalytic applications. The size of His@MNPs was larger than that of pristine MNPs (roughly 2–3 nm larger), which can plausibly be attributed to the histidine coating layer. To validate that, EDS analysis was deployed to identify the contributing elements of His@MNPs. From Figure 1c, it can be observed that all the components are well distributed. The occurrence of a nitrogen element in His@MNPs instead of unmodified MNPs proved that the histidine layer was well deposited. Additionally, the difference in zeta potential between His@MNPs and MNPs could also be deduced from the presence of a histidine layer on the surface of MNPs (Figure S1). The signature XRD patterns of pristine MNPs are shown in Figure S2a, which was well aligned with the standard JCPDS database (JCPDS 00-019-0629). These peaks are strong and distinct, implying the good crystallinity of bare MNPs. In respect to His@MNPs, though the peaks pattern is analogous to that of pristine MNP, the intensities are apparently decreased, which indicates the presence of a histidine coating layer. In the FT-IR spectra (Figure S2b), the characteristic peaks at 540 cm^{-1} and at around 3300–3400 cm^{-1}, which derived from Fe-O bond and O-H stretching vibration, respectively, were observed in both MNPs and His@MNPs, proving the occurrence of iron oxide particles [28]. Furthermore, the appearance of a peak at 1120 and 1386 cm^{-1} represented C-O stretching and COO- group originating from histidine residues. The broad peak at 1630 cm^{-1} in the His@MNPs pattern also indicated the presence of C = O stretching frequency which derived from histidine [29]. According to these results, the coating of histidine on the surface of MNPs was consolidated.

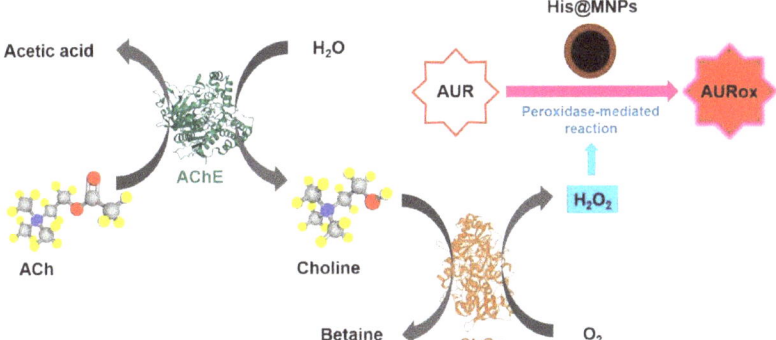

Scheme 1. Schematic illustration of ACh detection utilizing multiple enzymes and peroxidase-mimicking His@MNPs.

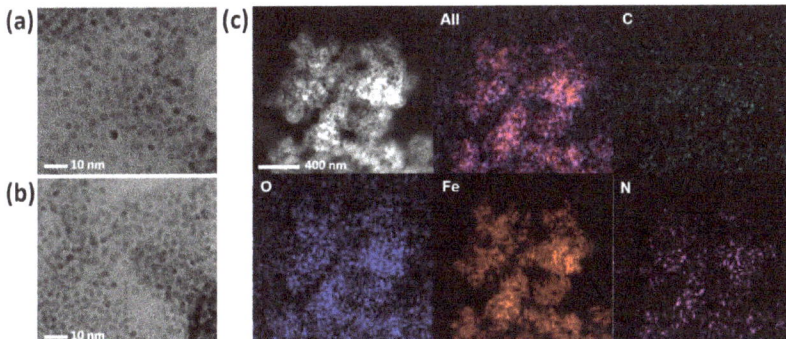

Figure 1. TEM analysis of (**a**) His@MNPs and (**b**) MNPs. (**c**) EDS analysis with the TEM images of His@MNPs.

3.2. Enhanced Peroxidase-Like Activity of His@MNPs

The catalytic activity of His@MNPs was investigated via the peroxidase-facilitated oxidation of colorimetric reagent TMB as substrate in the presence of H_2O_2. Due to the peroxidase-like effect of His@MNPs, reactive oxygen species (ROS) were created which further react with TMB to produce oxidized TMB, and thus the color of the reaction solution changed from colorless to blue. For comparison, the peroxidase-imitating activities of other amino acids coated MNPs, including Met@MNPs, Arg@MNPs, Ala@MNPs, and Lys@MNPs, as well as the pristine MNPs, were also examined. According to the results, the catalytic activity of His@MNPs was 10-fold higher compared to bare MNPs and was significantly higher than other amino acids coated MNPs (Figure 2). This result was also equivalent to the previously reported study [30], which describes the importance of mimicking the natural active site of HRP. Ferric and ferrous ions present on the His@MNPs might catalyze the peroxidase reaction, and the histidine residues additionally contributed to the catalysis, like the natural active site of HRP. Preliminary optimizations related to temperature and pH effects on the peroxidase-like activity of His@MNPs were also conducted. The results, henceforth, revealed that pH 4.0 and temperature at 40~50 °C were optimum conditions for the catalytic activity of His@MNPs (Figure S3).

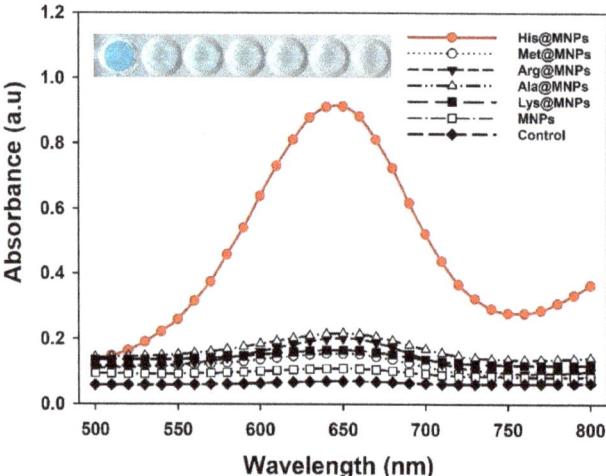

Figure 2. Investigation for the peroxidase-like activities of various amino acids coated MNPs and pristine MNPs. In the real image, the order is corresponding to the order of graph from top to bottom.

To further understand the affinity between His@MNPs and substrates (TMB and H_2O_2), steady-state kinetic values including K_m, V_{max}, and k_{cat} were computed and compared with those obtained from pristine MNPs. Figure 3 showed the typical Michaelis-Menten curves of initial velocities against various concentrations of TMB and H_2O_2 which were further used to draw Lineweaver-Burk plots. The Lineweaver-Burk plots then were applied to determine K_m and V_{max}. Table 1 summarized the catalytic parameters of His@MNPs, unmodified pristine MNPs, HRP, and other peroxidase-like nanozymes which were recently reported [2,3,11,16,31]. According to the Table 1, it is noticeable that the K_m value (TMB) for His@MNPs was approximately 3 times higher than that of natural HRP. However, the K_m value of His@MNPs toward H_2O_2 was roughly half that of bare MNPs, indicating that the histidine modification efficiently increased the affinity toward H_2O_2 rather than TMB, as similarly reported previously [30]. Though His@MNP showed marginally lower affinity toward TMB, its catalytic efficiency (k_{cat}) was determined to be higher than those obtained from other nanozymes and HRP. The high k_{cat} value could be attributed to the presence of many efficient active sites by the histidine modifications on MNPs.

Table 1. Comparison of kinetic constants of His@MNPs, pristine MNPs, HRP, and other peroxidase-mimicking nanozymes with H_2O_2 and TMB. K_m is the Michaelis-Menten constant, k_{cat} is derived from $k_{cat} = V_{max}/[E]$, where V_{max} is the maximal reaction velocity, and [E] is the iron concentration measured by ICP-AES.

	Substrate	[E] (mM)	V_{max} ($\mu M\ s^{-1}$)	K_m (mM)	k_{cat} (s^{-1})	Ref.
MIL-101(Fe)	TMB	-	1.0083	0.585	-	[2]
	H_2O_2	-	0.5138	1.89	-	
BSA-$Cu_3(PO_4)_2$ nanoflower	TMB	-	0.099	0.1709	-	[3]
	H_2O_2	-	0.063	1.00	-	
HRP	TMB	2.5×10^{-11}	0.1	0.434	4×10^3	[11]
	H_2O_2	2.5×10^{-11}	0.087	3.70	3.48×10^3	
NSP-carbon quantum dot	TMB	-	0.188	0.47	-	[16]
	H_2O_2	-	69.50	32.61	-	
FeS_2/SiO_2	TMB	-	0.31	0.948	2.22	[31]
	H_2O_2	-	1.181	0.0126	1.29	

Table 1. Cont.

	Substrate	[E] (mM)	V_{max} ($\mu M\ s^{-1}$)	K_m (mM)	k_{cat} (s^{-1})	Ref.
MNP	TMB	9.7×10^{-6}	0.18	0.142	1.88	This work
	H_2O_2	9.7×10^{-6}	567.83	689.38	5.9×10^4	
His@MNP	TMB	9.7×10^{-6}	0.51	0.149	5.29	This work
	H_2O_2	9.7×10^{-6}	676.45	381.62	3.9×10^4	

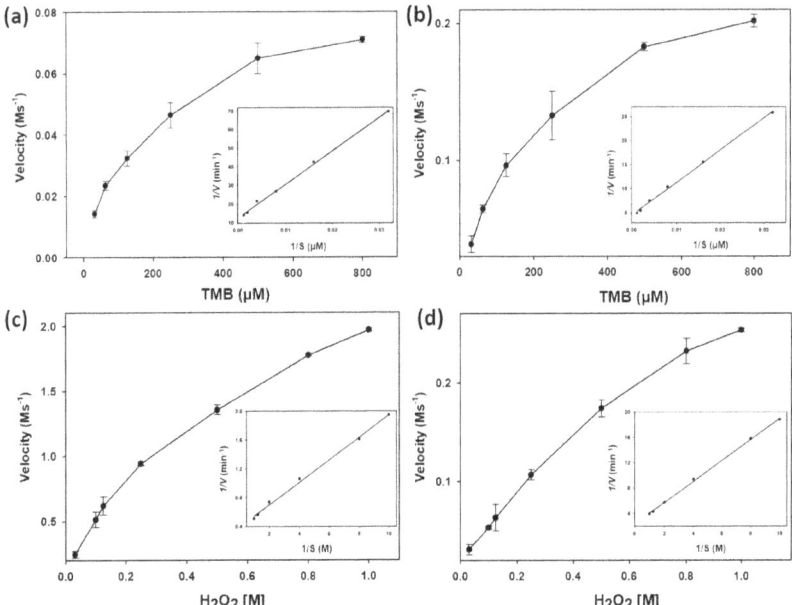

Figure 3. Steady-state kinetic assays of (**a**) His@MNPs and (**b**) pristine MNPs with TMB as substrate and their corresponding double reciprocal (Lineweaver-Burk) plots of activity. H_2O_2 substrate was used in the steady-state kinetic assays performed by (**c**) His@MNPs and (**d**) pristine MNPs.

Unlike natural enzymes, which are susceptible to harsh conditions, nanozymes are highly stable, withstanding even high temperatures or extreme pH conditions, in which most enzymes are denatured and lose their activities. To demonstrate the robustness of His@MNPs, various pH and temperature conditions were applied and the results were depicted in Figure S4. In particular, His@MNPs conserved their activities at various pH conditions, even at acidic or basic conditions (where it remained over 90% active), whereas HRP lost roughly 60% activity within basic pH conditions. Likewise, HRP began to plunge when the temperature was higher than 30 °C, losing approximately 70% of activity at 37 °C, while His@MNPs still retained acceptable activity over 70%. The minor loss of activity in His@MNPs at high temperature could conceivably be explained due to the histidine organic component. However, the activity loss is marginal, proving the high thermal resistance of the as-developed His@MNPs.

In general, the peroxidase-mimicking activity is demonstrated by confirming the generation of free hydroxyl (·OH) radicals from H_2O_2 [32,33]. To test this, a non-fluorescent TA probe was deployed to examine the formation of free OH radicals, which are produced during the decomposition of H_2O_2. Upon the formation of OH, the non-fluorescent TA molecules transform into highly fluorescent products (2-hydroxy terephthalic acid), which emit a unique fluorescence signal at around 435 nm. It is worth mentioning that, compared to bare MNPs, the amount of free OH radicals generated from His@MNPs was considerably increased as demonstrated in Figure 4, indicating their higher catalytic activity.

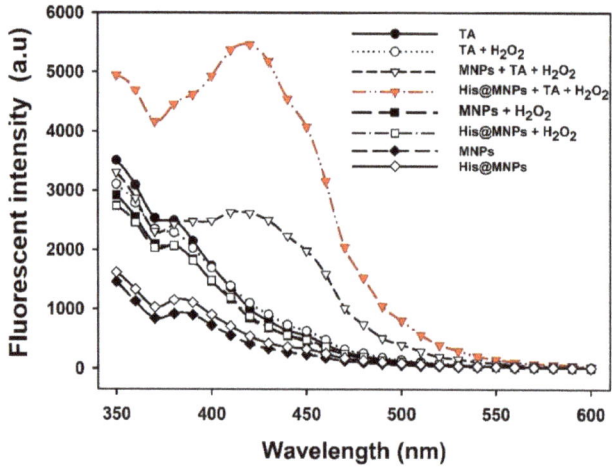

Figure 4. Fluorescence spectra for detecting free hydroxyl (·OH) radicals produced during the peroxidase-mediated decomposition of H_2O_2, utilizing TA as a probe.

3.3. Highly Sensitive Fluorescent Detection of Choline and ACh

The level of ACh in clinical samples was determined via cascade catalytic reactions starting from AChE coupled with ChOx, followed by a peroxidase-facilitated reaction performed by His@MNPs as follows.

$$Acetylcholine \xrightarrow{AChE} Choline + acetic\ acid \quad (1)$$

$$Choline \xrightarrow{ChOx} betaine + H_2O_2 \quad (2)$$

$$H_2O_2 + AUR \xrightarrow{His@MNP} H_2O + AURox \quad (3)$$

According to the above chain reactions, it can be observed that in the occurrence of ACh, enzyme AChE catalyzes its hydrolysis to generate choline, which is the specific substrate for choline oxidase. After choline molecules were cleaved by ChOx, H_2O_2 molecules were formed and subsequently consumed by the peroxidase-mediated reaction catalyzed by His@MNPs. This reaction produces free ·OH radicals that interact with AUR to create oxidized AUR. Based on previous reports, the concentration of ACh in body fluid is extremely low, generally in the nanomolar range [34,35]. Thus, AUR was exploited instead of other peroxidase substrates to monitor the presence of ACh since it can produce a very sensitive, highly fluorescent product (AURox). By recruiting AUR as a peroxidase substrate, the enzymatic reaction catalyzed by His@MNPs was performed at neutral pH (pH 7.0).

Under the aforementioned conditions, the peroxidase activity of His@MNPs to detect H_2O_2 with AUR as a substrate was examined. The results, which were shown in Figure S5, demonstrated that within a suitable range, the fluorescence signals derived from oxidized AUR increased as the concentrations of H_2O_2 raised, which established a good linear correlation ($R^2 > 0.99$). Accordingly, we further investigated the sensitivity of this system towards choline and ACh. In detail, as the concentration of choline increases from 0.3 μM to 5.0 μM, a positively proportional increase was observed in the respective fluorescence intensity, thus constructing a highly linear correlation ($R^2 > 0.99$) (Figures 5a and S6). Calculated from the linear regression equation (y = 0.9449x + 294.63), the limit of detection (LOD) was determined to be 200 nM. The LOD value was calculated based on the formula: LOD = 3 × δ/slope, where δ is the standard deviation of blank and slope is the slope of calibration curve. Subsequently, a convenient, one-step operation was carried out to detect ACh via the multiple enzymatic reactions (AChE, ChOx, and peroxidase-mimicking His@MNPs). Figure 5b as well as Figure S7 showed a qualified linear relationship ($R^2 > 0.99$) between the fluorescent signal and concentration of ACh (0.25 µM to 5 µM). Particularly, a linear equation was established (y = 5.9983x + 539.76) when the concentration of ACh increases from 0.25 µM to 5 µM. Based on this equation, LOD down to 100 nM was achieved, which was lower than those of recent studies [3,8,36].

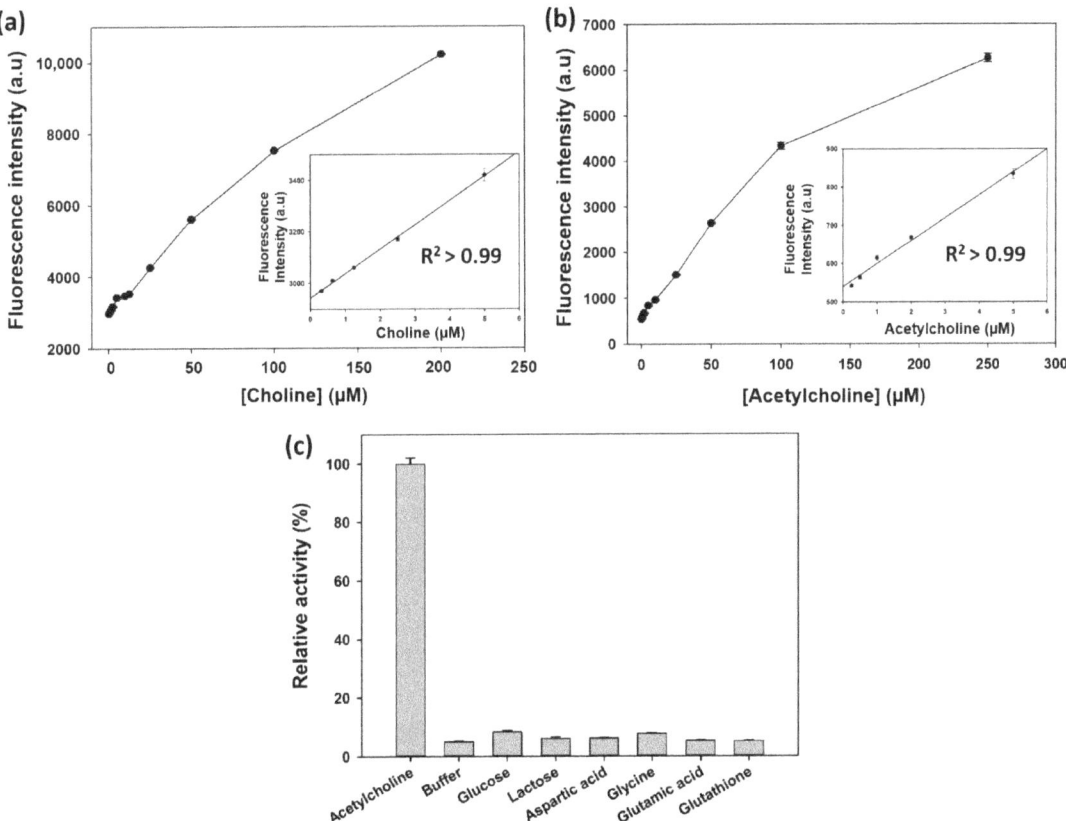

Figure 5. Dose-response curve for the detection of (**a**) choline and (**b**) ACh with their corresponding linear calibration plots using His@MNPs based assay. (**c**) Selectivity of His@MNPs based assay towards ACh. In the assay, concentration of ACh is 10 µM while other interfering substances are all 100 µM.

For biosensors, selectivity is an important criterion to assess the practical property, especially in the diagnostic area [3]. To confirm the selectivity of the developed His@MNPs based sensing regime, several interfering substances were utilized, including glucose, lactose, aspartic acid, glycine, glutamic acid and glutathione. The results illustrated in Figure 5c indicated that only in the presence of Ach can a significant fluorescent signal be undoubtedly recognized, while other substances did not generate any considerable signals, despite the fact that their concentrations were used at 10-times higher compared to the target ACh.

The practical application of the developed regime was further verified by detecting the level of ACh in human serum (10%) via a standard addition method. The obtained results exhibited an excellent precision, yielding CVs in the range of 1.16 to 7.54% and recovery rates from 100.20% to 101.92%, demonstrating a good agreement with the spiked amount of ACh (Table 2). Thus, the results confirm that the developed sensing systems utilizing multiple enzymes and His@MNPs as peroxidase mimics are promising to be applied for the real detection of ACh in clinical settings. Additionally, our His@MNPs based method was compared with a commercial Choline/Acetylcholine Assay Kit (ab65345) for quantitatively determining target acetylcholine. The observed correlation between the two methods was $R^2 > 0.99$ (Figure S8). This high concordance between the two methods ensured the validity of the His@MNPs based strategy for reliable quantitative determination of acetylcholine.

Table 2. Detection precision of the His@MNPs based assay for the determination of ACh levels in spiked human serum sample.

Sample	Added Value (μM)	Measured Value [a] (μM)	SD [b]	CV [c] (%)	Recovery [d] (%)
1	3	3.01	0.23	7.54	100.20
2	5	5.07	0.06	1.16	101.47
3	8	8.15	0.31	3.85	101.92

[a] The average value of six independent measurements. [b] The standard deviation (SD) of six successive measurements. [c] Coefficient of variation. [d] Recovery = (Measured value/Expected value) × 100.

4. Conclusions

In conclusion, we have successfully synthesized very small-sized His@MNPs via a facile one-pot method, yielding their significant enhancement in the peroxidase-like activity. Compared to pristine MNPs, the newly synthesized His@MNPs, which were designed to mimic the structure of active site of natural HRP, exhibited approximately 10-fold higher peroxidase-like activity. The activity enhancement was mainly due to the increased affinity toward H_2O_2 and the presence of many efficient active sites by the histidine modifications on MNPs. By coupling with appropriate enzymes including AChE and ChOx, the developed nanozymes then were successfully exploited to detect ACh with high specificity and sensitivity in which the LOD was recorded to be as low as 100 nM. With the satisfied outcomes, our developed nanozymes with good stability and enhanced peroxidase-mimicking activity are sufficient to be utilized for biological sensing applications.

Supplementary Materials: The following are available online at https://www.mdpi.com/article/10.3390/nano11051207/s1, Methods for investigating the mechanism for the peroxidase-like activity of His@MNPs, Figure S1: Zeta potential values of His@MNPs and pristine MNPs in sodium acetate buffer (0.1M, pH 4.0), Figure S2: The graph (a) XRD and (b) FT-IR spectra of His@MNPs and pristine MNPs, Figure S3: Effect of (a) pH and (b) temperature on the catalytic activity for oxidation of TMB catalyzed by His@MNPs, Figure S4: Catalytic stabilities of His@MNPs and pristine MNPs against (a) pH and (b) temperature, Figure S5: Dose-response curve for the detection of H_2O_2 and their corresponding linear calibration plot using His@MNPs as peroxidase mimics and AUR substrate, Figure S6: Fluorescence emission spectra for the detection of choline, Figure S7: Fluorescence emission spectra for the detection of ACh, Figure S8: Correlation between His@MNPs based method (x) and Choline/Acetylcholine detection kit (y) for the quantitative determination of ACh.

Author Contributions: Conceptualization, investigation, writing—original draft preparation, H.J.C.; validation, writing—review and editing, Q.H.N. and conceptualization, supervision, writing—review and editing, M.I.K. All authors have read and agreed to the published version of the manuscript.

Funding: This work was supported by a National Research Foundation of Korea (NRF) grant funded by the Korea government (Ministry of Science and ICT (NRF-2019R1A2C1087459) and by the Korean National Police Agency (Project name: Development of visualization technology for biological evidence in crime scenes based on nano-biotechnology/Project Number: PA-K000001-2019-401).

Data Availability Statement: Not applicable.

Conflicts of Interest: The authors declare no conflict of interest.

References

1. Bolat, E.Ö.; Tığ, G.A.; Pekyardımcı, Ş. Fabrication of an amperometric acetylcholine esterase-choline oxidase biosensor based on MWCNTs-Fe$_3$O$_4$ NPs-CS nanocomposite for the determination of aetylcholine. *J. Electroanal. Chem.* **2017**, *758*, 241–248. [CrossRef]
2. Gou, J.; Wu, S.; Wang, Y.; Zhao, M.A. A label-free fluorescence biosensor based on a bifunctional MIL-101 (Fe) nanozyme for sensitive detection of choline and acetylcholine at nanomolar level. *Sens. Actuator B Chem.* **2020**, *312*, 128021.
3. Kong, D.; Jin, R.; Zhao, X.; Li, H.; Yan, X.; Liu, F.; Sun, P.; Gao, Y.; Liang, X.; Lin, Y.J.; et al. Protein–inorganic hybrid nanoflower-rooted agarose hydrogel platform for point-of-care detection of acetylcholine. *ACS Appl. Mater. Interfaces* **2019**, *11*, 11857–11864. [CrossRef] [PubMed]
4. Hampel, H.; Mesulam, M.; Cuello, A.C.; Farlow, M.R.; Giacobini, E.; Grossberg, G.T.; Khachaturian, A.S.; Vergallo, A.; Cavedo, E.; Snyder, P.J.; et al. The cholinergic system in the pathophysiology and treatment of Alzheimer's disease. *Brain* **2018**, *141*, 1917–1933. [CrossRef] [PubMed]
5. Castro, D.M.; Dillon, C.; Machnicki, G.; Allegri, R.F. The economic cost of Alzheimer's disease: Family or public-health burden? *Dement. Neuropsychol.* **2010**, *4*, 262–267. [CrossRef] [PubMed]
6. Zhao, L.H. Alzheimer's disease facts and figures. *Alzheimers Dement.* **2020**, *16*, 391–460.
7. Dunphy, R.; Burrinsky, D.J. Detection of choline and acetylcholine in a pharmaceutical preparation using high performance liquid chromatography/electrospray ionization mass spectrometry. *Analysis* **2003**, *31*, 905–915. [CrossRef]
8. He, S.-B.; Wu, G.-W.; Deng, H.-H.; Liu, A.-L.; Lin, X.-H.; Xia, X.-H.; Chen, W. Choline and acetylcholine detection based on peroxidase-like activity and protein antifouling property of platinum nanoparticles in bovine serum albumin scaffold. *Biosens. Bioelectron.* **2014**, *62*, 331–336. [CrossRef]
9. Craig, L.A.; Hong, N.S.; McDonald, R.J. Revisiting the cholinergic hypothesis in the development of Alzheimer's disease. *Neurosci. Biobehav. Rev.* **2011**, *35*, 1397–1409. [CrossRef]
10. Huang, Y.; Ren, J.; Qu, X. Nanozymes: Classification, catalytic mechanisms, activity regulation, and applications. *Chem. Rev.* **2019**, *119*, 4357–4412. [CrossRef]
11. Gao, L.; Zhuang, J.; Nie, L.; Zhang, J.; Zhang, Y.; Gu, N.; Wang, T.; Feng, J.; Yang, D.; Perrett, S.J. Intrinsic peroxidase-like activity of ferromagnetic nanoparticles. *Nat. Nanotechnol.* **2007**, *2*, 577–583. [CrossRef]
12. Liu, Y.; Zhu, S.; Wei, Y.; Liu, X.; Jiao, S.; Yang, J. Ultrasensitive detection of miRNA-155 based on controlled fabrication of AuNPs@MoS$_2$ nanostructures by atomic layer deposition. *Biosens. Bioelectron.* **2019**, *144*, 111660. [CrossRef]
13. Wang, L.; Hu, Z.; Wu, S.; Pan, J.; Xu, X.; Niu, X. A peroxidase-mimicking Zr-based MOF colorimetric sensing array to quantify and discriminate phosphorylated proteins. *Anal. Chim. Acta* **2020**, *1121*, 26–34. [CrossRef] [PubMed]
14. Loynachan, C.N.; Thomas, M.R.; Gray, E.R.; Richards, D.A.; Kim, J.; Miller, B.S.; Brookes, J.C.; Agarwal, S.; Chudasama, V.; McKendry, R.A. Platinum nanocatalyst amplification: Redefining the gold standard for lateral flow immunoassays with ultrabroad dynamic range. *ACS Nano* **2018**, *12*, 279–288. [CrossRef] [PubMed]
15. Nguyen, P.T.; Ahn, H.T.; Kim, M.I. Reagent-free colorimetric assay for galactose using agarose gel entrapping nanoceria and galactose oxidase. *Nanomaterials* **2020**, *10*, 895. [CrossRef]
16. Tripathi, K.M.; Ahn, H.T.; Chung, M.; Le, X.A.; Saini, D.; Bhati, A.; Sonkar, S.K.; Kim, M.I.; Kim, T, N, S, and P-Co-doped carbon quantum dots: Intrinsic peroxidase activity in a wide pH range and its antibacterial applications. *ACS Biomater. Sci. Eng.* **2020**, *6*, 5527–5537. [CrossRef] [PubMed]
17. Duan, D.; Fan, K.; Zhang, D.; Tan, S.; Liang, M.; Liu, Y.; Zhang, J.; Zhang, P.; Liu, W.; Qiu, X.; et al. Nanozyme-strip for rapid local diagnosis of Ebola. *Biosens. Bioelectron.* **2015**, *74*, 134–141. [CrossRef] [PubMed]
18. Shin, H.Y.; Kim, B.-G.; Cho, S.; Lee, J.; Na, H.B.; Kim, M.I. Visual determination of hydrogen peroxide and glucose by exploiting the peroxidase-like activity of magnetic nanoparticles functionalized with a poly (ethylene glycol) derivative. *Microchim. Acta* **2017**, *184*, 2115–2122. [CrossRef]
19. Srinivasan, B.; Li, Y.; Jing, Y.; Xu, Y.; Yao, X.; Xing, C.; Wang, J.-P. A detection system based on giant magnetoresistive sensors and high-moment magnetic nanoparticles demonstrates zeptomole sensitivity: Potential for personalized medicine. *Angew. Chem. Int. Ed.* **2009**, *121*, 2802–2805. [CrossRef]
20. Wang, Y.; Dostalek, J.; Knoll, W. Magnetic nanoparticle-enhanced biosensor based on grating-coupled surface plasmon resonance. *Anal. Chem.* **2011**, *83*, 6202–6207. [CrossRef]
21. Yang, L.; Ren, X.; Tang, F.; Zhang, L. A practical glucose biosensor based on Fe$_3$O$_4$ nanoparticles and chitosan/nafion composite film. *Biosens. Bioelectron.* **2009**, *25*, 889–895. [CrossRef] [PubMed]
22. Cardoso, V.F.; Francesko, A.; Ribeiro, C.; Bañobre-López, M.; Martins, P.; Lanceros-Mendez, S. Advances in magnetic nanoparticles for biomedical applications. *Adv. Healthc. Mater.* **2018**, *7*, 1700845. [CrossRef] [PubMed]
23. Slimani, Y.; Hannachi, E. Magnetic nanosensors and their potential applications. In *Nanosensors for Smart Cities*; Han, B., Nguyen, T.A., Singh, P.K., Eds.; Elsevier: London, UK, 2020; Chapter 9; pp. 143–155.
24. Dhavale, R.P.; Dhavale, R.; Sahoo, S.C.; Kollu, P.; Jadhav, S.U.; Patil, P.S.; Dongale, T.D.; Chougale, A.D.; Patil, P.B. Chitosan coated magnetic nanoparticles as carriers of anticancer drug Telmisartan: pH-responsive controlled drug release and cytotoxicity studies. *J. Phys. Chem. Solids* **2021**, *148*, 109749. [CrossRef]

25. Sahin, S.; Ozmen, I. Determination of optimum conditions for glucose-6-phosphate dehydrogenase immobilization on chitosan-coated magnetic nanoparticles and its characterization. *J. Mol. Catal. B Enzym.* **2016**, *133*, S25–S33. [CrossRef]
26. Komersová, A.; Kovářová, M.; Komers, K.; Lochař, V.; Čegan, A. Why is the hydrolytic activity of acetylcholinesterase pH dependent? Kinetic study of acetylcholine and acetylthiocholine hydrolysis catalyzed by acetylcholinesterase from electric eel. *Z. Naturforsch. C J. Biosci.* **2018**, *73*, 345–351. [CrossRef] [PubMed]
27. Silman, H.; Karline, A. Effect of local pH changes caused by substrate hydrolysis on the activity of membrane-bound acetylcholinesterase. *Proc. Natl. Acad. Sci. USA* **1967**, *58*, 1664–1668. [CrossRef] [PubMed]
28. Silva, V.A.; Andrade, P.L.; Silva, M.P.C.; Bustamante, A.; Valladares, L.D.L.S.; Aguiar, J.A. Synthesis and characterization of Fe_3O_4 nanoparticles coated with fucan polysaccharides. *J. Magn. Magn. Mater.* **2013**, *343*, 138–143. [CrossRef]
29. Zhang, W.; Niu, X.; Meng, S.; Li, X.; He, Y.; Pan, J.; Qiu, F.; Zhao, H.; Lan, M. Histidine-mediated tunable peroxidase-like activity of nanosized Pd for photometric sensing of Ag+. *Sens. Actuator B Chem.* **2018**, *273*, 400–407. [CrossRef]
30. Fan, K.; Wang, H.; Xi, J.; Liu, Q.; Meng, X.; Duan, D.; Gao, L.; Yan, X. Optimization of Fe_3O_4 nanozyme activity via single amino acid modification mimicking an enzyme active site. *Chem. Commun.* **2017**, *53*, 424–427. [CrossRef]
31. Huang, X.; Xia, F.; Nan, Z. Fabrication of FeS_2/SiO_2 double mesoporous hollow spheres as an artificial peroxidase and rapid determination of H_2O_2 and glutathione. *ACS Appl. Mater. Interfaces* **2020**, *12*, 46539–46548. [CrossRef]
32. Huang, Z.; He, W.; Shen, H.; Han, G.; Wang, H.; Su, P.; Song, J.; Yang, Y. $NiCo_2S_4$ microflowers as peroxidase mimic: A multi-functional platform for colorimetric detection of glucose and evaluation of antioxidant behavior. *Talanta* **2021**, *230*, 122337. [CrossRef]
33. Karim, M.D.; Singh, M.; Weerathunge, P.; Bian, P.; Zheng, R.; Dekiwadia, C.; Ahmed, T.; Walia, S.; Gaspera, E.D.; Singh, S.; et al. Visible-light-triggered reactive oxygen species mediated antibacterial activity of peroxidase mimic CuO nanorods. *ACS Appl. Nano Mater.* **2018**, *1*, 1694–1704. [CrossRef]
34. Kim, M.S.; Cho, S.; Joo, S.H.; Lee, J.; Kwak, S.K.; Kim, M.I.; Lee, J. N-and B-codoped graphene: A strong candidate to replace natural peroxidase in sensitive and selective bioassays. *ACS Nano* **2019**, *13*, 4312–4321. [CrossRef]
35. Valekar, A.H.; Batule, B.S.; Kim, M.I.; Cho, K.-H.; Hong, D.-Y.; Lee, U.-H.; Chang, J.-S.; Park, H.G.; Hwang, Y.K. Novel amine-functionalized iron trimesates with enhanced peroxidase-like activity and their applications for the fluorescent assay of choline and acetylcholine. *Biosens. Bioelectron.* **2018**, *100*, 161–168. [CrossRef] [PubMed]
36. Su, L.; Yu, X.; Qin, W.; Dong, W.; Wu, C.; Zhang, Y.; Mao, G.; Feng, S. One-step analysis of glucose and acetylcholine in water based on the intrinsic peroxidase-like activity of Ni/Co LDHs microspheres. *J. Mater. Chem. B* **2017**, *5*, 116–122. [CrossRef] [PubMed]

Article

Boosting Magnetoelectric Effect in Polymer-Based Nanocomposites

Alexander Omelyanchik [1,2], Valentina Antipova [1], Christina Gritsenko [1], Valeria Kolesnikova [1], Dmitry Murzin [1], Yilin Han [3], Andrei V. Turutin [4,5], Ilya V. Kubasov [4], Alexander M. Kislyuk [4], Tatiana S. Ilina [4], Dmitry A. Kiselev [4], Marina I. Voronova [4], Mikhail D. Malinkovich [4], Yuriy N. Parkhomenko [4], Maxim Silibin [6,7,8], Elena N. Kozlova [3], Davide Peddis [2,9], Kateryna Levada [1], Liudmila Makarova [1,10], Abdulkarim Amirov [1,11,*] and Valeria Rodionova [1,*]

[1] REC Smart Materials and Biomedical Applications, Immanuel Kant Baltic Federal University, 236041 Kaliningrad, Russia; asomelyanchik@kantiana.ru (A.O.); vantipova1@kantiana.ru (V.A.); christina.byrka@gmail.com (C.G.); vgkolesnikova1@kantiana.ru (V.K.); dvmurzin@yandex.ru (D.M.); elevada@kantiana.ru (K.L.); la.loginova@physics.msu.ru (L.M.)
[2] Department of Chemistry and Industrial Chemistry (DCIC), University of Genova, 16146 Genova, Italy; davide.peddis@unige.it
[3] Biomedical Centre, Department of Neuroscience, Uppsala University, 751 24 Uppsala, Sweden; yilin.han@neuro.uu.se (Y.H.); elena.kozlova@neuro.uu.se (E.N.K.)
[4] Laboratory of Physics of Oxide Ferroelectrics and Department of Materials Science of Semiconductors and Dielectrics, National University of Science and Technology MISiS, 119049 Moscow, Russia; aturutin92@gmail.com (A.V.T.); kubasov.ilya@gmail.com (I.V.K.); akislyuk94@gmail.com (A.M.K.); ilina.tatina@gmail.com (T.S.I.); dm.kiselev@gmail.com (D.A.K.); mvoron@bk.ru (M.I.V.); malinkovich@yandex.ru (M.D.M.); parkh@rambler.ru (Y.N.P.)
[5] Department of Physics and I3N, University of Aveiro, 3810-193 Aveiro, Portugal
[6] Institute of Advanced Materials and Technologies, National Research University of Electronic Technology "MIET", 124498 Moscow, Russia; sil_m@mail.ru
[7] Institute for Bionic Technologies and Engineering, I.M. Sechenov First Moscow State Medical University, 119991 Moscow, Russia
[8] Scientific-Manufacturing Complex "Technological Centre" Shokin Square, House 1, Bld. 7, Zelenograd, 124498 Moscow, Russia
[9] Institute of Structure of Matter–CNR, Monterotondo Stazione, 00016 Rome, Italy
[10] Faculty of Physics, Lomonosov Moscow State University, 1-2 Leninskie Gory, 119234 Moscow, Russia
[11] Amirkhanov Institute of Physics of Dagestan Federal Research Center, Russian Academy of Sciences, 367003 Makhachkala, Russia
* Correspondence: amiroff_a@mail.ru (A.A.); vvrodionova@kantiana.ru (V.R.)

Citation: Omelyanchik, A.; Antipova, V.; Gritsenko, C.; Kolesnikova, V.; Murzin, D.; Han, Y.; Turutin, A.V.; Kubasov, I.V.; Kislyuk, A.M.; Ilina, T.S.; et al. Boosting Magnetoelectric Effect in Polymer-Based Nanocomposites. Nanomaterials 2021, 11, 1154. https://doi.org/10.3390/nano11051154

Academic Editor: Paolo Arosio

Received: 27 March 2021
Accepted: 26 April 2021
Published: 28 April 2021

Publisher's Note: MDPI stays neutral with regard to jurisdictional claims in published maps and institutional affiliations.

Copyright: © 2021 by the authors. Licensee MDPI, Basel, Switzerland. This article is an open access article distributed under the terms and conditions of the Creative Commons Attribution (CC BY) license (https://creativecommons.org/licenses/by/4.0/).

Abstract: Polymer-based magnetoelectric composite materials have attracted a lot of attention due to their high potential in various types of applications as magnetic field sensors, energy harvesting, and biomedical devices. Current researches are focused on the increase in the efficiency of magnetoelectric transformation. In this work, a new strategy of arrangement of clusters of magnetic nanoparticles by an external magnetic field in PVDF and PFVD-TrFE matrixes is proposed to increase the voltage coefficient (α_{ME}) of the magnetoelectric effect. Another strategy is the use of 3-component composites through the inclusion of piezoelectric $BaTiO_3$ particles. Developed strategies allow us to increase the α_{ME} value from ~5 mV/cm·Oe for the composite of randomly distributed $CoFe_2O_4$ nanoparticles in PVDF matrix to ~18.5 mV/cm·Oe for a composite of magnetic particles in PVDF-TrFE matrix with 5%wt of piezoelectric particles. The applicability of such materials as bioactive surface is demonstrated on neural crest stem cell cultures.

Keywords: multiferroics; magnetoelectric effect; nanoparticles; cobalt ferrite; barium titanate; PVDF; PVDF-TrFE

1. Introduction

Multiferroics are a class of material where magnetism and ferroelectricity coexist in coupling and synergy. The development of new composite multiferroic materials with better properties than in single-phase multiferroics, having the interrelated piezoelectric and ferromagnetic properties once again take a lot of attention [1–3]. Coupled electrical polarization and magnetization give rise to their mutual control. For example, the direct magnetoelectric (ME) effect is the magnetically tunable polarization, change of the value or direction of electrical polarization under the applied magnetic field. Those unique properties are advance for the application of ME composites in energy transfer/harvesting [4–6], magnetic field sensors [5,7,8] and biomagnetic field sensors [9,10].

Magnetorheological smart materials are a class of composite materials having both rheological and magnetic properties [11]. This kind of material is usually composed of ferro (i-) magnetic micro- or nanofiller and elastic polymer matrix [12]. One of the advantages of the elastic polymer composites is that they can be easily shaped for a specific application, for instance, via using a 3d-printer [13]. It gives rise to interest in the utilization of magnetorheological composites in different applications as mechanical manipulators, actuators, tunable dampers, as well as soft robots, etc. [14].

If the above properties (multiferroics and magnetorheological) are met in one continuity, these materials will merge attributes and advantages from both families. An interesting example is represented by the magnetoelectric polymeric composites—materials consisting of magnetic/magnetostrictive filler (e.g., magnetic nanoparticles (NPs)) and piezopolymer matrix or polymer-bonded composites of ferroelectric and magnetic particles [15,16]. In this class of materials, ME coupling occurs through strain interactions (elastic coupling) of magnetic filler and piezoelectric particles or matrix [12,17–19]. As we know, the magnitude of the magnetoelectric effect in elastic polymer-bonded composites is by an order of magnitude larger (~700 mV/cm·Oe [20]) compared to the composites based on a piezoelectric polymeric matrix and magnetic nanoparticles (~40 mV/cm·Oe [15,18]). This fact can be explained by a different mechanism of coupling: in bonded composites, elastic coupling was explained by the mutual movement of two kinds of particles (magnetic in a magnetic field and ferroelectric in an electric field, respectively) [21], while in the piezopolymeric matrix is due to magnetostriction of magnetic fillers [15]. The highest value of ME effect in polymer-bonded composites however was achieved in composites of micron-sized particles of lead zirconate titanate (PZT) and neodymium iron boron (NdFeB) [20], which do not meet the requirements of biocompatibility. The goal of this work is to keep the relatively high biocompatibility of composites based on a piezoelectric polymeric matrix, to decrease the amount of inorganic inclusions and to achieve a high value of ME effect at the same time.

Despite the magnetoelectric effect in polymeric nanocomposites (NCs) is still smaller than in ceramic or laminar structures, they have advantages in simple fabrication, flexibility, and easy shaping [15]. Additionally, polymeric interfaces can show good biocompatibility, which together with multiferroic properties make them a unique tool for a set of bioapplications (e.g., cultivation surfaces with remotely controlled electric surface charge and mechanical stresses by applying an external magnetic field [22,23]). Application of both stimuli—charge and mechanical stress—may promote cell responses such as a controlled differentiation of stem cells. The physical stimulation of stem cell differentiation can replace the biochemical methods that are being used at the current time in stem cell-based therapy of neurodegenerative disorders [24]. Differentiation of stem cells into osteocytes [25], cardiomyocytes [26], and neural cells [27], initiated by electrical and mechanical stimulus has been studied. Neural cells are more sensitive to electrical stimulation because of their electric activities. Electrical stimulus induced by piezoelectric polymers leads to targeted axonal growth, inducing directed cell migration and promoting neurogenesis [24]. The first step of the investigation of physical factors' effects on neuronal stem cell differentiation is their cultivation on piezoelectric polymers that are the tests for biocompatibility with followed targeted differentiation. The biocompatibility of PVDF was demonstrated earlier on neuronal stem cells, isolated at the later stage of embryonic development [28]. Using the neural stem cells isolated in the early embryonic period can increase the ability to direct their differentiation for future applications [29]. The biocompatibility effect of

PVDF on the neural stem cells isolated in the early embryonic period should be studied additionally. In vitro cell or organ growing for further transplantation is an attractive stem cell-based therapy for the treatment of neurodegenerative disorders such as Parkinson's disease, Huntington's disease, Alzheimer's, amyotrophic lateral sclerosis [30], spinal cord injury [28], and brain damage [31].

In this work, we prepared NCs based on two types of polymers, poly(vinylidene fluoride) (PVDF) and its copolymers with trifluoroethylene (PVDF-TrFE) [32]. NCs based on PVDF-TrFE demonstrated a higher magnetoelectric performance and thus were chosen for further experiments. The highly crystalline $CoFe_2O_4$ NPs were prepared via a sol-gel auto-combustion method [33] and they were used for the preparation of rheological magnetoelectric materials [19]. Further, new strategies to increase magnetoelectric response were involved: (i) application of magnetic field during crystallization of polymer to align clusters of magnetic NPs and (ii) creation of 3-component composite with ferroelectric $BaTiO_3$ particles. We tested the piezoelectric polymers for future application as biointerfaces for activation and targeted differentiation of neuronal stem cells: neuronal stem cells isolated at the early embryonic stage cultivated on PVDF-based surface were able to proliferate and differentiate into main types of neural cells (neurons and glial cells).

2. Materials and Methods

2.1. Synthesis of $CoFe_2O_4$ (CFO), $Zn_{0.25}Co_{0.75}Fe_2O_4$ (ZCFO) and $BaTiO_3$ (BTO) Particles

Samples of $CoFe_2O_4$ NPs were prepared by the self-combustion method described in detail elsewhere [33]. The $Fe(NO_3)_3 \cdot 9H_2O$ (Carlo Erba Reagenti SpA, Cornaredo, Italy), $Co(NO_3)_2 \cdot 6H_2O$ (Scharlab S.L, Barcelona, Spain), citric acid (Scharlab S.L., Barcelona, Spain), and of 30% ammonia solution (Carlo Erba Reagenti SpA, Cornaredo, Italy) were used without further purification. In this process, 1-molar iron and cobalt nitrate aqueous solutions in a 2:1 ratio, respectively, and citric acid with 1:1 molar ratio of metals to citric acid were prepared. The pH level was adjusted to the value of ~7 by dropwise addition of aqueous ammonia. Obtained sol was placed on a hotplate at 150 °C to form a gel for 2 h. The gels underwent successively a thermal treatment at 300 °C for 15 min, where the auto-combustion reaction took place. Additionally, the Zn substituted cobalt ferrite ($Zn_{0.25}Co_{0.75}Fe_2O_4$, ZCFO) NPs were prepared with the same sol-gel auto-combustion method. A more detailed characterization of ZCFO NPs used here is reported earlier [34].

$BaTiO_3$ (BTO) particles were prepared by the solid-phase reaction method, followed by sintering using conventional ceramic technology described in detail elsewhere [35]. Briefly, $BaCO_3$ and TiO_2 powder with a purity of at least 99.95% were used as precursors. Then, $BaTiO_3$ particles were prepared by solid-state reaction method in two stages: at $T_1 = 1150$ °C during time $\tau_1 = 4$ h (1st stage) and $T_2 = 1170$ °C in during $\tau_2 = 4$ h (2nd stage).

2.2. Fabrication of Magnetoelectric Nanocomposites (NCs)

For composite fabrication, the two different types of polymers, poly(vinylidene fluoride) (PVDF) and its copolymer with trifluoroethylene (PVDF-TrFE) were used as a polymer matrix. For the preparation of the polymer-precursor solution, PVDF (Alfa Aesar, Kandel, Germany) or PVDF-TrFE 55/45 (Piezotech, King of Prussia, PA, USA) granules were dissolved in dimethylformamide (DMF) (Sigma-Aldrich, Darmstadt, Germany) at 40 °C followed by mixing until complete dissolution of polymer granules. The concentrations were about 1:4 in weight ratio for PVDF/DFM and 1:6 for PVDF-TrFE/DMF solution. The dissolution time was about 45 min for PVDF and 90 min for PVDF-TrFE. The total concentrations of PVDF/DMF and PVDF-TrFE/DMF were 1:8 and 1:12, respectively, since at the next step an additional amount of DMF was introduced together with NPs.

Nanocomposites of NPs embedded in the piezoelectric polymer matrix were fabricated by the solvent evaporation method assisted by a *doctor blade* technique [36]. The so-called doctor blade or blade coating method is one of the simple methods for lab-scale production of thin polymer composites. In this method, the polymer solution is placed on the substrate in front of the moving blade and is smoothed out by it. The thickness of the layer is controlled by adjusting a gap between the knife (blade) and substrate. The thickness of the final evaporated layer depends on the gap between knife and substrate, speed of coating, the temperature of the substrate and physical properties of solution (viscosity, density, etc.). The technological protocol of composite fabrication is strongly dependent on the type of polymer, fillers, and type of solvent. The CFO or ZCFO NPs were ground, mixed with the second part of DMF solvent and dispersed in preliminary prepared polymer-precursor solutions in an ultrasonic bath for 2 h. The mixing of fillers in DMF solutions was applied to decrease the particle agglomerations and their more homogeneous distribution in polymer solutions. In the next set of samples, in the system demonstrated higher magnetoelectric properties (oriented CFO/PVDF-TRfE), 5% and 10% weight content of BaTiO$_3$ (BTO) particles was added at the same step as CFO particles.

The solution of particles and polymers precursor was spread on a clean glass substrate using a coating blade at a fixed distance between the substrates. The solvent was evaporated by heating the composites in an oven at 75 °C for 15 min. Then, for the fabrication of the ordered samples, this protocol was modified as follows: clusters of magnetic NPs were aligned in the magnetic field before evaporation of the polymer's solvent. The magnetic field was applied in-plane of the dish with the precursor solution during evaporation (Figure 1a). After evaporation, the particles were immobilized in a polymer matrix, when the magnetic field was removed, aligned samples were obtained.

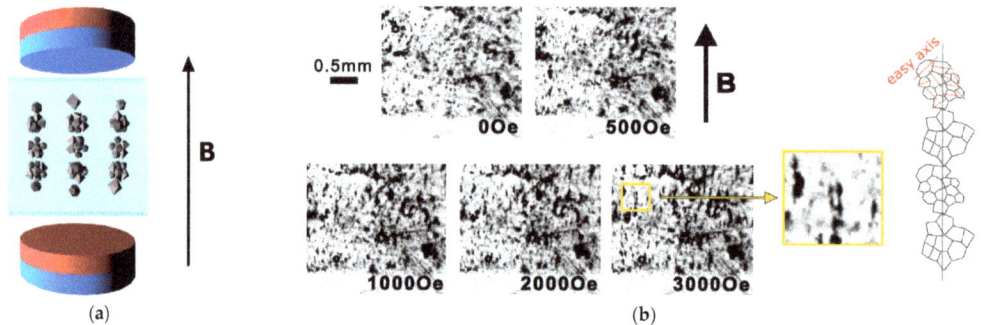

Figure 1. (a) Illustration of the alignment of CFO NPs in PVDF polymer in a magnetic field; (b) optical image of the formation of ordered chains of CFO NPs clusters in the liquid precursor of PVDF-TrFE polymer under external magnetic fields of different inductions. The sketch represents a structure of the chain as an elongated assembly of NPs clusters with the random distribution of easy axes of individual particles inside each cluster (red lines).

Figure 1b shows the alignment process of clusters composed of CFO NPs into chains in a gradually increasing magnetic field up to 3 kOe. Electromagnets of the magnetometer (7400 System VSM; Lake Shore Cryotronics Inc., Westerville, OH, USA) were used to generate a homogeneous magnetic field ($\pm 0.1\%$) in a volume 10 mm^3, which is bigger than the volume of the samples (typical shape of sample is a square with edges of 4 mm and thickness in the range of 30–60 µm). After switching of the magnetic field, optical images were obtained with a 5.3 MP monochrome camera PixeLINK PL-D725MU-T (Edmund Optics Inc., Barrington, NJ, USA) placed between two coils of the electromagnet. At a field of about 500 Oe, clusters of particles start to move, forming aligned structures. At a field of about ~3 kOe, those structures achieved the final state and the further increase of field does not change the shape of the clusters' chains. The clusters of CFO NPs in PVDF-TrFE-based solutions showed better alignment in the magnetic field, because of higher viscosity and lower time of drying in comparison with PVDF-based (Figure S5). Moreover, the difference between the two polymers was in the structure of the surface. According to the atomic force microscopy (AFM), the pore size was 30 ± 12 nm in PVDF-TrFE-based and 100 ± 64 nm in PVDF-based NCs (see explanation in SI, Figure S6). The same experiment of registration of movement and reorganization of particle aggregations, performed on samples after evaporation, showed that the particles were rigidly fixed in the polymer matrix (no displacement was detected within experimental error).

Finally, magnetoelectric NCs were obtained by detaching the glass substrate. All samples were prepared with 15% weight content (wt.%) of CFO or ZCFO NPs, because according to literature data around the enhanced formation of ferroelectric β- and γ-phases of PVDF polymers is expected [37].

All samples were poled using direct contact poling in a custom-designed chamber for 40 min at 40 °C [38]. The chamber was constructed as the adiabatic camera from thermo-insulated material (polystyrene foam) and equipped with a thermoregulation system. The maximal poled electric field was 50 MV/m.

2.3. Structural and Magnetic Characterization

The X-ray diffraction (XRD) studies were performed with a DaVinci2 diffractometer (Bruker, MA, USA) using Cu Kα (λ = 1.54056 Å) in the 2θ geometry in a range of 10–70 degrees. The average size of crystallites d_{XRD} was calculated for (440) peak with Scherrer's equation:

$$d_{XRD} = 0.94 \cdot \lambda / (B \cdot \cos \theta) \quad (1)$$

where B is full width at half maximum (FWHM) and θ is a position of XRD reflections.

The size distribution of NPs was investigated by using an S-5500 Transmission Electron Microscopy (TEM; Hitachi, Japan).

The magnetic properties were studied with a vibrating sample magnetometer 7400 System (VSM; Lake Shore Cryotronics Inc., Westerville, OH, USA) in the magnetic field up to 1.1 T at room temperature (295 K). Since the maximal acquired field was not sufficient to fully saturate the sample, the value of saturation magnetization was extrapolated with the fitting of the high-field region using the Law of Approach to Saturation (LAS):

$$M(H) = M_S(1 - A/H - B/H^2) \quad (2)$$

where A and B are fitting parameters [39]. The Equation (2) was applied previously to estimate M_S in different ferrite nanoparticles [34,40].

A deeper investigation of the magnetic properties was conducted by FORC analysis (first-order reversal curve [41,42]). To measure the FORC, the sample was first saturated, then the applied field was decreased to the value of the return field (H_r). The curve measured from the H_r to the saturation field is a single FORC. The cycle of at least 100 repetitions by decreasing the value of the H_r was recorded. The measurement of such curves provides information from different paths of magnetization and interaction fields for all the phases that contribute to the hysteresis loop. FORC method was recently applied to study the magnetomechanical properties of magnetic elastomers (intrinsic magnetic hysteresis of magnetic filler and mechanical compliance of the matrix) [43]. The FORC diagram interpretation is based on its comparison with the Classical Preisach Model of hysteresis [44]. In this model, the hysteron is a mathematical operator that acts on the magnetic field and produces a square hysteresis loop characterized by a coercive field H_c (half-width) and an interaction field H_u (horizontal bias) [45]. Each magnetic phase in the material could be described by a single hysteron, while a set of hysterons will describe the macroscopic hysteresis cycle of the entire sample. The hysteron's distribution $\rho(H_c, H_u)$ is represented on the two-dimensional Preisach plane with H_c and H_u axis profiles. By comparing the H_c and H_u axis profiles the information about the magnetic interactions in the system can be obtained [46]. The FORC-curves were obtained via 7400 VSM FORC Utility (Lake Shore Cryotronics Inc., Westerville, OH, USA).

2.4. Magnetoelectric Properties

The ME studies were carried out using a custom-designed setup for measuring the magnetoelectric voltage ΔV with a lock-in amplifier (Model SR830, Stanford Research,

Sunnyvale, CA, USA) at frequencies of 1 Hz–100 kHz. The input impedance of the lock-in amplifier is 10 MOhm. The ME coefficient α_{ME} was defined using the following equation:

$$\alpha_{ME} = \frac{\Delta V}{b \, \Delta H} \qquad (3)$$

where ΔV is the amplitude of the induced ME voltage, b is the thickness of the sample, and ΔH is the amplitude of the AC field H_{AC}. The accuracy of ME signal measurements was less than 1%. The amplitude of H_{AC} was about 10 Oe and the DC field was varied up to 10 kOe. The H_{AC} and H_{DC} fields were applied across the plane of the sample, that is $H_{AC} \| H_{DC} \| \Delta V$ (Figure 2). The Helmholtz coils were used for the generation of AC field, the DC bias field was applied using an adjustable Halbach type magnet system (AMT & C LLC, Troitsk, Russia). Electric contacts were made by the coating of aluminum foils on the larger surface of composite films, thus, the ME coefficient was measured in α_{33} mode.

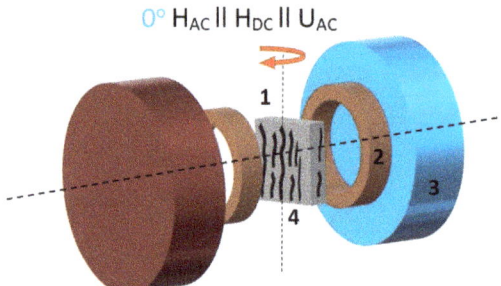

Figure 2. Scheme of the experiment for direct magnetoelectric (ME) measurements (**1**—sample, **2**—Helmholtz coils, **3**—DC magnetic field source, **4**—aligned chains of particle clusters); the red arrow indicates the direction in which sample was rotated.

2.5. Magnetic and Piezoresponse Force Microscopy

Piezoresponse force microscopy (PFM) and local polarization switching spectroscopy measurements were carried out with MFP-3D (Asylum Research, Goleta, CA, USA) commercial scanning probe microscope using the CSG30/Pt (Tipsnano, Tallinn, Estonia) conductive probe with the spring constant of 0.6 N/m. The PFM out-of-plane images were scanned in the single frequency PFM mode at 3 V and a frequency of ~7 kHz. An alternating current (AC) voltage (3 V) was superimposed onto a triangular square-stepping wave (f = 0.5 Hz, with writing and reading times 25 ms, and bias window up to ±150 V) during the remnant piezoelectric hysteresis loops measurements. PFM images were also measured with applying DC magnetic field. The magnetic field ($B_{ext.}$ = 1.4 kOe) was applied perpendicular to the plane of the samples. To estimate the effective d_{33} piezoelectric constants, the deflections and vibration sensitivity of the cantilever alignment were calibrated by GetReal procedure using the IgorPro software. For the quantification of switching and piezoelectric coefficient of samples, a dual AC resonance tracking piezoresponse force microscopy (DART-PFM) was employed, which allowed us to probe the piezoresponse that originated within a single domain with a spatial resolution up to submicrometers [47]. DART-PFM is comparatively a reliable technology to probe the piezoresponse from thin polymer samples, because it uses dual AC resonance tracking to quantify the shift of resonance to avoid the noise effects of the surface height topography and suppress the contributions from electrostatic effects [48].

Besides, effective piezoelectric coefficient d_{33}(Voltage) hysteresis loops were investigated for further understanding of the magnetic field influence on piezoelectric response. The hysteresis loops (PFM Amplitude (pm) and PFM phase) were acquired using the simple harmonic oscillator (SHO) fit with Asylum Research software to exclude the magnification effect of Q factor of the contact resonance. Effective longitudinal piezoelectric response ("effective d_{33}") was calculated by Equation (4):

$$d_{33} \text{ (pm/V)} = (\text{PFM Amplitude (pm)} \times \cos(\text{PFM Phase}))/\text{Applied AC voltage (V)}. \quad (4)$$

Magnetic force microscopy (MFM) images were obtained using the ASYMFM HC magnetic probe (Asylum Research, Goleta, CA, USA). For MFM scans commercial cantilevers (ASYMFM HC) coated with a magnetic layer of CoPt/FePt (tip apex radius 45 nm; H_C > 5 kOe) were utilized. The lift height is 300 nm.

2.6. Biological Tests of Polymeric Interfaces

Boundary cap neural crest stem cells (bNCSC) culture is a transient neural crest-derived group of cells located at the dorsal root entry zone. Previous experiments have shown that bNCSCs can differentiate into sensory neurons and glial cells in vitro [29] and in vivo after transplantation [49]. The bNCSC were generated from E11.5 days mouse embryo constitutively expressing red fluorescence signal (RFP) under actin promoter [50]. The bNCSCs were cultured as neurospheres in propagation medium: N2 medium containing bFGF (basic fibroblast growth factor) and EGF (epidermal growth factor) (20 ng/mL, RnD Systems, Minneapolis, MN, USA), and B27 supplement (Gibco, Waltham, MA, USA). bNCSCs were dissociated to single cells with 3PlE and plated on sterilized PVDF substrate (70% alcohol and UV-light for 2 h) covered surface on the bottom of 4-well dishes (D = 16.5 mm). bNCSCs were cultured in proliferation medium (stem cell medium) for 45 min. After that, the medium was replaced with a differentiation medium (DMEM-F12/Neurobasal medium supplemented with N2, B27, 0.1 mM non-essential amino acids and 2 mM sodium pyruvate).

After 72 hours' incubation time neurospheres on the PVDF substrate were fixed for 15 min with 4% phosphate-buffered paraformaldehyde (PFH, Merk, Darmstadt, Germany) at room temperature (RT) and washed with phosphate buffered saline (PBS, Gibco, Carlsbad, CA, USA) three times for 10 min. Then, the cells were left overnight in PBS at +4 °C. Then in 12 h, cells were washed and incubated in preincubation solution (1% bovine serum albumin (BSA, Thermo Scientific, Waltham, MA, USA), 0.3% Triton X-100 (Invitrogen, Carlsbad, CA, USA), and 0.1% sodium azide (NaN3, Merk, Darmstadt, Germany) in PBS) for 60 min, and incubated with primary antibodies (GFAP; Rabbit, 1:500, Merk, Darmstadt, Germany; III β-Tubulin; Rabbit, 1:500, BioSite, Täby, Sweden) overnight at 4 °C, followed by the appropriate secondary antibodies (Alexa Fluor 488 goat anti-rabbit IgG (H + L; 1:250, Life Technologies, Carlsbad, CA, USA)) for 4 h at RT. Subsequently, the PVDF substrate was washed in PBS, and cells were incubated with Hoechst (1:10,000; Invitrogen, Carlsbad, CA, USA) to label cell nuclei, and then mounted on a glass slide for analysis.

After the immunostaining procedure, PVDF substrate with cells was examined using a fluorescence microscope Eclipse E800 (Nikon, Tokyo, Japan). Image analysis was carried out using ImageJ.

3. Results and Discussions

3.1. Characterization of CFO and BTO Particles

The XRD pattern of the powder $CoFe_2O_4$ (CFO) sample indicates the high crystallinity of nanoparticles without any amorphous content (Figure 3 and more detailed in Figure S1a). The observed reflections were indexed to a cubic spinel lattice according to card No.591-0063 for cobalt ferrite. The size of crystallites calculated by Equation (1) d_{XRD} = 17 ± 2 nm was close to the mean size of the particles observed with TEM microanalysis (Figure S1b, d = 15 ± 1 nm with standard deviation σ = 8 ± 1 nm). This fact indicates the high crystallinity of synthesized NPs. Field dependence of magnetization recorded at 300 K for powder NPs shows hysteretic behavior typical for ferrimagnetic nanoparticles in the blocked state (Figure S2a). The coercivity field (H_C) was ~1.3 kOe, saturation magnetization (M_S) was ~66 emu/g, and reduced remanence (M_R/M_S) of about 0.44. More detailed characterization of the magnetic and structural properties of particles was already reported elsewhere [34]. Magnetic interparticle interactions were evaluated by measuring the remanence curves and plotting of ΔM-plots (see SI and refs. [51,52] for more details). CFO powder sample shows a negative value of ΔM (Figure S2b) with the maximum intensity of about ~0.1 that suggests that the interparticle dipolar interactions are dominant. The $Zn_{0.25}Co_{0.75}Fe_2O_4$ (ZCFO) NPs have a similar average crystal size 16 ± 2 nm (Figure S1). Substitution of diamagnetic Zn^{2+} ions in the spinel structure of cobalt ferrite results in a decreased value of magnetic anisotropy (K_{CFO} = 1.6 × 10^6 erg/cm^3; K_{ZCFO} = 0.95 × 10^6 erg/cm^3 [34]) and a slightly higher value of the saturation magnetization (~74 emu/g) concerning pure CFO sample (Figure S1a). ZCFO powder sample also shows a negative value of ΔM (Figure S2b) of higher magnitude due to a larger saturation magnetization, which led to the stronger dipolar interactions.

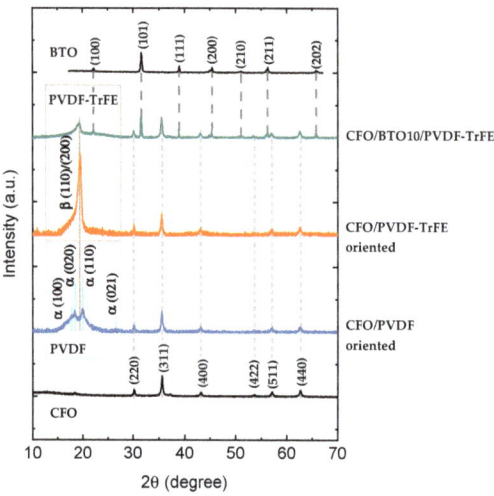

Figure 3. XRD patterns of CFO and BTO nanoparticles, CFO/PVDF, CFO/PVDF-TrFE, and CFO/BTO10/PVDF-TrFE nanocomposites. The Miller indexes specified for pure CFO, BTO particles and PVDF (PVDF-TrFE) polymer are guided to corresponding reflections in composites via dashed lines.

The XRD analysis of $BaTiO_3$ (BTO) particles indicates the presence of a perovskite tetragonal structure (Figure 3 and more detailed in Figure S3). Positions of main reflections were indexed according to card No.152-5437. Cell parameters were a = 3.995 Å, c = 4.030 Å; V = 64.31Å; c/a = 1.0088, typical for $BaTiO_3$. The size of crystallites (d_{XRD}) calculated with Equation (1) was 26 ± 8 nm.

3.2. Characterization of NCs

In Figure 3, XRD patterns for composites samples are presented. In all samples, diffraction peaks allocated in 30–70° 2θ-range and attributed to spinel ferrite $CoFe_2O_4$ and $Zn_{0.25}Co_{0.75}Fe_2O_4$ are indexed. The intensity of diffraction peaks is reduced compared to the pure powder sample (Figure S1a) due to high polymer content in the samples. In the low field region allocated diffraction peaks related to the PVDF and PVDF-TrFE polymers. The higher relative intensity of diffraction peak at ~20° in PVDF-TrFE than in PVDF indicates the higher crystallinity of PVDF-TrFE. Oriented and random nanocomposites showed similar XRD patterns (Figure S4a). In 3-component NCs there are three distinguished phases attributed to β-phase of PVDF-TrFE polymer, perovskite structure of BTO, and spinel structure of CFO particles (Figure S4b). The pattern for ZCFO/PVDF-TrFE is similar to the patterns for CFO-based NCs (Figure S4c).

Patterns for PVDF are almost identical to the reported ones in [53]. Where the major phase was the monoclinic α-phase crystal, confirmed by two intensive diffraction peaks at 18.4° and 20.0° and a low intense peak at 26.6°, corresponding to (020), (110), and (021) reflections. Characteristic diffraction peak at 20.6° of β-phase is also presented but it is merged with 110 reflections of dominant α-phase. In PVDF-TrFE polymer, β-phase is more pronounced because it is to form in this modification as follows from the literature [32].

Macroscopic magnetic properties of composites samples were studied with VSM at 300 K (Figure 4). M_S values of composite samples were reduced concerning CFO powder due to the presence of diamagnetic polymer content. The coercivity field of PVDF-based composites was almost equal to the same value of CFO powder ~1.3 kOe (Figure 4a). Thus matrix stiffness was relatively high, preventing mechanical rotation of particles [54]. PVDF-TrFE-based composites demonstrated a slightly higher coercive field of ~1.5 kOe (Figure 4b). Probably, it is related to slightly lower magnetic interparticle interactions, that were better dispersed in PVDF-TrFE (see explanation below). Interestingly, that the samples ordered in the magnetic field have almost the same magnetic properties as randomly oriented samples. Figure 4c,d shows the angular dependence of M-H loops recorded for oriented CFO/PVDF-TrFE sample in two different orientations of the magnetic field and sample axis (along chains of CFO NPs clusters). In the first case, when the orientation of the sample was always in-plane the hysteresis loops did not depend on the orientation of the field. In the second case, when the direction of the field changed from in-plane to out of plane orientation, a small difference was observed in both random and oriented samples but this was mainly due to the geometrical change of measuring configuration (mutual position of a sample and pick-up coils of VSM).

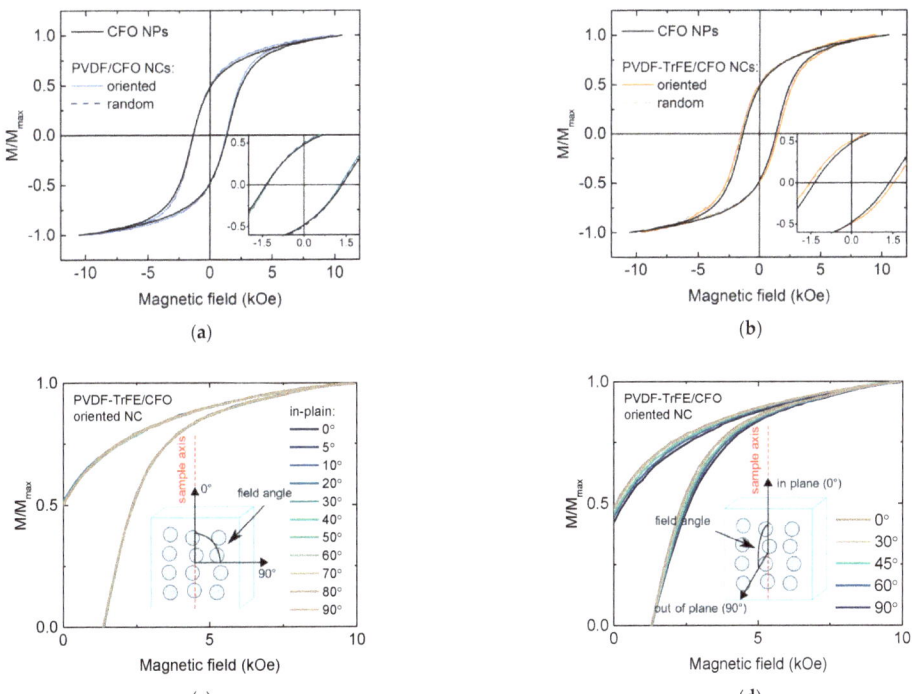

Figure 4. In-plane M-H loops reordered at 300 K for random and ordered (**a**) PVDF/CFO and (**b**) PVDF-TrFE/CFO nanocomposite compared with CFO NPs; M-H loops for ordered PVDF-TrFE/CFO sample as a function of sample axis and field direction in (**c**) in-plane and (**d**) from in-plane to out-of-plane orientations.

Notably, the formation of those ordered chains does not induce any magnetic anisotropy of composite samples. This fact can be explained by the dominant role of intra-aggregate interparticle magnetic interactions on macroscopic magnetic reversal processes. In other words, the arrangement of the clusters' chains in the magnetic field orients aggregates of several particles but inside those aggregates, the easy axes of magnetic anisotropy of individual NPs are still distributed randomly [54,55]. Individually single-domain magnetic NPs behave according to the Stoner–Wohlfarth model, thus angular dependence of magnetization was expected to change from rectangular to sloped line for easy and hard axes respectively. Indeed, according to TEM investigation (Figure 5b), the produced powder is formed by submicron-size aggregates of densely compacted particles. The dipolar nature of interparticle interactions was confirmed with ΔM-plots (Figure S2b) and FORC diagrams.

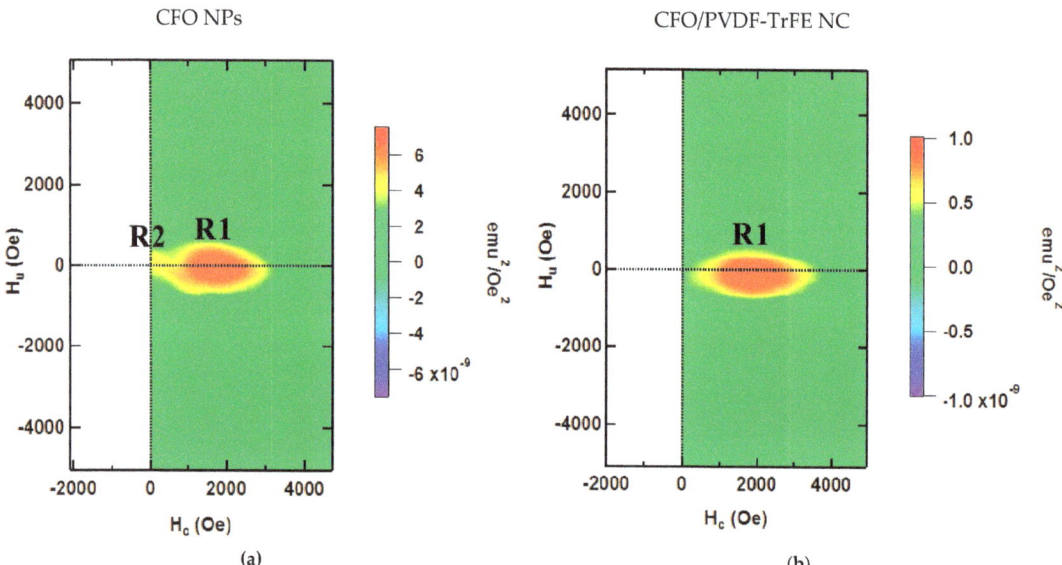

Figure 5. First order reversal curves (FORC) diagram for (**a**) CFO NPs and (**b**) PVDF-TrFE/CFO nanocomposite sample.

The H_r profile of FORC diagrams reflects the distribution of coercivities of the particle ensembles (Figure 5a). For close-packed CFO NPs, two main maxima can be observed in the FORC distribution (R1 and R2) [56]. The dominating region R1 reflects the behavior of individual particle clusters. The minor spread in the H_u profile indicates that the dipolar magnetic interaction between the particles is dominant: according to ref. [45], magnetic single-domain NPs in clusters is characterized by wider FORC distribution than for individual particles due to a strong and localized interaction. The region R2 results from the interaction among clusters. Indeed, in the case of particles distributed in PVDF-TrFE polymer (Figure 5b), the smaller region R2 is hindered by the sensitivity limit of the VSM device due to the larger distance between clusters [46]; in the case of powder sample (Figure 5a), clusters of NPs are close to each other: two distinct regions, R1 and R2, can be distinguished. R1 regions are identical in Figure 5a,b, it means the dispersion of individual particles in polymer does not affect the assembled clusters: interaggregate interaction has a minor effect, while macroscopic magnetic properties of the samples are determined mainly by their magnetocrystalline anisotropy and interactions in their assemblies (clusters of several NPs). For elastomers in presence of magnetic field during evaporation and without it (ordered and random samples, respectively), FORC-analysis does not show the differences in $H_c - H_u$ planes. This observation together with results of angular M-H measurements (Figure 4c,d) indicates that even in the case of ordered clusters of NPs, samples are magnetically isotropic; distribution of individual easy axis of particles is random.

The micromagnetic structure of prepared composite samples was studied using MFM in zero external magnetic fields (Figure 6). MFM images show magnetically active regions of about 0.1–1 µm formed by NPs aggregates (clusters of several particles) magnetized in the same direction and arranged in chains [19]. The formation of those chains is caused by the applied external magnetic field during polymerization. The contrast spots render magnetized regions that do not match the signal of MFM magnitude with the topology. The appearance of the magnetized regions on topologically flat areas confirms that particles were immersed into the polymer and not exposed on the surface. In composites, evaporated in the absence of a magnetic field, the magnetic contrast from clusters of magnetic particles shows no preferential orientation of their magnetic moments (Figure S7).

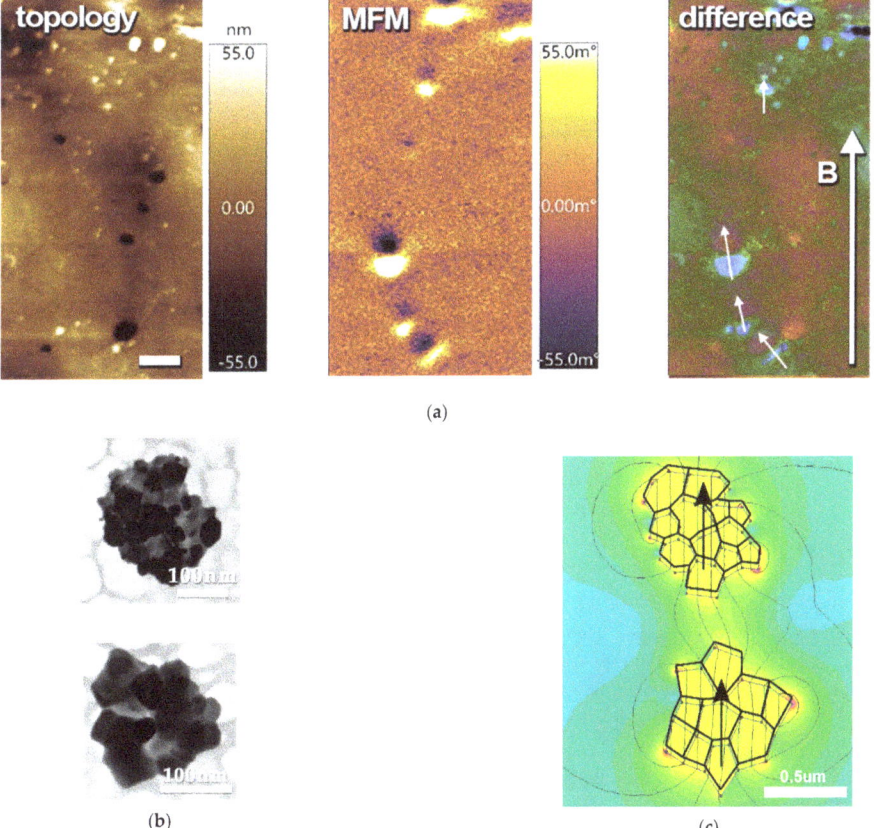

Figure 6. (**a**) Magnetic force microscopy (MFM) images of PVDF-TrFE/CFO nanocomposite: topology, MFM signals, and their difference. Arrow B indicates the direction of the applied magnetic field during polymerization. Scale bar is 2 µm; (**b**) TEM image of separated aggregates of powder CFO NPs; (**c**) illustration of a possible configuration of two aggregates and simulated magnetic field distribution for this configuration.

To quantitatively evaluate the impact of two sorts of interactions (intra- and interaggregate) on magnetization state, the finite element method was performed utilizing the FEMM software. To fulfill the simulation, a hypothetical case of two aggregates with sizes close to those estimated from MFM (Figure 6a), the shapes close to observed with TEM (Figure 6b), and measured magnetic properties of CFO powder was reproduced (Figure 6c). A situation of collinearly magnetized aggregates (all magnetic moments of individual particles formed those aggregates aligned in a head-to-tail manner) is rendered in Figure 6c. Magnetostatics energy of this configuration was minimal among other considered cases (see more data in Figure S8). For example, configuration with the head-to-head magnetization of aggregates has the maximal energy of interaggregate interactions with a total energy of about 40% higher than in the previous configuration. If one or both aggregates have a close structure with minor stray field and negligible interaggregate interactions, the total energy increased by one and two orders of magnitude respectively. This finding confirms that despite the appearance of magnetic microstructure observed with MFM, its impact on macroscopic magnetic properties is still minor, while magnetic interactions inside aggregates and particles themselves are dominant [57].

3.3. Random and Oriented NCs Based on CFO NPs in PVDF and PVDF-TrFE NCs

The dependence of ME voltage coefficients (α_{ME}) versus DC magnetic field (H_{DC}) for all composites has a peak-like behavior (Figure 7). The non-monotonous ME response with a maximum at ~4 kOe is related to the magnetization processes of the CFO nanoparticles (see Section 3.5). All measurements were performed at a fixed frequency of 10 kHz, which is below resonant frequency for the samples. The lower frequency was chosen because of the limitations of further biological experiments (long-time treatment at higher frequencies will induce unwanted heating of the system).

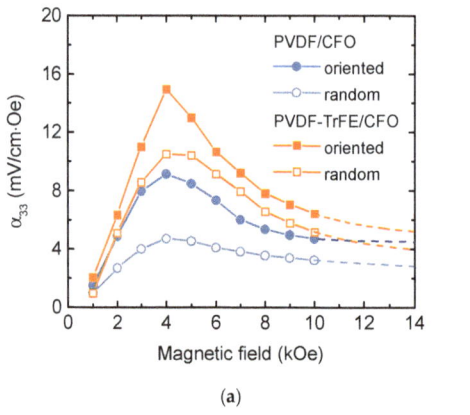
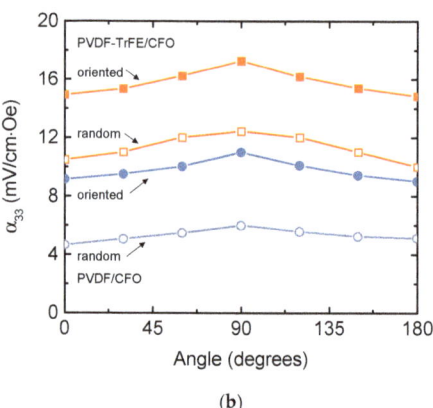

Figure 7. (a) Field and (b) angular dependencies of the ME voltage coefficient (α_{ME}) on DC bias magnetic field for ordered and random PVDF/CFO and PVDFTrFE/CFO composites at AC field frequency of 10 kHz.

The ME voltage coefficients α_{ME} depend on the type of piezoelectric matrix: samples with PVDF-TrFE piezoelectric matrix demonstrates larger ME coefficient in comparison with PVDF-based (Figure 7a), which can be associated with better piezoelectric properties of PVDF-TrFE polymer (d_{33} = −38 pC/N) in comparison with PVDF (d_{33} = −25.8 pC/N) [58]. Observed values of ME coefficients are in the range of reported data on ME polymer composites [15]. However, for an accurate comparison of ME effect, such factors as types of magnetic and ferroelectric components, their magnetic, ferroelectric, piezomagnetic and piezoelectric properties, phase coexistence, mechanical coupling, fabrication techniques should be taken into account. For example, P. Martins et al. observed linear response of ME coefficient versus biasing DC magnetic field in the field up to 5 kOe in

Ni$_{0.5}$Zn$_{0.5}$Fe$_2$O$_4$/PVDF-TrFE nanocomposites [17]. They observed a maximal α_{33} value of 1.35 mV/cm·Oe for composite with 15 wt% Ni$_{0.5}$Zn$_{0.5}$Fe$_2$O$_4$ nanoparticles (NPs) in 40 kHz AC field with the amplitude of about 1 Oe and DC biasing field of 5 kOe. J. Zhang et al. [59] found a higher α_{33} value of about 40 mV/cm·Oe in CoFe$_2$O$_4$/PVDF-TrFE measured in the same field condition. In later work, P. Martins et al. compared the values of ME coefficient of X^{2+}Fe^{3+}$_2$O$_4$/PVDF-TrFE, where X^{2+} = Zn/Mn, Co, and Fe [18]. The higher value of ME coefficient was observed in nanocomposite with CoFe$_2$O$_4$ NPs of about 15 nm prepared via the hydrothermal route. Those particles had a coercivity of ~2.5 kOe and magnetization of ~61 emu/g at 10 kOe while the other studied particles exhibit superparamagnetic behavior at room temperature. Measured values of α_{ME} coefficient are lower than in some cases listed above (~17 and ~11 mV/cm·Oe for PVDF-TrFE/CFO and PVDF/CFO samples, respectively), the further improvement can be achieved by the optimization of concentration, composition, and better dispersion of magnetic particles.

The orientation of CFO particles during polymerization enhanced α_{ME} coefficient (~50% for PVDF/CFO and ~30% for PVDF-TrFE/CFO polymer composites). Interesting to note, that even if samples have isotropic magnetic properties, they demonstrated anisotropy magnetoelectric properties. This fact is due to geometrical features of samples: in the in-plane orientation of the field it is more difficult to deform the sample, while the out-of-plane orientation when rectangular sample placed perpendicular to the field direction, is easier. In cases of oriented samples, the angular dependence of the magnetoelectric coefficient becomes slightly sharper ($\alpha_{ME}(0°)/\alpha_{ME}(90°)$ is ~77% for random and ~83% for oriented PVDF/CFO, ~84% for random and ~86% for oriented PVDF-TrFE/CFO samples).

To study the piezoelectrical domain switching behavior of the PVDF/CFO and PVDF-TrFE/CFO samples, the "mini" chessboard structures were written without a magnetic field (Figure 8a,c). The dark and bright square areas (7.5 × 7.5 μm^2) correspond to applied +150 V and −150 V DC biases, respectively. Strong PFM contrast confirms the complete switching process in composite polymer samples under poling. The domains created are rather stable in time. It was found that the magnitude of the PFM signal of polarized regions for sample PVDF-TrFE/CFO is 4 times higher than for PVDF/CFO sample. Then the samples were placed in a magnetic field (B$_{ext.}$ = 1.4 kOe) and the same polarized area was scanned again (Figure 8b,d). It is experimentally shown that for PVDF/CFO sample, an external magnetic field increases (~30%) the signal of the remnant piezoelectric response or effective d$_{33}$. For comparison, the PVDF-TrFE/CFO sample shows a ~64% reduction in the effective d$_{33}$. Besides, the values of the effective piezoelectric coefficient d$_{33}$ (Figure 8e,f) for PVDF/CFO composite are higher than for PVDF-TrFE/CFO. These experiments demonstrate locally induced ME coupling on composites, which is associated with the deformation of the piezoelectric matrix by magnetic CFO NPs. Thus, this method additionally confirms the magnetoelectric nature of the composites [60].

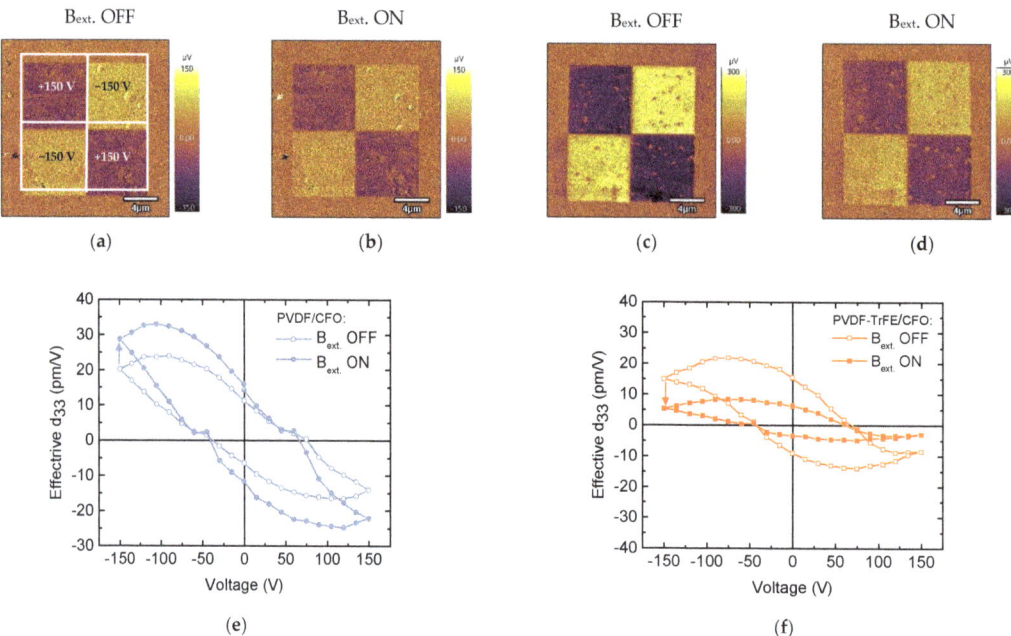

Figure 8. Piezoresponse force microscopy (PFM) images (a–d) and remnant local piezoelectric hysteresis loops (e,f) at applied ($B_{ext.}$ ON) and switched off ($B_{ext.}$ OFF) the magnetic field ($B_{ext.}$ = 1.4 kOe) for (a,b,e) PVDF/CFO and (c,d,f) PVDF-TrFE/CFO samples.

3.4. Further Improvements of ME Efficiency

Further improvement of ME efficiency was achieved via the inclusion of one additional component, with 5% and 10% weight content of piezoelectric $BaTiO_3$ (BTO) particles in more efficient ordered PVDF-TrFE/CFO NC. Thus, 3-component NC samples, ordered PVDF-TrFE/BTO5/CFO and PVDF-TrFE/BTO10/CFO, were obtained. The adding of BTO fillers led to a small enhancement in the ME effect (Figure 9a), which is related to the contribution of BTO with higher piezoresponse in comparison with PVDF-based polymers (d_{33} = 191 pC/N [61]). When the concentration of BTO increased from 5 to 10% the ME voltage coefficients α_{ME} decreased from 18.5 to 17 mV/cm·Oe. That was attributed to a reduced quality of the crystallization of polymer when it is overfilled. Nonetheless, those values are sensory higher than this value for ordered PVDF-TrFE/CFO NC. The local piezoelectric hysteresis loop measured utilizing PFM demonstrates the response of d_{33} BTO and PVDF-TrFE components to the applied magnetic field as a result of magnetoelectric interaction (Figure 9b). The introduction of diamagnetic BTO particles into composites does not affect their magnetic properties and they are almost identical within experimental error with 2-component PVDF-TrFE/CFO NC.

Additionally, the Zn substituted cobalt ferrite ($Zn_{0.25}Co_{0.75}Fe_2O_4$, ZCFO) NPs with the same size as CFO NPs, the slightly higher saturation magnetization, and lower magnetic anisotropy were used to tune a magnetoelectric response. M-H loop of PVDF-TrFE/ZCFO is shown in Figure 9c. PVDF-TrFE/ZCFO NC has much lower coercivity (H_C~0.6 kOe) and irreversibility fields (H_{irr}~3.8 kOe) than those for the family of PVDF-TrFE/CFO NCs (H_C~1.5 kOe; H_{irr}~7 kOe). NC with ZCFO particles shows slightly higher ME performance, which more likely can be attributed to the higher M_S value of ZCFO NPs. The higher value of M_S leads to the stronger magnetostatic interactions and thus to the stronger interactions of particle clusters. A more significant change was detected in the position of peak (H_{peak}) in the field dependence of the ME voltage coefficients α_{ME}. The H_{peak}

was reduced at a factor 0.56–0.71, which is quite close to the ratio of magnetic anisotropy constants $K_{ZCFO}/K_{CFO} \sim 0.6$.

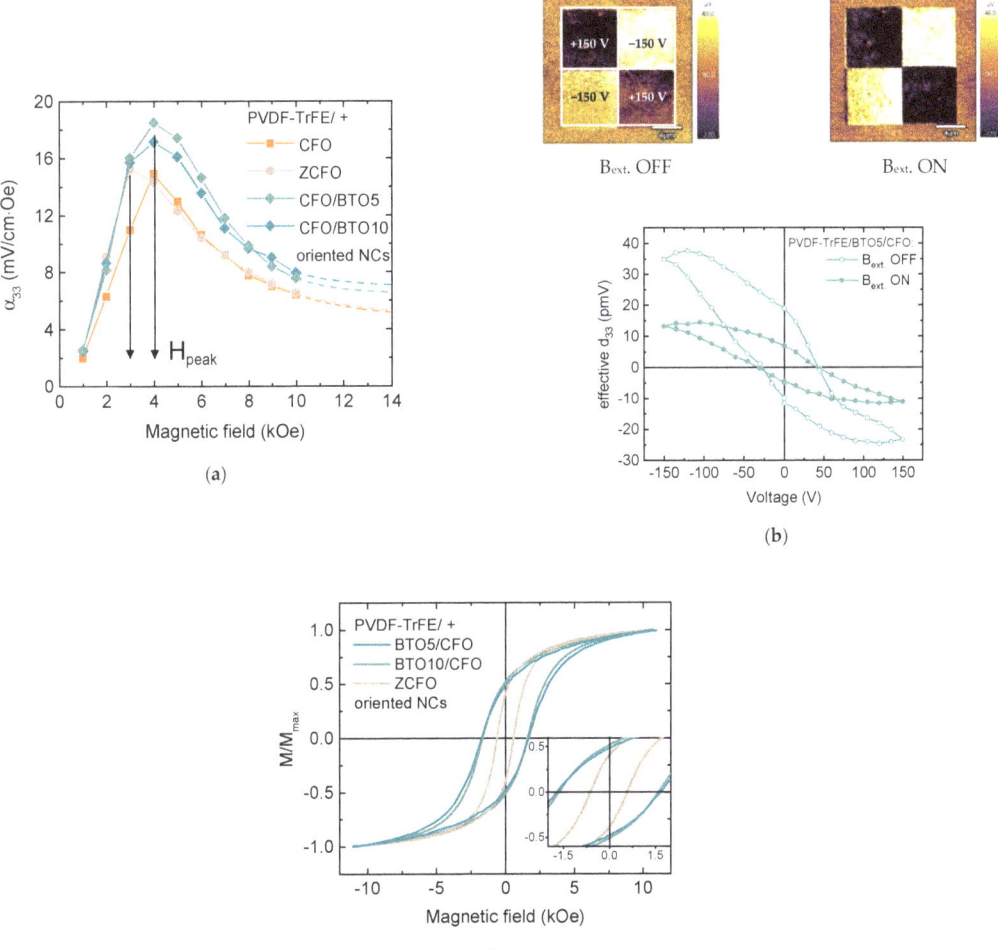

Figure 9. (**a**) Dependencies of the ME voltage coefficient (α_{ME}) on DC bias magnetic field composites at AC field frequency of 10 kHz for PVDF-TrFE -based samples with ZCFO, BTO5/CFO and BTO10/CFO fillers (PVDF-TrFE/CFO is shown again for comparison); (**b**) piezoresponse force microscopy (PFM) images and remnant local piezoelectric hysteresis loops at applied and switched off the magnetic field ($B_{ext.}$ = 1.4 kOe) for PVDF-TrFE/BTO5/CFO sample; (**c**) M-H loops at 300 K for PVDF-TrFE/ZCFO, PVDF-TrFE/BTO5/CFO and PVDF-TrFE/BTO10/CFO NCs.

3.5. The Origin of ME Effect

The magnetoelectric effect was explained through magnetostatic interactions. In a zero magnetic field in an equilibrium state, magnetic moments of aggregates are parallel (C1) or antiparallel (C2) depending on the initial location (Figure 10, symmetrical cases are not shown). When an external magnetic field higher than the magnetic anisotropy of particles is applied (H > H_A), a high-energy configuration (C3) is formed because of magnetization of particles. In this case, dipolar forces will invoke a rearrangement of particles to reach another low-energy configuration (C4) if the viscosity of the matrix will allow (as in liquid precursor of polymer in Figure 1b, when chains were formed) or to arising of mechanical stresses if the matrix is rigid [54,55]. In the case of 3-component NCs, the polymer matrix transfers the mechanical stresses from matrix to BTO particles by elastic coupling that causes their electric polarization due to piezoelectric effect. This model can explain the reason for the increase of magnetoelectric response in oriented samples. Indeed, in samples with the random distribution of magnetic particles, configurations C1 and C2 have the same probability, but the low-energy C2 case in an external magnetic field will form just another low-energy case C4. Thus only 50% of particles after a magnetic field is ON will be in the high-energy configuration generating stronger mechanical stresses on the piezopolymer matrix and causing a stronger magnetoelectric response. This model also can explain, why the field-dependence of magnetoelectric response has peak-like behavior. Reaching a certain field close to the magnetic anisotropy field of CFO particles (H_A), almost all local magnetic moments of magnetic particles are oriented along the external magnetic field. For the case of ZCFO NPs with a lower anisotropy field (H_A for single-domain magnetic particles is proportional to H_{irr} [34]), the value of H_{peak} is proportionally lower. Further increase of the magnetic field only slightly rotates those moments in direction of the magnetic field but, at the same time, the external magnetic field hides the gradient of the local field. The magnetic force acting on each magnetic dipole (in this contest, an aggregate) is proportional to the multiplication of the value of magnetic moment on the gradient of the magnetic field (M·grad(B)) formed by external field and neighbor sources of the magnetic field (such as neighbor aggregates). Thus, the reduction of the gradient of the magnetic field will lead to the reduction of energy of interparticle (or interaggregate) interactions and reduce the strain of the piezoelectric matrix.

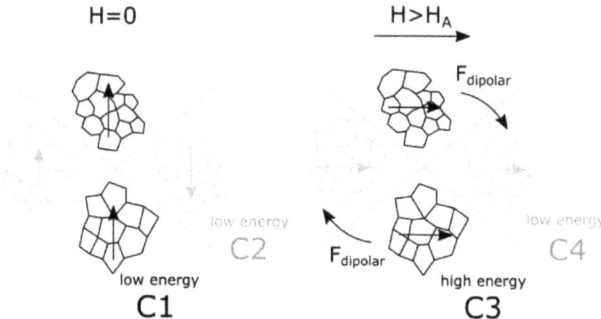

Figure 10. Schematic diagram explaining the mechanism of arising of dipolar interactions leading to the displacement of aggregates of CFO NPs: in zero magnetic fields (explanation is in the text).

3.6. Biological Tests

Previously it has been shown that PVDF can be used for attachment and differentiation of cells from different tissues: cardiovascular [26], osteogenic [25], muscle [62], and neuronal cells [63]. The neural crest stem cells (bNCSC) adhesion, survival, and differentiation were compatible with the PVDF material in line with previous findings on hippocampal neurospheres [27]. However, the effect of electromagnetic activities on neurogenesis remains controversial. The effect of PVDF on neural stem cells isolated in the later time of embryonic development was shown previously [63]. However, isolation of cells at an earlier stage of embryonic development gives the ability to direct their differentiation is more addressed, but biocompatibility of materials should be studied additionally. To test the biocompatibility of PVDF, we used bNCSC that reliably demonstrated their ability to generate neurons and glial cells in vitro [64,65]. This type of cells is characterized by highly proliferative activity (approximate cell population doubling time ~15–18 h) and was isolated at the early stage of embryonic development (11.5 days mouse embryos). After 24 h of bNCSCs culture on PVDF substrate, the number of cells was increased, with some of them attached to the polymer and forming neurospheres, as evidence of strong proliferation activity (Figure 11). After 48 h the size of neurospheres was markedly increased. In addition, a small number of elongated bNCSCs were present, which suggested the onset of migrated and differentiated cells (Figure 11). After 72 h, the processes of cell migration from neurospheres on the polymer substrate became more prominent. This process resulted in a reduced density of intercellular connections and possibly due to increased adhesion to the surface of the substrate. bNCSC placed on the material strongly attached and differentiated during 3 days, which is in agreement with the previously published protocol [66,67].

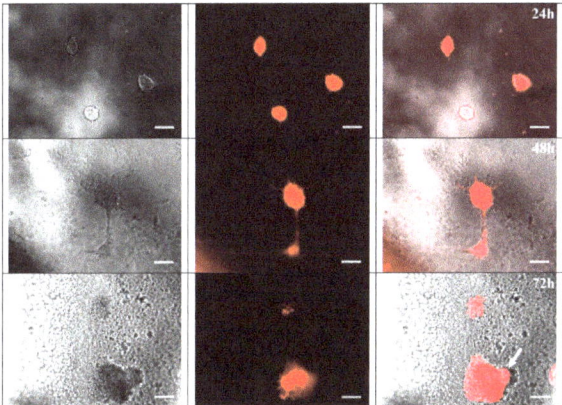

Figure 11. Images of neural crest stem cells (bNCSC), expressing red fluorescence signal (RFP) after 24 h, 48 h, 72 h of culture on PVDF substrate (×20); arrow indicates cells migrating from the neurosphere. Scale bar is 25 µm.

Two markers of cell differentiation were used for the immunofluorescent staining procedure: neuronal—β3-tubulin and glial—GFAP (glial fibrillary acidic protein). Both β3-tubulin positive cells and GFAP positive cells were present in the stained samples (Figure 12). These data can be interpreted as the cells retaining their potential for differentiation into neurons and glial cells in PVDF substrate. In addition, the stained samples showed traces of the migratory activity of bNCSC cells on the PVDF surface (Figure 12).

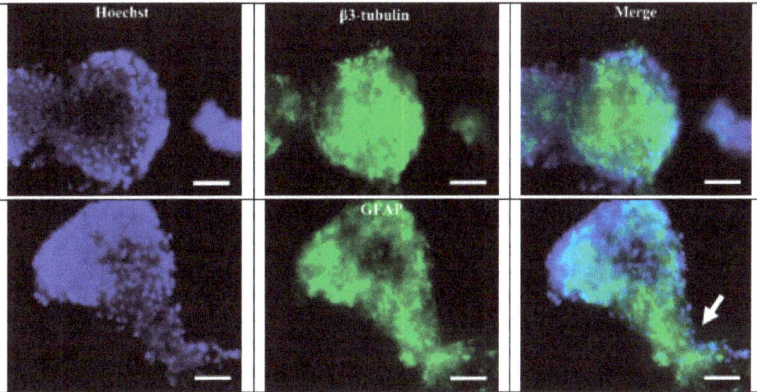

Figure 12. βIII-tubulin and GFAP (green) expression bNCSCs growing on PVDF substrate (×40). Blue–Hoechst nuclear stain; arrow indicates cells migrating from the neurosphere. Scale bar = 25 µm.

Neuronal stem cells isolated at the early embryonic stage cultivated on PVDF-based surface were able to proliferate and differentiate into main types of neural cells (neurons and glial cells). This finding together with an expected possibility to stimulate and modulate those processes remotely via magnetic field invoking local electric polarization of interface suggests that the developed materials can be used as bio-interfaces. For example, as a tool for neuronal stem cell cultivation and targeted deafferentation for future application in the treatment of neurodegenerative disorders and spinal cord injury and brain damage.

4. Conclusions

We have prepared and studied a set of samples of polymer-based nanocomposites having magnetic and magnetoelectric properties owing to the inclusion of magnetic nanoparticles in piezopolymers. New strategies to increase the magnetoelectric performance of PVDF- and PVDF-TrFE-based nanocomposites by the orientation of NPs clusters in chains in the polymer matrix and by the creation of 3-component nanocomposites by adding one additional component (ferroelectric particles) have been demonstrated. In our study, the magnetoelectric voltage coefficient (α_{ME}) of oriented 3-component PVDF-TrFE/BTO5/CFO composites was about 18.5 mV/cm·Oe that is four times higher than it is in the randomly oriented 2-component PVDF/CFO composite. A model based on magnetostatic interactions of clusters of magnetic nanoparticles with randomly distributed easy axes for the explanation of the ME transformation in 3-component composites has been suggested. Local magnetic and piezoelectric properties have been studied employing scanning probe microscopy. Further researches will be aimed to increase the magnetoelectric performance by changing particle (both magnetic and ferroelectric) size, shape, and concentration in such composites. Currently, the work on the use of obtained magnetoelectric composites as a bioactive interface is in progress. The possibility to remotely vary the surface charge by applying the magnetic field can be used to modulate the live process of neuronal stem cells, e.g., to control and induce their differentiation. Our results suggest that PVDF-based substrates are biocompatible for neuronal stem cells isolated at the early embryonic stage and thus they can be used as a matrix for cell cultivation for future application in the treatment of neurodegenerative disorders and spinal cord injury and brain damage. Furthermore, the functionalization of the PVDF substrate may contribute to the guidance of stem cell differentiation toward the required type of cells and may be used for the development of new in vitro differentiation strategies.

Supplementary Materials: The following are available online at https://www.mdpi.com/article/10.3390/nano11051154/s1. Figure S1: (a) XRD pattern of $CoFe_2O_4$ and $Zn_{0.25}Co_{0.75}Fe_2O_4$ NPs; (b) TEM micrograph and size distribution of $CoFe_2O_4$ NPs. Figure S2: (a) Room temperature (~300 K) M-H loop (b) ΔM-plot of $CoFe_2O_4$ and $Zn_{0.25}Co_{0.75}Fe_2O_4$ NPs in form of powder. Figure S3: XRD pattern of $BaTiO_3$ particles. Figure S4: Comparison of XRD patterns for (a) random and oriented CFO/PVDF and CFO/PVDF-TrFE nanocomposites; (b) nanocomposites with 5 and 10% of $BaTiO_3$ particles; (c) ZCFO/PVDF-TrFE nanocomposite. Figure S5: Optical images of aligned clusters of CFO NPs in (a) PVDF-TrFE and (b) PVDF polymers. Figure S6: Atomic force microscopy (AFM) images of (a) PVDF-TrFE/CFO and (b) PVDF/CFO NCs; histograms of the pore depth distribution obtained using the WSxM program for (c) PVDF-TrFE/CFO and (d) PVDF/CFO NCs. The median value and standard deviation (σ) of log-normal distribution are presented. Figure S7: Magnetic force microscopy (MFM) images of PVDF-TrFE/CFO nanocomposite evaporated in the absence of magnetic field: (a) topology and (b) MFM signals. Figure S8: Results of FEMM simulation of magnetic induction (B) for different configurations of initial magnetization of particles aggregates: (a) "head-to-tail" magnetization state; (b) "head-to-head" magnetization state; (c) one aggregate has a close structure and the second is uniformly magnetized; (d) two aggregates have close structures.

Author Contributions: Conceptualization, A.O., L.M., and V.R.; methodology, A.O., A.A., V.R., K.L.; validation, A.O., A.A., V.R., K.L.; formal analysis, A.O., M.S., A.A., V.R.; investigation, A.O., A.V.T., I.V.K., A.M.K., T.S.I., D.A.K., M.I.V., M.D.M., C.G., V.K., Y.H., V.A., A.A., D.M., M.S.; resources, V.R., Y.N.P.; data curation, A.O., A.A., V.A.; writing—original draft preparation, A.O., V.A.; writing—review and editing, A.A., V.R., K.L., A.V.T.; visualization, A.O., A.A.; supervision, E.N.K., V.R., D.P., K.L.; project administration, V.R.; funding acquisition, V.R. All authors have read and agreed to the published version of the manuscript.

Funding: This work was supported by the Russian Foundation for Basic Research (RFBR) according to the research project No. 18-32-20219. Partially work was supported by the Russian Science Foundation No. 21-72-30032 (interparticle and interaggregate interactions for all samples and NCs). The discussion of results has been made possible through the mobility grant provided by the 5 top 100 Russian Academic Excellence Project at the Immanuel Kant Baltic Federal University. L.M. acknowledges funding from the Grant of the President the Russian Federation No. MK-716.2020.2. T.S.I., A.M.K. and D.A.K. acknowledge financial support of the Ministry of Science and Higher Education of the Russian Federation as a part of the State Assignment (basic research, Project No. 0718-2020-0031 "New magnetoelectric composite materials based on oxide ferroelectrics having an ordered domain structure: production and properties") in part of the XRD study of NCs. Y.N.P. and I.V.K. acknowledge the financial support of the Ministry of Education and Science of the Russian Federation in the framework of the Increase Competitiveness Program of NUST «MISiS» (No. K2-2019-015) in part of AFM study. A.V.T. and M.D.M. acknowledge financial support of the Russian Science Foundation, project No. 18-79-10265 as part of MFM and PFM measurements of ME nanocomposite samples. E.N.K. acknowledges the financial support of the Swedish National Space Agency (reg. No. 2020-00163).

Institutional Review Board Statement: Not applicable.

Informed Consent Statement: Not applicable.

Data Availability Statement: The data presented in this study are available on request from the corresponding author.

Acknowledgments: BTO particles were kindly provided by Larisa Reznichenko from Southern Federal University (Rostov-on-Don, Russia). The authors thank Inna Rakova for the improvement of English.

Conflicts of Interest: The authors declare no conflict of interest.

References

1. Pereira, N.; Lima, A.C.; Lanceros-Mendez, S.; Martins, P. Magnetoelectrics: Three Centuries of Research Heading towards the 4.0 Industrial Revolution. *Materials* **2020**, *13*, 4033. [CrossRef]
2. Spaldin, N.A. Materials Science: The Renaissance of Magnetoelectric Multiferroics. *Science* **2005**, *309*, 391–392. [CrossRef]
3. Spaldin, N.A.; Ramesh, R. Advances in magnetoelectric multiferroics. *Nat. Mater.* **2019**, *18*, 203–212. [CrossRef]
4. Makarova, L.A.; Alekhina, J.; Isaev, D.A.; Khairullin, M.F.; Perov, N.S. Tunable layered composites based on magnetoactive elastomers and piezopolymer for sensors and energy harvesting devices. *J. Phys. D Appl. Phys.* **2020**. [CrossRef]
5. Vidal, J.V.; Turutin, A.V.; Kubasov, I.V.; Kislyuk, A.M.; Kiselev, D.A.; Malinkovich, M.D.; Parkhomenko, Y.N.; Kobeleva, S.P.; Sobolev, N.A.; Kholkin, A.L. Dual Vibration and Magnetic Energy Harvesting with Bidomain LiNbO3-Based Composite. *IEEE Trans. Ultrason. Ferroelectr. Freq. Control* **2020**, *67*, 1219–1229. [CrossRef] [PubMed]
6. Turutin, A.V.; Vidal, J.V.; Kubasov, I.V.; Kislyuk, A.M.; Malinkovich, M.D.; Parkhomenko, Y.N.; Kobeleva, S.P.; Kholkin, A.L.; Sobolev, N.A. Low-frequency magnetic sensing by magnetoelectric metglas/bidomain LiNbO$_3$ long bars. *J. Phys. D Appl. Phys.* **2018**, *51*. [CrossRef]
7. Turutin, A.V.; Vidal, J.V.; Kubasov, I.V.; Kislyuk, A.M.; Kiselev, D.A.; Malinkovich, M.D.; Parkhomenko, Y.N.; Kobeleva, S.P.; Kholkin, A.L.; Sobolev, N.A. Highly sensitive magnetic field sensor based on a metglas/bidomain lithium niobate composite shaped in form of a tuning fork. *J. Magn. Magn. Mater.* **2019**, *486*, 165209. [CrossRef]
8. Pereira, N.; Lima, A.C.; Correia, V.; Perinka, N.; Lanceros-Mendez, S.; Martins, P. Magnetic proximity sensor based on magneto-electric composites and printed coils. *Materials* **2020**, *13*, 1729. [CrossRef]
9. Murzin, D.; Mapps, D.J.; Levada, K.; Belyaev, V.; Omelyanchik, A.; Panina, L.; Rodionova, V. Ultrasensitive Magnetic Field Sensors for Biomedical Applications. *Sensors* **2020**, *20*, 1569. [CrossRef]
10. Roy, K.; Ghosh, S.K.; Sultana, A.; Garain, S.; Xie, M.; Bowen, C.R.; Henkel, K.; Schmeiβer, D.; Mandal, D. A Self-Powered Wearable Pressure Sensor and Pyroelectric Breathing Sensor Based on GO Interfaced PVDF Nanofibers. *ACS Appl. Nano Mater.* **2019**, *2*, 2013–2025. [CrossRef]

11. SKang, S.; Choi, K.; Nam, J.D.; Choi, H.J. Magnetorheological elastomers: Fabrication, characteristics, and applications. *Materials* **2020**, *13*, 4597. [CrossRef]
12. Martins, P.; Silva, M.; Reis, S.; Pereira, N.; Amorín, H.; Lanceros-Mendez, S. Wide-range magnetoelectric response on hybrid polymer composites based on filler type and content. *Polymers* **2017**, *9*, 62. [CrossRef] [PubMed]
13. Bastola, A.K.; Hoang, V.T.; Li, L. A novel hybrid magnetorheological elastomer developed by 3D printing. *Mater. Des.* **2017**, *114*, 391–397. [CrossRef]
14. Ze, Q.; Kuang, X.; Wu, S.; Wong, J.; Montgomery, S.M.; Zhang, R.; Kovitz, J.M.; Yang, F.; Qi, H.J.; Zhao, R. Magnetic Shape Memory Polymers with Integrated Multifunctional Shape Manipulation. *Adv. Mater.* **2020**, *32*, 1–8. [CrossRef] [PubMed]
15. Martins, P.; Lanceros-Méndez, S. Polymer-based magnetoelectric materials. *Adv. Funct. Mater.* **2013**, *23*, 3371–3385. [CrossRef]
16. Palneedi, H.; Annapureddy, V.; Priya, S.; Ryu, J. Status and Perspectives of Multiferroic Magnetoelectric Composite Materials and Applications. *Actuators* **2016**, *5*, 9. [CrossRef]
17. Martins, P.; Moya, X.; Phillips, L.C.; Kar-Narayan, S.; Mathur, N.D.; Lanceros-Mendez, S. Linear anhysteretic direct magnetoelectric effect in Ni 0.5Zn0.5Fe2O4/poly(vinylidene fluoride-trifluoroethylene) 0-3 nanocomposites. *J. Phys. D Appl. Phys.* **2011**, *44*, 482001. [CrossRef]
18. Martins, P.; Kolen'Ko, Y.V.; Rivas, J.; Lanceros-Mendez, S. Tailored Magnetic and Magnetoelectric Responses of Polymer-Based Composites. *ACS Appl. Mater. Interfaces* **2015**, *7*, 15017–15022. [CrossRef] [PubMed]
19. Makarova, L.A.; Alekhina, Y.A.; Omelyanchik, A.S.; Peddis, D.; Spiridonov, V.V.; Rodionova, V.V.; Perov, N.S. Magnetorheological foams for multiferroic applications. *J. Magn. Magn. Mater.* **2019**, *485*, 413–418. [CrossRef]
20. Makarova, L.; Alekhina, Y.; Kramarenko, E.; Omelyanchik, A.; Rodionova, V.; Malyshkina, O.; Perov, N. Composite multiferroic materials consisting of NdFeB and PZT particles embedded in elastic matrix: The appearance of electrical polarization in a constant magnetic field. *EPJ Web Conf.* **2018**, *185*, 07008. [CrossRef]
21. Makarova, L.A.; Alekhina, Y.A.; Omelyanchik, A.S.; Rodionova, V.V.; Malyshkina, O.V.; Perov, N.S. Elastically coupled ferromagnetic and ferroelectric microparticles: New multiferroic materials based on polymer, NdFeB and PZT particles. *J. Magn. Magn. Mater.* **2019**, *470*, 89–92. [CrossRef]
22. Tang, B.; Zhuang, J.; Wang, L.; Zhang, B.; Lin, S.; Jia, F.; Dong, L.; Wang, Q.; Cheng, K.; Weng, W. Harnessing Cell Dynamic Responses on Magnetoelectric Nanocomposite Films to Promote Osteogenic Differentiation. *ACS Appl. Mater. Interfaces* **2018**, *10*, 7841–7851. [CrossRef] [PubMed]
23. Castro, N.; Fernandes, M.M.; Ribeiro, C.; Correia, V.; Minguez, R.; Lanceros-Méndez, S. Magnetic Bioreactor for Magneto-, Mechano- and Electroactive Tissue Engineering Strategies. *Sensors* **2020**, *20*, 3340. [CrossRef]
24. Zhu, R.; Sun, Z.; Li, C.; Ramakrishna, S.; Chiu, K.; He, L. Electrical stimulation affects neural stem cell fate and function in vitro. *Exp. Neurol.* **2019**, *319*, 112963. [CrossRef] [PubMed]
25. Mirzaei, A.; Saburi, E.; Enderami, S.E.; Bagherabad, M.B.; Enderami, S.E.; Chokami, M.; Moghadam, A.S.; Salarinia, R.; Ardeshirylajimi, A.; Mansouri, V.; et al. Synergistic effects of polyaniline and pulsed electromagnetic field to stem cells osteogenic differentiation on polyvinylidene fluoride scaffold. *Artif. Cells Nanomed. Biotechnol.* **2019**. [CrossRef] [PubMed]
26. Hitscherich, P.; Wu, S.; Gordan, R.; Xie, L.H.; Arinzeh, T.; Lee, E.J. The effect of PVDF-TrFE scaffolds on stem cell derived cardiovascular cells. *Biotechnol. Bioeng.* **2016**. [CrossRef]
27. Xia, L.; Shang, Y.; Chen, X.; Li, H.; Xu, X.; Liu, W.; Yang, G.; Wang, T.; Gao, X.; Chai, R. Oriented Neural Spheroid Formation and Differentiation of Neural Stem Cells Guided by Anisotropic Inverse Opals. *Front. Bioeng. Biotechnol.* **2020**. [CrossRef] [PubMed]
28. Li, J.; Lepski, G. Cell transplantation for spinal cord injury: A systematic review. *Biomed. Res. Int.* **2013**. [CrossRef]
29. Hjerling-Leffler, J.; Marmigère, F.; Heglind, M.; Cederberg, A.; Koltzenburg, M.; Enerbäck, S.; Ernfors, P. The boundary cap: A source of neural crest stem cells that generate multipe sensory neuron subtypes. *Development* **2005**. [CrossRef] [PubMed]
30. Sakthiswary, R.; Raymond, A.A. Stem cell therapy in neurodegenerative diseases: From principles to practice. *Neural Regen. Res.* **2012**, *7*, 1822–1831. [CrossRef]
31. Zhou, Y.; Shao, A.; Xu, W.; Wu, H.; Deng, Y. Advance of Stem Cell Treatment for Traumatic Brain Injury. *Front. Cell. Neurosci.* **2019**, *13*. [CrossRef]
32. Meng, N.; Zhu, X.; Mao, R.; Reece, M.J.; Bilotti, E. Nanoscale interfacial electroactivity in PVDF/PVDF-TrFE blended films with enhanced dielectric and ferroelectric properties. *J. Mater. Chem. C* **2017**, *5*, 3296–3305. [CrossRef]
33. Cannas, C.; Falqui, A.; Musinu, A.; Peddis, D.; Piccaluga, G. CoFe$_2$O$_4$ nanocrystalline powders prepared by citrate-gel methods: Synthesis, structure and magnetic properties. *J. Nanopart. Res.* **2006**, *8*, 255–267. [CrossRef]
34. Omelyanchik, A.; Levada, K.; Pshenichnikov, S.; Abdolrahim, M.; Baricic, M.; Kapitunova, A.; Galieva, A.; Sukhikh, S.; Astakhova, L.; Antipov, S.; et al. Green Synthesis of Co-Zn Spinel Ferrite Nanoparticles: Magnetic and Intrinsic Antimicrobial Properties. *Materials* **2020**, *13*, 5014. [CrossRef] [PubMed]
35. Shilkina, L.A.; Talanov, M.V.; Shevtsova, S.I.; Grin', P.G.; Kozakov, A.T.; Dudkina, S.I.; Nikol'skii, A.V.; Reznichenko, L.A. Isomorphism problems in lead-barium titanate. *J. Alloys Compd.* **2020**, *829*, 154589. [CrossRef]
36. Ribeiro, C.; Costa, C.M.; Correia, D.M.; Nunes-Pereira, J.; Oliveira, J.; Martins, P.; Gonçalves, R.; Cardoso, V.F.; Lanceros-Méndez, S. Electroactive poly(vinylidene fluoride)-based structures for advanced applications. *Nat. Protoc.* **2018**, *13*, 681–704. [CrossRef] [PubMed]
37. Andrew, J.S.; Clarke, D.R. Enhanced ferroelectric phase content of polyvinylidene difluoride fibers with the addition of magnetic nanoparticles. *Langmuir* **2008**. [CrossRef]

38. Oh, W.J.; Lim, H.S.; Won, J.S.; Lee, S.G. Preparation of PVDF/PAR composites with piezoelectric properties by post-treatment. *Polymers* **2018**, *10*, 1333. [CrossRef]
39. Morrish, A.H. *The Physical Principles of Magnetism*; IEEE Press: Piscataway, NJ, USA, 1965. [CrossRef]
40. Omelyanchik, A.; Singh, G.; Volochaev, M.; Sokolov, A.; Rodionova, V.; Peddis, D. Tunable magnetic properties of Ni-doped CoFe2O4 nanoparticles prepared by the sol–gel citrate self-combustion method. *J. Magn. Magn. Mater.* **2019**, *476*. [CrossRef]
41. Stancu, A.; Pike, C.; Stoleriu, L.; Postolache, P.; Cimpoesu, D. Micromagnetic and Preisach analysis of the First Order Reversal Curves (FORC) diagram. *J. Appl. Phys.* **2003**. [CrossRef]
42. Kolesnikova, V.; Martínez-garcía, J.C.; Rodionova, V.; Rivas, M. Study of bistable behaviour in interacting Fe-based microwires by First Order Reversal Curves. *J. Magn. Magn. Mater.* **2020**, *2*, 166857. [CrossRef]
43. Vaganov, M.V.; Raikher, Y.L. Effect of mesoscopic magnetomechanical hysteresis on magnetization curves and first-order reversal curve diagrams of magnetoactive elastomers. *J. Phys. D Appl. Phys.* **2020**, *53*. [CrossRef]
44. Mayergoyz, I.D.; Friedman, G. Generalized preisach model of hyst eresis (invited). *IEEE Trans. Magn.* **1988**, *24*, 212–217. [CrossRef]
45. Pike, C.R.; Roberts, A.P.; Verosub, K.L. Characterizing interactions in fine magnetic particle systems using first order reversal curves. *J. Appl. Phys.* **1999**, *85*, 6660–6667. [CrossRef]
46. Vaganov, M.V.; Linke, J.; Odenbach, S.; Raikher, Y.L. Model FORC diagrams for hybrid magnetic elastomers. *J. Magn. Magn. Mater.* **2017**, *431*, 130–133. [CrossRef]
47. Jesse, S.; Baddorf, A.P.; Kalinin, S.V. Switching spectroscopy piezoresponse force microscopy of ferroelectric materials. *Appl. Phys. Lett.* **2006**. [CrossRef]
48. Hong, S.; Woo, J.; Shin, H.; Jeon, J.U.; Pak, Y.E.; Colla, E.L.; Setter, N.; Kim, E.; No, K. Principle of ferroelectric domain imaging using atomic force microscope. *J. Appl. Phys.* **2001**. [CrossRef]
49. Aldskogius, H.; Berens, C.; Kanaykina, N.; Liakhovitskaia, A.; Medvinsky, A.; Sandelin, M.; Schreiner, S.; Wegner, M.; Hjerling-Leffler, J.; Kozlova, E.N. Regulation of boundary cap neural crest stem cell differentiation after transplantation. *Stem Cells* **2009**, *27*, 1592–1603. [CrossRef]
50. Vintersten, K.; Monetti, C.; Gertsenstein, M.; Zhang, P.; Laszlo, L.; Biechele, S.; Nagy, A. Mouse in red: Red fluorescent protein expression in mouse ES cells, embryos, and adult animals. *Genesis* **2004**. [CrossRef]
51. Geshev, J.; Mikhov, M.; Schmidt, J.E. Remanent magnetization plots of fine particles with competing cubic and uniaxial anisotropies. *J. Appl. Phys.* **1999**, *85*, 7321–7327. [CrossRef]
52. García-Otero, J.; Porto, M.; Rivas, J. Henkel plots of single-domain ferromagnetic particles. *J. Appl. Phys.* **2000**, *87*, 7376. [CrossRef]
53. Cai, X.; Lei, T.; Sun, D.; Lin, L. A critical analysis of the α, β and γ phases in poly(vinylidene fluoride) using FTIR. *RSC Adv.* **2017**, *7*, 15382–15389. [CrossRef]
54. Vaganov, M.V.; Borin, D.Y.; Odenbach, S.; Raikher, Y.L. Training effect in magnetoactive elastomers due to undermagnetization of magnetically hard filler. *Phys. B Condens. Matter* **2020**, *578*, 411866. [CrossRef]
55. Vaganov, M.V.; Borin, D.Y.; Odenbach, S.; Raikher, Y.L. Modeling the magnetomechanical behavior of a multigrain magnetic particle in an elastic environment. *Soft Matter* **2019**, *15*, 4947–4960. [CrossRef] [PubMed]
56. Oroujizad, S.; Almasi-Kashi, M.; Alikhanzadeh-Arani, S. A FORC investigation into the effect of Cu additive on magnetic characteristics of Co-Ni alloy nanoparticles. *J. Magn. Magn. Mater.* **2019**, *473*, 169–175. [CrossRef]
57. Sánchez, E.H.; Vasilakaki, M.; Lee, S.S.; Normile, P.S.; Muscas, G.; Murgia, M.; Andersson, M.S.; Singh, G.; Mathieu, R.; Nordblad, P.; et al. Simultaneous individual and dipolar collective properties in binary assemblies of magnetic nanoparticles. *arXiv* **2019**, arXiv:1909.13500. [CrossRef]
58. Omote, K.; Ohigashi, H.; Koga, K. Temperature dependence of elastic, dielectric, and piezoelectric properties of "single crystalline" films of vinylidene fluoride trifluoroethylene copolymer. *J. Appl. Phys.* **1997**, *81*, 2760–2769. [CrossRef]
59. Zhang, J.X.; Dai, J.Y.; So, L.C.; Sun, C.L.; Lo, C.Y.; Or, S.W.; Chan, H.L.W. The effect of magnetic nanoparticles on the morphology, ferroelectric, and magnetoelectric behaviors of CFO/P(VDF-TrFE) 0-3 nanocomposites. *J. Appl. Phys.* **2009**, *105*, 054102. [CrossRef]
60. Gheorghiu, F.; Stanculescu, R.; Curecheriu, L.; Brunengo, E.; Stagnaro, P.; Tiron, V.; Postolache, P.; Buscaglia, M.T.; Mitoseriu, L. PVDF–ferrite composites with dual magneto-piezoelectric response for flexible electronics applications: Synthesis and functional properties. *J. Mater. Sci.* **2020**, *55*, 3926–3939. [CrossRef]
61. Gao, J.; Xue, D.; Liu, W.; Zhou, C.; Ren, X. Recent progress on BaTiO3-based piezoelectric ceramics for actuator applications. *Actuators* **2017**, *6*, 24. [CrossRef]
62. Ribeiro, S.; Gomes, A.C.; Etxebarria, I.; Lanceros-Méndez, S.; Ribeiro, C. Electroactive biomaterial surface engineering effects on muscle cells differentiation. *Mater. Sci. Eng. C* **2018**. [CrossRef] [PubMed]
63. Esmaeili, E.; Soleimani, M.; Ghiass, M.A.; Hatamie, S.; Vakilian, S.; Zomorrod, M.S.; Sadeghzadeh, N.; Vossoughi, M.; Hosseinzadeh, S. Magnetoelectric nanocomposite scaffold for high yield differentiation of mesenchymal stem cells to neural-like cells. *J. Cell. Physiol.* **2019**. [CrossRef] [PubMed]
64. Trolle, C.; Konig, N.; Abrahamsson, N.; Vasylovska, S.; Kozlova, E.N. Boundary cap neural crest stem cells homotopically implanted to the injured dorsal root transitional zone give rise to different types of neurons and glia in adult rodents. *BMC Neurosci.* **2014**, *15*, 60. [CrossRef] [PubMed]

65. Konig, N.; Trolle, C.; Kapuralin, K.; Adameyko, I.; Mitrecic, D.; Aldskogius, H.; Shortland, P.J.; Kozlova, E.N. Murine neural crest stem cells and embryonic stem cell-derived neuron precursors survive and differentiate after transplantation in a model of dorsal root avulsion. *J. Tissue Eng. Regen. Med.* **2017**. [CrossRef] [PubMed]
66. Kozlova, E.N.; Jansson, L. Differentiation and migration of neural crest stem cells are stimulated by pancreatic islets. *Neuroreport* **2009**, *20*, 833–838. [CrossRef] [PubMed]
67. Kozlova, E.N. The Effect of Mesoporous Silica Particles on Stem Cell Differentiation. *J. Stem Cell Res. Ther.* **2017**, *2*, 73–78. [CrossRef]

Article

The Heating Efficiency and Imaging Performance of Magnesium Iron Oxide@tetramethyl Ammonium Hydroxide Nanoparticles for Biomedical Applications

Mohamed S. A. Darwish [1,2], Hohyeon Kim [1], Minh Phu Bui [1], Tuan-Anh Le [1], Hwangjae Lee [3], Chiseon Ryu [3], Jae Young Lee [3,*] and Jungwon Yoon [1,*]

[1] School of Integrated Technology, Gwangju Institute of Science and Technology, Gwangju 61005, Korea; msa.darwish@gmail.com (M.S.A.D.); duplisona@gm.gist.ac.kr (H.K.); buiminhphu@gist.ac.kr (M.P.B.); tuananhle@gist.ac.kr (T.-A.L.)

[2] Egyptian Petroleum Research Institute, 1 Ahmed El-Zomor Street, El Zohour Region, Nasr City, Cairo 11727, Egypt

[3] School of Materials Science and Engineering, Gwangju Institute of Science and Technology, Gwangju 500-712, Korea; hj31p@gist.ac.kr (H.L.); ryuchiseon@gist.ac.kr (C.R.)

* Correspondence: jaeyounglee@gist.ac.kr (J.Y.L.); jyoon@gist.ac.kr (J.Y.)

Abstract: Multifunctional magnetic nanomaterials displaying high specific loss power (SLP) and high imaging sensitivity with good spatial resolution are highly desired in image-guided cancer therapy. Currently, commercial nanoparticles do not sufficiently provide such multifunctionality. For example, Resovist® has good image resolution but with a low SLP, whereas BNF® has a high SLP value with very low image resolution. In this study, hydrophilic magnesium iron oxide@tetramethyl ammonium hydroxide nanoparticles were prepared in two steps. First, hydrophobic magnesium iron oxide nanoparticles were fabricated using a thermal decomposition technique, followed by coating with tetramethyl ammonium hydroxide. The synthesized nanoparticles were characterized using XRD, DLS, TEM, zeta potential, UV-Vis spectroscopy, and VSM. The hyperthermia and imaging properties of the prepared nanoparticles were investigated and compared to the commercial nanoparticles. One-dimensional magnetic particle imaging indicated the good imaging resolution of our nanoparticles. Under the application of a magnetic field of frequency 614.4 kHz and strength 9.5 kA/m, nanoparticles generated heat with an SLP of 216.18 W/g, which is much higher than that of BNF (14 W/g). Thus, the prepared nanoparticles show promise as a novel dual-functional magnetic nanomaterial, enabling both high performance for hyperthermia and imaging functionality for diagnostic and therapeutic processes.

Keywords: hyperthermia; magnesium iron oxide; magnetic particle imaging; nanoparticle

1. Introduction

Iron oxide nanoparticles (IONPs) are widely used in magnetic separation, drug-delivery, imaging, and hyperthermia cancer treatments, due to their biocompatibility, magnetic-imaging capability, and hypothermic characteristics [1–4]. In particular, one of the most developed IONP techniques is magnetic hyperthermia in which heat is generated under an alternating magnetic field (AMF) to destroy tumor cells. IONPs that can elevate the temperature of the target tissue above 45 °C are suitable for tumor treatment [4]. For effective treatment, the cancerous tissue temperature should reach 42 to 45 °C, whereas temperatures over 50 °C destroy the cancer cells via thermoablation.

A short treatment time is highly desirable for safe and effective hyperthermia treatment. The specific loss power (SLP) is used as a parameter to investigate how much energy is absorbed per magnetic nanoparticles mass (NPs) under AMF [2,3]. Hence, extensive efforts have been made to produce magnetic nanomaterials possessing high SLP values.

Concurrently, there is increasing interest in using NPs as imaging agents to monitor the location and distribution of nanomaterials precisely and stably for magnetic-particle imaging (MPI) applications [5–7].

MPI detects the positions and concentrations of IONPs via time-varying magnetic fields. Thus, the IONPs act as MPI tracer agents, because they can be magnetized instantaneously by an external magnetic field and can stimulate a nonlinear response in a near-zero magnetic field [6]. Compared with existing imaging modalities, MPI is recognized as a favorable tool for cancer diagnosis, as it offers the advantages of zero background signal, zero signal reduction with increasing tissue depth, quantitative linearity [7], and high sensitivity. Additionally, for MPI, there is no need for ionizing radiation.

Various techniques have been developed for IONP fabrication, including co-precipitation, thermal decomposition, and hydro-/solvothermal processes [8–13]. Among them, the thermal decomposition technique has been widely studied. Thermal decomposition allows for fine control over the morphology and size of IONPs by changing the synthetic parameters (e.g., precursors, surfactants, pH value, and reaction time) [12,13]. The surface properties of IONPs critically influence the overall performance of the materials. Hence, surface modification of IONPs is usually carried out either during or after synthesis [9,10].

The presence of coatings on the surface of IONPs can protect the magnetic core from oxidative environments, enhance stability and compatibility, and improve tumor-targeting efficiency. Generally, IONPs are biocompatible and non-toxic; thus, they are suitable for biomedical applications. It should be noted that biocompatibility is also strongly influenced by the coating material [14–17]. To date, two types of iron oxide NPs have been applied in clinical treatment: (i) shelled with polysaccharide layer and (ii) shelled with silica layer [14].

The current challenge in applying IONPs for diagnostic and therapeutic processes is the development of a smart theranostic agent with high performance for hyperthermia treatment and MPI. Currently, RES NPs is promising as an MPI contrast agent; however, it has a low SLP value that restricts its use in therapeutic hyperthermia cancer treatments. On the other hand, BNF has a relatively high SLP value in hyperthermia systems, but a low MPI sensitivity. Therefore, IONPs that can combine, in a single material, the unique advantages of commercially and clinically used IONPs will be promising theranostic agents. As ideal theranostic agents, magnetic IONPs should be non-toxic and biocompatible with a high magnetization value.

Controlling the size and morphology of NPs is an effective technique for improving SLP, image resolution, and biocompatibility [5–7]. Bauer et al. prepared zinc-doped magnetite cubic NPs and achieved good MPI performance and a high SLP of 1019.2 W/g under the action of an alternating magnetic field of strength 16 kA/m and frequency 380 kHz; however, these magnetic-field parameters exceed the safety limit of SLP testing, i.e., 5×10^9 A m^{-1}s^{-1} [5]. Recently, Dadfar et al. prepared citrate-coated superparamagnetic IONPs and achieved good MPI performance and a high SLP of 350 W/g, with a magnetic field of strength 46 kA/m and frequency 186 kHz; however, the main limitation of the study was also that the magnetic field parameters exceeded the safety limit [6]. In another report, Song et al. prepared carbon-coated FeCo NPs and achieved good MPI performance and a high SLP of 406 W/g with a magnetic field of strength 100 kA/m and frequency 30 kHz [7]; however, they did not clearly address the potential biocompatibility issue, as the dissolution of the FeCo NPs would lead to the release of toxic Co^{2+} ions inside the living organism [17].

In this study, we synthesized and characterized high-performance nanomaterials that have high SLP and high sensitivity with good spatial resolution, based on magnesium iron oxide@tetramethyl ammonium hydroxide (MgIONPs@TMAH) NPs. To our knowledge, the efficiency of MgIONPs@TMAH as a theranostic agent has not been investigated. In this study, we focused on the dual features of high SLP performance and high MPI sensitivity necessary for the application of MgIONPs@TMAH in practice. To this end, MgIONPs@TMAH nanoparticles were compared to the commercially and clinically used IONPs.

2. Experimental Work

2.1. Materials

Iron (III) acetylacetonate (97% purity), magnesium acetate tetrahydrate ($\geq$99% purity), benzyl ether (98% purity), tetramethyl ammonium hydroxide (10 wt% in H_2O) and oleic acid ($\geq$99% purity) were purchased from Sigma Aldrich (St. Louis, MO, USA). Commercial BNF NPs (iron oxide nanoparticles coated with dextran (Micromod Partikeltechnologie GmbH, Rostock, Germany) and Ferucarbotran (RES NPs) iron oxide NPs shelled with carboxydextran (Meito Sangyo Co., Ltd., Nagoya, Japan)) were used in the study.

2.2. Synthesis of Magnesium Iron Oxide@Tetramethyl Ammonium Hydroxide Nanoparticles (MgIONPs@TMAH)

MgIONPs@TMAH nanoparticles were prepared by the following two steps consisting of the synthesis of hydrophobic MgIONPs and surface modification with TMAH, as described below.

2.2.1. Preparation of Hydrophobic Nanoparticles

MgIONPs were fabricated by a thermal decomposition process in benzyl ether. Iron (III) acetylacetonate (0.7063 g), magnesium acetate tetrahydrate (0.02787 g), and oleic acid (0.3389 g) were mixed with benzyl ether (20 mL). The solution was transferred to a three-neck flask with a temperature sensor, nitrogen, a condenser, and a magnetic bar stirrer set on a heating mantel. The solution was then mixed well for 60 min at room temperature for degassing. The temperature was increased to 200 °C at a rate of 8 °C/min with vigorous stirring. Afterwards, the solution was maintained at 200 °C for 50 min. The temperature was then increased to 295 °C at a rate of heating of 5 °C/min with vigorous stirring and maintained at 295 °C for 60 min, before cooling to room temperature. The prepared NPs were washed twice with ethanol and separated using a magnet.

2.2.2. Surface Modification of MgIONPs with Tetramethyl Ammonium Hydroxide

To produce hydrophilic NPs for water dispersion, the prepared MgIONPs were coated with TMAH. In brief, 10 mL TMAH was dispersed in DI water (10 mL). The TMAH solution was then added to the prepared MgIONPs in an ultrasonic bath and further mixed for 30 min. The coated NPs were separated using a magnetic bar, washed with water, and then dried in an oven at 80 °C for 2 days. The obtained NPs powder was well dispersed in water and used for further studies.

2.3. Characterization

X-ray diffraction (XRD) of the NPs was carried out using an X-ray diffractometer (Rigaku, Tokyo, Japan). The crystallite sizes (D_p) of the NPs were calculated using Scherrer's formula, as follows [18]:

$$\text{Crystallite size } (D_p) = K\lambda/(B\cos\theta), \quad (1)$$

D_p: the average crystallite size (nm), B: the full width at half maximum of the XRD peak, λ: the X-ray wavelength (1.5406 Å), K: Scherrer's constant (shape parameter: 0.89), and θ: the XRD peak position.

$$\frac{1}{d_{hkl}^2} = \frac{h^2 + k^2 + l^2}{a^2} \quad (2)$$

The zeta potential and particle size analyzer (ELSZ-2000; Photal Otsuka Electronics Co., Osaka, Japan) was used to evaluate the absorbance properties using ultraviolet-visible spectroscopy. The magnetic behaviors of NPs were investigated using vibrating sample magnetometry (VSM; Lake Shore 7400 series; Lake Shore Cryotronics, Inc., Westerville, OH, USA). ICP-OES (Optima 8300, Perkin Elmer, Waltham, MA, USA) was performed to determine the metal content of the NPs. The morphology of NPs was examined by TEM (Tecnai G2S Twin; Philips, USA) at 300 keV. For the magnetic particle imaging (MPI) experiment, a gradient field of 2.6 T/m was applied to generate the field-free point, and an excitation field of 200 μT at 25 kHz was used to magnetize the particles.

Heating profile was evaluated using a custom-designed laboratory system. A function generator was used to generate a sinusoidal voltage signal, which was further amplified to the desired power through an alternating current power amplifier (AE Techron 7224, Elkhart, IN, USA). The SLP value was calculated using the following equation. Note that we used data obtained from the first 10 s.

$$SLP = (C_p/m) \times (dT/dt) \tag{3}$$

dT/dt: time dependent temperature, C_p: 4.184 for water, and m: the mass of elements per volume.

2.4. In Vitro Cytocompatibility Test

Cytocompatibility of the nanoparticles was evaluated through in vitro cytotoxicity tests with murine NIH-3T3 fibroblasts cells. First, the cytotoxicity of the nanoparticles was tested with a WST assay [19], which can quantitatively measure the metabolic activity of mitochondria in live cells. The culture medium (1 mL) containing the different NP contents (0.125, 0.25, 0.5, 1, or 2 mg/mL) was added to the cells and incubated for 24 h. Then, the cells were washed with sterile Dulbecco's phosphate-buffered saline. Finally, fresh culture medium (0.5 mL) and WST assay solution (0.05 mL) were added to each well and incubated for an additional 2 h. Then, the absorbance of each sample solution at 450 nm was measured using a plate reader. We reported cell viability using the following equation:

$$\text{Cell Viability (\%)} = (A_c - A_s)/A_c \times 100, \tag{4}$$

A_c: the absorbance of the control sample and A_s: the absorbance of the sample solution.

In addition, we examine the cytocompatibility of nanoparticles by using a Live/Dead staining kit (Invitrogen, Carlsbad, CA, USA) according to the manufacturer's protocol. This assay results in the staining of live cells and dead cells with green and red colors, respectively. The percentage of live cells in the total cells (live and dead cells) after the exposure to the nanoparticles can indicate cytocompatibility. Fluorescence images were acquired using a fluorescence microscope (DMI3000B; Leica, Wetzlar, Germany).

3. Results and Discussion

3.1. Synthesis and Characterization of MgIONPs@TMAH

MgIONPs@TMAH were prepared and their efficiency as a theranostic agent for hyperthermia and MPI was investigated and compared with commercial iron oxide NPs. MgIONPs@TMAH were first prepared using a thermal decomposition process, followed by coating with TMAH to obtain relatively uniform water-dispersable NPs (Figure 1).

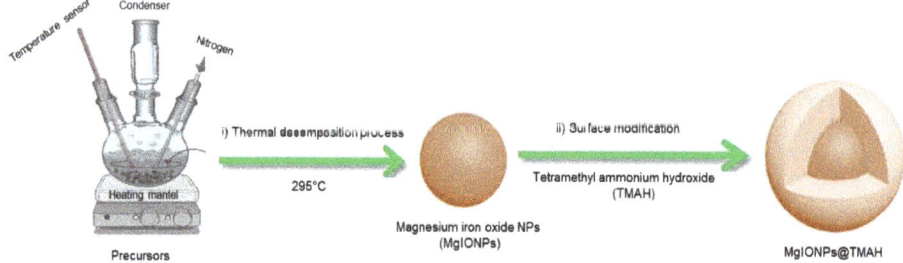

Figure 1. Schematic representation of the steps in the synthesis of MgIONPs@TMAH.

The prepared MgIONPs@TMAH were relatively monodisperse, with a narrow size distribution (15.0 ± 5.0 nm) (Figure 2A). For comparison, we acquired TEM images of two commercial IONPs. RES NPs had an average particle size of 3.0–8.5 nm, whereas BNF NPs showed an average particle size (14.0 ± 3.5 nm) (Figure 2B,C). The crystalline nature of the prepared MgIONPs@TMAH was examined using high-resolution TEM/XRD; the results of which are discussed in the following section. The corresponding SAED image of the NPs displayed ring characteristics, which is proportional to a structure of small domains. The SAED pattern also showed widespread rings with a low intensity, indicating the reflection planes of the NPs. Generally, the presence of aggregation behavior between NPs is a result of high surface energy and magnetic dipole–dipole force. This detrimental aggregation was reduced by introducing a shell layer of polymer or organic material on the surface of the NPs [20,21].

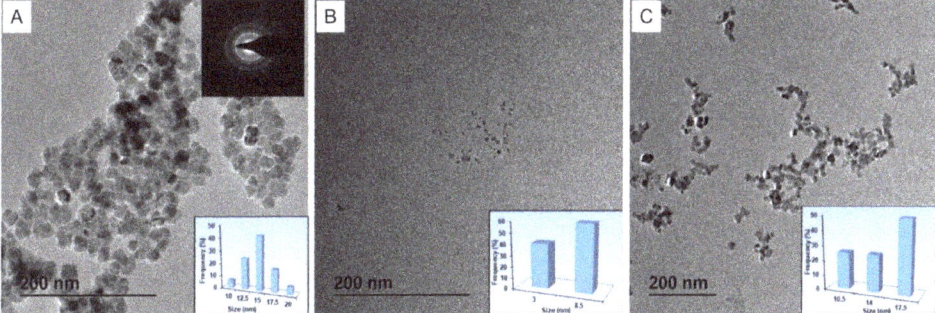

Figure 2. (A) TEM and SAED images of the prepared hydrophilic magnesium iron oxide@tetramethyl ammonium hydroxide nanoparticles (MgIONPs@TMAH), (B) RES NPs, and (C) BNF NPs.

Elemental compositions of the prepared MgIONPs@TMAH were quantified by ICP-OES. Element weight represented 58.52% of the entire NP content, in which Fe and Mg atoms were 58.3% and 0.22% in the composition, respectively. The crystalline properties of MgIONPs@TMAH were investigated and compared with the standard Joint Committee on Powder Diffraction Standards data (JCPDS: 88-1935) using XRD (Figure 3A) [22]. The indexed peaks were (220), (311), (400), (511), and (440), and magnesium ferrite structure was confirmed. The crystallite sizes and the associated lattice parameters for MgIONPs@TMAH were 13.89 nm and 8.38 Å, respectively. The crystalline properties of NPs were calculated based on maximum intensity peak. Peak broadening depends on various factors, such as instrumental effects, strain effects, and a finite crystallite size. The peaks were rather broad and weak, likely due to disorder and small crystallite effects. The crystalline properties of NPs are important for heating efficiency [23]. When the MNP size is maintained below a critical volume/size during the nanoparticle synthesis, the MNPs tend to behave as single magnetic domain structures, and at the smallest sizes, they exhibit superparamagnetic behavior under standard conditions. As results, the value of Ms increases with the size of the MNPs until it reaches a maximum that is close to the bulk magnetization value. As the size of the MNPs increases, they eventually possess pseudo single-domains and then multi-domain structures, in which the moment of each domain may not be oriented in the same direction [24]. As the relaxation time and magnetization of the NPs have an effect on the SLP value, a specific NPs size can exhibit effective heating [25]. Vreeland et al. recorded that approximately 22 nm is the effective size to enhance the SLP value for superparamagnetic NPs [25]. The diameter and shape of magnetic NPs are also important factors associated with heating efficiency. Core-shell NP cubes exhibit a high heating profile, due to low anisotropy and spin disorder reduction [23].

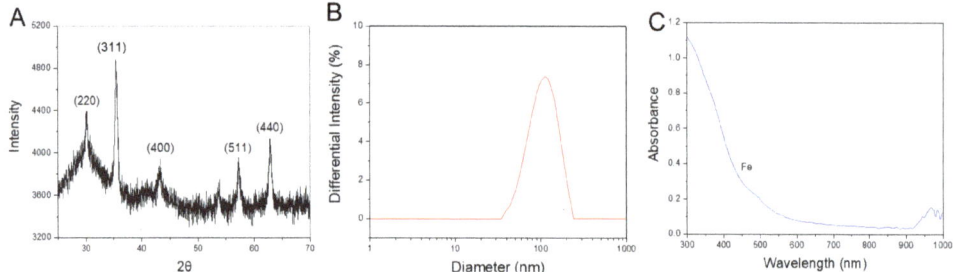

Figure 3. (**A**) XRD spectrum, (**B**) size distribution, and (**C**) absorbance spectrum of the prepared MgIONPs@TMAH.

The mean hydrodynamic size of MgIONPs@TMAH obtained by DLS was 167.0 ± 0.8 nm (Figure 3B). Note that this value is slightly higher than that obtained from TEM analysis, which is typical for hydrophilic NPs due to water–material interactions.

The zeta potential (ζ) of the MgIONPs@TMAH was investigated, as it is closely related to the stability of nanoparticles. The magnetic NPs stability is significant in biomedical applications [26]. A low zeta potential value (−12.0 ± 0.1 mV) implies that the NP may show poor stability in aqueous solutions. ±30 mV is considered the limit value for setting stability in colloidal systems, due to the formation of a high repulsion force between nanoparticles at this value [27].

The absorbance of the prepared NPs was measured by UV-Vis spectroscopy. MgIONPs@TMAH spectra showed broad absorption over the visible range (300–600 nm), as a result of the presence of d-orbital in Fe^{3+} (Figure 3C). In particular, the absorption peak at 490 nm corresponded to Fe^{3+} in a tetrahedral coordination [28–30].

The magnetic behavior of MgIONPs@TMAH was studied using VSM at room temperature (~23 °C). The magnetization (M)–field strength (H) curves showed hysteresis loops, indicating the ferromagnetism of the prepared MgIONPs@TMAH with a magnetization saturation (M_s) at 55.1 emu/g (Figure 4). MgIONPs@TMAH showed very low coercivity (Hc) and a remanence magnetization (M_r) of 30.2 Oe and 5.5 emu/g, respectively, due to the soft magnetic nature of the NPs. The squareness (SQ) is defined as the ratio of remanence magnetization to magnetization saturation. A single magnetic domain structure is observed when SQ ≥ 0.5, whereas a material having SQ < 0.5 is considered to have a multi-domain structure. In our study, MgIONPs@TMAH samples showed the presence of multi-domain structures [31]. Performing a ZFC-FC analysis is useful to obtain the blocking temperature, and thus the information about the superparamagnetic state at RT. However, unfortunately, ZFC-FC analysis is not available at the moment. So, it is a limitation for this research that will be performed as future work.

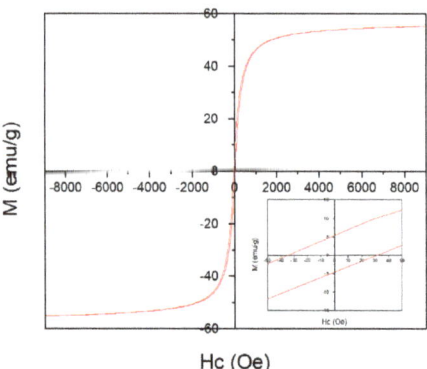

Figure 4. VSM of the prepared MgIONPs@TMAH.

3.2. Hyperthermia Performance

SLP was used as an indicator to evaluate the absorbed energy amount per metal NP mass under the action of an AMF [2]. In general, the SLP (heat generation) of NPs under an external AMF depends on Néel and Brownian relaxations [3]. Importantly, for human exposure, it is pivotal to maintain the product of the magnetic field strength (H) and its frequency (f) below a threshold safety value known as the Brezovich criterion. Based on this safety limit, the product of the frequency and the field amplitude ($C = H \times f$) should remain below 5×10^9 A m^{-1}s^{-1} to minimize any collateral effects of alternating magnetic fields on the human body [2,3]. Heating efficiency of the prepared nanoparticles was investigated by applying AC magnetic fields of various strengths (40 or 50 kA/m) at a frequency of 97 kHz. The values of C in our experiments were calculated to be 3.8×10^9 and 4.8×10^9 A m^{-1}s^{-1} for 40 and 50 kA/m, respectively, which did not exceed the safety limit. On the other hand, at a constant field strength of 13.5 kA/m, while varying the frequency from 159.8 to 269.9 kHz, the values of C in our experiments were calculated to be 2.1×10^9 and 3.6×10^9 A m^{-1}s^{-1} for 159.8 and 269.9 kHz, respectively. These values also did not exceed the safety limit. However, in the case of using the low field strength of 9.5 kA/m and a high field frequency of 614.4 kHz, the value of C in our experiments was calculated to be 5.8×10^9 A m^{-1}s^{-1}, which slightly exceeds the safety limit. For the hyperthermia system in this study, the SLP values were obtained by employing the effects of field frequency and field variation to investigate and optimize the heating effects.

3.2.1. Effects of Magnetic-Field Strength Variation with a Constant Frequency

The heating efficiency of MgIONPs@TMAH was investigated at a constant frequency of 97 kHz, while varying the strength (40–50 kA/m), as shown in Figure 5A. The increase in temperature against time was mostly linear, and its rate gradually slowed. The rate of heating increased with the strength (Figure 5A). Accordingly, the heating efficiency increased as the strength increased from 40 to 50 kA/m. For example, the increases in temperature were 6 and 20 °C after 200 s of AMF applications of 40 and 50 kA/m, respectively.

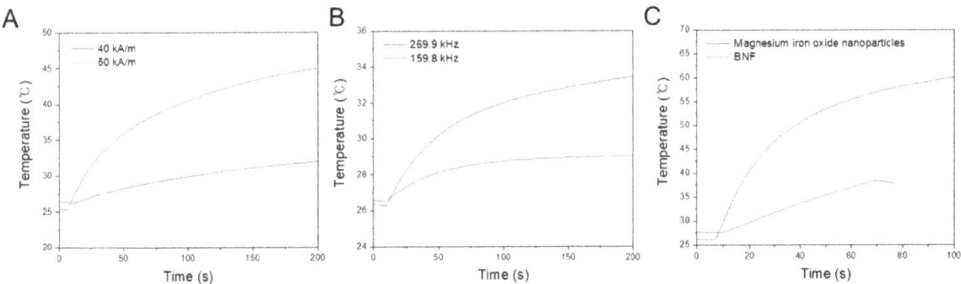

Figure 5. Temperature elevation by (**A**) MgIONPs@TMAH in response to the application of AMF with a constant frequency of 97 kHz, while varying the strength from 40 to 50 kA/m. (**B**) MgIONPs@TMAH in response to the application of an AMF with a constant strength of 13.5 kA/m, while varying the frequency from 159.8 to 269.9 kHz. (**C**) MgIONPs@TMAH (36.5 mg/mL) and BNF NPs (60 mg/mL) with the application of an AMF having a frequency of 614.4 kHz and a strength of 9.5 kA/m.

3.2.2. Effects of Field Frequency Variation with a Constant Magnetic Field Strength

The heating efficiency of MgIONPs@TMAH was investigated at constant field strength of 13.5 kA/m, while varying the frequency from 159.8 to 269.9 kHz, as shown in Figure 5B. The heating efficiency increased with the variation of field frequency from 159.8 to 269.9 kHz. As a result, the temperature elevations were 2.5 °C and 7 °C after 200 s of AMF applications at 159.8 and 269.9 kHz, respectively.

3.2.3. Effects of a High-Field Frequency with a Low Magnetic Field Strength

The heating efficiency of MgIONPs@TMAH was investigated and compared with BNF under the conditions of low field strength of 9.5 kA/m and a high field frequency of 614.4 kHz, as shown in Figure 5C. During the first 70 s of AMF application, a temperature change of 26 °C was observed for the prepared NPs, whereas a temperature change of only 10.5 °C was observed for BNF. This indicates that the heating profile of the MgIONPs@TMAH is substantially better than that of the commercial NPs.

Under the aforementioned conditions, the heating profile of the fabricated NPs was enhanced under the field of frequency 614.4 kHz and strength 9.5 kA/m. Comparisons of the SLP and intrinsic loss parameter (ILP) of the MgIONPs@TMAH and the commercial NPs are shown in Table 1. The difference in ILP is high, which indicates that the sample is dependent on applied field and agglomerating, which is probably caused by the low stability of the suspension.

Table 1. Comparisons of the SLP and ILP of the prepared NPs and the commercial NPs (BNF).

Conditions	f = 614.4 kHz; H = 9.5 kA/m	f = 97 kHz; H = 40 kA/m
MgIONPs@TMAH	SLP: 216.18 W/g ILP: 3.8 nHm^2/kg	SLP: 10.84 W/g ILP: 0.069 nHm^2/kg
BNF	SLP: 14 W/g ILP: 0.25 nHm^2/kg	SLP: 195.2 W/g ILP: 1.25 nHm^2/kg

We successfully obtained magnetic NPs exhibiting a high SLP under specific conditions with the given magnetic field parameters and a saturation temperature of 45 °C, suggesting that MgIONPs@TMAH can be used in cancer treatment. The heating time needed for the temperature to reach 45 °C is affected by various factors, including magnetic field conditions, the viscous medium, and the diameter and sample concentration of the magnetic NPs. Localized hyperthermia using NPs under low magnetic fields is favorable for treating small or deeply set tumors. It was reported that zinc-doped magnetite cubic nanoparticles show a 5-fold improvement in the specific absorption rate (SAR) in magnetic hyperthermia while providing a good MPI signal, when comparing with un-doped spherical nanoparticles [5].

3.3. Magnetic Particle Imaging Performance

It is notable that, due to the exploitation of magnetization, the tracer used in MPI is similar to that used in magnetic hyperthermia. The mechanism of heat generation in magnetic hyperthermia depends on the Brownian and Néel relaxation principles for magnetization reversal, which is similar to the signal generation mechanism in MPI. To generate a signal, MPI tracers must undergo magnetization reversal, which requires the overcoming of inertial torques and viscous and thermal effects using the Néel and Brownian relaxations. Modulation of the diameter and particle shape of NPs can tune the magnetic behavior, induce saturation magnetization, and further enhance magnetic hyperthermia properties and MPI [5–7]. SLP and the spatial resolution of MPI are influenced by the saturation magnetization, as shown by Equation (5) for SLP and Equation (6) for MPI:

$$SLP = \frac{\pi \cdot \mu_0 \cdot \chi''(f) \cdot H^2 \cdot f}{\rho_{MNPs} \cdot \phi}, \text{ where } \chi''(f) = \frac{\mu_0 M_s^2 V}{3 k_B T} \frac{\omega \tau_R}{(1 + \omega^2 \tau^2_R)}, \quad (5)$$

f and H: the frequency and magnetic field strength, ρ_{MNPs}: the magnetic NP density, ϕ: the magnetic NP volume fraction, μ_0: the permeability, and χ'': the susceptibility. M_s: the magnetization, V: the volume of the magnetic NP, $\omega = 2\pi f$: the AMF sweep rate, τ_R: the relaxation time, and k_B: Boltzmann's constant.

$$\text{MPI : spatial resolution} \approx \frac{24 k_B T}{\mu_0 \pi M_s} G^{-1} d^{-3}, \quad (6)$$

G: selection field gradient, *d*: magnetic diameter NP, *T*: temperature of the magnetic NP.

MPI allows for the monitoring of the locations and distributions of magnetic NPs. The performance of the MgIONPs@TMAH (36.5 mg/mL) as a contrast agent for MPI was compared with that of RES NPs (36.5 mg/mL). RES NPs (100 µL) and MgIONPs@TMAH (100 µL) were subjected to an excitation field to generate the MPI signals, of which images were acquired from two positions using 1D-MPI, as shown in Figure 6B. MgIONPs@TMAH produced a signal approximately the same as that of RES NPs, as shown in Figure 6A; thus, the imaging capability of MgIONPs@TMAH was recognized by 1D-MPI.

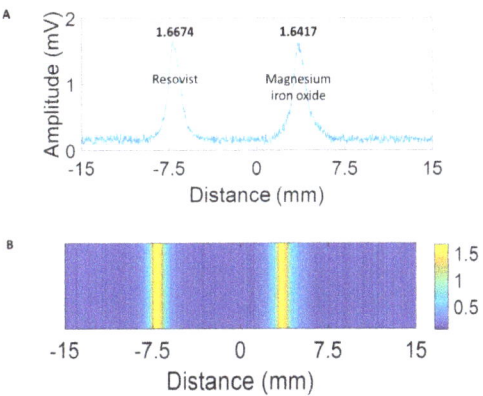

Figure 6. (**A**) Signals and (**B**) magnetic particle imaging (MPI) images of RES NPs and MgIONPs@TMAH.

3.4. In Vitro Cytotoxicity Tests

The cytotoxicity of MgIONPs@TMAH was examined with NIH-3T3 fibroblasts as shown in (Figure 7). For in vitro cytotoxicity tests, we performed both metabolic activity measurements using a WST assay (Figure 7B) and live/dead staining (Figure 7A). A WST assay can indicate cell viability based on colorimetric products by measuring the absorbance of the medium. It is important to note that there can be a variation in measurement (and error), which sometimes results in values over 100% (higher absorbance than the control). The relative cell viability of the sample (0.125 mg/mL) was not significantly different from that of the control sample. The cytotoxicity of the MgIONPs@TMAH showed dose dependence. The viability of the cells somewhat decreased with the increasing concentrations (1 and 2 mg/mL). WST results were in agreement with those of live/dead staining (Figure 7A). With 0–0.5 mg/mL of MgIONPs@TMAH, most of the cells were stained live, with only a small number of red-stained (dead) cells. These results indicate that the MgIONPs@TMAH have low cytotoxicity (<1 mg/mL).

The major issue in the developed nanomaterial is the high tendency of particles to agglomerate, even after the functionalization. It is fundamentally one of the most important factors for biomedical applications and still a challenging target [32–36]. Magnetic nanoparticles have large surface energy that causes the colloidal instability and aggregation. To mitigate this issue, layering of various coating materials, such as surfactants, has been commonly performed either during or after the synthesis of the nanoparticles. In general, post synthesis techniques can be introduced to modify nanoparticles to avoid aggregation or to stimulate disaggregation. The formation of a layer on the particle can prevent aggregation. If the synthetic route allows for the inserting of surface coating molecules before aggregation takes place, the particles can remain well dispersible in target media straight away. Additionally, surface molecules can be exchanged through chemical synthesis after the particle synthesis. Such post synthetic technique offers the possibility for a variety of surface modifications, which can be adjusted to the required applications. In some cases, especially in water, stability of nanoparticles is strongly affected by their surface charge.

This behavior is influenced by the ion content of the water and the amount of charged ions on the surface of the nanoparticles. Surface modification and surface charge can have a major impact on the biological response to particles, including phagocytosis and genotoxicity; hence, these parameters need to be controlled. However, if the surface properties cannot be controlled, nanoparticles quickly form large particles due to agglomeration. Also, the heating efficiency of nanoparticles is affected by the size of the nanoparticles and the coating layer. Hence, the magnetic nanoparticles should be prepared with a proper method to reach the target temperature [37,38].

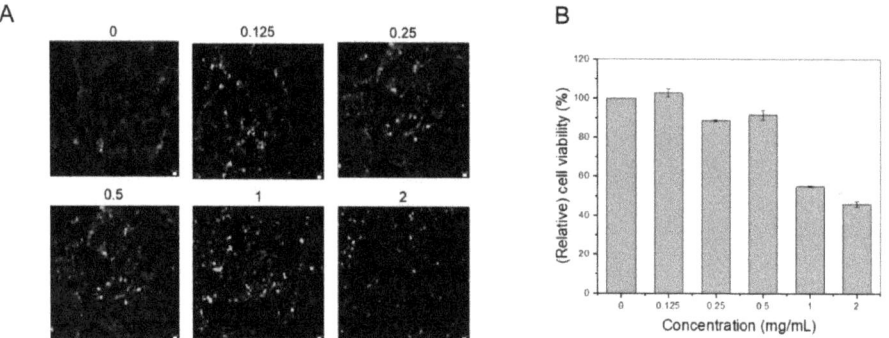

Figure 7. (**A**) Live/dead fluorescence imaging of the cultured cells and (**B**) cytocompatibility examined using a WST assay for various NP concentrations.

The average particle diameter in the single domain (from 20 to 70 nm) along with a narrow size distribution was reported to exhibit good heating performance [39]. The reported values for SLP in the literature are in the range of 10–100 W/g for magnetic field conditions of $H = 10$ kA/m and $f \approx 400$ kHz [40]. Nigam et al. reported that magnetite nanoparticles coated with citrate had a Ms value of 57 emu/g with high SLP to reach the target temperature (43 °C) in a short time [41]. High SLP (150 W/g) was reached by Reyes-Ortega et al. under a magnetic field condition of $H = 20$ mT and $f = 205$ kHz [42]. Magnetite nanoparticles stabilized with chitosan (15 nm) showed higher SLP (119 W/g) than uncoated magnetite nanoparticles likely due to the enhancement in the particle dispersion, which was obtained by the coating of the hydrophilic shell layer [43]. The magnetic fluid MFL AS (Nanotherm®, MagForce Nanotechnologies AG, Berlin, Germany), which consists of superparamagnetic iron oxide nanoparticles dispersed in water at an iron concentration of 112 mg/mL, has been approved for clinical trials. The iron oxide core is covered by an aminosilane-type shell and its size is approximately 15 nm in diameter. The required magnetic field to reach the target temperature with a frequency of 100 kHz and a variable field strength of 0–18 kA/m is generated in the MFH®300F applicator. We compared the results of this study to those reported in the literature, as summarized in Table 2.

Table 2. Comparison between particles reported in the literature and the current study.

Sample	Size (nm)	M_s (emu/g)	SLP	Alternating Current (AC) Field Condition	ILP	MPI Good Performance
Chitosan-coated Fe_3O_4 [44]	37	71.5	595	$H: 14, f: 335$	9	-
PEG-coated Fe_3O_4 [45]	31	54	355	$H: 27, f: 400$	1.22	-
Fe_3O_4 [46]	37	67	213	$H: 23.9, f: 571$	0.72	-
Fe_3O_4,Sph [5]	19.2 ± 1.3	101.5	189.6	$H: 16, f: 380$	1.9	+
Fe_3O_4,Cube	15.5 ± 1.1	107.3	356.2		3.6	
$Zn_{0.4}Fe_{2.6}O_4$,Zn-Sph	19.1 ± 1.0	125.7	438.6		4.5	
$Zn_{0.4}Fe_{2.6}O_4$,Zn-Cube	15.4 ± 1.1	130.4	1019.2		10.47	
FeCo@C [7]	40 nm	192	406	$H: 100, f: 30$	1.3	+
Citrate-coated IONPs [6]	10.6 ± 1.8	73.8	230	$H: 46, f: 186$	0.58	+
	13.1 ± 2.2	82.5	350		0.88	
Fe_3O_4@ZnO [47]	10	31.2	80	$H: 25.12, f: 250$	0.49	-
MgIONPs@TMAH	15.0 ± 5.0	55.1	216.18	$H: 9.5, f: 614.4$	3.8	Current study

4. Conclusions

In summary, we have successfully reported the synthesis of hydrophilic MgIONPs@TMAH through a two-step process consisting of thermal decomposition and the subsequent coating with TMAH to obtain relatively uniform water-dispersable NPs (15.0 ± 5.0 nm). For the hyperthermia system, the heating efficiencies were investigated and optimized by differing field frequency and field strength. The heating efficiency of the prepared nanoparticles was enhanced significantly more by the increase in the field strength (from 40 to 50 kA/m) than by the increase in the field frequency (from 159.8 to 269.9 kHz). Under magnetic field conditions of frequency 614.4 kHz and strength 9.5 kA/m, MgIONPs@TMAH nanoparticles show a 15-fold improvement in the specific loss power (SLP) in magnetic hyperthermia as compared to the BNF nanoparticles. MgIONPs@TMAH produced an MPI signal that was approximately the same as that produced by Resovist nanoparticles when subjected to an excitation field. The imaging capability of the prepared nanoparticles was recognized by 1D-MPI. We successfully obtained magnetic nanoparticles that exhibit a high SLP under specific conditions with the given magnetic field parameters, a saturation temperature of 45 °C and a good MPI signal, suggesting that the MgIONPs@TMAH can be useful for various biomedical applications, such as image-guided cancer treatment.

Author Contributions: Conceptualization, M.S.A.D.; Methodology, M.S.A.D., H.K., M.P.B., T.-A.L., H.L. and C.R.; Supervision, J.Y.L. and J.Y.; Writing—original draft, M.S.A.D.; Writing—review & editing, J.Y.L. and J.Y. All authors have read and agreed to the published version of the manuscript.

Funding: This work was supported in part by the National Research Foundation of Korea under Grant 2019M3C1B8090798, in part by the Korea Medical Device Development Fund grant funded by the Korea government (Project No. 202012E12) and in part by the Korea Evaluation Institute of Industrial Technology under Grant 20003822.

Data Availability Statement: The data is contained within the article.

Conflicts of Interest: The authors declare no conflict of interest.

References

1. Sánchez-Cabezas, S.; Montes-Robles, R.; Gallo, J.; Sancenón, F.; Martínez-Máñez, R. Combining magnetic hyperthermia and dual T1/T2 MR imaging using highly versatile iron oxide nanoparticles. *Dalton Trans.* **2019**, *48*, 3883–3892. [CrossRef]
2. Carvalho, A.; Gallo, J.; Pereira, D.; Valentão, P.; Andrade, P.; Hilliou, L.; Ferreira, P.; Bañobre-López, M.; Martins, J. Magnetic Dehydrodipeptide-Based Self-Assembled Hydrogels for Theragnostic Applications. *Nanomaterials* **2019**, *9*, 541. [CrossRef] [PubMed]

3. Le, T.; Phu Bui, M.; Yoon, J. Theoretical Analysis for Wireless Magnetothermal Deep Brain Stimulation Using Commercial Nanoparticles. *Int. J. Mol. Sci.* **2019**, *20*, 2873. [CrossRef]
4. Appa Rao, P.; Srinivasa Rao, K.; Pydi Raju, T.R.K.; Kapusetti, G.; Choppadandi, M.; Chaitanya Varma, M.; Rao, K.H. A systematic study of cobalt-zinc ferrite nanoparticles for self-regulated magnetic hyperthermia. *J. Alloys Compd.* **2019**, *794*, 60–67. [CrossRef]
5. Bauer, L.; Situ, S.; Griswold, M.; Samia, A. High-performance iron oxide nanoparticles for magnetic particle imaging-guided hyperthermia (hMPI). *Nanoscale* **2016**, *8*, 12162–12169. [CrossRef]
6. Dadfar, S.; Camozzi, D.; Darguzyte, M.; Roemhild, K.; Varvara, P.; Metselaar, J.; Banala, S.; Straub, M.; Guvener, N.; Engelmann, U.; et al. Size-isolation of superparamagnetic iron oxide nanoparticles improves MRI, MPI and hyperthermia performance. *J Nanobiotechnol.* **2020**, *18*, 22. [CrossRef] [PubMed]
7. Song, G.; Kenney, M.; Chen, Y.; Zheng, X.; Deng, Y.; Chen, Z.; Wang, S.; Gambhir, S.; Dai, H.; Rao, J. Carbon-coated FeCo nanoparticles as sensitive magnetic-particle-imaging tracers with photothermal and magnetothermal properties. *Nat. Biomed. Eng.* **2020**, *4*, 325–334. [CrossRef] [PubMed]
8. Niu, Z.P.; Wang, Y.; Li, F.S. Magnetic properties of nanocrystalline Co–Ni ferrite. *J. Mater. Sci.* **2006**, *41*, 5726–5730. [CrossRef]
9. Wang, X.; Zhuang, J.; Peng, Q.; Li, Y. A general strategy for nanocrystal synthesis. *Nature* **2005**, *437*, 121–124. [CrossRef]
10. Sorescu, M.; Grabias, A.; Tarabasanu-Mihaila, D.; Diamandescu, L. Influence of cobalt and nickel substitutions on populations, hyperfine fields, and hysteresis phenomenon in magnetite. *J. Appl. Phys.* **2002**, *91*, 8135–8137. [CrossRef]
11. Rani, S.; Varma, G.D. Superparamagnetism and metamagnetic transition in Fe3O4 nanoparticles synthesized via co-precipitation method at different pH. *Phys. B Condens. Matter* **2015**, *472*, 66–77. [CrossRef]
12. Sun, S.; Zeng, H.; Robinson, D.B.; Raoux, S.; Rice, P.M.; Wang, S.X.; Li, G. Monodisperse MFe$_2$O$_4$ (M = Fe Co, Mn) nanoparticles. *J. Am. Chem. Soc.* **2004**, *126*, 273–279. [CrossRef] [PubMed]
13. Jana, N.R.; Chen, Y.; Peng, X. Size- and shape-controlled magnetic (Cr, Mn, Fe, Co, Ni) oxide nanocrystals via a simple and general approach. *Chem. Mater.* **2004**, *16*, 3931–3935. [CrossRef]
14. Arami, H.; Khandhar, A.; Liggitt, D.; Krishnan, K.M. In vivo delivery, pharmacokinetics, biodistribution and toxicity of iron oxide nanoparticles. *Chem. Soc. Rev.* **2015**, *44*, 8576–8607. [CrossRef]
15. Petersen, E.J.; Nelson, B.C. Mechanisms and measurements of nanomaterial-induced oxidative damage to DNA. *Anal. Bioanal. Chem.* **2010**, *398*, 613–650. [CrossRef] [PubMed]
16. Ling, D.; Hyeon, T. Chemical design of biocompatible iron oxide nanoparticles for medical applications. *Small* **2013**, *9*, 1450. [CrossRef]
17. Leyssens, L.; Vinck, B.; Van Der Straeten, C.; Wuyts, F.; Maes, L. Cobalt toxicity in humans—A review of the potential sources and systemic health effects. *Toxicology* **2017**, *387*, 43–56. [CrossRef]
18. Guardia, P.; Di Corato, R.; Lartigue, L.; Wilhelm, C.; Espinosa, A.; Garcia-Hernandez, M.; Gazeau, F.; Manna, L.; Pellegrino, T. Water-Soluble Iron Oxide Nanocubes with High Values of Specific Absorption Rate for Cancer Cell Hyperthermia Treatment. *ACS Nano* **2012**, *6.4*, 3080–3091. [CrossRef]
19. Thirunavukkarasu, G.K.; Cherukula, K.; Lee, H.; Yeon Jeong, Y.; Park, I.; Young Lee, J. Magnetic field-inducible drug-eluting nanoparticles for image-guided thermo-chemotherapy. *Biomaterials* **2018**, *180*, 240–252. [CrossRef]
20. Ayyappan, S.; Mahadevan, S.; Chandramohan, P.; Srinivasan, M.P.; Philip, J.; Raj, B. Influence of Co^{2+} Ion Concentration on the Size, Magnetic Properties, and Purity of CoFe$_2$O$_4$ Spinel Ferrite Nanoparticles. *J. Phys. Chem. C* **2010**, *114*, 6334–6341. [CrossRef]
21. Chinnasamy, C.N.; Senoue, M.; Jeyadevan, B.; Perales-Perez, O.; Shinoda, K.; Tohji, K. Synthesis of size-controlled cobalt ferrite particles with high coercivity and squareness ratio. *J. Colloid Interface Sci.* **2003**, *263*, 80–83. [CrossRef]
22. Maensiri, S.; Sangmanee, M.; Wiengmoon, A. Magnesium Ferrite (MgFe$_2$O$_4$) Nanostructures Fabricated by Electrospinning. *Nanoscale Res. Lett.* **2009**, *4*, 221–228. [CrossRef]
23. Noh, S.H.; Na, W.; Jang, J.T.; Lee, J.H.; Lee, E.J.; Moon, S.H.; Lim, Y.; Shin, J.S.; Cheon, J. Nanoscale magnetism control via surface and exchange anisotropy for optimized ferrimagnetic hysteresis. *Nano Lett.* **2012**, *12*, 3716–3721. [CrossRef] [PubMed]
24. Jun, Y.W.; Huh, Y.M.; Choi, J.S.; Lee, J.H.; Song, H.T.; Kim, K.; Yoon, S.; Kim, K.S.; Shin, J.S.; Suh, J.S.; et al. Nanoscale size effect of magnetic nanocrystals and their utilization for cancer diagnosis via magnetic resonance imaging. *J. Am. Chem. Soc.* **2005**, *127*, 5732–5733. [CrossRef]
25. Rosensweig, R.E. Heating Magnetic Fluid with Alternating Magnetic Field. *J. Magn. Magn. Mater.* **2002**, *252*, 370–374. [CrossRef]
26. Xu, Y.; Qin, Y.; Palchoudhury, S.; Bao, Y. Water-Soluble Iron Oxide Nanoparticles with High Stability and Selective Surface Functionality. *Langmuir* **2011**, *27*, 8990–8997. [CrossRef]
27. Kmita, A.; Lachowicz, D.; Żukrowski, J.; Gajewska, M.; Szczerba, W.; Kuciakowski, J.; Zapotoczny, S.; Sikora, M. One-Step Synthesis of Long Term Stable Superparamagnetic Colloid of Zinc Ferrite Nanorods in Water. *Materials* **2019**, *12*, 1048. [CrossRef]
28. O'Leary, S.K.; Lim, P.K. On determining the optical gap associated with an amorphous semiconductor: A generalization of the Tauc model. *Solid State Commun.* **1997**, *104*, 17–21. [CrossRef]
29. Mallick, P.; Dash, B.N. X-ray diffraction and UV-Visible characterizations of γ–Fe$_2$O$_3$ nanoparticles annealed at different temperature. *Nanosci. Nanotechnol.* **2013**, *3*, 130–134.
30. El Ghandoor, H.; Zidan, H.M.; Khalil, M.M.; Ismail, M.I.M. Synthesis and some physical properties of magnetite (Fe$_3$O$_4$) nanoparticles. *Int. J. Electrochem. Sci.* **2012**, *7*, 5734–5745.
31. Prabhakaran, T.; Hemalatha, J. Combustion synthesis and characterization of cobalt ferrite nanoparticles. *Ceram. Int.* **2016**, *42*, 14113–14120. [CrossRef]

32. Engelmann, U.; Buhl, E.; Draack, S.; Viereck, T.; Ludwig, F.; Schmitz-Rode, T.; Slabu, I. Magnetic Relaxation of Agglomerated and Immobilized Iron Oxide Nanoparticles for Hyperthermia and Imaging Applications. *IEEE Magn. Lett.* **2018**, *9*, 1–5. [CrossRef]
33. Myrovali, E.; Maniotis, N.; Samaras, T.; Angelakeris, M. Spatial focusing of magnetic particle hyperthermia. *Nanoscale Adv.* **2020**, *2*, 408–416. [CrossRef]
34. Brero, F.; Albino, M.; Antoccia, A.; Arosio, P.; Avolio, M.; Berardinelli, F.; Bettega, D.; Calzolari, P.; Ciocca, M.; Corti, M.; et al. Hadron Therapy, Magnetic Nanoparticles and Hyperthermia: A Promising Combined Tool for Pancreatic Cancer Treatment. *Nanomaterials* **2020**, *10*, 1919. [CrossRef] [PubMed]
35. Al-Musawi, S.; Albukhaty, S.; Al-Karagoly, H.; Almalki, F. Design and Synthesis of Multi-Functional Superparamagnetic Core-Gold Shell Nanoparticles Coated with Chitosan and Folate for Targeted Antitumor Therapy. *Nanomaterials* **2021**, *11*, 32. [CrossRef]
36. Salimi, M.; Sarkar, S.; Hashemi, M.; Saber, R. Treatment of Breast Cancer-Bearing BALB/c Mice with Magnetic Hyperthermia using Dendrimer Functionalized Iron-Oxide Nanoparticles. *Nanomaterials* **2020**, *10*, 2310. [CrossRef]
37. Ma, M.; Wu, Y.; Zhou, J.; Sun, Y.; Zhang, Y.; Gu, N. Size dependence of specific power absorption of Fe_3O_4 particles in AC magnetic field. *J. Magn. Magn. Mater.* **2004**, *268*, 33–39. [CrossRef]
38. Hergt, R.; Hiergeist, R.; Zeisberger, M.; Glockl, G.; Weitschies, W.; Ramirez, L.P.; Hilger, I.; Kaiser, W.A. Enhancement of AC-losses of magnetic nanoparticles for heating applications. *J. Magn. Magn. Mater.* **2004**, *280*, 358–368. [CrossRef]
39. Hergt, R.; Dutz, S. Magnetic Particle Hyperthermia—Biophysical Limitations of a Visionary Tumour Therapy. *J. Magn. Magn. Mater.* **2007**, *311*, 187–192. [CrossRef]
40. Hergt, R.; Dutz, S.; Muller, R.; Zeisberger, M. Magnetic particle hyperthermia: Nanoparticle magnetism and materials development for cancer therapy. *J. Phys. Condens. Matter* **2006**, *18*, S2919. [CrossRef]
41. Nigam, S.; Barick, K.; Bahadur, D. Development of citrate-stabilized Fe_3O_4 nanoparticles: Conjugation and release of doxorubicin for therapeutic applications. *J. Magn. Magn. Mater.* **2011**, *323*, 237–243. [CrossRef]
42. Reyes-Ortega, F.; Delgado, Á.V.; Schneider, E.K.; Checa Fernández, B.L.; Iglesias, G.R. Magnetic Nanoparticles Coated with a Thermosensitive Polymer with Hyperthermia Properties. *Polymers* **2018**, *10*, 10. [CrossRef]
43. Shete, P.B.; Patil, R.M.; Thorat, N.D.; Prasad, A.; Ningthoujam, R.S.; Ghosh, S.J.; Pawar, S.H. Magnetic chitosan nanocomposite for hyperthermia therapy application: Preparation, characterization and in vitro experiments. *Appl. Surf. Sci.* **2014**, *288*, 149. [CrossRef]
44. Chauhan, A.; Midha, S.; Kumar, R.; Meena, R.; Singh, P.; Jhab, S.; Kuanr, K. Rapid tumor inhibition via magnetic hyperthermia regulated by caspase 3 with time-dependent clearance of iron oxide nanoparticles. *Biomater. Sci.* **2021**. [CrossRef] [PubMed]
45. Liu, X.L.; Fan, H.M.; Yi, J.B.; Yang, Y.; Choo, E.S.G.; Xue, J.M.; Ding, J. Optimization of surface coating on Fe_3O_4 nanoparticles for high performance magnetic hyperthermia agents. *J. Mater. Chem.* **2012**, *22*, 8235–8244. [CrossRef]
46. Fuentes-García, J.A.; Carvalho Alavarse, A.; Moreno Maldonado, A.C.; Toro-Córdova, A.; Ibarra, M.R.; Goya, G.F. Simple Sonochemical Method to Optimize the Heating Efficiency of Magnetic Nanoparticles for Magnetic Fluid Hyperthermia. *ACS Omega* **2020**, *5*, 26357–26364. [CrossRef]
47. Gupta, J.; Hassan, P.A.; Barick, K.C. Core-shell Fe_3O_4@ZnO nanoparticles for magnetic hyperthermia and bio-imaging applications. *AIP Adv.* **2021**, *11*, 025207. [CrossRef]

Article

Evaluation of Physicochemical Properties of Amphiphilic 1,4-Dihydropyridines and Preparation of Magnetoliposomes

Oksana Petrichenko [1,*], Aiva Plotniece [2,3], Karlis Pajuste [2], Martins Rucins [2], Pavels Dimitrijevs [2,3], Arkadij Sobolev [2], Einars Sprugis [4] and Andrejs Cēbers [1]

[1] Laboratory of Magnetic Soft Materials, Faculty of Physics, Mathematics and Optometry, University of Latvia, 3 Jelgavas str., LV-1004 Riga, Latvia; andrejs.cebers@lu.lv
[2] Latvian Institute of Organic Synthesis, 21 Aizkraukles Str., LV-1006 Riga, Latvia; aiva@osi.lv (A.P.); kpajuste@osi.lv (K.P.); rucins@osi.lv (M.R.); p.dimitrijevs@osi.lv (P.D.); arkady@osi.lv (A.S.)
[3] Department of Pharmaceutical Chemistry, Faculty of Pharmacy, Riga Stradiņš University, 21 Dzirciema Str., LV-1007 Riga, Latvia
[4] Laboratory of Chemical Technologies, Institute of Solid State Physics, University of Latvia, 8 Kengaraga Str., LV-1063 Riga, Latvia; esprugis@cfi.lu.lv
* Correspondence: oksana.petricenko@lu.lv

Citation: Petrichenko, O.; Plotniece, A.; Pajuste, K.; Rucins, M.; Dimitrijevs, P.; Sobolev, A.; Sprugis, E.; Cēbers, A. Evaluation of Physicochemical Properties of Amphiphilic 1,4-Dihydropyridines and Preparation of Magnetoliposomes. *Nanomaterials* 2021, *11*, 593. https://doi.org/10.3390/nano11030593

Academic Editor: Paolo Arosio

Received: 27 January 2021
Accepted: 22 February 2021
Published: 27 February 2021

Publisher's Note: MDPI stays neutral with regard to jurisdictional claims in published maps and institutional affiliations.

Copyright: © 2021 by the authors. Licensee MDPI, Basel, Switzerland. This article is an open access article distributed under the terms and conditions of the Creative Commons Attribution (CC BY) license (https://creativecommons.org/licenses/by/4.0/).

Abstract: This study was focused on the estimation of the targeted modification of 1,4-DHP core with (1) different alkyl chain lengths at 3,5-ester moieties of 1,4-DHP (C_{12}, C_{14} and C_{16}); (2) N-substituent at position 1 of 1,4-DHP (N-H or N-CH_3); (3) substituents of pyridinium moieties at positions 2 and 6 of 1,4-DHP (H, 4-CN and 3-Ph); (4) substituent at position 4 of 1,4-DHP (phenyl and napthyl) on physicochemical properties of the entire molecules and on the characteristics of the obtained magnetoliposomes formed by them. It was shown that thermal behavior of the tested 1,4-DHP amphiphiles was related to the alkyl chains length, the elongation of which decreased their transition temperatures. The properties of 1,4-DHP amphiphile monolayers and their polar head areas were determined. The packing parameters of amphiphiles were in the 0.43–0.55 range. It was demonstrated that the structure of 1,4-DHPs affected the physicochemical properties of compounds. "Empty" liposomes and magnetoliposomes were prepared from selected 1,4-DHP amphiphiles. It was shown that the variation of alkyl chains length or the change of substituents at positions 4 of 1,4-DHP did not show a significant influence on properties of liposomes.

Keywords: 1,4-dihydropyridine amphiphiles; iron oxide nanoparticles; magnetoliposomes; lipid monolayers; physicochemical properties

1. Introduction

Scientists worldwide have made many efforts to expand the invention and development of broad range nanoparticle delivery systems. Liposomes have been extensively studied as promising delivery systems due to their efficiency, biocompatibility and dual character, i.e., the ability to entrap either hydrophobic or hydrophilic drugs, improving their pharmacokinetic and pharmacodynamic properties [1–4]. Magnetic iron oxide nanoparticles are used for different scientific and technological purposes due to their peculiar properties. The main feature of these particles is to enable movement in a magnetic field. Magnetic nanoparticles (MNPs), due to their biocompatibility and functionality, have been considered to be promising for applications in medicine. Their unique magnetic and electric properties allow for their application in magnetic resonance imaging as contrast agents [5] for treatment in hyperthermia [6,7]. Currently, MNPs are studied for cell labelling and separation [2,8]. Magnetic iron oxide nanoparticles have been used, for example, as a trigger drug release from magnetoliposomes (MLs), through a magneto–nanomechanical approach [9], for magnetically guided cells in tissue engineering [2,10,11] when the MNPs can be coated with a polymer or encapsulated inside liposomes producing MLs. Magnetoliposomes are also used for magnetofection and drug delivery by magnetic targeting [12–14].

The investigation of MLs' morphology and physical properties is an important issue. The chemical structure and shape of cationic compounds determine their self-assembling and DNA complexation properties, and hence the gene delivery activity [15].

Synthetic nanoparticle-forming cationic lipid-like compounds have been developed as delivery agents for the transfer of genetic materials, including plasmid DNA (pDNA) molecules into cells [16,17] and for therapy and diagnostic applications [18]. In general, among the synthetic cationic delivery systems, quaternary ammonium surfactants are more toxic than their analogues, with the cationic charge delocalized in a heterocyclic ring [19–21]. It is important to evaluate liposome forming lipid properties for the development of new liposome systems [22].

Multiple amphiphilic 1,4-dihydropyridine (1,4-DHP) derivatives with various lengths of the alkyl chain at positions 3 and 5 of the 1,4-DHP ring were studied earlier as membranotropic compounds. These amphiphiles were found to condense and efficiently deliver plasmid DNA (pDNA) into different cell lines in vitro [23,24]. It was demonstrated that dodecyloxycarbonyl substituents at positions 3 and 5 of the 1,4-DHP molecule were optimal for gene transfection efficacy in the group of these synthetic lipid-like amphiphiles [23].

1,4-DHPs are important heterocyclic scaffolds with exceptional biological properties, and they take an important position in synthetic, medicinal and bioorganic chemistry [25]. Representatives of 4-aryl-1,4-DHPs are known as calcium channel blockers and have been widely used for the treatment of hypertension [26]. After the discovery of excellent therapeutic benefits of 1,4-DHP derivatives as calcium antagonists, the number of other activities of 1,4-DHPs, such as neuroprotective [27], radioprotective [28], antimutagenic [29], antioxidative [30], anticancer [31] and antimicrobial [32,33] have been reported.

Over the last 20 years, studies of pyridinium moieties containing compounds based on a 1,4-DHP core have revealed that they possess a number of unique properties. Cationic amphiphiles derived from polyfunctional 1,4-DHPs possess self-assembling properties that are sufficient to form nanoaggregates spontaneously without surfactants in an aqueous environment as lipid-like compounds due to the presence of both hydrophobic and hydrophilic parts in the molecule. These compounds based on the 1,4-DHP core are attractive because, along with self-assembling properties, they contain 1,4-DHP as an active linker [34,35], which is an intrinsic structural part of many pharmacologically active compounds and drugs with highly specific physiological activities (cardiovascular, anticancer, antimutagenic, as L-type calcium channel blockers, etc.) [36,37]. Additionally, these compounds have been reported to possess antiradical [38], and in vitro cell growth modulating activities [39]. Derivatives with N-dodecylpyridinium or N-hexadecylpyridinium moiety at the 1,4-DHP cycle exhibit cytotoxicity on tumor cell lines [3]. 4-(N-Dodecylpyridinium)-1,4-DHP has been reported to efficiently cross the blood–brain barrier and improve memory by enhancing the GABAergic and synapticplasticity processes [40]. Liposomes formed by these 1,4-DHPs are a promising tool for the delivery of DNA into cells [23,38].

The main goal of this work was to evaluate the influence of 1,4-dihydropyridine substituents on amphiphile physicochemical properties and the formation of magnetoliposomes (MLs). The following physicochemical parameters such as thermal behavior (TGA/DTA and DSC) of new amphiphiles and properties of monolayers composed by tested amphiphiles were studied. Preparation of magnetoliposomes in this study is a method to evaluate the influence of the 1,4-DHP structure on ML formation. The reverse-phase evaporation (REV) method along with the use of 1,4-DHP amphiphiles has proved its applicability to produce MLs [41–43]. The obtained results may add knowledge for the further comprehension of the structure–activity and liposome parameter relationships of these tested compounds.

2. Materials and Methods

2.1. Chemicals

All chemical reagents for the synthesis of the lipid-like 1,4-DHP amphiphiles were purchased from Acros Organics (Geel, Belgium), Sigma-Aldrich/Merck KGaA (Darmstadt, Germany), or Alfa Aesar (Lancashire, UK) and used without further purification. For the production of a ferrofluid containing maghemite (γ-Fe_2O_3) nanoparticles and iron salts (Fluka/Merck KGaA (Darmstadt, Germany), namely $FeCl_2 \cdot 4H_2O$, $FeCl_3 \cdot 6H_2O$ and $Fe(NO_3)_3$, as well as nitric acid and ammonium hydroxide (Scharlau Chemie S.A., Barcelona, Spain) were used.

2.2. Magnetic Nanoparticles Synthesis and Characterization

Magnetic nanoparticles were synthesized following the Massart method [44] by co-precipitation of anionic magnetite (Fe_3O_4) from aqueous solutions of Fe^{2+} and Fe^{3+} chlorides using ammonium hydroxide with the following oxidation of Fe_3O_4 with $Fe(NO_3)_3$. As a result, positively charged γ-Fe_2O_3 MNPs were produced. An acidic ferrofluid (FF) was produced by peptizing the collected MNPs in an aqueous medium. To produce FF–citr with a pH~6.4, obtained γ-Fe_2O_3 MNPs were coated by citrate ions [45]. Trisodium citrate dihydrate was used to stabilize the magnetic nanoparticles.

The magnetic characteristics and size distribution of the synthesized nanoparticles were determined using a vibrating sample magnetometer (Lake Shore Cryotronics, Inc., model 7404 VSM, Westerville, OH, USA) and the software for processing the magnetization data.

2.3. Synthesis of 1,4-DHP Amphiphiles

1,1'-[(3,5-Didodecyloxycarbonyl-4-phenyl-1,4-dihydropyridine-2,6-diyl)dimethylen] bispyridinium dibromides (**1**, **4–6**) and 1,1'-[(3,5-dialkoxycar-bonyl-4-phenyl-1,4-dihydropyridine-2,6-diyl) dimethylen]bispyridinium dibromides (**2**, **3**) were synthesized by the previously reported methods [15,24].

1,4-DHP derivative **7** (1'-[(3,5-didodecyloxycarbonyl-4-(2-napthyl)-1,4-dihyd-ropyridine-2,6-diyl)dimethylen]bispyridinium dibromide) was synthesized in analogy with other compounds. Briefly, the developed synthesis of the cationic 1,4-DHP **7** includes three sequential steps. The first step is the synthesis of corresponding 2,6-dimethyl 1,4-DHP derivative in a two-component Hantzsch-type cyclization; the second step involves the bromination of the methyl groups of 2,6-dimethyl-1,4-DHP derivative with N-bromosuccinimide; and the third step is the nucleophilic substitution of bromine of 2,6-dibromomethylene-1,4-DHP with pyridine yielding the target compound **7**. The synthesis and characterization of the original compounds are presented in more detail in Supplementary data.

Purities of the compounds were analyzed by HPLC using the Waters Alliance 2695 system and Waters 2485 UV/Vis detector equipped with a SymmetryShield RP_{18} column (5 µm, 4.6 × 150 mm, Waters corporation, Milford, Massachusetts, USA) for parent 1,4-DHP or an Alltima CN column (5 µm, 4.6 × 150 mm, Grace, Columbia, MD, USA) for cationic moieties containing 1,4-DHP amphiphiles **1–7** using a gradient elution with acetonitrile/water containing 0.1% phosphoric acid as the mobile phase (v/v), at a flow rate of 1 mL/min. Peak areas were determined electronically using a Waters Empower 2 chromatography data system.

2.4. Thermal Analysis of 1,4-DHP Amphiphiles

2.4.1. Thermogravimetric and Differential Thermal Analysis for the Tested 1,4-DHP Amphiphiles

Thermogravimetric (TGA) and differential thermal (DTA) analyses for the tested 1,4-DHP amphiphiles **1–7** were performed for a 3–5 mg sample with a Shimadzu DTG-60 instrument in an Ar atmosphere (Ar 5.0 from Linde Gas SIA, Riga, Latvia) with a 50 mL/min flow in a temperature range from 30 °C to 300 °C at a heating rate of 5 °C/min. Data files were transformed into an ASCII file for further analysis using TA60 ver. 2.10 software (Shimadzu Corporation, Kyoto, Japan).

2.4.2. Differential Scanning Calorimetry for the Tested 1,4-DHP Amphiphiles

Dry samples of the tested 1,4-DHP amphiphiles **1–7** were characterized by differential scanning calorimetry (DSC). The samples were analyzed using a DSC131 evo instrument from Setaram (Caluire, France). Each sample (generally 5 to 10 mg) was weighed using an analytical scale and then cautiously placed in a 30 µL aluminum crucible. The crimped crucible was then placed in the sample compartment of a DSC instrument along with a crimped reference aluminum crucible. Argon, at a rate of 30 mL/min, was used as a purge gas. Each experiment included 3 heating-cooling cycles to determine different phase transitions. The temperature was increased at a heating rate of 10 °C/min from room temperature to approx. 130 °C depending on the sample decomposition temperature. Upon reaching the target temperature, the system was allowed to cool down to 50 °C, and an additional 2 cycles were performed in a similar way. The setup of the experiments and the data obtained were analyzed using the Calisto Data Acquisition software, ver. 1.493.

2.5. Characterization of Monolayers Formed by 1,4-DHP Amphiphiles or Surface Pressure–Area (π–A) Isotherms

The properties of monolayers composed of 1,4-DHP amphiphiles and their polar head areas were determined from π–A isotherms, which were obtained using the Langmuir–Blodgett trough. The surface pressure–molecular area (π–A) compression isotherms were measured using a computer-controlled Langmuir trough (Medium trough, KSV NIMA Instruments, Finland; A_{total} = 243 cm^{-2}) made of Teflon and equipped with two compression barriers. The surface pressure of the monolayer was monitored with a Wilhelmy plate made of platinum, which was cleaned by flushing it with ethanol and Milli-Q water, and then burned by a Bunsen burner.

Prior to measurements, the trough and barriers were thoroughly rinsed with ethanol and Milli-Q water. Cleanliness of the aqueous surface was ensured by sweeping the barriers across the surface, and the aqueous surface was considered clean when $\pi \leq 0.1$ mN/m. Monolayers were formed by carefully spreading an appropriate volume of the lipid solution in chloroform dropwise on the deionized water surface at 23 $\pm$ 1 °C using a Hamilton micro-syringe. The carrying solvent (CHCl$_3$) was allowed to evaporate for 10 min before compressions began. The monolayers were compressed at a constant rate of 10 mm/min. Measurements were made at 23 $\pm$ 1 °C and repeated at least three times to ensure the reproducibility of the results. The experimentally detected standard deviations of the molecular area and surface pressure did not exceed 2%.

2.6. Magnetoliposome Preparation

To produce MLs by the REV method, the first step is to obtain of an organic phase emulsion containing a fixed amount of 1,4-DHP amphiphile as a chloroform solution, diethyl ether (3 mL) and ferrofluid (FF–citr, pH~6.4) (1 mL). This mixture was sonicated in an ultrasonic bath (Sonorex Type RK-100, Bandelin electronic GmbH, Berlin, Germany) for 20 min. Then the organic solvents were evaporated under reduced pressure (350–400 mBar) using a rotavapor (Büchi 215/V-700, Büchi Labortechnik AG, Flawil, Switzerland) and bath temperature around 30 °C. After the removal of the most of the organic solvent, viscous gel was formed, which became an aqueous suspension. Then 3 mL of deionized H$_2$O was added after which a resulting mixture was evaporated under reduced pressure in the same conditions for an additional 20 min to remove traces of the solvent. The obtained suspension was filtered through a 0.45 µm syringe filter and purified by magnetic decantation to remove all non-encapsulated magnetic nanoparticles.

2.7. Characterization of Liposomes by Dynamic Light Scattering (DLS) and Transmission Electron Microscopy

DLS measurements of the particle hydrodynamic size distributions in the aqueous medium formed by the examined amphiphiles were performed by a Zetasizer Nano ZS instrument (Malvern Instruments Ltd., Malvern, UK) with Malvern Instruments Ltd.

Software 7.12. Nanoparticles were analyzed with the following specifications: medium, water; refractive index: 1.330; viscosity: 0.8872 cP; temperature, 25 °C; dielectric constant, 78.5. Nanoparticles: liposomes; refractive index of materials: 1.60. Detection angle was 173°, with a wavelength of 633 nm. The data were analyzed using the multimodal number distribution software equipped with the instrument. The measurements were repeated three times in order to check their reproducibility.

For transmission electron microscopy studies, one drop of the sample was adsorbed to a formvar carbon-coated copper grid and negatively stained with 1% aqueous solution of uranyl acetate. The grids were examined with a JEM-1230 TEM (Jeol, Tokyo, Japan) at 100 kV.

3. Results and Discussion

3.1. Magnetic Nanoparticle Synthesis and Characterization

Magnetic nanoparticles were synthesized by co-precipitating anionic magnetite (Fe_3O_4) from aqueous solutions of Fe^{2+} and Fe^{3+} chlorides using ammonium hydroxide with the following oxidation of Fe_3O_4 with $Fe(NO_3)_3$. As a result, positively charged γ-Fe_2O_3 MNPs were produced. Ferrofluid containing γ-Fe_2O_3–citr was obtained with an MNP coating by citrate ions, FF–citr pH~6.4. The polydispersity index of the obtained FF–citr was 0.181 ± 0.006; the ζ-potential was -38.0 ± 2.3 mV according to DLS data. The volume fraction $\Phi_{FF-citr} = 1.5\%$ was determined by iron concentration colorimetric analysis (with 5-sulfosalicilic acid dehydrate, absorbance at wavelength $\lambda = 425$ nm). Fe concentration in the obtained FF–citr was 0.95 M, which corresponded to the γ-Fe_2O_3 nanoparticle content (76 mg/mL), and the FF density was determined to be 1.06 g/cm^3. The magnetization curve and size distribution of the FF–citr nanoparticles are shown in the Figure 1. Magnetic properties of the synthesized MNPs were determined using a vibrating sample magnetometer and software for processing magnetization curves. The magnetic diameter of the nanoparticles in the main population was determined to be 15 nm by adjusting the magnetization curve of the ferrofluid to a Langevin formalism weighted by the size distribution of the γ-Fe_2O_3–citr MNPs.

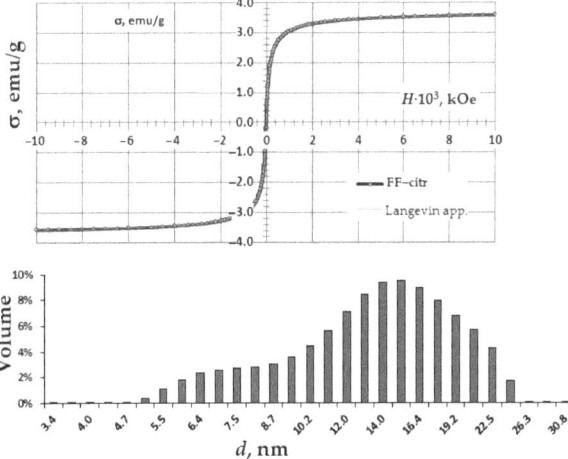

Figure 1. Magnetization curve and size distribution histogram of the obtained superparamagnetic γ-Fe_2O_3–citr magnetic nanoparticles.

3.2. Synthesis of 1,4-DHP Derivatives

The synthesis of the selected 1,4-DHP amphiphiles **1–7** varying in substituents at the 1,4-DHP ring was carried out by the previously described methods [23,38]. The studied 1,4-DHP derivatives were divided into four groups to evaluate the influence of the structure elements on the physicochemical properties of the compounds and on the properties of MLs (Figure 2):

- 1,4-DHPs with different alkyl chain lengths at 3,5-ester moieties of 1,4-DHP (C_{12}, C_{14} and C_{16}) (comps. 1–3);
- variation of the N-substituent at position 1 of 1,4-DHP (N-H or N-CH$_3$) (comps. 1 and 4);
- variation of the substituents at pyridinium moieties as cationic head groups at positions 2 and 6 of 1,4-DHP (H, 4-CN and 3-Ph) (comps. 1, 5 and 6);
- variation of the substituent at position 4 of 1,4-DHP (Ph and Nh) (comps. 1 and 7).

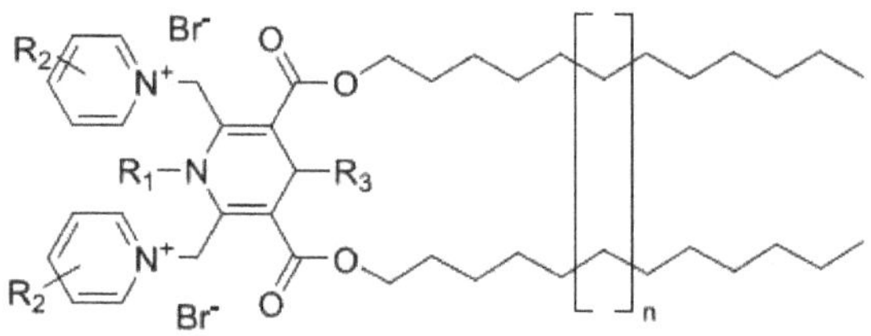

Comp.	R$_1$	R$_2$	R$_3$	n
1	H	H	Ph	1
2	H	H	Ph	3
3	H	H	Ph	5
4	CH$_3$	H	Ph	1
5	H	4-CN	Ph	1
6	H	3-Ph	Ph	1
7	H	H	2-Nh	1

Figure 2. Structures of the studied 1,4-DHP amphiphiles 1–7.

1,1′-[(3,5-Bisdodecyloxycarbonyl-4-phenyl-1,4-dihydropyridine-2,6-diyl)dimethylen]bis(pyridin-1-ium) (or substituted pyridinium) dibromides (**1, 5, 6**), 1,1′-[(3,5-dialkoxycarbonyl-4-phenyl-1,4-dihydropyridine-2,6-diyl)dimethylen]bis(pyridine-1-ium) dibromides (**2, 3**) and 1,1′-[(3,5-didodecyloxy-carbonyl-4-(2-napthyl)-1,4-dihydropyridine-2,6-diyl)dimethylen]bis(pyridin-1-ium) dibromide (**7**) were obtained according to Scheme S1 (Supplementary data). Briefly, the corresponding parents 3,5-bis(alkoxycarbonyl)-2,6-dimethyl-4-aryl-1,4-dihydropyridines were obtained by the classical Hantzsch synthesis from the corresponding acetoacetic ester, the corresponding aldehyde and ammonium acetate [38].

1,1′-[(3,5-Bis((dodecyloxy)carbonyl)-1-methyl-4-phenyl-1,4-dihydropyridine-2,6-diyl)dimethylene]-bis(pyridin-1-ium) dibromide (**4**) was obtained according to Scheme S2 (see the Supplementary data). Briefly, the parent 3,5-didodecyloxy-carbonyl-4-phenyl-1,2,6-trimethyl-1,4-dihydropyridine was synthesized from dodecyl acetoacetate, benzaldehyde and methylamine hydrochloride as a nitrogen source in pyridine by refluxing the reaction mixture for 6 h.

Bromination of 2,6 methyl groups of parent 1,4-DHP was performed by N-bromosuccinimide in methanol giving 2,6-di(bromomethyl)-3,5-bis(alkoxy-carbonyl)-4-aryl-1,4-dihydropyridine, which without purification was treated by the corresponding pyridine derivative resulting in formation of the target 1,4-DHP amphiphiles 1–7.

^{1}H-NMR spectra data and other physicochemical parameters of compounds **1–6** were in agreement with those reported in the literature [23,38,46]. Characterization of the original compounds—1,4-dihydropyridine (1,4-DHP) amphiphiles 1–3, 5–7—is given in the Supplementary data. Measured by LC–MS mass-to-charge (m/z) values of the re-synthesized compounds were in good agreement with the calculated values and also with the previously reported ones. In addition, the characteristic signals of 2,6-methylene group protons in ^{1}H-NMR spectra were observed as an AB-system, which confirmed the diastereotopic properties of CH_2X protons in the molecules of 1,4-DHP amphiphiles and confirmed their structures [38]. The purities of the studied compounds were at least 98% according to high-performance liquid chromatography data.

It is known from the literature that some 1,4-DHP molecules exhibit a significant sensitivity to light, leading to the complete loss of pharmacological activity [47,48]. It is necessary to emphasize that the tested cationic moieties containing 1,4-DHP are more stable than the corresponding parent compounds without cationic moieties. Our previous studies of electrochemical oxidation of 1,4-DHP derivatives containing cationic pyridinium methylene groups in position 2 and 6, by cyclic voltammetry on a stationary glassy carbon electrode in dry acetonitrile, demonstrated that they had electrooxidation potentials of 1.57–1.58 V [49]. These data were also in agreement with our previous results, where the electrochemical oxidation potential of a similar cationic 1,4-DHP was determined as 1.7 V, and the electrochemical oxidation of this compound was characterized as a two-electron process [50], whereas the parent compounds—1,4-DHP derivatives without cationic moieties demonstrated lower electrooxidation potentials. Thus, 4-phenyl substituted Hantzsch 1,4-dihydropyridine had a potential of 1.08 V [51], and other different 4-aryl substituted 1,4-DHPs had potentials around 1.1 V [52], but 4-monoalkyl substituted 1,4-DHPs had oxidation potentials of 1.01–1.03 V, respectively [53].

Additionally, it was demonstrated that pyridinium moieties containing 3,5-didodecyloxycarbonyl-4-phenyl-1,4-dihydropyridine derivatives showed 25–60% radical scavenging activity, which was comparable with the antiradical activity (ARA) of Diludin (40%)—a widely known antioxidant. Other 1,4-DHP amphiphiles containing saturated heterocyclic moieties—N-methylmorpholinium or N-methylpyrrolidinium derivatives—demonstrated more pronounced ARA, namely 95% and 54%, respectively [38]. The choice of cationic moiety containing 1,4-DHP amphiphiles **1–7** for the evaluation of their physicochemical properties in order to determine stable and safe lipids for formation of magnetoliposomes is based on the above mentioned data.

3.3. Thermal Analysis of 1,4-DHP Amphiphiles

It is known that the temperature of phase transition depends on the structure of the hydrocarbon chains in lipid molecules and also on the nature of their polar heads. 1,4-DHP amphiphiles **1–7** were tested using the TGA/DTA technique to determine the thermal stability and phase transitions of the compounds. Along with the TGA/DTA technique, the compounds also were tested by DSC to clarify in detail phase transitions that occur before the decomposition of the substance starts. Thermal studies were carried out in order to assess how the structures of the amphiphiles influence their thermal stability and phase transition. The values obtained by analyzing the TGA/DTA curves are presented in Table S1 (Supplementary data).

As shown in Figures 3A,C,E, 4 and 5A,C,E, the last transitions of all compounds correspond to the compounds' decomposition. Analysis of the TGA curves of the samples and comparison with the DTA curves show that when approaching the temperatures of the last transition, the sample noticeably started losing weight, which means the beginning of the compound decomposition process. The decomposition process is accompanied by significant heat absorption. Figure 3A,C,E shows the dynamics of comps. **1–3** curves as a function of the heating temperature obtained by the TGA/DTA technique. The curve of comp. **1** exhibited one weak and two distinct endothermic peaks (Figure 3A). The first transition temperature peak for comp. **1** (Figure 3A) was at 56 °C, but it had low intensity. TGA data for comp. **1** were in a good agreement with our previously published results [54]. With an increase in the length of the lipophilic chains, "broadening" of the peaks was observed (Figure 3C,E, comps. **2** and **3** and Table S1). It should be admitted that N-CH_3-substituted 1,4-DHP showed a similar trajectory of the DTA curves (Figure 4 and Table S1) as unsubstituted 1,4-DHP. Both compounds had a distinct first order endothermic transition, but the transition of comp. 4 had a wider temperature range, the so-called "broadening" transition.

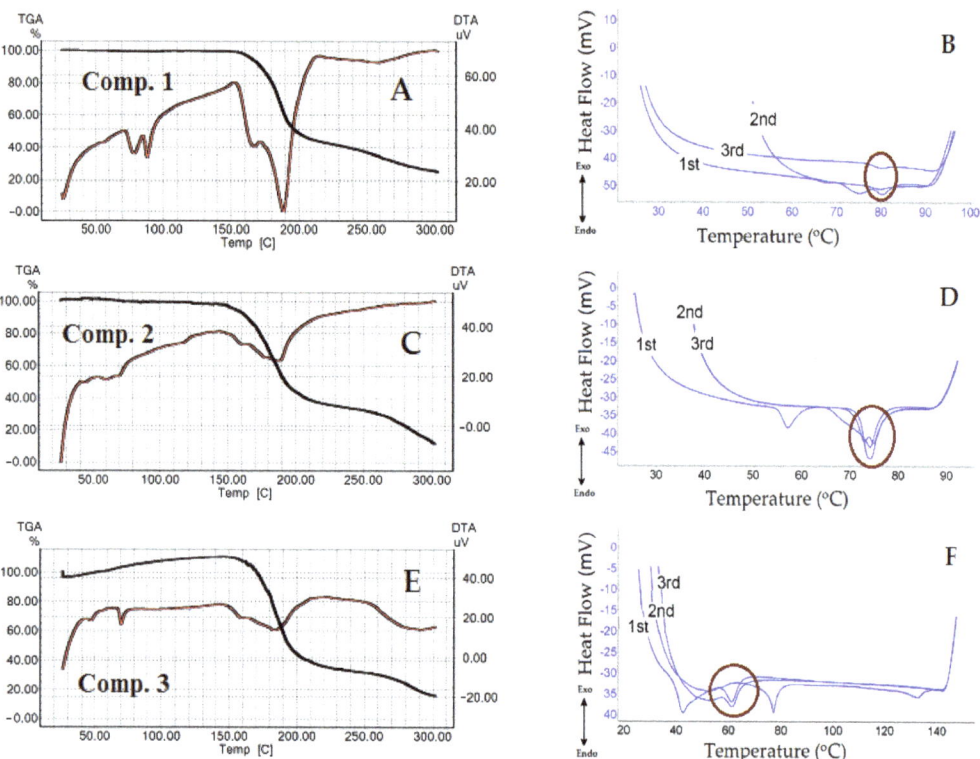

Figure 3. (**A,C,E**): TGA/DTA curves for comps. **1**–**3**. (**B,D,F**): DSC triple heating curves of comps. **1**–**3** in the temperature ranges before compound decomposition. The circles show endothermic peaks reproduced during the three heating cycles.

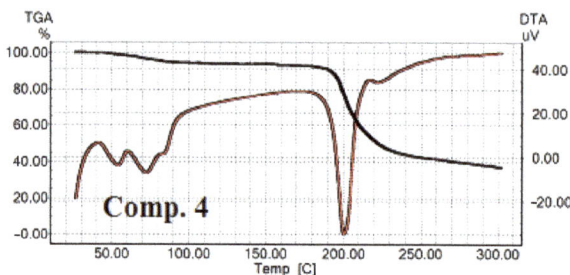

Figure 4. TGA/DTA curves for comp. **4**.

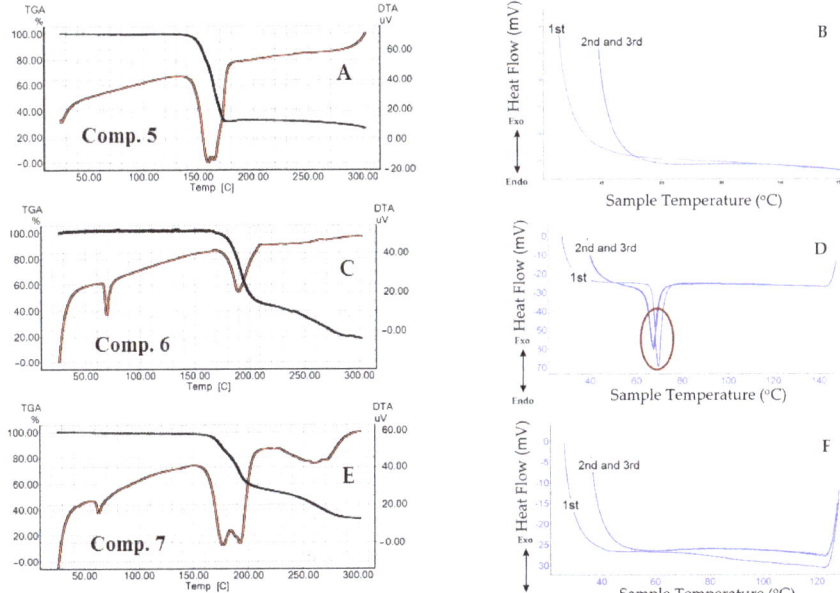

Figure 5. (**A,C,E**): TGA/DTA curves for comps. **5–7**. (**B,D,F**): DSC triple heating curves of comps. **5–7** in the temperature range before compound decomposition. Endothermic peaks that occurred after triple heating are marked by circles.

In Figure 3B,D,F, united curves of comps. **1–3** after triple heating obtained by the DSC technique are shown. The samples of the 1,4-DHP amphiphiles to be analyzed by DSC technique were heated up to temperatures below the compound decomposition temperatures, which were determined by TGA for each substance minus 10–20 °C. This process was repeated three times to examine also the thermal stability of the compounds and the repeatability trajectory of the curves. The DSC curves for comps. **1–3** demonstrated common features: the curves at the first heating showed several transitions, but the curves at the second and third heating displayed only one transition. The results of the heating and cooling curves analysis are listed in Table 1. In addition, Table 1 lists the results of sample cooling (exothermic processes), which correspond to those of the endothermic process when the compound was heated.

Table 1. Values obtained from DSC curves. "↓" denotes endothermic and "↑" denotes exothermic transition. Compounds **5** and **7** did not show any transitions before the decomposition temperature. Cooling curves are presented in the Supplementary data (Figure S2).

Comp.	Transition	Transitions Temperatures Range, °C			
		1st Heating (Process)	2nd Heating (Process)	3rd Heating (Process)	All Coolings (Process)
1	1st	65.22–70.05 ↓			
	2nd	70.05–78.08 ↓	77.21–82.96 ↓	77.00–82.70 ↓	65.00–60.00 (weak signal)
	3rd	78.10–82.70 ↓			
2	1st	55.20–60.15 ↓	71.97–76.83 ↓	70.00–78.00 ↓	75.27–72.02 ↑
	2nd	69.90–76.42 ↓			
3	1st	37.31–57.42 ↓	57.42–66.44 ↓	56.82–66.71 ↓	63.53–55.89 (weak signal)
	2nd	70.86–81.57 ↓			
6	1st	64.20–76.32 ↓	59.14–74.98 ↓	59.58–74.54 ↓	56.53–44.63 ↑

The curves of compounds **1–3** and **6** demonstrated exothermic peaks after triple heating and cooling. For comps. **1** and **2**, the exothermic process showed weak signals (see the cooling curves in Figure S1 in the Supplementary data).

TGA/DTA analysis showed that comp. **5** (4-CN substituent in the pyridinium at cationic head groups) has no endothermic transition (Figure 5A) at a temperature lower than the compound decomposition temperature. This fact was confirmed by DSC (Figure 5B). Comp. **6** (3-Ph substituent at cationic head groups) had a pronounced endothermic peak, which was observed in the same temperature range in the DSC curves (Figure 5C for DTA and Figure 5D for DSC curves). A comparison of comp. **5** and comp. **6** makes it possible to observe the effect of the substituent on the thermal behavior and thermal stability of these compounds. Variation of the substituent at position 4 of 1,4-DHP, namely phenyl for comp. **1** and napthyl for comp. **7**, also confirmed this influence (Figures 3A and 5E). The curve of comp. **7** showed a rather narrow first order endothermic transition temperature range when tested by TGA/DTA; this first transition had a rather small absorbed heat value of 57.75 J/g. However, this transition was not confirmed by DSC for comp. **7** (Figure 4F).

It is underlined in the literature that the properties of the liposomes are mainly dependent on the physicochemical characteristics of lipids. It is known that the length and the degree of saturation of the lipid chain influence the phase transition temperature, such as gel to liquid crystalline state. The phase transition temperature depends on the length of the fatty acid chains, their degree of saturation, charge and head group types [55,56]. Usually a longer alkyl chain has a higher transition temperature, and introduction of double bonds decreases the transition temperature [56–58]. All studied 1,4-DHP amphiphiles have saturated lipophilic chains. For comps. **1–3** with increasing length of alkyl chains for two CH_2 groups, the values of the first transition temperature according to TGA data were in the same range of 56 °C, 44 °C and 54 °C (Table S1), respectively. This could be explained by the influence of pyridinium as cationic head group or 1,4-DHP core as an active linker. Our previous data regarding thermogravimetric analysis of structurally related pyridine amphiphiles with various heterocycles as cationic head groups demonstrated a wider transition phase range for the first transition state [54]. In agreement with DSC data, these temperatures for comps. **1–3** were 80 °C, 75 °C and 62 °C, respectively. Similar phenomena were observed for liposomal compositions of DPPC with monocationic 1,4-DHP amphiphiles—4-(N-alkylpyridinium-1,4-DHP)—as additives where with the increase of alkyl chain length, the phase transition temperature of compositions was decreased [59].

3.4. Surface Pressure–Area Isotherms, Mechanical Properties of Monolayers

Surface pressure (π) is defined as the decrease in surface tension of the aqueous medium when surfactant is added, $\pi = \gamma_0 - \gamma$, where γ_0 is the surface tension of water and γ is the surface tension of water with the surfactant monolayer. As seen from the $\pi-A$ isotherms, all studied compounds were able to form stable monolayers in an aqueous medium. Graphs of the surface pressure for monolayers composed of 1,4-DHP amphiphiles **1–7** were plotted. The critical surface pressure is defined as the surface pressure at which the monolayer collapses.

As is shown in Figure S2 (Supplementary data), all compounds had a similar collapsing surface pressure around 45 mN/m (Table 2) except for comp. **5** (4-CN), which had a critical pressure of 53 mN/m. All of the studied compounds had a similar $\pi-A$ isotherm pattern without an apparent liquid expanded (LE) to liquid condensed (LC) phase transition.

Table 2. Mechanical properties of monolayers composed of 1,4-DHP amphiphiles **1–7**. P (mN/m) is the critical pressure of the monolayers; C_s^{-1} mN/m) is the compressibility modulus.

Comps.	1	2	3	4	5	6	7
P ± SD, mN/m	46.62 ± 0.02	46.53 ± 0.75	46.33 ± 0.58	48.53 ± 0.68	53.06 ± 0.68	44.81 ± 0.44	47.13 ± 0.74
C_s^{-1} ± SD, mN/m	156.65 ± 1.24	170.46 ± 2.37	160.18 ± 1.58	170.44 ± 3.95	209.98 ± 1.69	120.71 ± 3.11	189.12 ± 2.92

Compressibility moduli C_s^{-1}, mN/m (Figure S3, Supplementary data) were calculated for comps. **1–7** (Table 2) from the $\pi-A$ data obtained from the monolayer compressions using the following Equation (1) [60]:

$$C_s^{-1} = -A\,(\delta\pi/\delta A) \qquad (1)$$

where $\delta\pi/\delta A$ is the slope of the monolayer, and the area, A, corresponds to the mean molecular area (MMA) at the indicated surface pressure, π. According to the literature [60], the values of the compressibility modulus ranging from 0 to 12.5 mN/m refer to the gas phase of the films, from 12.5 to 50 mN/m for the liquid-expanded (LE) films, from 100 to 250 mN/m for the liquid-condensed (LC) films, and above 250 mN/m for the solid films (S) [60,61].

According to the values of C_s^{-1}, the monolayers collapse in the LC phase, and none of the compounds reach the solid phase. A pronounced decrease in C_s^{-1} occurs at a surface pressure about 5–8 mN/m below the critical pressure acquired from the π–A isotherms, which means that the collapse of the monolayer does not spontaneously happen at the critical pressure point, but starts earlier, which is in accordance with a different behavior of the surfactant monolayers [38].

The MMA of the compound of interest can be extracted from the $\pi-A$ isotherms in the liquid condensed phase, and the obtained values for comps. **1–7** are listed in Table 3. It is a matter of long debate whether the hydrophobic tail has an influence on the surfactant area per molecule or not. Obtained results demonstrate that the alkyl chain length in the ester moieties varying from C_{12} to C_{16} (comps. **1, 2** and **3**) did not significantly influence the mean molecular area. On the other hand, the methylation of the dihydropyridine nitrogen atom at position 1 enlarged the area of the molecule almost by 15%. As could be expected, the introduction of a large hydrophobic phenyl group to pyridinium moieties in the positions 2 and 6 of the 1,4-DHP cycle enlarged the MMA, but a relatively small cyano group did not. In contrast, the introduction of a bulky naphthyl moiety in position 4 of the 1,4-DHP core slightly decreased the MMA. This means that the tail length and moieties in the position 4 of 1,4-DHP are not important in terms of the MMA, in contrast to the 1,4-DHP N-substituents and hydrophobic substituents at the pyridinium moiety as the cationic part of the amphiphile.

Table 3. Mean molecular areas (MMA, Å^2) of the 1,4-DHP amphiphiles **1–7**, and calculated values of the packing parameter (p) of these compounds.

Comps.	1	2	3	4	5	6	7
MMA $\pm$ SD, Å^2	82.77 $\pm$ 0.60	82.48 $\pm$ 1.19	86.91 $\pm$ 1.50	95.01 $\pm$ 0.94	82.91 $\pm$ 0.61	96.64 $\pm$ 1.08	76.65 $\pm$ 1.59
p *	0.51	0.51	0.48	0.44	0.51	0.43	0.55

* For all compounds SD < 0.02.

The packing parameter of compounds **1–7** was calculated using the obtained molecular areas (see Table 3) from the following Equation (2) [62,63]:

$$p = v_0/al_0, \qquad (2)$$

where v_0 and l_0 are the volume and the length of the surfactant tail [62], and a is the equilibrium area per molecule. The packing parameter values determine the shape of the micelle formed in the aqueous medium ($p = 0 \leq 1/3$ for the sphere, $1/3 \leq 1/2$ for the cylinder, and $1/2 \leq 1$ for the flexible bilayer) vesicles [62,63]. Every studied compound had p values near 1/2 or higher (see Table 3), which led to an assumption that comps. **1–7** form bilayer structures in the aqueous medium. According to the elaborated theoretical considerations, the formation of vesicles occurs in the systems with the packing parameters between 1/2 and 1 [63].

3.5. Magnetoliposome Preparation, Evaluation and Characterization

Cationic moiety containing 1,4-DHP amphiphiles **1–7** were chosen as lipid-like compounds for evaluation to produce magnetoliposomes (MLs) and characterization of liposome properties. According to our previous work [38], comp. 1 displays more pronounced delivery activity. Therefore we used this compound as the standard for characterization of other amphiphiles. The MLs shown in Figure 6B were obtained by the REV method from 1,4-DHP amphiphile **1** and FF containing negatively charged MNPs coated by citrate anions (γ-Fe$_2$O$_3$-citr) with pH~6.4 shown in Figure 6A. Transmission electron microscopy images (Figure 6) showed MLs with diameters in the range of 50–100 nm, while diameters of pure MNPs were around 10–15 nm (see MNP size distributions in Figure 1 and the TEM image in Figure 6A). It was confirmed that 1,4-DHP amphiphile **1** and MNPs coated with citrate anions formed magnetoliposomes with sizes that are applicable in biomedicine purposes.

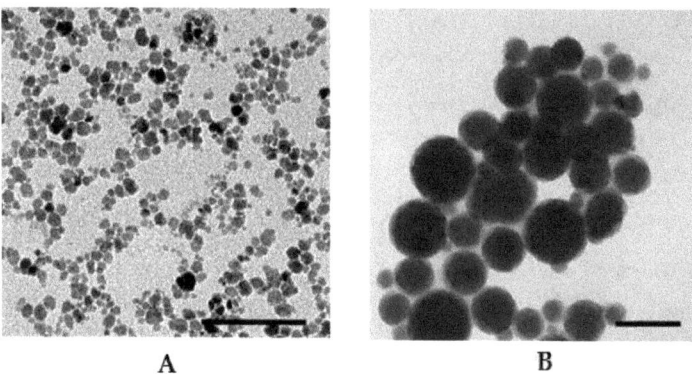

Figure 6. TEM images of (**A**) γ-Fe$_2$O$_3$ nanoparticles and (**B**) magnetoliposomes (MLs) formed by magnetic nanoparticles (MNPs) coated with citrate ions (γ-Fe$_2$O$_3$–citr 55 mg/mL) and 1,4-DHP **1**; $n_{DHP}/n_{\gamma\text{-Fe2O3}}$ = 0.04. Scale bar is 100 nm. MLs obtained by the reverse-phase evaporation (REV) method.

For further studies, 1,4-DHP amphiphiles **1–3** and **7** were chosen. Initially "empty" liposomes of pure 1,4-DHP amphiphiles **1–3** and **7** without FF additive were prepared by the REV method. DLS data of "empty" liposomal samples are summarized in Table 4 and Figure S4 in the Supplementary data.

Table 4. DLS data of "empty" liposome dispersions of 1,4-DHP amphiphiles **1–3** and **7** obtained by the REV method without ferrofluid (FF) additive. PdI is the polydispersity index; Z-ave D_H is the diameter that represents the mean hydrodynamic diameter of all liposomes in the distribution. The mean hydrodynamic diameter, D_H, depicts the hydrodynamic size of the main population of the tested sample.

"Empty" Liposomes	PdI	Z-ave D_H, nm	Distr. Peaks (max) Mean D_H, nm (%)		
			Peak 1 (%)	Peak 2 (%)	Peak 3 (%)
Comp. 1 *	0.263 ± 0.050	146.4 ± 1.4	188 (97)	4757 (3)	–
Comp. 1	0.263 ± 0.013	150.0 ± 8.1	190 (97)	3246 (3)	–
Comp. 2	0.486 ± 0.007	106.9 ± 8.3	216 (79)	34 (18)	4491 (3)
Comp. 3	0.431 ± 0.013	219.2 ± 5.4	347 (95)	46 (5)	–
Comp. 7	0.428 ± 0.017	290.6 ± 4.6	557 (93)	34 (2)	2829 (5)

* spontaneous swelling (SpSw).

The results for comp. 1 confirmed that the sizes of liposomes prepared by SpSw and REV methods were comparable. The dispersions for comp. 1 prepared by both the REV and SpSw methods had the lowest PdI value of 0.263, which indicated that the samples were more homogeneous than those of the other compounds, while the other samples were identified as moderately homogeneous samples, with PdI values of 0.428–0.486. DLS data analysis showed that the average diameter of the "empty" liposomes formed by amphiphiles **1–3** and **7** were in the 107–291 nm range. All the tested amphiphiles formed mainly one liposome population, namely 93–97% for comps. **1, 3,** and **7** and 79% for comp. **2**.

For detailed studies, 1,4-DHP amphiphiles **1–3** and **7** were chosen as membrane-forming agents to produce magnetoliposomes by the reverse-phase evaporation (REV) method. After encapsulation, the mixture was purified by magnetic decantation to remove all non-encapsulated MNPs. As underlined in the literature for similar liposomal systems, considering the strong difference in maximum magnetization of magnetic nanoparticles and magnetoliposomes in aqueous media, due to the diamagnetic contribution of water, the aqueous magnetoliposomes are not attracted to the magnet. Therefore, only non-encapsulated magnetic nanoparticles are separated in this way, keeping the magnetoliposomes in the supernatant phase. The lipid phase remains unchanged upon decantation, with the initial and final lipid concentrations being the same in the sample [64].

The total iron content encapsulated in MLs was around 20% from the starting concentration, which was confirmed by iron detection calorimetric analysis of the magnetoliposomal samples obtained by REV. The DLS data of magnetoliposomal samples are summarized in Table 5 and Figure S5 in the Supplementary data.

Table 5. DLS data of liposomes of 1,4-DHP amphiphiles **1–3** and **7** obtained with FF–citr. PdI is the polydispersity index; Z-ave D_H is the diameter that represents the mean hydrodynamic diameter of all liposomes in the distribution. The mean hydrodynamic diameter, D_H, depicts the hydrodynamic size of the main population of the tested sample. Dispersions were prepared by the REV method.

Liposomes	PdI	Z-ave D_H, nm	Distr. Peaks (max) Mean D_H, nm (%)		
			Peak 1 (%)	Peak 2 (%)	Peak 3 (%)
Comp. 1	0.291 ± 0.016	299.3 ± 1.8	432 (93)	58 (7)	–
Comp. 2	0.246 ± 0.017	149.6 ± 1.0	204 (94)	38 (6)	–
Comp. 3	0.286 ± 0.001	137.6 ± 0.1	201 (99)	4896 (1)	–
Comp. 7	0.521 ± 0.003	143.8 ± 0.7	316 (70)	54 (25)	4567 (5)
FF–citr	0.181 ± 0.006	38.4. ± 0.4	47 (100)	–	–

The obtained results demonstrated that main characteristic parameters for liposomal samples were comparable for "empty" liposomes and MLs. PdI values for MLs samples of comps. **1–3** were 0.246–0.291, indicating homogeneity of samples, while the PdI value of the sample of comp. **7** was 0.521. DLS data analysis showed that the average diameter of the MLs formed by amphiphiles **1–3** and **7** were in the 138–299 nm range. All tested amphiphiles formed mainly one liposome population, namely 93–99% for comps. **1–3** and 70% for comp. **7**.

4. Conclusions

Targeted modification of the 1,4-DHP core with different substituents both in the polar and in the non-polar parts of the molecule was performed, resulting in four groups of amphiphiles for the evaluation of the influence of structural elements on the physicochemical properties of compounds and on the properties of magnetoliposomes.

The obtained results by TGA/DTA demonstrate that with increasing length of the ester chains for 1,4-DHP amphiphiles **1–3**, the transition temperatures were shifted to lower temperatures. By TGA/DTA, transition temperatures are first transition, 56 °C; and second transition, 79 °C for comp.**1**; first transition, 44 °C; and second transition, 60 °C for comp.

2; and first transition, 54 °C; and second transition, 64 °C for comp. **3**. By DSC, these temperatures were 80 °C (comp. **1**), 75 °C (comp. **2**) and 62 °C (comp. **3**). Due to the relatively higher phase transition temperature, these amphiphiles may be potentially used as additives in composition with other synthetic lipids for the development of nanovectors.

It was shown that the variation of the alkyl chain length or the change of substituents at position 4 of 1,4-DHP did not show a significant influence on the mean molecular areas of the tested compound monolayers. In contrast, the introduction of the N-methyl substituent at position 1 of the 1,4-DHP molecule enlarges the mean molecular area almost by 15%, and also an addition of hydrophobic sterically hindered phenyl substituents at pyridinium moieties at positions 2 and 6 of the 1,4-DHP molecule slightly decreases the mean molecular area of the compound. The transition to the LC phase of comps. **1–7** is not clearly distinguished and occurred at 20–25 mN/m. Additionally, it was suggested that the tested 1,4-DHP amphiphiles **1–7** form bilayer structures in the aqueous medium because the calculated packing parameter values range from 0.43 to 0.55, which is in agreement with the theoretical considerations that the formation of vesicles occurs in the systems with the packing parameters between 1/2 and 1 [62].

It was demonstrated that the variation of the alkyl chain length or the change of substituents at position 4 of 1,4-DHP did not show a significant influence on the properties of liposomes.

Supplementary Materials: The following are available online at https://www.mdpi.com/2079-4991/11/3/593/s1. Scheme S1. Synthesis of 1,4-dihydropyridine (1,4-DHP) amphiphiles **1–3, 5–7**; Scheme S2. Synthesis of 1,4-dihydropyridine (1,4-DHP) amphiphile **4**; Table S1. Temperatures characteristics of tested compounds **1–7**, obtained by analysing TGA and DTA curves. Figure S1. Cooling curves obtained by DSC after heating process for tested compounds **1–3, 5–7**; Figure S2. 1,4-DHP amphiphiles **1–7** surface pressure—mean molecular area isotherms at 23 ± 1 °C; Figure S3. Compressibility modulus-surface pressure dependences obtained for the 1,4-DHP amphiphiles **1–7** monolayers; Figure S4. Hydrodynamic size distribution of the 'empty' liposomes formed by 1,4-DHP amphiphiles **1–3** and **7**. Liposomes obtained by REV; Figure S5. Hydrodynamic size distribution of the magnetoliposomes formed by 1,4-DHP amphiphiles **1–3** and **7**.

Author Contributions: Conceptualization, O.P.; methodology, O.P. and A.P.; validation, K.P.; investigation, O.P., M.R., P.D., E.S. and K.P.; data curation, O.P., M.R.; writing—original draft preparation, O.P. and P.D.; writing—review and editing, O.P., A.P. and A.S.; supervision, A.C. All authors have read and agreed to the published version of the manuscript.

Funding: This work was financially supported by PostDocLatvia Project No 1.1.1.2/VIAA/1/16/018 (O. Petrichenko: TGA and DTA analysis, magnetic nanoparticles synthesis, preparation and characterization of magnetoliposomes), PostDocLatvia Project No 1.1.1.2/VIAA/2/18/371 (M. Rucins: synthesis and characterization of 1,4-DHP amphiphiles), LIOS internal grant IG-2018-13 (P. Dimitrijevs: LB measurements) and M.era-net project FMF No.1.1.1.5/ERANET/18/04.

Data Availability Statement: The data presented in this study are available within this article and in Supplementary Data.

Acknowledgments: The authors are indebted to M. Maiorov from the Institute of Physics of the University of Latvia for the measurements of magnetic properties of NPs. The authors also are grateful to V. Ose from the Latvian Biomedical Research and Study Centre for TEM studies.

Conflicts of Interest: The authors declare no conflict of interest. The funders had no role in the design of the study; in the collection, analyses, or interpretation of data; in the writing of the manuscript; or in the decision to publish the results.

References

1. Laouini, A.; Jaafar-Maalej, C.; Limayem-Blouza, I.; Star, S.; Charcosset, C.; Fessi, H. Preparation, chararacterization and aplication of liposomes. *J. Colloid Sci. Biotehnol.* **2012**, *1*, 147–168. [CrossRef]
2. Monteiro, N.; Martins, A.; Reis, R.L.; Neves, N.M. Liposomes in tissue engineering and regenerative medicine. *J. R. Soc. Interface* **2014**, *11*, 20140459. [CrossRef] [PubMed]

3. Rucins, M.; Dimitrijevs, P.; Pajuste, K.L.; Petrichenko, O.; Jackevica, L.; Gulbe, A.; Kibilda, S.; Smits, K.; Plotniece, M.; Tirzite, D.; et al. Contribution of molecular structure to self-assembling and biological properties of bifunctional lipid-like 4-(N-alkylpyridinium)-1,4-dihydropyridines. *Pharmaceutics* **2019**, *11*, 115. [CrossRef] [PubMed]
4. Deshpande, P.P.; Biswas, S.; Torchilin, V.P. Current trends in the use of liposomes for tumor targeting. *Nanomedicine* **2013**, *8*, 1509–1528. [CrossRef]
5. Skouras, A.; Mourtas, S.; Markoutsa, E.; De Goltstein, M.C.; Wallon, C.; Sarah, C.; Antimisiaris, S.G. Magnetoliposomes with high USPIO entrapping efficiency, stability and magnetic properties. *Nanomedicine: NBM* **2011**, *7*, 572–579. [CrossRef] [PubMed]
6. Hofmann-Amtenbrink, M.; Hofmann, H.; Montet, X. Superparamagnetic nanoparticles—A tool for early diagnostics. *Swiss Med. Wkly.* **2010**, *140*, w13081. [CrossRef] [PubMed]
7. Babincová, N.; Sourivong, P.; Babinec, P.; Bergemann, C.; Babincová, M.; Durdík, Š. Application of magnetoliposomes with encapsulated doxorubicin for integrated chemotherapy and hyperthermia of rat C6 glioma. *Z. Naturforsch. C J. Biosci.* **2018**, *73*, 265–271. [CrossRef]
8. Gordon, R.; Hogan, C.E.; Neal, M.L.; Anantharam, V.; Kanthasamy, A.G.; Kanthasamy, A. A simple magnetic separation method for high-yield isolation of pure primary microglia. *J. Neurosci. Methods* **2011**, *194*, 287–296. [CrossRef]
9. Nardoni, M.; della Valle, E.; Liberti, M.; Relucenti, M.; Casadei, M.A.; Paolicelli, P.; Apollonio, F.; Petralito, S. Can pulsed electromagnetic fields trigger on-demand drug release from high-Tm magnetoliposomes? *Nanomaterials* **2018**, *8*, 196. [CrossRef]
10. Ding, H.; Sagar, V.; Agudelo, M.; Polakka-Kanthikeel, S.; Subba Rao Atluri, V.; Raymond, A.; Samikkannu, T.; Nair, M.P. Enhanced blood–brain barrier transmigration using a novel transferrin embedded fluorescent magneto-liposome nanoformulation. *Nanotechnology* **2014**, *25*, 055101. [CrossRef]
11. Fan, Z.; Fu, P.P.; Yu, H.; Raya, P.C. Theranostic nanomedicine for cancer detection and treatment. *J. Food Drug Anal.* **2014**, *22*, 3–17. [CrossRef] [PubMed]
12. Thomsen, L.B.; Thomsen, M.S.; Moos, T. Targeted drug delivery to the brain using magnetic nanoparticles. *Ther. Deliv.* **2015**, *6*, 1145–1155. [CrossRef]
13. Garcia-Pinel, B.; Jabalera, Y.; Ortiz, R.; Cabeza, L.; Jimenez-Lopez, C.; Melguizo, C.; Prados, J. Biomimetic magnetoliposomes as oxaliplatin nanocarriers: In vitro study for potential application in colon cancer. *Pharmaceutics* **2020**, *12*, 589. [CrossRef]
14. Cardoso, B.D.; Rodrigues, A.R.O.; Almeida, B.G.; Amorim, C.O.; Amaral, V.S.; Castanheira, E.M.S.; Coutinho, P.J.G. Stealth magnetoliposomes based on calcium-substituted magnesium ferrite nanoparticles for curcumin transport and release. *Int. J. Mol. Sci.* **2020**, *21*, 3641. [CrossRef] [PubMed]
15. Paul, B.; Bajaj, A.; Indi, S.S.; Bhattacharya, S. Synthesis of novel dimeric cationic lipids based on an aromatic backbone between the hydrocarbon chains and headgroup. *Tetrahedron Lett.* **2006**, *47*, 8401–8405. [CrossRef]
16. De Smedt, S.C.; Demeester, J.; Hennik, W.E. Cationic polymer based gene delivery systems. *Pharm. Res.* **2000**, *17*, 113–126. [CrossRef]
17. Ibraheem, D.; Elaissari, H.; Fessi, H. Gene therapy and DNA delivery systems. *Int. J. Pharm.* **2014**, *459*, 70–83. [CrossRef] [PubMed]
18. Pattni, B.S.; Chupin, V.V.; Torchilin, V.P. New developments in liposomal drug delivery. *Chem. Rev.* **2015**, *115*, 10938–10966. [CrossRef]
19. Lv, H.; Zhang, S.; Wang, B.; Cui, S.; Yan, J. Toxicity of cationic lipids and cationic polymers in gene delivery. *J. Control. Release* **2006**, *114*, 100–109. [CrossRef]
20. Ilies, M.A.; Johnson, B.H.; Makori, F.; Miller, A.; Seitz, W.A.; Thompson, E.B.; Balaban, A.T. Pyridinium cationic lipids in gene delivery: An in vitro and in vivo comparison of transfection efficiency versus a tetraalkylammonium congener. *Arch. Biochem. Biophys.* **2005**, *435*, 217–226. [CrossRef]
21. Damen, M.; Groenen, A.J.J.; van Dongen, S.F.M.; Nolte, R.J.M.; Scholte, B.J.; Feiters, M.C. Transfection by cationic gemini lipids and surfactants. *MedChemComm* **2018**, *9*, 1404–1425. [CrossRef]
22. Inglut, C.T.; Sorrin, A.J.; Kuruppu, T.; Vig, S.; Cicalo, J.; Ahmad, H.; Huang, H.-C. Immunological and toxicological considerations for the design of liposomes. *Nanomaterials* **2020**, *10*, 190. [CrossRef]
23. Hyvönen, Z.; Plotniece, A.; Reine, I.; Chekavichus, B.; Duburs, G.; Urtti, A. Novel cationic amphiphilic 1,4-dihydropyridine derivatives for DNA delivery. *Biochim. Biophys. Acta* **2000**, *1509*, 451–466. [CrossRef]
24. Hyvönen, Z.; Ruponen, M.; Rönkkö, S.; Suhonen, P.; Urtti, A. Extracellular and intracellular factors influencing gene transfection mediated by 1,4-dihydropyridine amphiphiles. *Eur. J. Pharm. Sci.* **2002**, *15*, 449–460. [CrossRef]
25. Mishra, A.P.; Bajpai, A.; Rai, A.K. 1,4-Dihydropyridine: A dependable heterocyclic ring with the promising and most anticipable therapeutic effects. *Mini-Rev. Med. Chem.* **2019**, *19*, 1219–1254. [CrossRef] [PubMed]
26. Godfraind, T. Discovery and development of calcium channel blockers. *Front. Pharmacol.* **2017**, *29*, 286. [CrossRef] [PubMed]
27. Klusa, V. Atypical 1,4-dihydropyridine derivatives, an approach to neuroprotection and memory enhancement. *Pharmacol. Res.* **2016**, *113*, 754–759. [CrossRef]
28. Zhang, Y.; Wang, J.; Li, Y.; Wang, F.; Yang, F.; Xu, W. Synthesis and radioprotective activity of mitochondria targeted dihydropyridines in vitro. *Int. J. Mol. Sci.* **2017**, *25*, 2233. [CrossRef]
29. Leonova, E.; Ošiņa, K.; Duburs, G.; Bisenieks, E.; Germini, D.; Vassetzky, Y.; Sjakste, N. Metal ions modify DNA-protecting and mutagen-scavenging capacities of the AV-153 1,4-dihydropyridine. *Mutat. Res. Genet. Toxicol. Environ. Mutagen.* **2019**, *845*, 403077. [CrossRef]

30. Milkovic, L.; Vukovic, T.; Zarkovic, N.; Tatzber, F.; Bisenieks, E.; Kalme, Z.; Bruvere, I.; Ogle, Z.; Poikans, J.; Velena, A.; et al. Antioxidative 1,4-dihydropyridine derivatives modulate oxidative stress and growth of human osteoblast-like cells in vitro. *Antioxidants* **2018**, *7*, 123. [CrossRef]
31. Manna, D.; Akhtar, S.; Maiti, P.; Mondal, S.; Kumar Mandal, T.; Ghosh, R. Anticancer activity of a 1,4-dihydropyridine in DMBA-induced mouse skin tumor model. *Anticancer Drugs* **2020**, *31*, 394–402. [CrossRef] [PubMed]
32. Lentz, F.; Reiling, N.; Spengler, G.; Kincses, A.; Csonka, A.; Molnár, J.; Hilgeroth, A. Dually acting nonclassical 1,4-dihydropyridines promote the anti-tuberculosis (Tb) activities of clofazimine. *Molecules* **2019**, *24*, 2873. [CrossRef] [PubMed]
33. González, A.; Casado, J.; Chueca, E.; Salillas, S.; Velázquez-Campoy, A.; Angarica, V.E.; Bénejat, L.; Guignard, J.; Giese, A.; Sancho, J.; et al. Repurposing dihydropyridines for treatment of helicobacter pylori infection. *Pharmaceutics* **2019**, *11*, 681. [CrossRef] [PubMed]
34. Triggle, D.J. 1,4-Dihydropyridine as calcium channel ligands and privileged structures. *Cell. Mol. Neurobiol.* **2003**, *23*, 293–303. [CrossRef]
35. Triggle, D.J. The 1,4-dihydropyridine nucleus: A pharmacophoric template part. 1. Actions at ion channels. *Mini-Rev. Med. Chem.* **2003**, *3*, 215–223. [CrossRef] [PubMed]
36. Duburs, G.; Vīgante, B.; Plotniece, A.; Krauze, A.; Sobolevs, A.; Briede, J.; Kluša, V.; Velēna, A. Dihydropyridine derivatives as bioprotectors. *Chem. Today* **2008**, *26*, 68–70.
37. Cindric, M.; Cipak, A.; Serly, J.; Plotniece, A.; Jaganjac, M.; Mrakovcic, L.; Lovakovic, T.; Dedic, A.; Soldo, I.; Duburs, G.; et al. Reversal of multidrug resistance in murine lymphoma cell by amphiphilic dihydropyridine antioxidant derivative. *Anticancer Res.* **2010**, *30*, 4063–4070.
38. Pajuste, K.; Hyvönen, Z.; Petrichenko, O.; Kaldre, D.; Rucins, M.; Cekavicus, B.; Ose, V.; Skrivele, B.; Gosteva, M.; Morin-Picardat, E.; et al. Gene delivery agents possessing antiradical activity: Self-assembling cationic amphiphilic 1,4-dihydropyridine derivatives. *New J. Chem.* **2013**, *37*, 3062–3075. [CrossRef]
39. Bruvere, I.; Bisenieks, E.; Poikans, J.; Uldrikis, J.; Plotniece, A.; Pajuste, K.; Rucins, M.; Vigante, B.; Kalme, Z.; Gosteva, M.; et al. Dihydropyridine derivatives as cell growth modulators in vitro. *Oxid. Med. Cell. Longev.* **2017**, *2017*, 4069839. [CrossRef]
40. Jansone, B.; Kadish, I.; van Groen, T.; Beitnere, U.; Moore, D.R.; Plotniece, A.; Pajuste, K.; Klusa, V. A Novel 1,4-Dihydropyridine derivative improves spatial learning and memory and modifies brain protein expression in wild type and transgenic APPSweDI mice. *PLoS ONE* **2015**, *10*, e0127686. [CrossRef]
41. Beaune, G.; Dubertret, B.; Clément, O.; Vayssettes, C.; Cabuil, V.; Ménager, C. Giant vesicles containing magnetic nanoparticles and quantum dots: Feasibility and tracking by fiber confocal fluorescence microscopy. *Angew. Chem. Int. Ed.* **2007**, *46*, 5421–5424. [CrossRef] [PubMed]
42. Petrichenko, O.; Plotniece, A.; Pajuste, K.; Ose, V.; Cebers, A. Formation of magnetoliposomes using self-assembling 1,4-dihydropyridine derivative and maghemite γ-Fe$_2$O$_3$ nanoparticles. *Chem. Heterocycl. Compd.* **2015**, *51*, 672–677. [CrossRef]
43. Petrichenko, O.; Erglis, K.; Cebers, A.; Plotniece, A.; Pajuste, K.; Bealle, G.; Menager, C.; Dubois, E.; Perzynski, R. Bilayer properties of giant magnetic liposomes formed by cationic pyridine amphiphile and probed by active deformation under magnetic forces. *Eur. Phys. J. E* **2013**, *36*, 9. [CrossRef]
44. Bee, A.; Massart, R.; Neveu, S. Synthesis of very small fine maghemite particles. *JMMM* **1995**, *149*, 6–9. [CrossRef]
45. Răcuciu, M.; Creangă, D.E.; Airinei, A. Citric-acid-coated magnetite nanoparticles for biological applications. *Eur. Phys. J. E* **2006**, *21*, 117–121. [CrossRef]
46. Pajuste, K.; Plotniece, A.; Kore, K.; Intenberga, L.; Cekavicus, B.; Kaldre, D.; Duburs, G.; Sobolev, A. Use of pyridinium ionic liquids as catalysts for the synthesis of 3,5-bis(dodecyloxycarbonyl)-1,4-dihydropyridine derivative. *CEJC* **2011**, *9*, 143–148. [CrossRef]
47. De Luca, M.; Ioele, G.; Ragno, G. 1,4-Dihydropyridine antihypertensive drugs: Recent advances in photostabilization strategies. *Pharmaceutics* **2019**, *11*, 85. [CrossRef]
48. De Luca, M.; Ioele, G.; Spatari, C.; Ragno, G. Photodegradation of 1,4-dihydropyridine antihypertensive drugs: An updated review. *Int. J. Pharm. Pharm. Sci.* **2018**, *10*, 8–18. [CrossRef]
49. Rucins, M.; Smits, R.; Sipola, A.; Vigante, B.; Domracheva, I.; Turovska, B.; Muhamadejev, R.; Pajuste, K.; Plotniece, M.; Sobolev, A.; et al. Pleiotropic properties of amphiphilic dihydropyridines, dihydropyridones and aminovinylcarbonyl compounds. *Oxid. Med. Cell. Longev.* **2020**, *2020*, 8413713. [CrossRef]
50. Plotniece, A.; Pajuste, K.; Kaldre, D.; Cekavicus, B.; Vigante, B.; Turovska, B.; Belyakov, S.; Sobolev, A.; Dubur, G. Oxidation of cationic 1,4-dihydropyridine derivatives as model compounds for putative gene delivery agents. *Tetrahedron* **2009**, *65*, 8344–8349. [CrossRef]
51. Pardo-Jiménez, V.; Barrientos, C.; Pérez-Cruz, K.; Navarrete-Encina, P.A.; Olea-Azar, C.; Nuñez-Vergara, L.J.; Squella, J.A. Synthesis and electrochemical oxidation of hybrid compounds: Dihydropyridinefused coumarins. *Electrochim. Acta* **2014**, *125*, 457–464. [CrossRef]
52. Stradyn', Y.P.; Beilis, Y.I.; Uldrikis, Y.R.; Dubur, G.Y.; Sausin', A.E.; Chekavichus, B.S. Voltamperometry of 1,4-dihydropyridine derivatives—II. Electronic and steric effects in the electrooxidation of 4-substituted 1,4-dihydropyridines. *Chem. Heterocycl. Compd.* **1975**, *11*, 1299–1303. [CrossRef]
53. Turovska, B.; Goba, I.; Turovskis, I.; Grinberga, S.; Belyakov, S.; Stupnikova, S.; Liepinsh, T.; Stradins, J. Electrochemical oxidation of 4-monoalkyl-substituted 1,4-dihydropyridines. *Chem. Heterocycl. Compd.* **2008**, *44*, 1483–1490. [CrossRef]

54. Petrichenko, O.; Rucins, M.; Vezane, A.; Timofejeva, I.; Sobolev, A.; Cekavicus, B.; Pajuste, K.; Plotniece, M.; Gosteva, M.; Kozlovska, T.; et al. Studies of the physicochemical and structural properties of self-assembling cationic pyridine derivatives as gene delivery agents. *Chem. Phys. Lipids* **2015**, *191*, 25–37. [CrossRef]
55. Li, J.; Wang, X.; Zhang, T.; Wang, C.; Huang, Z.; Luo, X.; Deng, Y. A review on phospholipids and their main applications in drug delivery systems. *Asian J. Pharm. Sci.* **2015**, *10*, 81–98. [CrossRef]
56. Beltrán-Gracia, E.; López-Camacho, A.; Higuera-Ciapara, I.; Velázquez-Fernández, J.B.; Vallejo-Cardona, A.A. Nanomedicine review: Clinical developments in liposomal applications. *Cancer Nano* **2019**, *10*, 11. [CrossRef]
57. Lin, X.; Gu, N. Surface properties of encapsulating hydrophobic nanoparticles regulate the main phase transition temperature of lipid bilayers: A simulation study. *Nano Res.* **2014**, *7*, 1195–1204. [CrossRef]
58. Ballweg, S.; Sezgin, E.; Doktorova, M.; Covino, R.; Reinhard, J.; Wunnicke, D.; Hänelt, I.; Levental, I.; Hummer, G.; Ernst, R. Regulation of lipid saturation without sensing membrane fluidity. *Nat. Commun.* **2020**, *11*, 756. [CrossRef] [PubMed]
59. Tirzite, D.; Koronova, J.; Plotniece, A. Influence of some quaternised 1,4-dihydropyridines derivatives on liposomes and erythrocyte membranes. *Biochem. Mol. Biol. Int.* **1998**, *45*, 849–856. [PubMed]
60. Capuzzi, G.; Fratini, E.; Pini, F.; Baglioni, P.; Casnati, A.; Teixeira, J. Counterion complexation by calixarene ligands in cesium and potassium dodecyl micelles. A small angle neutron scattering study. *Langmuir* **2000**, *16*, 188–194. [CrossRef]
61. Vitovic, P.; Subjakova, V.; Hianik, T. The physical properties of lipid monolayers and bilayers containing calixarenes sensitive to cytochrome c. *Gen. Physiol. Biophys.* **2013**, *32*, 189–200. [CrossRef]
62. Nagarajan, R. Molecular packing parameter and surfactant self-assembly: The neglected role of the surfactant tail. *Langmuir* **2002**, *18*, 31–38. [CrossRef]
63. Šegota, S.; Težak, D. Spontaneous formation of vesicles. *Adv. Colloid Interface Sci.* **2006**, *121*, 51–75. [CrossRef] [PubMed]
64. Pereira, D.S.M.; Cardoso, B.D.; Rodrigues, A.R.O.; Amorim, C.O.; Amaral, V.S.; Almeida, B.G.; Queiroz, M.-J.R.P.; Martinho, O.; Baltazar, F.; Calhelha, R.C.; et al. Magnetoliposomes containing calcium ferrite nanoparticles for applications in breast cancer therapy. *Pharmaceutics* **2019**, *11*, 477. [CrossRef] [PubMed]

Article

Inactivation of Bacteria Using Bioactive Nanoparticles and Alternating Magnetic Fields

Vitalij Novickij [1,*], Ramunė Stanevičienė [2], Rūta Gruškienė [3], Kazimieras Badokas [4], Juliana Lukša [2], Jolanta Sereikaitė [3], Kęstutis Mažeika [5], Nikolaj Višniakov [6], Jurij Novickij [1] and Elena Servienė [2,3,*]

1. Faculty of Electronics, Vilnius Gediminas Technical University, 03227 Vilnius, Lithuania; jurij.novickij@vgtu.lt
2. Laboratory of Genetics, Nature Research Centre, 08412 Vilnius, Lithuania; ramune.staneviciene@gamtc.lt (R.S.); juliana.luksa@gamtc.lt (J.L.)
3. Faculty of Fundamental Sciences, Vilnius Gediminas Technical University, 10223 Vilnius, Lithuania; ruta.gruskiene@vgtu.lt (R.G.); jolanta.sereikaite@vgtu.lt (J.S.)
4. Institute of Photonics and Nanotechnology, Vilnius University, 10257 Vilnius, Lithuania; kazimieras.badokas@ff.vu.lt
5. Center for Physical Sciences and Technology, 02300 Vilnius, Lithuania; kestutis.mazeika@ftmc.lt
6. Faculty of Mechanics, Vilnius Gediminas Technical University, 03224 Vilnius, Lithuania; nikolaj.visniakov@vgtu.lt
* Correspondence: vitalij.novickij@vgtu.lt (V.N.); elena.serviene@gamtc.lt (E.S.)

Abstract: Foodborne pathogens are frequently associated with risks and outbreaks of many diseases; therefore, food safety and processing remain a priority to control and minimize these risks. In this work, nisin-loaded magnetic nanoparticles were used and activated by alternating 10 and 125 mT (peak to peak) magnetic fields (AMFs) for biocontrol of bacteria *Listeria innocua*, a suitable model to study the inactivation of common foodborne pathogen *L. monocytogenes*. It was shown that *L. innocua* features high resistance to nisin-based bioactive nanoparticles, however, application of AMFs (15 and 30 min exposure) significantly potentiates the treatment resulting in considerable log reduction of viable cells. The morphological changes and the resulting cellular damage, which was induced by the synergistic treatment, was confirmed using scanning electron microscopy. The thermal effects were also estimated in the study. The results are useful for the development of new methods for treatment of the drug-resistant foodborne pathogens to minimize the risks of invasive infections. The proposed methodology is a contactless alternative to the currently established pulsed-electric field-based treatment in food processing.

Keywords: electromagnetic fields; food processing; sterilization; nisin; *L. innocua*

1. Introduction

Foodborne diseases associated with bacteria represent a serious health problem, which may be fatal in some cases [1,2]. As a result, significant efforts are made to ensure food safety and adequate management of the bacterial contaminants. Thus, processing may involve thermal sterilization [3,4], pulsed electric field treatment [5], cold-plasma treatment [6], natural bacteriocins [7,8] or nanotechnological methods [9,10]. The synergistic approaches when several different methodologies are combined frequently deliver promising results in food control [11–13].

Many of the methodologies rely on the chemical interactions between the bacteria and the target molecule, which sooner or later may result in occurrence of the treatment-resistant microorganisms [14–16]. It is especially true in the context of antibiotics and bacteriocins, when bacteria through modifications of their cell envelope (i.e., charge and thickness) can develop resistance to the chemical treatment [17,18]. The pharmaceutical industry and health care systems have been combating antibiotic-resistant strains of bacteria for more than 60 years [19], and nisin is currently the most popular bacteriocin used in the food sector [20]. It is efficient against many Gram-positive bacteria, while still being unstable in alkaline environment or during exposure to special protease [21,22]. Therefore, nisin nanoencapsulation is one of the best validated methods in order to prolong its bactericidal time, improve storage performance or obtain a sustained release [22–24]. Nevertheless, nanoencapsulation does not solve all the problems, and nisin still remains ineffective against Gram-negative bacteria or bacteria with thick cytoplasmic membrane and cell wall [25–27]. As a result, other food processing methods such as hydrostatic pressure techniques [28], cold plasma [29], ultraviolet light [30], ultrasound [31] or pulsed electric fields [32] have been extensively focused on for their use in the improvement of food safety. Nevertheless, the best effects are expected when a synergism between various methods is acquired [33]. Therefore, in order to improve the efficacy of nisin and induce a synergistic response, methods affecting the cell wall and plasma membrane are required.

One of the solutions is to use physical methods for cell permeabilization and thus allow nisin to incorporate itself in the bacterial cell membrane by binding to essential precursors for cell wall biosynthesis, which ultimately leads to formation of pores, loss of solutes in bacteria and subsequent cell death [34]. In our previous works we have successfully shown that the efficacy of nisin nanoparticles can be effectively improved by electroporation [35,36]. Combination with mild thermal treatment further improves the efficacy even against Gram-negative bacteria in stationary growth phase [37]. However, application of pulsed electric fields (PEFs) involves metal electrodes being in direct contact with the treated sample, which results in ion release, electrode degradation due to electrolysis, various electrochemical reactions, pH gradients or voltage breakdowns [38–42]. All these factors to a certain extent can affect the quality of food and are considered as a limitation of PEF-based techniques.

One of the solutions to overcome mentioned above limitations could be the application of magnetic fields, which do not require a direct contact with the sample. The phenomenon of contactless electroporation using pulsed magnetic fields and conductive nanoparticles has been confirmed recently [43]. However, in such a case, high power setups are required, and the methodology is still far from the capability to scale it industrially. Therefore, application of magnetic nanomaterial can be introduced to achieve a profound inactivation effect in low amplitude, but high frequency magnetic fields [44] enabling a multi-factorial treatment including magnetic hyperthermia. The parametrical flexibility and efficient removability of the nanoparticles from food can be highlighted as the main advantages of such methodology [45]. In order to further improve the efficacy, the magnetic nanoparticles can be functionalized and serve as drug carriers for targeted treatment [46]. Previously we tried to develop nisin-functionalized magnetic nanoparticles [47] for controlled release using high frequency alternating magnetic fields (AMFs) [48], but it was shown that the 10 mT, 100 kHz magnetic field is ineffective for potentiation of nisin-based treatment when short exposure times are used (2 min). Therefore, in this work, we employed a more powerful 125 mT, 200 kHz magnetic setup and increased the exposure times. The 10 mT, 100 kHz setup was used as a reference, and equivalent parametric protocols (exposure time-wise) were derived.

The results are useful for the development of new methods for treatment of the drug-resistant foodborne pathogens to minimize the risks of invasive infections. We show that it is not solely the thermal stress influencing bioactivity of nisin nanoparticles, but also the alternating magnetic fields significantly improve the efficacy. The actuality of the proposed methodology lies in the contactless nature of the treatment, which is advantageous in terms of contamination and/or electrochemical reactions, which are typical for widely-implemented PEF-based treatment chambers.

2. Materials and Methods
2.1. Alternating Magnetic Fields

The experimental setup consisted of two generators: (1) a low magnetic field (LMF) 10 mT, 100 kHz generator and (2) a high magnetic field (HMF) 125 mT, 200 kHz generator. The coil of the LMF generator was made from hollow (5 mm diameter) copper wire resulting in a 1 layer 8 winding solenoid structure (inner effective diameter of 23 mm). The coil of the HMF generator was made from hollow (8 mm diameter) copper wire resulting in a 1 layer 3 winding solenoid structure (inner effective diameter of 10 mm). Liquid cooling was used to prevent heating of the coils. Both generators were compatible with 0.2 mL polymerase chain reaction (PCR) sterile tubes (Quali Electronics Inc., Columbia, SC, USA).

The measured waveforms of both AMF generators are presented in Figure 1. The waveforms were acquired using a calibrated loop sensor (VGTU, Vilnius, Lithuania) and post-processed in OriginPro 8.5 Software (OriginLab, Northampton, MA, USA).

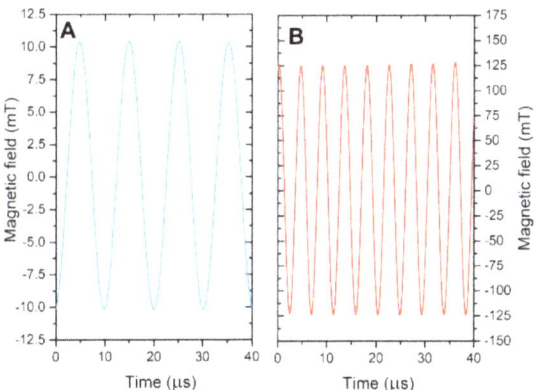

Figure 1. Measured alternating magnetic field waveforms. (**A**) 100 kHz low magnetic field generator and (**B**) 200 kHz high magnetic field generator.

The exposure time was controlled to establish a dose-dependent cellular response. For the LMF generator the 5, 30 and 60 min protocols were employed, while for the HMF the 5, 15 and 30 min protocols were used.

2.2. Thermal Influence

The thermal influence was estimated using a compact Pt1000 sensor (Innovative Sensor Technology, Wattwil, Switzerland). Temperature was measured with a varied time step (5–30 min) to grasp the moment of temperature saturation. In order to prevent the influence of eddy currents on the sensor response during AMF pulses, the pulse sequences were stopped for several seconds to acquire the measurement. After that, the pulsing was resumed for a further time step. For the LMF protocols the temperature did not exceed 37 °C, while for HMF it was below 45 °C. In order to estimate the influence of thermal stress and to allow adequate evaluation of the magnetic field-mediated methodology, a separate experiment where the bacterial cells were incubated in 37 and 45 °C for 5, 15, 30 and 60 min was performed. The experimental scheme included cells separately and with nisin, nisin nanoparticles (CaN) and nisin-free nanoparticles (CaO) to mirror the experiment with magnetic fields.

2.3. Nisin-Loaded Magnetic Nanoparticles

Nisin-loaded iron oxide magnetic nanoparticles (NPs) were prepared as previously described [47]. Briefly, under vigorous stirring, 0.587 g $FeCl_3 \cdot 6H_2O$ and 0.278 g $FeSO_4 \cdot 7H_2O$ were mixed in 10 mL of water and heated to 80 °C under nitrogen in a three-necked flask. Then, 3.5 mL of NH_4OH (10%) was dropped into the solution. After reaction for 30 min at 80 °C, 0.3 g of calcium citrate in 0.6 mL water was added directly into the reaction solution. The temperature was increased to 95 °C, and stirring continued for an additional 90 min. Then, the solution was cooled down to room temperature naturally. The iron oxide particles were separated by a magnet from reaction mixture, washed with deionized water several times and dried at 45 °C for 12 h. The prepared dried powder was stored in the refrigerator. Before use, the required amount of iron oxide was re-dissolved in water using an ultrasonic water bath for 3 h and centrifuged at 6400× g for 2 h. The final iron oxide nanoparticle solution was used for the following nisin loading. Synthesized iron oxide nanoparticles capped with citric acid corresponded to Fe_2O_3 phase (Maghemite-C, ICDD Card No. 00-039-1346) with a possible small quantity of magnetite as judged with the X-ray diffraction method.

A vibrating sample magnetometer was used for the magnetization measurement of the sample. The lock-in amplifier SR510 (Stanford Research Systems, Sunnyvale, CA, USA) was applied for the detection of the signal from the sense coils generated by vibrating sample. A Gauss/Teslameter FH-54 (Magnet Physics Dr. Steingroever GmbH, Cologne, Germany) was used to measure the magnetic field strength between the poles of the laboratory magnet, which was supplied by the power source SM 330-AR-22 (Delta Elektronika, ND Zierikzee, Netherlands). The Mössbauer spectrum was measured using the ^{57}Co (Rh) source in the transmission geometry with a Mössbauer spectrometer (Wissenschaftliche Elektronik GmbH, Starnberg, Germany). The Mössbauer spectrum was fitted to hyperfine field distributions applying WinNormos software (version 3.0 by R.A.Brand, Wissenschaftliche Elektronik GmbH, Starnberg, Germany).

For the preparation of nisin-loaded particles, a nisin solution in water at the concentration of 10 μg/mL was added dropwise to the iron oxide nanoparticle solution (0.05 mg/mL) at the ratio 1/4 (v/v) under constant stirring at room temperature. The final concentration of nisin in the product was 2 μg/mL, and that of iron oxide nanoparticles in the solution was 0.04 mg/mL. For the preparation of control, instead of nisin solution, water was used. The solution of prepared nisin-loaded NPs was stored at +4 °C. Nisin loading on the particles was confirmed by Fourier transform infrared spectroscopy and thermogravimetric analysis. The diameter of nisin-loaded particles determined by atomic force microscopy and the hydrodynamic diameter determined by dynamic light scattering method was 11.3 ± 1.4 and 26.4 ± 2.0 nm, respectively [44]. The size distribution (hydrodynamic radius) of nisin-loaded particles is presented in Figure 2.

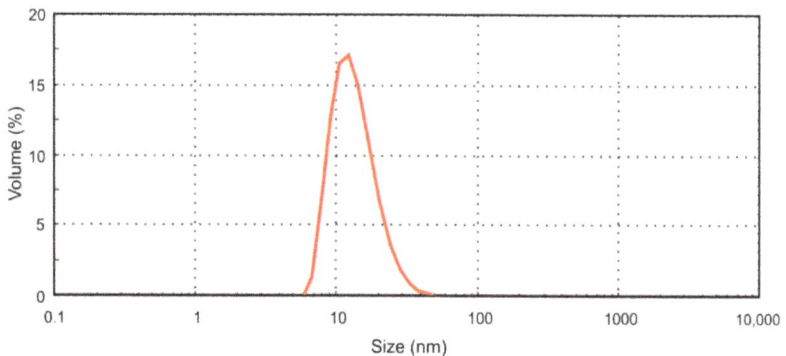

Figure 2. Size distribution of nisin-loaded iron oxide magnetic nanoparticles.

2.4. Bacterial Cells

The bacteria *Listeria innocua* CECT 910T (kindly provided by Maria Joao Fraqueza, University of Lisbon, Lisbon, Portugal) was cultivated in brain heart infusion (BHI) medium (1.25% brain extract, 0.5% heart extract, 1% peptone, 0.2% dextrose, 0.5% NaCl, 0.25% Na_2HPO_4) for 16–18 h with continuous shaking at 37 °C. Overnight grown cells were collected by centrifugation at 6000× g for 5 min, washed 3 times with 1 mol/L sorbitol, re-suspended in 1 mol/L sorbitol at a final concentration of about 1×10^9 cells/mL and used in AMF treatment experiments.

For the analysis of bacteria viability, the cells of *L. innocua* in 1 mol/L sorbitol (50 μL) were mixed with equal volume of nisin-unloaded/loaded magnetic nanoparticle solution

(v/v 1:1) and treated with AMFs. The final concentration of the nisin in the mixture was 1 µg/mL, and the concentration of iron oxide nanoparticles was ~0.02 mg/mL. After the treatment (5–60 min), the samples were incubated at room temperature (20 °C) for total time of (treatment time + incubation = 1 h, e.g., 5 min AMF treatment + 55 min incubation) without agitation, serial dilutions were performed in sterile 0.9% NaCl and 50 µL of each solution was spread onto BHI-agar plates with incubation following overnight at 37 °C. After the incubation, colonies were counted as colony forming units (CFU), and the mean value of CFU/mL was calculated. In subsequent bacteria viability experiments, the volume of nisin-unloaded/loaded magnetic nanoparticles was doubled (v/v 2:1) and tripled (v/v 3:1). As a reference, nisin-only solution was also implemented in the study.

2.5. Scanning Electron Microscopy

For the preparation of scanning electron microscopy (SEM), L. innocua cells (1×10^9 CFU/mL) were incubated with nanoparticles and/or treated by AMFs and followed serial dilutions in sterile water. Then, 5 µL measurements of solutions were dropped onto the specimen stubs covered with copper foil tape and gently dried at room temperature. For the microscopy preparation, samples were sputter coated with a 25 nm gold layer using a Q150T ES sputter coater (Quorum Technologies, Laughton, UK). Twenty or more images per cell treatment were obtained using an Apollo 300 (CamScan, Cambridge, UK) scanning electron microscope operating at 15 kV.

2.6. Statistical Analysis

One-way analysis of variance (ANOVA; $p < 0.05$) was used to compare results. If ANOVA indicated a statistically significant result ($p < 0.05$), Tukey's honest significance difference (HSD) multiple comparison test for evaluation of the difference was used. The data were post-processed in OriginPro software (OriginLab, Northhampton, MA, USA). All experiments were performed with at least three repetitions, and the treatment efficiency is expressed as mean ± standard deviation.

3. Results

Before the experiments with cells, Mössbauer spectroscopy and magnetization measurements of iron oxide magnetic nanoparticles were performed. As shown in Figure 3, the magnetization dependence has no hysteresis or remnant magnetization and thus is characteristic of superparamagnetic nanoparticles. The saturation magnetization of maghemite/magnetite nanoparticles of ≈48 emu/g is lower in comparison with 92 emu/g of bulk magnetite. For nanosized materials, saturation magnetization decreases because of the magnetically dead layer, magnetic disorder of the surface layer and the presence of nonmagnetic adsorption on surface materials.

The broadening of spectral lines of the Mössbauer spectrum (Figure 4) showed superparamagnetism of nanoparticles. Two hyperfine field distributions were used to fit to the spectrum (Table 1). The two hyperfine field distributions $P(B)$, which differ by isomer shift, give better quality of fitting than only one distribution as shown by the dashed line. The hyperfine field distribution with fixed smaller isomer shift δ = 0.28 mm/s was attributed to the contribution of magnetite tetrahedral sublattice A or to maghemite [49]. Another hyperfine distribution had larger fixed isomer shift of δ = 0.66 mm/s. The isomer shift of the contribution of magnetite octahedral B sublattice is larger because of the presence of both Fe^{2+} and Fe^{3+}. It is noteworthy that using the fixed isomer shift of hyperfine field distribution, the contribution of B sublattice can be separated for magnetite but not for maghemite γ-Fe_2O_3 without Fe^{2+}. For bulk magnetite, hyperfine field B_0 = 49 T for A and ≈ 46 T for B sublattice [49]. Average hyperfine fields <B> of both distributions were 26% lower than those of bulk magnetite, indicating superparamagnetic relaxation of nanoparticles (Table 1). The characteristic size of maghemite/magnetite nanoparticles ≈11.6 nm was obtained applying hyperfine field dependence on the size of nanoparticles $B = B_0 (1 - kT/2 KV)$, where k is the Boltzmann constant, T is temperature, $K \approx 10^4$ J/m^3 is magnetic anisotropy of magnetite and V is average volume of nanoparticles [50]. According to the sub-spectra ratio, there was about 14% Fe^{2+} of all iron.

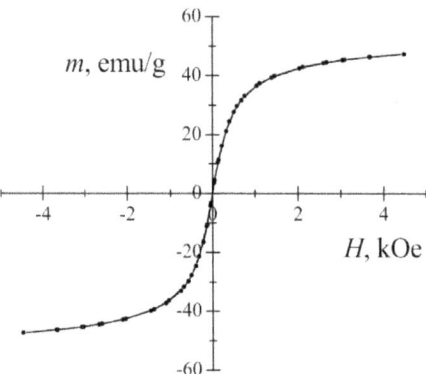

Figure 3. Magnetization dependence of maghemite/magnetite nanoparticles on the applied magnetic field.

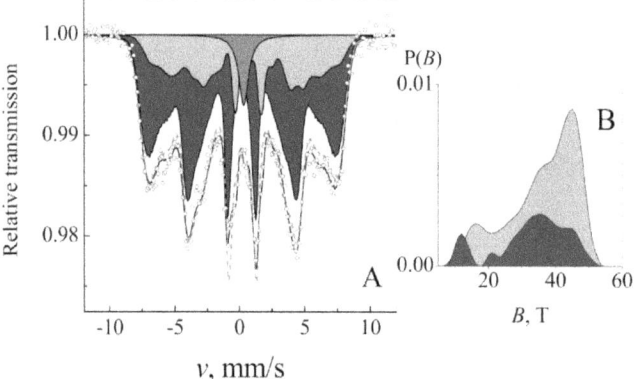

Figure 4. Mössbauer spectrum (**A**) of maghemite/magnetite nanoparticles at 21 °C temperature and two hyperfine field distributions P(B) (**B**) applied for fitting. Dashed gray line (**A**) given for comparison indicates fitting using one hyperfine field distribution.

Table 1. Parameters of fitting to Mössbauer spectrum: isomer shift δ, quadrupole shift 2ε, average hyperfine field <B> of distribution and relative area A attributed to magnetite [§].

Sub-spectrum	δ, mm/s	2ε, mm/s	<B>, T	A, %
Magnetite A (Fe^{3+}) or maghemite	0.28 *	0.02 ± 0.01	36.2	69
Magnetite B ($Fe^{2+} + Fe^{3+}$)	0.66 *	−0.23 ± 0.03	33.2	27
Superparamagnetic singlet	0.43 ± 0.04	-	-	4

[§] Isomer shift is given with respect to α-Fe. * fixed.

Considering that the developed NPs have both the biological activity and a thermal physical stressor due to superparamagnetic relaxation, the increase of temperature and the resulting hyperthermia was evaluated for each concentration of NPs. The results are summarized in Figure 5.

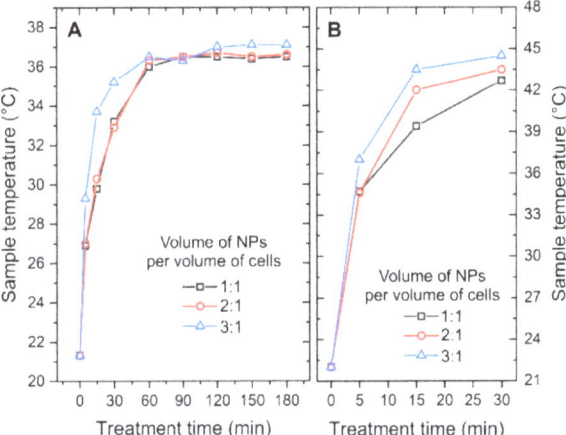

Figure 5. Measured dependence of sample temperature on concentration of nanoparticles and treatment time. (**A**) 100 kHz low magnetic field generator and (**B**) 200 kHz high magnetic field generator.

As it can be seen in Figure 5A, the low AMFs resulted in a saturated temperature after 60 min, while the dependence on the NP concentration was weak and not statistically significant. The highest temperature of 37 °C was reached. In the case of high AMFs (Figure 5B), the differences were more profound. The temperature reached saturation after 30 min and was almost 45 °C.

Further, experiments with cells incubated with or without nanoparticles for up to 1 h at different equivalent temperatures were performed. The results are summarized in Figure 6.

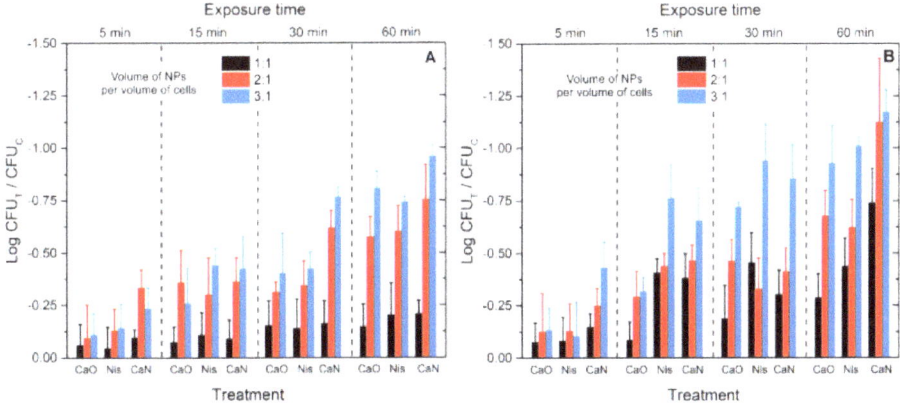

Figure 6. The viability response of *L. innocua* to thermal stress and nisin-loaded (CaN), nisin-free (CaO) magnetic nanoparticles or nisin-only solution (Nis). (**A**) 37 °C experiments; (**B**) 45 °C experiments. In all cases the treatment data (CFU_T) are normalized to untreated control samples (CFU_C).

As expected, the efficacy of nisin or CaO and CaN nanoparticles depends on concentration and temperature. Nevertheless, cell exposure to 1 h 37 °C thermal stress with CaN NPs potentiates the efficacy of the treatment to reach 1 log inactivation only (Figure 6A). Similar efficacies are more rapidly reached (15–30 min) when 45 °C thermal stress is applied (Figure 6B); however, in all cases the efficacies are below 1.25 log reduction. It should be also noted that in the case of AMF treatments, the NPs heat the samples gradually (refer to

Figure 5). However, the data presented in Figure 6 are for stable temperature incubation; thus, more thermal stress is experienced by the cells compared to AMF treatment. Considering the acquired data and weak effects of temperature, further experiments targeting cells in AMFs were performed.

In order to preserve a predominantly non-thermal treatment the exposure time was limited to 60 min for the low AMF generator and 30 min for the high AMF generator. The results after low AMF treatment are summarized in Figure 7.

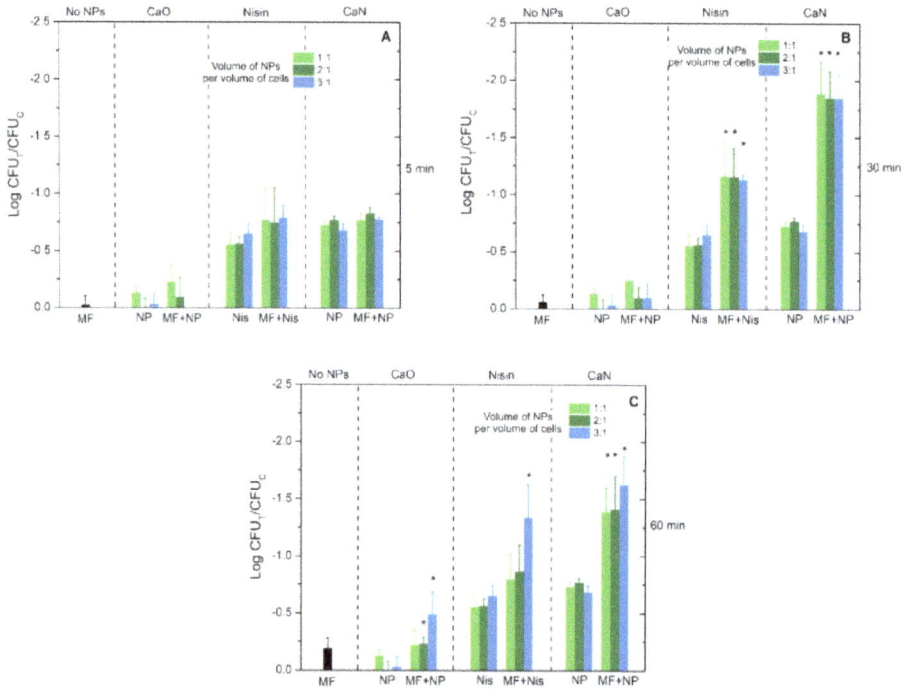

Figure 7. The viability response of *L. innocua* to low alternating magnetic field (10 mT, 100 kHz) and activated nisin-loaded (CaN), nisin-free (CaO) magnetic nanoparticles or nisin-only solution (Nisin). (**A**) After 5 min exposure; (**B**) 30 min exposure and (**C**) 60 min exposure. In all cases the treatment (CFU_T) is normalized to untreated control samples (CFU_C). Asterisk (*) corresponds to statistically significant difference ($p < 0.05$) versus NP-only or nisin-only treatment.

It can be seen that the exposure time for the MF-only treatment has a weak influence on the cell viability. In the case of 5 and 30 min treatments, the differences are not statistically significant, while for the 60 min protocol (Figure 7C) a minor <0.25 log viability reduction was detected. Exposure of the cells to magnetic nanoparticles only (CaO without heating) did not result in any effect, or it fell within the standard deviation of data independent of the applied concentration of NPs. However, in the case of nisin-loaded magnetic nanoparticles (CaN without heating), a detectable drop in cell viability was observed. On average, the CaN treatment resulted in a better treatment efficacy compared to nisin-only treatment—the highest difference was close to 1 log reduction of cell viability.

Finally, the combination of the treatments with low alternating magnetic fields potentiated the inactivation efficacy during the 30 and 60 min protocols, while for the 5 min procedure the effect was non-detectable. Similarly to MF-free treatment, the dependence on the NP concentration was not profound in most of cases, indicating a saturated treatment efficacy. It was concluded that application of low AMFs (10 mT, 100 kHz) with the proposed nanoparticles has no practical application in a food processing context due to weak inactivation efficacy. The difference between thermal stress response (Figure 6A) and the best low AMF protocol (Figure 7B) is only 1–1.5 log of cell viability reduction.

Similar analysis was performed for the high AMF (125 mT, 200 kHz) treatment. The results are summarized in Figure 8.

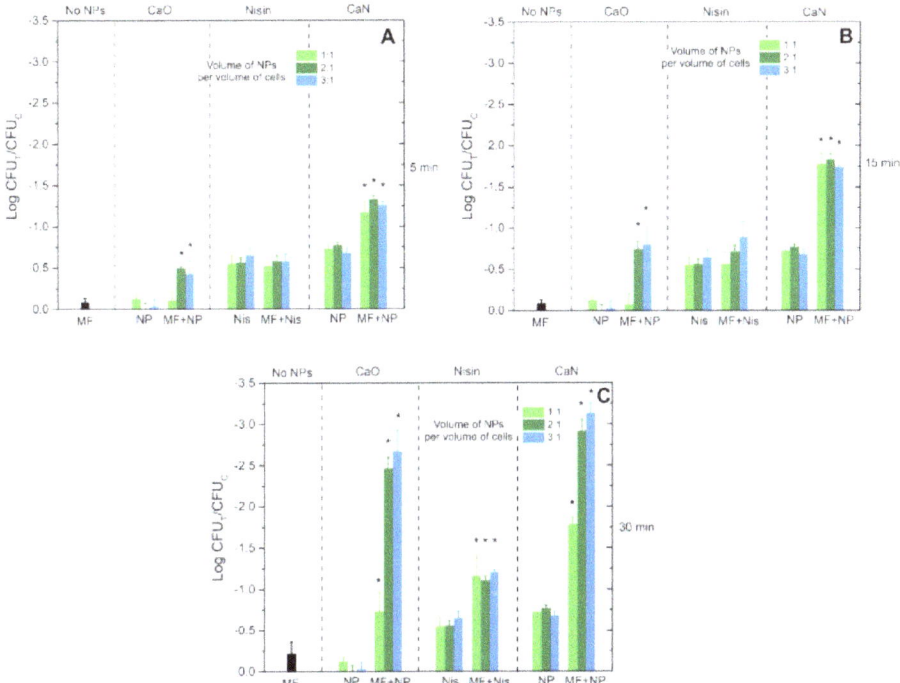

Figure 8. The viability response of *L. innocua* to high alternating magnetic field (125 mT, 200 kHz) and activated nisin-loaded (CaN), nisin-free (CaO) magnetic nanoparticles or nisin-only solution (Nisin). (**A**) After 5 min exposure; (**B**) 15 min exposure and (**C**) 30 min exposure. In all cases the treatment (CFU_T) is normalized to untreated control samples (CFU_C). Asterisk (*) corresponds to statistically significant difference ($p < 0.05$) versus NP-only or nisin-only treatment.

Similarly to the low AMF treatment, the MF-only (125 mT, 200 kHz) procedure resulted in a minor decrease of the cell viability (independent of the exposure time). Nevertheless, in the case of high AMFs, the 5 min protocol (Figure 8A) triggered a statistically significant difference in the inactivation rate when CaO nisin-free NPs were used, contrary to the 10 mT treatment.

An exposure time-dependent response was acquired with the high AMF treatment, i.e., the inactivation efficacy gradually increased, and the highest inactivation of 3 log reduction was observed (Figure 8C), which is a significant result. If compared to the response mediated by temperature (Figure 6B), additional 2 log reduction was associated with the AMF component.

The SEM analysis of the applied protocols was further introduced in the study. The representative exemplary images are shown in Figure 9.

It can be seen that untreated bacteria and cells after MF-only treatment morphologically look the same. However, when CaO is applied with MF treatment, differences can be seen (i.e., Figure 9F, CaO after 125 mT treatment)—the rigid structure of the cell is ruined, and the integrity of the membrane is questionable. A similar result is observed when CaN-NP-only treatment is used (Figure 9A,B, CaN). Lastly, when the nisin magnetic NPs are combined with AMF treatment, the cellular damage is irreversible, and the observed effects

(Figure 9F, CaN) are in good agreement with the viability data (Figure 8C). To summarize, the SEM images adequately represent the viability data overviewed in Figures 7 and 8.

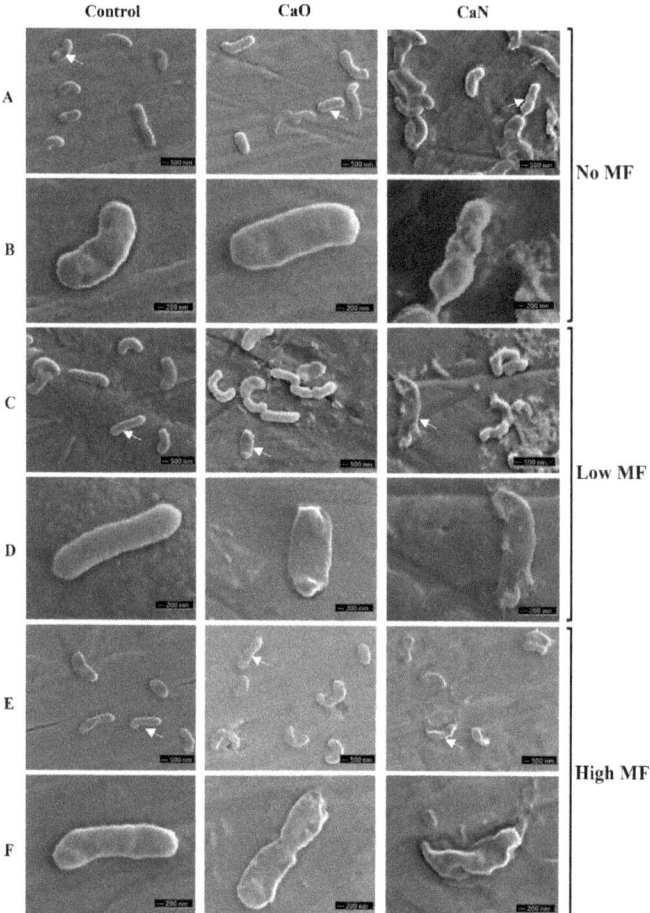

Figure 9. Representative scanning electron microscopy images of various protocols with and without nanoparticles. (**A**) Untreated and samples with CaO and CaN only; (**C**) 10 mT, 100 kHz/60 min treated and samples after CaO and CaN + 10 mT AMF procedure; (**E**) 125 mT, 200 kHz/30 min treated and samples after CaO and CaN + 125 mT AMF procedure; magnified images of bacteria highlighted with white arrows are shown in (**B,D,F**).

4. Discussion

It was shown that the application of alternating magnetic fields with functionalized magnetic nanoparticles can effectively manipulate the initial resistance of bacteria to nisin. Due to the pathogenicity of the environmentally persistent *L. monocytogenes*, it is challenging to safely conduct large-scale experiments using this bacteria. However, nonpathogenic *L. innocua*, widespread in the environment and in food bacteria, have similar genotypic and phenotypic characteristics to *L. monocytogenes* [51]; thus, this species was employed as a surrogate bacteria giving a safety margin to protect researchers and preventing exposure to the pathogens.

We showed that 10 and 125 mT AMFs (without NPs) have no-to-weak effect on the viability of the *L. Innocua*. When combined with magnetic nanoparticles the inactivation efficacy improved significantly. The observed effects could be partly attributed to the thermal stress and mild thermal treatment [52,53], which is induced due to the relaxation losses in magnetic nanoparticles. The capability to potentiate the effects of nisin nanoparticles by mild thermal treatment was reported previously [37]. Nevertheless, the effects of thermal stress are limited to 1 log reduction of bacterial cell viability, which is insignificant in a food processing context. Application of 125 mT AMFs allowed up to 3 log reduction of cell viability to be reached.

Alternating magnetic fields of similar parameters (100–200 mT, kHz range) were reported to trigger inactivation of *E. coli* when longer exposure times (6–16 h) were used [54]. Such exposure rates limit the economic benefits and applicability of the methodology in industry. However, in our case the treatment was limited to 30 min, featuring a contactless and predominantly non-thermal methodology. Nevertheless, there are some reports on *Saccharomyces cerevisiae* yeast treated by a 5 mT magnetic field for 30 min that report inactivation in the 30% to 70% range [55], while others report growth stimulation [56]. It was also shown that magnetic fields in the mT range can affect enzyme activity and lipid peroxidation [57]. Our data indicate that there is a tendency of viability reduction after magnetic field treatment without nanoparticles. However, descriptive conclusions on the biophysics of the phenomena cannot be formed due to weak and mainly statistically insignificant effects.

When combined with functionalized NPs, the 10 mT treatment returned insignificant results (industry-wise) independent of exposure time and nanoparticles used since the induced inactivation was too low. Surprisingly, the 30 min treatment resulted in a better treatment efficacy compared to the 60 min treatment. We believe that the result could be attributed to an increased bacteria growth in higher temperature (37 °C). Such a phenomenon was reported previously [58,59].

The major improvements of the AMF methodology in our work lie in the potentiation of bioactive effects of functionalized and nisin-free ferromagnetic nanoparticles in the 125 mT magnetic field. Of course, partly, the inactivation efficacy was affected by mild temperature increase according to magnetic hyperthermia methodology [60]; however, it was not the dominant stressor. The non-thermal effect of NPs could be attributed to local field amplification due to conductive nanoparticles in close proximity with the cell membrane [43]. Combined with relaxation losses and resulting mild thermal treatment, the fluidity of the cell membrane could also have been altered [61]. Nevertheless, the proposed methodology offers precise control of the treatment parameters. In terms of temperature and considering the physics behind this phenomena, a further increase of the AMF frequency is required to produce more hysteresis losses [62] and thus more effective local heating and nisin activity improvement. Differently from microwave heating processing steps where the heat is generated within the food itself [63], functionalized magnetic nanoparticles enable activation of the nanoparticle itself in close proximity to target microorganisms, followed by heat diffusion in the whole volume. This may potentially allow mild thermal treatments (<45 °C) of the food with nisin exposed to higher temperature (due to magnetic NP functionalization) where it can act more rapidly in the optimal bactericidal mode [64]. Further multiparametric analysis is required including optimization of treatment time, NP concentration and increase of magnetic field frequency. Considering the effects of AMFs, further increase of the pulse amplitude is required, which according to our data may further improve the inactivation of microorganisms. A shorter period of pulses may also positively affect the treatment outcome due to the dB/dt-dependent induced electric field component [65].

5. Conclusions

To conclude, functionalized magnetic nanoparticles have high actuality in drug delivery systems [66], the cancer treatment context [67] or in the detection of bacterial pathogens [68]. Superparamagnetic nanoparticles do not have remnant magnetization in the absence of the external magnetic field and are considered to have no toxicity [69].

Moreover, iron oxide is already approved by FDA for medical and food applications, making nanoparticles good candidates for use in food pathogen biocontrol [70]. Our data indicate that there is a niche in mild hyperthermia-assisted food processing steps that can be utilized to overcome resistance to food-compatible bacteriocins and potentiate the inactivation effects of AMFs several-fold. The methodology is a competitive contactless alternative to currently established PEF-mediated treatments.

Author Contributions: Conceptualization, V.N. and E.S.; methodology, V.N., J.S., R.G., N.V., J.N., E.S.; validation, R.S., V.N., E.S., J.L.; investigation, R.S., V.N., E.S., K.B., J.L., K.M.; resources, V.N., J.N., N.V., J.S., E.S.; writing—original draft preparation, V.N., E.S., J.S. supervision, V.N., E.S., J.S. All authors have read and agreed to the published version of the manuscript.

Funding: This research received no external funding.

Data Availability Statement: Data available from the corresponding author V.N. on request.

Acknowledgments: We thank Deimantė Prušinskaitė for technical assistance.

Conflicts of Interest: The authors declare no conflict of interest.

References

1. Wilson, D.; Materón, E.M.; Ibáñez-Redín, G.; Faria, R.C.; Correa, D.S.; Oliveira, O.N. Electrical detection of pathogenic bacteria in food samples using information visualization methods with a sensor based on magnetic nanoparticles functionalized with antimicrobial peptides. *Talanta* **2019**, *194*, 611–618. [CrossRef] [PubMed]
2. Rajkovic, A.; Jovanovic, J.; Monteiro, S.; Decleer, M.; Andjelkovic, M.; Foubert, A.; Beloglazova, N.; Tsilla, V.; Sas, B.; Madder, A.; et al. Detection of toxins involved in foodborne diseases caused by Gram-positive bacteria. *Compr. Rev. Food Sci. Food Saf.* **2020**, *19*, 1605–1657. [CrossRef] [PubMed]
3. Martín-Vertedor, D.; Rodrigues, N.; Ítala, M.G.; Veloso, A.C.A.; Peres, A.M.; Pereira, J.A. Impact of thermal sterilization on the physicochemical-sensory characteristics of Californian-style black olives and its assessment using an electronic tongue. *Food Control.* **2020**, *117*, 107369. [CrossRef]
4. Auksornsri, T.; Bornhorst, E.R.; Tang, J.; Tang, Z.; Songsermpong, S. Developing model food systems with rice based products for microwave assisted thermal sterilization. *LWT* **2018**, *96*, 551–559. [CrossRef]
5. Zhu, N.; Wang, Y.-l.; Zhu, Y.; Yang, L.; Yu, N.; Wei, Y.; Zhang, H.; Sun, A.-D. Design of a treatment chamber for low-voltage pulsed electric field sterilization. *Innov. Food Sci. Emerg. Technol.* **2017**, *42*, 180–189. [CrossRef]
6. Mendes-Oliveira, G.; Jensen, J.L.; Keener, K.M.; Campanella, O.H. Modeling the inactivation of Bacillus subtilis spores during cold plasma sterilization. *Innov. Food Sci. Emerg. Technol.* **2019**, *52*, 334–342. [CrossRef]
7. Zhang, J.; Li, S.; Wang, W.; Pei, J.; Zhang, J.; Yue, T.; Youravong, W.; Li, Z. Bacteriocin assisted food functional membrane for simultaneous exclusion and inactivation of Alicyclobacillus acidoterrestris in apple juice. *J. Membr. Sci.* **2021**, *618*, 118741. [CrossRef]
8. Vijayakumar, P.; Muriana, P. Inhibition of Listeria monocytogenes on Ready-to-Eat Meats Using Bacteriocin Mixtures Based on Mode-of-Action. *Foods* **2017**, *6*, 22. [CrossRef]
9. Xu, J.; Zhang, M.; Bhandari, B.; Cao, P. Microorganism control and product quality improvement of Twice-cooked pork dish using ZnO nanoparticles combined radio frequency pasteurization. *LWT* **2018**, *95*, 65–71. [CrossRef]
10. Xu, J.; Zhang, M.; Cao, P.; Adhikari, B. Effect of ZnO nanoparticles combined radio frequency pasteurization on the protein structure and water state of chicken thigh meat. *LWT* **2020**, *134*, 110168. [CrossRef]
11. Xu, J.; Zhang, M.; Bhandari, B.; Kachele, R. ZnO nanoparticles combined radio frequency heating: A novel method to control microorganism and improve product quality of prepared carrots. *Innov. Food Sci. Emerg. Technol.* **2017**, *44*, 46–53. [CrossRef]
12. Rokbani, H.; Daigle, F.; Ajji, A. Combined Effect of Ultrasound Stimulations and Autoclaving on the Enhancement of Antibacterial Activity of ZnO and SiO_2/ZnO Nanoparticles. *Nanomaterials* **2018**, *8*, 129. [CrossRef] [PubMed]
13. Rai, M.; Paralikar, P.; Jogee, P.; Agarkar, G.; Ingle, A.P.; Derita, M.; Zacchino, S. Synergistic antimicrobial potential of essential oils in combination with nanoparticles: Emerging trends and future perspectives. *Int. J. Pharm.* **2017**, *519*, 67–78. [CrossRef] [PubMed]
14. Zhi, S.; Stothard, P.; Banting, G.; Scott, C.; Huntley, K.; Ryu, K.; Otto, S.; Ashbolt, N.; Checkley, S.; Dong, T.; et al. Characterization of water treatment-resistant and multidrug-resistant urinary pathogenic Escherichia coli in treated wastewater. *Water Res.* **2020**, *182*, 115827. [CrossRef]
15. Murphy, C.P.; Carson, C.; Smith, B.A.; Chapman, B.; Marrotte, J.; McCann, M.; Primeau, C.; Sharma, P.; Parmley, E.J. Factors potentially linked with the occurrence of antimicrobial resistance in selected bacteria from cattle, chickens and pigs: A scoping review of publications for use in modelling of antimicrobial resistance (IAM.AMR Project). *Zoonoses Public Health* **2018**, *65*, 957–971. [CrossRef]
16. Phillips, C.A. Bacterial biofilms in food processing environments: A review of recent developments in chemical and biological control. *Int. J. Food Sci. Technol.* **2016**, *51*, 1731–1743. [CrossRef]
17. Ben Lagha, A.; Haas, B.; Gottschalk, M.; Grenier, D. Antimicrobial potential of bacteriocins in poultry and swine production. *Vet. Res.* **2017**, *48*, 1–12. [CrossRef]

18. Schofs, L.; Sparo, M.D.; Bruni, S.F.S. Gram-positive bacteriocins: Usage as antimicrobial agents in veterinary medicine. *Vet. Res. Commun.* **2020**, 1–12. [CrossRef]
19. Field, D.; Blake, T.; Mathur, H.; O'Connor, P.M.; Cotter, P.D.; Paul Ross, R.; Hill, C. Bioengineering nisin to overcome the nisin resistance protein. *Mol. Microbiol.* **2019**, *111*, 717–731.
20. Bahrami, A.; Delshadi, R.; Jafari, S.M.; Williams, L. Nanoencapsulated nisin: An engineered natural antimicrobial system for the food industry. *Trends Food Sci. Technol.* **2019**, *94*, 20–31. [CrossRef]
21. Tan, Z.; Luo, J.; Liu, F.; Zhang, Q.; Jia, S. Effects of pH, Temperature, Storage Time, and Protective Agents on Nisin Antibacterial Stability. In *Advances in Applied Biotechnology*; Springer Nature: Berlin/Heidelberg, Germany, 2015; pp. 305–312.
22. Peng, X.; Zhu, L.; Wang, Z.; Zhan, X.-B. Enhanced stability of the bactericidal activity of nisin through conjugation with gellan gum. *Int. J. Biol. Macromol.* **2020**, *148*, 525–532. [CrossRef] [PubMed]
23. Guiotto, A.; Pozzobon, M.; Canevari, M.; Manganelli, R.; Scarin, M.; Veronese, F.M. PEGylation of the antimicrobial peptide nisin A: Problems and perspectives. *Il Farmaco* **2003**, *58*, 45–50. [CrossRef]
24. Guiga, W.; Swesi, Y.; Galland, S.; Peyrol, E.; Degraeve, P.; Sebti, I. Innovative multilayer antimicrobial films made with Nisaplin®or nisin and cellulosic ethers: Physico-chemical characterization, bioactivity and nisin desorption kinetics. *Innov. Food Sci. Emerg. Technol.* **2010**, *11*, 352–360. [CrossRef]
25. Crandall, A.D.; Montville, T.J. Nisin Resistance in Listeria monocytogenes ATCC 700302 Is a Complex Phenotype. *Appl. Environ. Microbiol.* **1998**, *64*, 231–237. [CrossRef] [PubMed]
26. Zhou, H.; Fang, J.; Tian, Y.; Lu, X.Y. Mechanisms of nisin resistance in Gram-positive bacteria. *Ann. Microbiol.* **2014**, *64*, 413–420. [CrossRef]
27. Zhou, L.; Van Heel, A.J.; Montalban-Lopez, M.; Kuipers, O.P. Potentiating the Activity of Nisin against Escherichia coli. *Front. Cell Dev. Biol.* **2016**, *4*, 7. [CrossRef]
28. Huang, H.W.; Hsu, C.P.; Wang, C.Y. Healthy expectations of high hydrostatic pressure treatment in food processing industry. *J. Food Drug Anal.* **2020**, *28*, 1–13. [CrossRef]
29. Gavahian, M.; Khaneghah, A.M. Cold plasma as a tool for the elimination of food contaminants: Recent advances and future trends. *Crit. Rev. Food Sci. Nutr.* **2019**, *60*, 1581–1592. [CrossRef]
30. Hinds, L.M.; O'Donnell, C.P.; Akhter, M.; Tiwari, B.K. Principles and mechanisms of ultraviolet light emitting diode technology for food industry applications. *Innov. Food Sci. Emerg. Technol.* **2019**, *56*, 102153. [CrossRef]
31. Bhargava, N.; Mor, R.S.; Kumar, K.; Sharanagat, V.S. Advances in application of ultrasound in food processing: A review. *Ultrason. Sonochem.* **2021**, *70*, 105293. [CrossRef]
32. Gómez, B.; Munekata, P.E.S.; Gavahian, M.; Barba, F.J.; Martí-Quijal, F.J.; Bolumar, T.; Campagnol, P.C.B.; Tomašević, I.; Lorenzo, J.M. Application of pulsed electric fields in meat and fish processing industries: An overview. *Food Res. Int.* **2019**, *123*, 95–105. [CrossRef] [PubMed]
33. Berdejo, D.; Pagán, E.; García-Gonzalo, D.; Pagán, R. Exploiting the synergism among physical and chemical processes for improving food safety. *Curr. Opin. Food Sci.* **2019**, *30*, 14–20. [CrossRef]
34. Wiedemann, I.; Benz, R.; Sahl, H.-G. Lipid II-Mediated Pore Formation by the Peptide Antibiotic Nisin: A Black Lipid Membrane Study. *J. Bacteriol.* **2004**, *186*, 3259–3261. [CrossRef] [PubMed]
35. Novickij, V.; Stanevičienė, R.; Grainys, A.; Lukša, J.; Badokas, K.; Krivorotova, T.; Sereikaitė, J.; Novickij, J.; Servienė, E. Electroporation-assisted inactivation of Escherichia coli using nisin-loaded pectin nanoparticles. *Innov. Food Sci. Emerg. Technol.* **2016**, *38*, 98–104. [CrossRef]
36. Novickij, V.; Zinkevičienė, A.; Stanevičienė, R.; Gruškienė, R.; Servienė, E.; Vepštaitė-Monstavičė, I.; Krivorotova, T.; Lastauskienė, E.; Sereikaitė, J.; Girkontaitė, I.; et al. Inactivation of Escherichia coli Using Nanosecond Electric Fields and Nisin Nanoparticles: A Kinetics Study. *Front. Microbiol.* **2018**, *9*, 3006. [CrossRef] [PubMed]
37. Novickij, V.; Stanevičienė, R.; Staigvila, G.; Gruškienė, R.; Sereikaitė, J.; Girkontaitė, I.; Novickij, J.; Servienė, E. Effects of pulsed electric fields and mild thermal treatment on antimicrobial efficacy of nisin-loaded pectin nanoparticles for food preservation. *LWT* **2020**, *120*, 108915. [CrossRef]
38. Maglietti, F.H.; Michinski, S.D.; Olaiz, N.M.; Castro, M.A.; Suárez, C.A.; Marshall, G.R. The Role of Ph Fronts in Tissue Electroporation Based Treatments. *PLoS ONE* **2013**, *8*, e80167. [CrossRef]
39. Guenther, E.; Klein, N.; Mikus, P.; Stehling, M.K.; Rubinsky, B. Electrical breakdown in tissue electroporation. *Biochem. Biophys. Res. Commun.* **2015**, *467*, 736–741. [CrossRef]
40. Schottroff, F.; Johnson, K.; Johnson, N.B.; Bédard, M.F.; Jaeger, H. Challenges and limitations for the decontamination of high solids protein solutions at neutral pH using pulsed electric fields. *J. Food Eng.* **2020**, *268*, 109737. [CrossRef]
41. Ruzgys, P.; Novickij, V.; Novickij, J.; Šatkauskas, S. Influence of the electrode material on ROS generation and electroporation efficiency in low and high frequency nanosecond pulse range. *Bioelectrochemistry* **2019**, *127*, 87–93. [CrossRef]
42. Rodaite-Riseviciene, R.; Saule, R.; Snitka, V.; Saulis, G. Release of Iron Ions From the Stainless Steel Anode Occurring During High-Voltage Pulses and Its Consequences for Cell Electroporation Technology. *IEEE Trans. Plasma Sci.* **2014**, *42*, 249–254. [CrossRef]
43. Miklavcic, D.; Novickij, V.; Kranjc, M.; Polajzer, T.; Meglic, S.H.; Napotnik, T.B.; Romih, R.; Lisjak, D. Contactless electroporation induced by high intensity pulsed electromagnetic fields via distributed nanoelectrodes. *Bioelectrochemistry* **2020**, *132*, 107440. [CrossRef]
44. Inbaraj, B.S.; Chen, B.-H. Nanomaterial-based sensors for detection of foodborne bacterial pathogens and toxins as well as pork adulteration in meat products. *J. Food Drug Anal.* **2016**, *24*, 15–28. [CrossRef] [PubMed]

45. Roger, J.d.A.; Magro, M.; Spagnolo, S.; Bonaiuto, E.; Baratella, D.; Fasolato, L.; Vianello, F. Antimicrobial and magnetically removable tannic acid nanocarrier: A processing aid for Listeria monocytogenes treatment for food industry applications. *Food Chem.* **2018**, *267*, 430–436. [CrossRef] [PubMed]
46. Yusoff, A.H.M.; Salimi, M.N. Superparamagnetic Nanoparticles for Drug Delivery. In *Applications of Nanocomposite Materials in Drug Delivery*; Elsevier BV: Amsterdam, The Netherlands, 2018; pp. 843–859. ISBN 9780128137581.
47. Gruskiene, R.; Krivorotova, T.; Stanevičienė, R.; Ratautas, D.; Serviene, E.; Sereikaite, J. Preparation and characterization of iron oxide magnetic nanoparticles functionalized by nisin. *Coll. Surf. B Biointerfaces* **2018**, *169*, 126–134. [CrossRef]
48. Novickij, V.; Stanevičiene, R.; Vepštaite-Monstavičе, I.; Gruškiene, R.; Krivorotova, T.; Sereikaite, J.; Novickij, J.; Serviene, E. Overcoming antimicrobial resistance in bacteria using bioactive magnetic nanoparticles and pulsed electromagnetic fields. *Front. Microbiol.* **2018**, *8*. [CrossRef]
49. Oh, S.J.; Cook, D.C.; Townsend, H.E. Characterization of Iron Oxides Commonly Formed as Corrosion Products on Steel. *Hyperfine Interact.* **1998**, *112*, 59–66. [CrossRef]
50. *Mössbauer Spectroscopy*; Cambridge University Press: Cambridge, UK, 1986; ISBN 9780521261012.
51. Mohan, V.; Wibisono, R.; De Hoop, L.; Summers, G.; Fletcher, G.C. Identifying Suitable Listeria innocua Strains as Surrogates for Listeria monocytogenes for Horticultural Products. *Front. Microbiol.* **2019**, *10*, 2281. [CrossRef]
52. Wang, P.; Zou, M.; Liu, K.; Gu, Z.; Yang, R. Effect of mild thermal treatment on the polymerization behavior, conformation and viscoelasticity of wheat gliadin. *Food Chem.* **2018**, *239*, 984–992. [CrossRef]
53. Zhou, L.; Liao, T.; Liu, J.; Zou, L.; Liu, C.; Liu, W. Unfolding and Inhibition of Polyphenoloxidase Induced by Acidic pH and Mild Thermal Treatment. *Food Bioprocess Technol.* **2019**, *12*, 1907–1916. [CrossRef]
54. Li, M.; Qu, J.-H.; Peng, Y.-Z. Sterilization of Escherichia coli cells by the application of pulsed magnetic field. *J. Environ. Sci.* **2004**, *16*, 348–352.
55. Bayraktar, V.N. Magnetic Field Effect on Yeast Saccharomyces Cerevisiae Activity at Grape Must Fermentation. *Biotechnol. Acta* **2013**, *6*, 125–137. [CrossRef]
56. Santos, L.O.; Alegre, R.M.; Garcia-Diego, C.; Cuellar, J. Effects of magnetic fields on biomass and glutathione production by the yeast Saccharomyces cerevisiae. *Process. Biochem.* **2010**, *45*, 1362–1367. [CrossRef]
57. Sahebjamei, H.; Abdolmaleki, P.; Ghanati, F. Effects of magnetic field on the antioxidant enzyme activities of suspension-cultured tobacco cells. *Bioelectromagnetics* **2006**, *28*, 42–47. [CrossRef] [PubMed]
58. Rowan, N.J.; Anderson, J.G. Effects of Above-Optimum Growth Temperature and Cell Morphology on Thermotolerance of Listeria monocytogenesCells Suspended in Bovine Milk. *Appl. Environ. Microbiol.* **1998**, *64*, 2065–2071. [CrossRef]
59. Li, Y.; Brackett, R.; Chen, J.; Beuchat, L.R. Mild heat treatment of lettuce enhances growth of Listeria monocytogenes during subsequent storage at 5 °C or 15 °C. *J. Appl. Microbiol.* **2002**, *92*, 269–275. [CrossRef]
60. Golneshan, A.A.; Lahonian, M. The effect of magnetic nanoparticle dispersion on temperature distribution in a spherical tissue in magnetic fluid hyperthermia using the lattice Boltzmann method. *Int. J. Hyperth.* **2011**, *27*, 266–274. [CrossRef]
61. Yoon, Y.; Lee, H.; Lee, S.; Kim, S.; Choi, K.-H. Membrane fluidity-related adaptive response mechanisms of foodborne bacterial pathogens under environmental stresses. *Food Res. Int.* **2015**, *72*, 25–36. [CrossRef]
62. Shaterabadi, Z.; Nabiyouni, G.; Soleymani, M. Physics responsible for heating efficiency and self-controlled temperature rise of magnetic nanoparticles in magnetic hyperthermia therapy. *Prog. Biophys. Mol. Biol.* **2018**, *133*, 9–19. [CrossRef]
63. Puligundla, P.; Abdullah, S.A.; Choi, W.; Jun, S.; Oh, S.-E.; Ko, S. Potentials of Microwave Heating Technology for Select Food Processing Applications—A Brief Overview and Update. *J. Food Process. Technol.* **2013**, *4*, 1–9. [CrossRef]
64. Prudêncio, C.V.; Mantovani, H.C.; Cecon, P.R.; Vanetti, M.C.D. Differences in the antibacterial activity of nisin and bovicin HC5 against Salmonella Typhimurium under different temperature and pH conditions. *J. Appl. Microbiol.* **2015**, *118*, 18–26. [CrossRef]
65. Umeki, S.; Watanabe, T.; Shimabukuro, H.; Kato, T.; Yoshikawa, N.; Taniguchi, S.; Tohji, K. Change in Zeta Potential by Alternating Electromagnetic Treatment as Scale Prevention Process. *Trans. Mater. Res. Soc. Jpn.* **2007**, *32*, 611–614. [CrossRef]
66. Wahajuddin, S.A. Superparamagnetic iron oxide nanoparticles: Magnetic nanoplatforms as drug carriers. *Int. J. Nanomed.* **2012**, *7*, 3445–3471.
67. Jurgons, R.; Seliger, C.; Hilpert, A.; Trahms, L.; Odenbach, S.; Alexiou, C. Drug loaded magnetic nanoparticles for cancer therapy. *J. Phys. Condens. Matter* **2006**, *18*, S2893–S2902. [CrossRef]
68. Hwang, J.; Kwon, D.; Lee, S.; Jeon, S. Detection of Salmonella bacteria in milk using gold-coated magnetic nanoparticle clusters and lateral flow filters. *RSC Adv.* **2016**, *6*, 48445–48448. [CrossRef]
69. Cao, M.; Li, Z.; Wang, J.; Ge, W.; Yue, T.; Li, R.; Colvin, V.L.; Yu, W.W. Food related applications of magnetic iron oxide nanoparticles: Enzyme immobilization, protein purification, and food analysis. *Trends Food Sci. Technol.* **2012**, *27*, 47–56. [CrossRef]
70. Al-Shabib, N.A.; Husain, F.M.; Ahmed, F.; Khan, R.A.; Khan, M.S.; Ansari, F.A.; Alam, M.Z.; Ahmed, M.A.; Khan, M.S.; Baig, M.H.; et al. Low Temperature Synthesis of Superparamagnetic Iron Oxide (Fe_3O_4) Nanoparticles and Their ROS Mediated Inhibition of Biofilm Formed by Food-Associated Bacteria. *Front. Microbiol.* **2018**, *9*, 2567. [CrossRef]

Article

Deviation of Trypsin Activity Using Peptide Conformational Imprints

Kiran Reddy Kanubaddi [1,†], Pei-Yu Huang [2,†], Ya-Lin Chang [2], Cheng Hsin Wu [2], Wei Li [2], Ranjith Kumar Kankala [1,3], Dar-Fu Tai [2,*] and Chia-Hung Lee [1,*]

1. Department of Life Science, National Dong Hwa University, Hualien 97401, Taiwan; 810513107@gms.ndhu.edu.tw (K.R.K.); ranjithkankala@hqu.edu.cn (R.K.K.)
2. Department of Chemistry, National Dong Hwa University, Hualien 97401, Taiwan; 610012029@gms.ndhu.edu.tw (P.-Y.H.); m9812003@ems.ndhu.edu.tw (Y.-L.C.); m9812022@gms.ndhu.edu.tw (C.H.W.); ga025361@yahoo.com.tw (W.L.)
3. College of Chemical Engineering, Huaqiao University, Xiamen 361021, China
* Correspondence: dftai@gms.ndhu.edu.tw (D.-F.T.); chlee016@mail.ndhu.edu.tw (C.-H.L.)
† These authors contributed equally to this work.

Abstract: In this study, a methodology utilizing peptide conformational imprints (PCIs) as a tool to specifically immobilize porcine pancreatic alpha-trypsin (PPT) at a targeted position is demonstrated. Owing to the fabrication of segment-mediated PCIs on the magnetic particles (PCIMPs), elegant cavities complementary to the PPT structure are constructed. Based on the sequence on targeted PPT, the individual region of the enzyme is trapped with different template-derived PCIMPs to show certain types of inhibition. Upon hydrolysis, N-benzoyl-L-arginine ethyl ester (BAEE) is employed to assess the hydrolytic activity of PCIMPs bound to the trypsin using high-performance liquid chromatography (HPLC) analysis. Further, the kinetic data of four different PCIMPs are compared. As a result, the PCIMPs presented non-competitive inhibition toward trypsin, according to the Lineweaver–Burk plot. Further, the kinetic analysis confirmed that the best parameters of PPT/PCIMPs $^{233-245+G}$ were $V_{max} = 1.47 \times 10^{-3}$ mM s^{-1}, $K_m = 0.42$ mM, $k_{cat} = 1.16$ s^{-1}, and $k_{cat}/K_m = 2.79$ mM^{-1} s^{-1}. As PPT is bound tightly to the correct position, its catalytic activities could be sustained. Additionally, our findings stated that the immobilized PPT could maintain stable activity even after four successive cycles.

Keywords: porcine pancreatic trypsin; molecularly-imprinted polymers; magnetic particles; conformational imprint; secondary structure

Citation: Kanubaddi, K.R.; Huang, P.-Y.; Chang, Y.-L.; Wu, C.H.; Li, W.; Kankala, R.K.; Tai, D.-F.; Lee, C.-H. Deviation of Trypsin Activity Using Peptide Conformational Imprints. *Nanomaterials* **2021**, *11*, 334. https://doi.org/10.3390/nano11020334

Academic Editor: Paolo Arosio
Received: 3 December 2020
Accepted: 22 January 2021
Published: 27 January 2021

Publisher's Note: MDPI stays neutral with regard to jurisdictional claims in published maps and institutional affiliations.

Copyright: © 2021 by the authors. Licensee MDPI, Basel, Switzerland. This article is an open access article distributed under the terms and conditions of the Creative Commons Attribution (CC BY) license (https://creativecommons.org/licenses/by/4.0/).

1. Introduction

Porcine pancreatic alpha-trypsin (PPT), a proteolytic enzyme, is a pancreatic serine protease (EC 3.4.21.4) with specificity for arginine or lysine substrate towards catalytic hydrolysis on esters and amides, under mild reaction conditions [1–3]. Owing to these facts, this proteolytic enzyme is often utilized in industrial and biomedical applications [1,4,5]. However, such enzymes are often immobilized into various substrates to improve stability and reusability without affecting their activity [2,6,7]. In this vein, various enzyme immobilization methods have been reported, such as covalent linkage, non-covalent adsorption, and encapsulation systems, among others [8–12]. Nevertheless, the quest for optimum performance is still on due to their conformational changes during immobilization [11,13]. Although an enzyme possesses a uniform structure, it often changes the conformation continuously. Consequently, the immobilized biocatalyst is organized randomly during/after immobilization, resulting in different constitutions and a less dynamic form.

In recent times, the utilization of molecularly imprinted polymers (MIPs) has become an emerging carrier-bound system for the immobilization of biomolecules, as they offer reversible orientation [14]. In this context, the incorporation of MIPs with magnetic particles

(MPs) was first demonstrated by Ansell and Mosbach [15]. Since then, several efforts have been dedicated to the utilization of MIPs for diverse applications. In a case, Tong and colleagues applied MPs to recognize Ribonuclease A [16]. In another case, Jing and coworkers used lysozyme as a template to adsorb blood specimens with a detection limit at 5 ng/mL [17]. Previously, trypsin was utilized as the template to generate MIPs for different assays and inhibition studies [18]. Nonetheless, the employment of such a template for MPs was restricted to the usage of the whole protein.

Considering these facts, herein, we demonstrate an elegant method to immobilize PPT. As segment-mediated MIPs were fabricated on MPs, cavities complementary to PPT structure were constructed on those nanomaterials. As the imprinting of a random coil peptide successfully generates the desired nano-cavity for the corresponding peptide/protein [19,20], recently, a helical peptide was utilized as a template to generate helical cavities with high affinity for the target protein, having achieved satisfactory results [21]. Accordingly, in this work, several PPT peptide segments were selected to generate helix cavities, using peptide conformational imprints decorated over the magnetic particles (PCIMPs). This study is divided into four main steps: (i) Fabricating peptide conformational imprint (PCIs) on MPs carriers; (ii) adsorbing trypsin to PCIMPs; (iii) analyzing the binding kinetics of immobilized PPT; and (iv) evaluating the reusability of immobilized PPT. The immobilization of PPT based on the PCIs was carried out as illustrated in Figure 1.

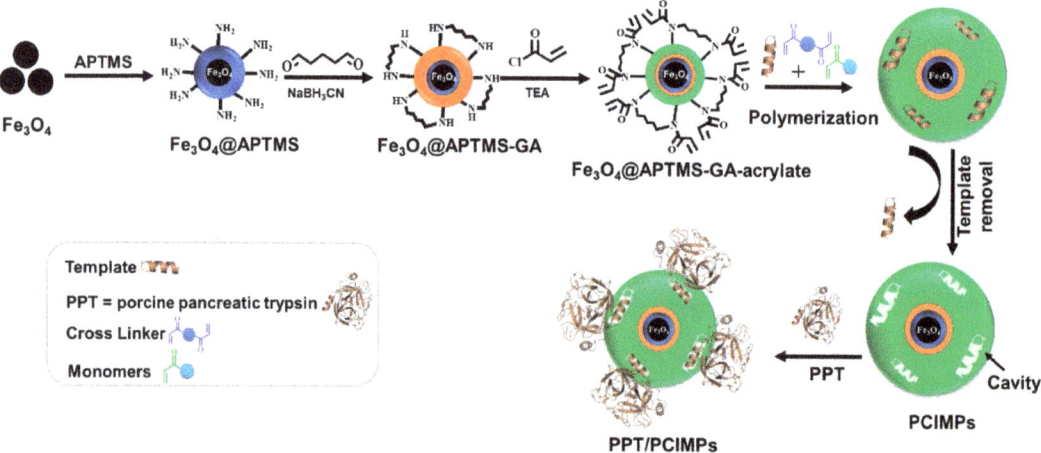

Figure 1. Scheme illustrating the fabrication of peptide conformational imprints (PCIs) on magnetic particles (MPs) and their binding to porcine pancreatic alpha-trypsin (PPT).

2. Materials and Methods

2.1. Reagents and Chemicals

3-(Aminopropyl) trimethoxysilane (APTMS) and ammonium acetate (NH$_4$Ac) were obtained from Acros Ltd. (Fair Lawn, New Jersey, United States). Iron (III) chloride hexahydrate (FeCl$_3$·6H$_2$O) and triethylamine (TEA) were purchased from Merck Ltd. (Darmstadt, Germany). N,N'-Ethylene bisacrylamide (EBAA) and N-benzyl acrylamide (BAA) were acquired from Lancaster (Lancashire, UK). Glutaraldehyde (GA) was obtained from Ferax (Berlin, Germany). All Fmoc-amino acids were purchased from BAChem (Bubendorf, Switzerland). Acrylamide (AM), acetic acid, N-benzoyl-L-arginine ethyl ester (BAEE), sodium cyanoborohydride (NaBH$_3$CN), N,N,N',N'-tetramethyl ethylene diamine (TEMED), tris (hydroxymethyl) amino-methane, porcine pancreatic trypsin (PPT), Tween®20, and urea were acquired from Sigma Co. Ltd. (St. Louis, MO, USA). Acetone, acetonitrile (ACN), dichloromethane (DCM), N,N-dimethyl formamide (DMF), piperidine, and toluene of High-Performance Liquid Chromatography (HPLC) grade were used. Purified distilled water acquired from a Milli-Q water purification system was used in all the experiments.

2.2. Template Synthesis

The peptide segments, such as PPT[107–116] (KLSSPATLNS), PPT[145–155] (KSSGSSYPSLL), PPT[169–178] (KSSYPGQITG), and PPT[233–245+G] (NYVNWIQQTIAANG), were produced through the Fmoc (fluorenylmethoxycarbonyl) solid-phase peptide synthesis approach using a Discover SPPS Microwave Peptide synthesizer (Kohan Co. Ltd., Taipei, Taiwan) available at the National Dong Hwa University (Hualien, Taiwan) [22].

2.3. Preparation of PCIs on MPs

2.3.1. Construction of Fe_3O_4@APTMS-GA

The synthesis of the Fe_3O_4 precursor, and subsequent immobilization of amine functionality, Fe_3O_4@APTMS, were performed as described previously [23,24]. Further, glutaraldehyde (GA) was coupled with Fe_3O_4@APTMS to construct stable secondary amine nanoparticles. Briefly, 100 mg of Fe_3O_4@APTMS was initially placed in 50 mL of ACN and subjected to ultrasonication for 30 min. Then, 162 µL of GA was added to the mixture. Further, a few drops of acetic acid were added to maintain the weakly acidic state of the reaction mixture, and stirring was performed for 2 h. Subsequently, 200 mg of $NaBH_3CN$ were added, and vigorous stirring was executed for another 2 h to make the reaction mixture weakly alkaline. Finally, the resultant particles were recovered with a strong magnet, washed several times with a solvent mixture of (H_2O:ACN = 1:1), and dried under vacuum.

2.3.2. Synthesis of Fe_3O_4@APTMS-GA-Acrylate

To prepare Fe_3O_4@APTMS-GA-acrylate, 300 mg of GA-modified MPs were initially dispersed in 25 mL of dry DCM and stirred for 15 min after adding TEA (0.48 mL). Then, acryloyl chloride (0.3 mL, 3.75 mmol) was added in a drop-wise manner to the mixture at 0 °C under N_2 purge and stirred for 24 h. Finally, the resultant product was washed with DCM and dried under vacuum.

2.3.3. Preparation of PCIMPs

To prepare PCIMPs, initially, 211.2 mg of N, N'-ethylene bisacrylamide (EBAA), 56.4 mg of benzyl acrylamide (BAA), and 25.2 mg of acrylamide (AA) were dissolved in a solvent mixture containing 16 mL of PBS (pH-7.6, 20 mM) and 2 mL of ethanol. Then, 7.5 µmol of template molecules (PPT[107–116], PPT[145–155], PPT[169–178], and PPT[233–245+G]) were dissolved separately in 20 mL of a solvent mixture of TFE and PBS at a ratio of 7:3 to exhibit the helical structure in the polymerization system. Further, the above two reaction mixtures were mixed after a while, and 90 mg of Fe_3O_4@APTMS-GA-Acrylate was added to make a pre-self-assembly reaction mixture. Then, 240 µL (10%, w/w) of ammonium persulfate and 90 µL (5%, w/v) of TEMED were added to the reaction and stirred for 24 h in the presence of N_2 at RT. The template removal was performed based on previous studies [25,26]. According to the following articles, acetic acid as a solvent disrupts the electrostatic interactions between the template and the polymer matrix, which can be separated. Notably, the template removal process could be achieved in few minutes. Finally, the polymer-MPs were obtained and washed with 25 mM urea (aq) containing 5% acetic acid and 0.5% tween-20 to remove the template. Subsequently, the pore structures formed after the removal of the four different templates were denoted as PCIMPs[107–116], PCIMPs[145–155], PCIMPs[169–178], and PCIMPs[233–245+G], respectively.

2.4. Determination of Binding Affinities of PCIMPs

Notably, the binding experiments were carried out in 10 min to avoid adsorption at non-specific binding sites on PCIMPs. Briefly, 10 mg of PCIMPs was added to PBS (pH-7.6, 20 mM) containing PPT at different concentrations (0.125, 0.25, 0.5, 1, and 1.5 mg/mL) and the resulting mixture was shaken for 10 min. Then, 200 µL of supernatant was collected and measured by Fluorescence Microplate Reader at $\lambda ex/\lambda em = 290$ nm/350 nm. Each experiment was repeated three times, and the results of the binding studies were evaluated using the Scatchard Equation (1) [27–29].

$$[RL]/[L] = (B_{max} - [RL])/K_d \tag{1}$$

where [L] is the concentration of PPT in the solution, [RL] is the concentration of bound PPT, B_{max} denotes the maximum number of binding sites, and K_d is the dissociation constant of the ligand.

2.5. Activity Assay of PPT and Immobilized PPT (PPT/PCIMPs)

The catalytic activity of PPT and PPT/PCIMPs was measured using the HPLC method. N-benzoyl-L-arginine ethyl ester (BAEE) was utilized as the starting material, while the product was N-Benzoyl-L-Arginine (BA), which was observed with time. The percentage of hydrolysis rate was calculated using the following Equation (2):

$$\text{Hydrolysis rate (\%)} = \frac{\text{Product area ratio}}{(\text{Starting area ratio + product area ratio})} \times 100 \tag{2}$$

For determining the catalytic activity of PPT, initially, 1 mL mixtures possessing different BAEE concentrations (0.5, 1.0, and 1.5 mM) were prepared using 50 mM of Tris-HCl buffer, with a pH equal to 7.6. Then, 20 µL of 1mM HCl containing 30 µg PPT was formulated. The assay was started by adding 20 µL of 1 mM HCl/30 µg PPT to 1 mL mixtures with the three BAEE concentrations mentioned above, respectively. For every min, 40 µL of the solution was collected from the reaction mixture and dissolved in 500 µL of ACN: buffer = 15:85, and 99.5 µL of the resultant solution was injected for HPLC detection until the end of the reaction.

2.6. PPT/PCIMPs Activity Assay

Briefly, 10 mg of each PPT/PCIMPs were separately added to 8.8 mL of the BAEE solutions (0.5, 1.0, and 1.5 mM), and for every min, 80 µL of that solution were separated from the mixture and dissolved in 1 mL of ACN: buffer = 15:85. From this, 99.5 µL of the solution was collected for HPLC detection until the end of the reaction. The same procedure was also carried out for the reusability test.

An intelligent, high-performance liquid chromatography (HPLC, model L7100, Hitachi, Tokyo, Japan) set-up equipped with a UV detector (Hitachi model L-2420, Tokyo, Japan), an autosampler (Hitachi L-2200, Tokyo, Japan), and a Vercopak-RP C18 column (Vercotech Corp., Taipei, Taiwan) was used to determine the purity of peptides and for performing the kinetic analysis of the immobilized enzyme. In the hydrolysis test, the mobile phase of HPLC was composed of 0.38 mL of phosphoric acid, 0.47 mL of triethylamine, and 1 L of DI-H_2O. The solution was then adjusted to pH 2.4 with NaOH and HCl. The ultraviolet wavelength was set at 214 nm.

2.7. Determination of Kinetic Constants of PPT and PPT/PCIMPs

The kinetic parameters of PPT and PPT/PCIMPs were evaluated from the Michaelis–Menten plot obtained from the following Equation (3),

$$v = \frac{V_{max}[S]}{(K_m + [S])} \tag{3}$$

where v is the reaction velocity at [S], V_{max} is the maximum rate of the reaction, K_m is the Michaelis half-saturation constant, and [S] is the concentration of the substrate.

The turnover number (k_{cat}) was calculated using the below Equation (4).

$$k_{cat} = V_{max}/[E] \qquad (4)$$

where [E] is the enzyme concentration [13].

3. Results and Discussions
3.1. Rational Selection of the Template

The template for the imprinting was chosen considering the following parameters: (i) Peptide segments from the flank part of the PPT spatial structure were selected as the template. Due to five disulfide linkages connected among PPT, the choice of peptide segments able to influence catalysis is limited. (ii) The length of the peptide segments in the template is a significant parameter. For instance, short peptide residues form flexible structures that can help the imprinting and protein-rebinding processes [19,30]. Therefore in this study, four PPT peptides, specifically $PPT^{107-116}$, $PPT^{145-155}$, $PPT^{169-178}$, and $PPT^{233-245}$, were chosen. The locations of these segments are shown in Figure 2. At one end of the $PPT^{233-245}$ peptide, a glycine (G) residue was added to make a stable peptide chain with flexibility [31,32].

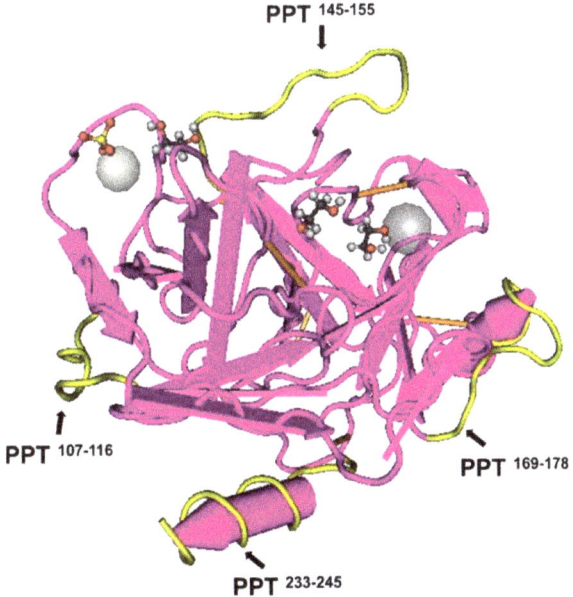

Figure 2. The structure of porcine pancreatic trypsin (cylinder: α-helix; arrow: β-sheet). The selected sequences are in yellow. These segments consist of the series: i.e., $PPT^{107-116}$, $PPT^{145-155}$, $PPT^{169-178}$, and $PPT^{233-245}$. The crystal structure of PPT was reproduced from http://www.ncbi.nlm.nih.gov/ and PDB ID: 1S81 [33].

3.2. Analysis of the Template

Furthermore, the template was synthesized using a CEM Discover Microwave Synthesizer (Kohan Co., Taipei, Taiwan) at National Dong Hwa University (Hualien, Taiwan). The peptide segments $PPT^{107-116}$, $PPT^{145-155}$, $PPT^{169-178}$, and $PPT^{233-245+G}$ were selected as templates. Initially, these segments were fabricated using the Fmoc solid-phase peptide synthesis [22]. Further, the purity of the template molecules was confirmed by HPLC equipped with an RP-18 (flow rate- 1 mL/min). Among the selected peptide segments, PPT$^{107-116}$ and PPT$^{169-178}$ showed a purity higher than 96%. Contrarily, the other two segments, PPT$^{145-155}$ and PPT$^{233-245+G}$, had a lower purity of around 88%, which could be attributed to their longer length, leading to a difficulty in the purification of those peptide segments [34]. Further, the molecular mass of the template was analyzed using a Matrix-Assisted Laser Desorption/Ionization-Time of Flight (MALDI/TOF) mass spectrometer (MS) (Bruker, Bremen, Germany) with a matrix consisting of 2,5-dihydroxybenzoic acid (DHB). The reported m/z values of PPT$^{107-116}$, PPT$^{145-155}$, PPT$^{169-178}$, and PPT$^{233-245+G}$ were observed at 1017.69, 1125.56, 1037.39, and 1614.24 [M + Na]$^+$, respectively. Further, the peptide segment PPT$^{233-245+G}$ was analyzed with a (JASCO, J-715, Tokyo, Japan) circular dichroism (CD) spectrometer to validate the helix structure in the mixtures of buffer and TFE (Figure 3). Usually, the helical peptides possess negative bands at 208 and 225 nm in a mixture of PBS and TFE, while the peptides with random coil structures show negative bands at 200 nm in PBS [35]. After that analysis, the selected PPT peptide segments were used to generate helix cavities using the PCIMPs-based approach.

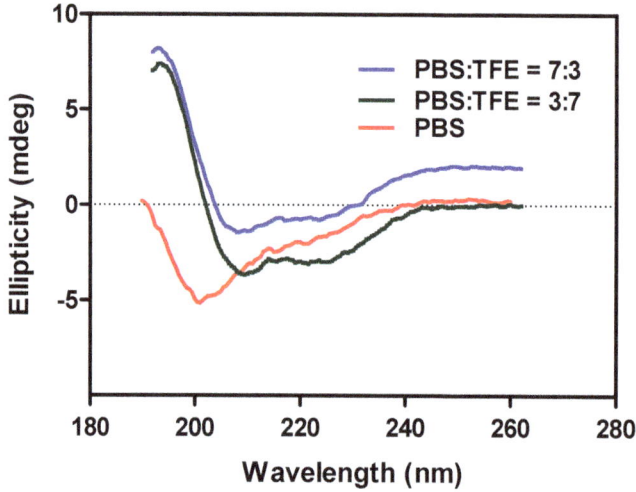

Figure 3. The circular dichroism (CD) spectrum of the PPT $^{233-245+G}$ segment in different solvent systems.

3.3. Characterization of MPs and PCIMPs

3.3.1. FTIR Analysis

A Fourier-transform infrared spectrometer (FTIR, Bruker TENSOR 27, Ettlingen, Germany) was employed to examine the successive surface modifications on MPs (Figure 4). The peaks at 586 cm^{-1} and 3444 cm^{-1} can be ascribed to Fe-O stretching vibration and O-H stretching of Fe$_3$O$_4$ (Figure 4a). The characteristic peaks of silanol groups (Si-O-H) on the surface of Fe$_3$O$_4$ at 1030 cm^{-1}, as well as at 1100 cm^{-1}, and the peak at 3421 cm^{-1} can represent the characteristic peaks of NH$_2$ (primary amine) of APTMS, indicating the successful modification of the Fe$_3$O$_4$ nanoparticles surface with amine groups (Figure 4b) [36–38]. Additionally, the peaks near of 3413 cm^{-1} can represent the existence of the N-H functional group, and no peak at 1739 cm^{-1} can indicate the C=O group at both ends of the glutaraldehyde molecule reacted with NH$_2$, attributed to the established stable secondary structure. The secondary amine-modified iron nanoparticles are more reactive than the primary amine because the inductive effect of secondary amine makes them more stable compared to the primary amine (Figure 4c). The peak at 1619 cm^{-1} can be ascribed to the characteristic peak of C=C, indicating successful acrylation of the MPs (Figure 4d) [39].

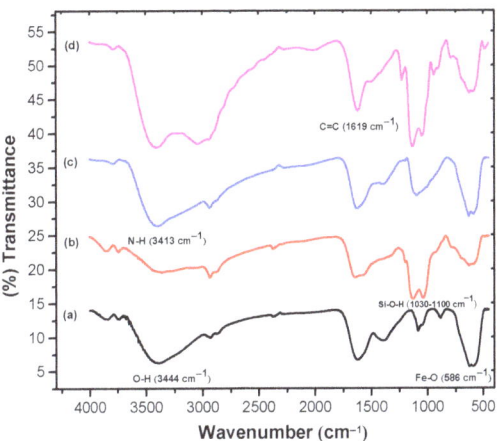

Figure 4. FTIR spectra of (**a**) Fe_3O_4, (**b**) Fe_3O_4@APTMS, (**c**) Fe_3O_4@APTMS-GA, and (**d**) Fe_3O_4@APTMS-GA-acrylate.

3.3.2. FE-SEM Analysis

The surface morphology of various MPs and PCIMPs was analyzed using a Field Emission Scanning Electron Microscope (FE-SEM, JEOL JSM-7000F/JEOL Ltd., Tokyo, Japan) (Figure 5). As a result, it was observed that the fabricated Fe_3O_4 particles were spherical, showing a uniform size distribution with an average size of ~237 nm (Figure 5a). Further, APTMS immobilization on Fe_3O_4 nanoparticles resulted in substantial changes in the size and shape of those MPs, having increased their average size to ~278 nm (Figure 5b). The subsequent immobilization of glutaraldehyde on the MPs resulted in an increase in their average size to ~309 nm (Figure 5c). Notably, a slight aggregation can be observed after the successive surface modification on the MPs. This could be because nanoparticles treated with different solvents and dry samples were collected after the surface modification. The dry power shows strong aggregation, as reported in previous studies [40]. Further, the acrylate monomer conjugation with MPs resulted in an average size of ~323 nm (Figure 5d). Amongst all PCIMPs, the PCIMPs[107–116], PCIMPs[145–155], and PCIMPs[169–178] have shown similar size at ~370 nm, whereas PCIMPs[233–245+G] were comparatively larger at ~408 nm (Figure 5e–h).

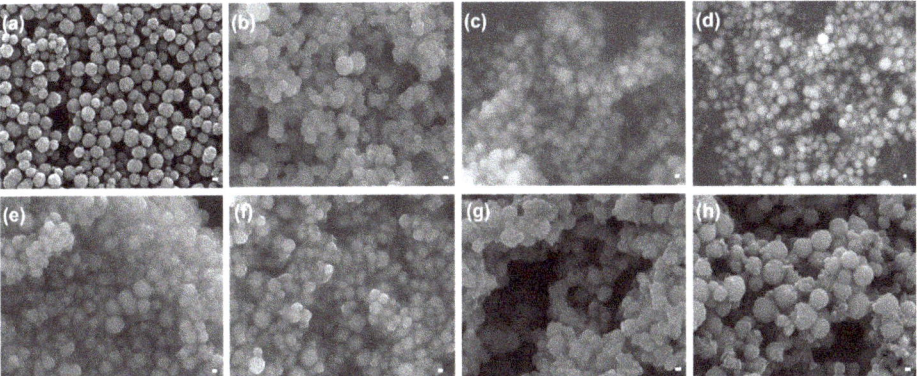

Figure 5. FE-SEM images of (**a**) Fe_3O_4, (**b**) Fe_3O_4@APTMS, (**c**) Fe_3O_4@APTMS-GA, (**d**) Fe_3O_4@APTMS-GA-acrylate, (**e**) PCIMPs[107–116], (**f**) PCIMPs[145–155], (**g**) PCIMPs[169–178], and (**h**) PCIMPs[233–245+G] (Scale bar: 100 nm).

3.4. Binding Studies of PCIMPs

For comparison, the binding affinities of the PPT to each PCIMPs were measured by the linear regression curve based on the Scatchard equation. As shown in Table 1, the PCIMPs $^{233-245+G}$ had the lowest K_d value (0.21 µM) of all the PCIMPs. It was observed from the results that the K_d values showed a decreasing trend with an increase in the number of peptide residues. Therefore, the higher the number of peptide segments in the template, the better the observed binding affinities. For instance, for the 14-mer peptide, the K_d was 0.21 µM, and it showed better affinity when compared to the 10 and 11-mer peptides [19]. Similarly, for PCIMPs$^{145-155}$, the K_d value was 0.38 µM, and it presented a better affinity than that of a 10-mer peptide. On the other hand, both PCIMPs$^{107-116}$ and PCIMPs$^{169-178}$ have shown a similar number of peptide residues in the template. In this case, affinities of the PPT to PCIMPs were more closely related to the molecular weight of the template residues. For example, the K_d value of the PCIMPs$^{169-178}$ was 0.55 µM, which showed a better binding affinity than PCIMPs$^{107-116}$ (0.65 µM).

Table 1. Binding affinity values of various PCIs on magnetic particles (PCIMPs) to PPT.

MPs	PCIMPs$^{107-116}$	PCIMPs$^{145-155}$	PCIMPs$^{169-178}$	PCIMPs$^{233-245+G}$
Residue	10	11	10	14
[K_d] µM	0.65	0.38	0.55	0.21
[B_{max}] nM	0.75	1.11	0.95	1.11
B_{max}/K_d	1.15	2.92	1.73	5.29

Previously, Griffete and colleagues developed a magnetic-protein imprinted polymer (M-PIP) by combining photopolymerization with a grafting approach onto surface-functionalized MPs. The authors demonstrated that the green fluorescent proteins were bound to MIPs in less than 2 h with a high affinity (K_d = 0.29 µM) [41]. In another study, MIPs were synthesized using a solid-phase approach on metal chelate functionalized glass-beads to immobilize trypsin using its surface histidine. Although less cross-reactivity with other proteins was observed, the dissociation constant value of the MIP-trypsin complex was 0.237 µM [42], with a lagging binding capacity. Notably, in this study, the PCIs developed on the surface of MPs create recognition sites that are complementary to the protein conformational structure and, therefore, significantly increase the specificity toward the targeted protein. The best binding performance of PCIMPs$^{233-245+G}$ occurred in 10 min with a high affinity (K_d = 0.21 µM). Upon a comprehensive evaluation of binding affinities and absorption time, it was apparent that conformational imprints on MPs acquired better results in these protein-imprinted particles. Together, our findings indicated a higher affinity of protein (PPT) to PCIMPs $^{233-245+G}$ (K_d = 0.21 µM), in comparison to the other MIPs grafting methods.

3.5. Kinetic Parameters of PPT and PPT/PCIMPs

In addition, the PCIMPs bound to PPT exhibited excellent catalytic activity. To demonstrate this aspects, the kinetic parameters of PPT and PPT/PCIMPs were explored by varying the BAEE substrate concentration (0.5–1.5 mM). They were then calculated using the Michaelis-Menten plot (Figure 6a). As shown in Table 2, among all the PPT/PCIMPs, PPT/PCIMPs $^{233-245+G}$ had the best kinetic parameters. The K_m value of PPT (0.36 mM) was almost similar to that of the PPT/PCIMPs $^{233-245+G}$ (0.42 mM), which could be due to the high feasibility of forming an enzyme-substrate complex, and also a lower diffusion restraint imposed on the flow of the substrate and product molecules from the grafted polymer matrix of the MPs [43,44]. The V_{max} values were found to be 3.2×10^{-3} mMs^{-1} and 1.47×10^{-3} mMs^{-1} for PPT and PPT/PCIMPs $^{233-245+G}$, in which the V_{max} was decreased for PPT/PCIMPs when compared to the free enzyme. The plausible reason might be due to the created steric hindrances that restrict the substrates' transport, enhance diffusional creation limitations, and decrease the enzyme's catalytic properties. These conclusions are in agreement with the results reported literature [45,46].

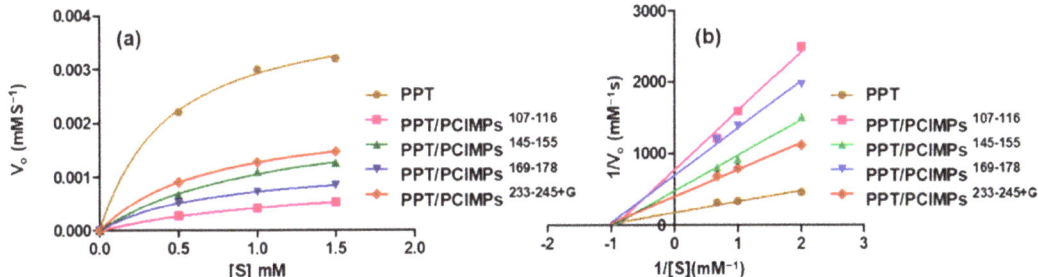

Figure 6. (a) Michaelis-Menten and (b) Lineweaver-Burk plots of PPT and PPT/PCIMPs obtained with various N-benzoyl-L-arginine ethyl ester (BAEE) solutions (1.5, 1, 0.5 mM). Note: The standard deviation for the Michaelis-Menten kinetics plot of PPT in different concentration (1.5 mM to 0.5 mM) is 1×10^{-4} whereas and PPT/PCIMPs$^{233-245+G}$ is 1.41067×10^{-5}, 7.2111×10^{-6}, and 5.50757×10^{-6} for (1.5, 1, and 0.5 mM).

Table 2. Kinetic parameters obtained from the Michaelis-Menten plot.

MPs	PPT	PPT/PCIMPs$^{107-116}$	PPT/PCIMPs$^{145-155}$	PPT/PCIMPs$^{169-178}$	PPT/PCIMPs$^{233-245+G}$
V_{max} (mM s^{-1})	3.2×10^{-3}	0.53×10^{-3}	1.25×10^{-3}	0.84×10^{-3}	1.47×10^{-3}
[K_m] mM	0.36	0.52	0.46	0.44	0.42
k_{cat} (s^{-1})	2.6	0.62	0.99	0.78	1.16
k_{cat}/K_m (mM^{-1} s^{-1})	7.32	1.19	2.15	1.77	2.79

Note: PPT= porcine pancreatic alpha-trypsin, PPT/PCIMPs = immobilized PPT.

Respectively, it was observed that the k_{cat} value of PPT/PCIMPs$^{233-245+G}$ was lower than that of PPT. The decrease of k_{cat} values upon immobilization of enzymes are frequently reported [13,47,48]. These findings suggest a limited diffusion of the substrate to the active site and higher structural rigidity of the immobilized PPT. Our results are quite comparable and in agreement with the ones reported in the literature [47–50]. Furthermore, the trypsin inhibition by PCIMPs was investigated by performing enzyme assays in the Tris-HCl buffer at pH 6.2, using BAEE as the substrate at various concentrations. The Lineweaver Burk plot (1/V$_o$ versus 1/S) is as shown in Figure 6b. It reveals that the PCIMPs exhibited non-competitive inhibition towards trypsin, in which the PCIMPs acted as inhibitors, while the BAEE functioned as a substrate. In non-competitive inhibition, the respective inhibitors bind to the free enzyme and the enzyme-substrate complex with the same affinity. Further, the inhibitor reduces the activity of the enzyme and binds equally well to the substrate [51,52].

3.6. Reusability

Additionally, the reusability of PPT/PCIMPs was examined. Initially, 10 mg of PPT/PCIMPs $^{145-155}$ was added to 8.8 mL of a 1.5 mM BAEE solution (50 mM Tris-HCl buffer, pH 7.6). The product concentration was monitored using HPLC. The test was conducted consecutively four times. It was observed from the results that the PPT/PCIMPs retained 90% of activity in 540 sec in the first cycle; however, in the subsequent cycles, it slightly dropped. The activity of the protein sustained after four cycles is as shown in Figure 7.

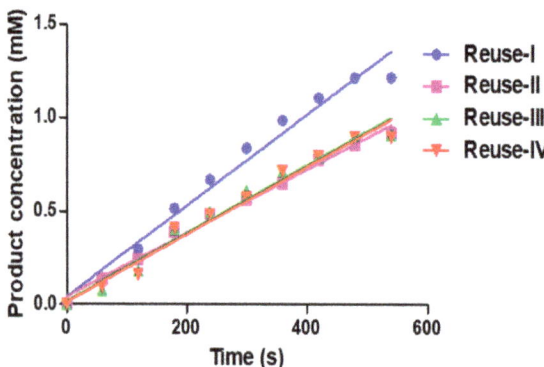

Figure 7. Reusability of PPT/PCIMPs[145–155].

3.7. Comparison Studies of the Proposed PPT/PCIMPs with Other Methods

The catalytic hydrolysis performance of the fabricated PPT/PCIMPs was compared to previous studies (Table 3). For example, Atacan and colleagues modified the surface of Fe_3O_4 nanoparticles with gallic acid. According to their research, K_m values of trypsin and immobilized trypsin were 5.1 and 7.88 mM, respectively, indicating that the immobilized trypsin has less affinity for the substrate, which might be attributed to the loss of enzyme flexibility. Although immobilized trypsin retained 92% of its initial activity after four months of storage at 4 °C, there was a dramatic decrease in its activity after being reused eight consecutive times [49]. In another study, trypsin was immobilized on polymer and grafted magnetic beads, in which the K_m for immobilized trypsin was found to be 13.6 mM, 1.4-fold higher than free trypsin, while V_{max} value was found to be 3946 U/mg, 1.5-fold lower than for the free trypsin, indicating that a change in the affinity of the enzyme towards the substrate occurred upon its immobilization [50]. In different work, by Bayramoglu and colleagues, polymer grafted magnetic beads were activated with glutaraldehyde for the immobilization of trypsin on affinity ligands attached to the beads' surface. Moreover, the reusability and activity were relatively good in this study when compared to the above work. The K_m and V_{max} values obtained for the immobilized trypsin were of 16.8 mM and 5115 U/mg, 1.8-fold higher and 1.5-fold lower than free trypsin, respectively. The K_m values could be explained by the fact that there existed conformational changes during enzyme immobilization [53].

Table 3. Comparison studies of proposed PPT/PCIMPs with other methods.

Trypsin/Immobilized Trypsin	K_m	V_{max}	k_{cat} (s^{-1})	k_{cat}/K_m (mM^{-1} s^{-1})	Reference
BPT/Immobilized BPT	5.1/7.88 mM	23/18.3 mM min^{-1}	-	-	[42]
BPT/Immobilized BPT	9.7/13.6 mM	5890/3946 U/mg	-	607/290	[43]
BPT/Immobilized BPT	9.3/16.8 mM	7345/5115 U/mg	-	-	[44]
PPT and PPT/PCIMPs$^{233-245+G}$	0.36/0.42 mM	$3.2 \times 10^{-3}/1.47 \times 10^{-3}$ mM s^{-1}	2.6/>1.16	7.32/2.79	This study

Abbreviations: Bovine Pancreas Trypsin (BPT), Porcine Pancreatic Trypsin (PPT), Note: U is defined as μmol.

Upon a comprehensive evaluation of kinetic parameters, it was evident that the elegant helical cavities imprinting strategy created recognition sites on the MPs surface, in which the enzymes were tightly bound. Moreover, it was achieved an improved catalytic hydrolysis in comparison to other previous studies. The best performance of PPT/PCIMPs for hydrolysis of BAEE had the following values for the kinetic parameters K_m, V_{max}, and k_{cat} values were 0.42 mM, 1.4 μM·s^{-1}, and 1.16·s^{-1}. Additionally, PPT/PCIMPs-imprinted materials exhibited stable catalytic activity and reusability.

4. Conclusions

In conclusion, a state-of-the-art method for point immobilization of enzymes on magnetic particles is accomplished. To maintain the catalytically competent state of an enzyme, an immobilized enzyme at a maximum degree of freedom is the ultimate choice. Our systems operate by binding enzyme partially and maintaining the remaining part of the enzyme free. The combination of site fixation with the use of conformation-specific PCIMPs could boost the catalytic process in many enzymes. Moreover, the experimental results also indicated the inhibition effect on capturing at the α-helix region to interfere with catalysis flexibility. The K_m of PPT/PCIMPs$^{233-245+G}$ was slightly higher than that of PPT, resulting in lower diffusion limitations of the substrate and product molecules from the polymer matrix to forming an enzyme-substrate complex. Consequently, this method is an appropriate choice for realizing the relationship between each segment's flexibility and catalytic activity. We thus believe the PCIMPs strategy can be more widely applied in green chemistry as a nano biocatalyst.

Author Contributions: Conceptualization, D.-F.T.; methodology, K.R.K., P.-Y.H., Y.-L.C., C.H.W. and W.L.; software, K.R.K. and P.-Y.H.; validation, K.R.K., P.-Y.H., Y.-L.C., C.H.W. and W.L.; formal analysis, K.R.K., P.-Y.H., Y.-L.C., C.H.W., W.L. and R.K.K.; resources, C.-H.L. and D.-F.T.; data curation, K.R.K., C.-H.L., D.-F.T., P.-Y.H., Y.-L.C., C.H.W., W.L. and R.K.K.; writing—original draft preparation, K.R.K., D.-F.T., R.K.K. and C.-H.L.; writing—review and editing, K.R.K., D.-F.T., R.K.K. and C.-H.L.; visualization, K.R.K., D.-F.T., R.K.K. and C.-H.L.; supervision, D.-F.T., and C.-H.L.; project administration, D.-F.T.; funding acquisition, D.-F.T. All authors have read and agreed to the published version of the manuscript.

Funding: This work is partially supported by the Taiwan Ministry of Science and Technology (MOST 106–2113-M-259–005).

Institutional Review Board Statement: Not applicable.

Informed Consent Statement: Not applicable.

Data Availability Statement: Not applicable.

Conflicts of Interest: The authors declare no conflict of interest.

References

1. Atacan, K.; Çakıroğlu, B.; Özacar, M. Covalent immobilization of trypsin onto modified magnetite nanoparticles and its application for casein digestion. *Int. J. Biol. Macromol.* **2017**, *97*, 148–155. [CrossRef]
2. Perutka, Z.; Šebela, M. Pseudotrypsin: A Little-Known Trypsin Proteoform. *Molecules* **2018**, *23*, 2637. [CrossRef] [PubMed]
3. Vorob'Ev, M.M. Proteolysis of β-lactoglobulin by Trypsin: Simulation by Two-Step Model and Experimental Verification by Intrinsic Tryptophan Fluorescence. *Symmetry* **2019**, *11*, 153. [CrossRef]
4. Siddiqui, I.; Husain, Q. Stabilization of polydopamine modified silver nanoparticles bound trypsin: Insights on protein hydrolysis. *Colloids Surf. B Biointerfaces* **2019**, *173*, 733–741. [CrossRef] [PubMed]
5. Sasai, Y.; Kanno, H.; Doi, N.; Yamauchi, Y.; Kuzuya, M.; Kondo, S.-I. Synthesis and Characterization of Highly Stabilized Polymer–Trypsin Conjugates with Autolysis Resistance. *Catalysts* **2016**, *7*, 4. [CrossRef]
6. Aslani, E.; Abri, A.; Pazhang, M. Immobilization of trypsin onto Fe_3O_4@SiO_2-NH_2 and study of its activity and stability. *Colloids Surf. B Biointerfaces* **2018**, *170*, 553–562. [CrossRef]
7. Sanchez, A.; Cruz, J.; Rueda, N.; Dos Santos, J.C.S.; Torres, R.; Ortiz, C.; Villalonga, R.; Fernandez-Lafuente, R. Inactivation of immobilized trypsin under dissimilar conditions produces trypsin molecules with different structures. *RSC Adv.* **2016**, *6*, 27329–27334. [CrossRef]
8. Stolarow, J.; Heinzelmann, M.; Yeremchuk, W.; Syldatk, C.; Hausmann, R. Immobilization of trypsin in organic and aqueous media for enzymatic peptide synthesis and hydrolysis reactions. *BMC Biotechnol.* **2015**, *15*, 77. [CrossRef]
9. Kankala, R.K.; Zhang, H.; Liu, C.; Kanubaddi, K.R.; Lee, C.; Wang, S.; Cui, W.; Santosaf, H.A.; Lin, K.; Chen, A. Metal Species–Encapsulated Mesoporous Silica Nanoparticles: Current Advancements and Latest Breakthroughs. *Adv. Funct. Mater.* **2019**, *29*, 1902652. [CrossRef]
10. Wahab, R.A.; Elias, N.; Abdullah, F.; Ghoshal, S.K. On the taught new tricks of enzymes immobilization: An all-inclusive overview. *React. Funct. Polym.* **2020**, *152*, 104613. [CrossRef]
11. Homaei, A.; Sariri, R.; Vianello, F.; Stevanato, R. Enzyme immobilization: An update. *J. Chem. Biol.* **2013**, *6*, 185–205. [CrossRef] [PubMed]

12. Reis, C.; Sousa, E.; Serpa, J.; Oliveira, R.; Santos, J. Design of immobilized enzyme biocatalysts: Drawbacks and opportunities. *Quím. Nova* **2019**, *42*, 768–783. [CrossRef]
13. Liu, C.; Saeki, D.; Matsuyama, H. A novel strategy to immobilize enzymes on microporous membranes via dicarboxylic acid halides. *RSC Adv.* **2017**, *7*, 48199–48207. [CrossRef]
14. Xing, R.; Ma, Y.; Wang, Y.; Wen, Y.; Liu, Z. Specific recognition of proteins and peptides via controllable oriented surface imprinting of boronate affinity-anchored epitopes. *Chem. Sci.* **2019**, *10*, 1831–1835. [CrossRef]
15. Ansell, R.J.; Mosbach, K. Magnetic molecularly imprinted polymer beads for drug radioligand binding assay. *Analyst* **1998**, *123*, 1611–1616. [CrossRef]
16. Tan, C.J. and Y.W. Tong, Preparation of Superparamagnetic Ribonuclease A Surface-Imprinted Submicrometer Particles for Protein Recognition in Aqueous Media. *Anal. Chem.* **2007**, *79*, 299–306. [CrossRef]
17. Jing, T.; Du, H.; Dai, Q.; Xia, H.; Niu, J.; Hao, Q.; Mei, S.; Zhou, Y. Magnetic molecularly imprinted nanoparticles for recognition of lysozyme. *Biosens. Bioelectron.* **2010**, *26*, 301–306. [CrossRef]
18. Guerreiro, A.; Poma, A.; Karim, K.; Moczko, E.; Takarada, J.; De Vargas-Sansalvador, I.P.; Turner, N.; Piletska, E.; De Magalhães, C.S.; Glazova, N.; et al. Influence of Surface-Imprinted Nanoparticles on Trypsin Activity. *Adv. Healthc. Mater.* **2014**, *3*, 1426–1429. [CrossRef]
19. Tai, D.-F.; Jhang, M.-H.; Chen, G.-Y.; Wang, S.-C.; Lu, K.-H.; Lee, Y.-D.; Liu, H.-T. Epitope-Cavities Generated by Molecularly Imprinted Films Measure the Coincident Response to Anthrax Protective Antigen and Its Segments. *Anal. Chem.* **2010**, *82*, 2290–2293. [CrossRef]
20. Tai, D.-F.; Ho, Y.-F.; Wu, C.-H.; Lin, T.-C.; Lu, K.-H.; Lin, K.-S. Artificial-epitope mapping for CK-MB assay. *Analyst* **2011**, *136*, 2230–2233. [CrossRef]
21. Chou, C.-Y.; Lin, C.-Y.; Wu, C.-H.; Tai, D.-F. Sensing HIV Protease and Its Inhibitor Using "Helical Epitope"—Imprinted Polymers. *Sensors* **2020**, *20*, 3592. [CrossRef] [PubMed]
22. Collins, J.M.; Porter, K.A.; Singh, S.K.; Vanier, G.S. High-Efficiency Solid Phase Peptide Synthesis (HE-SPPS). *Org. Lett.* **2014**, *16*, 940–943. [CrossRef] [PubMed]
23. Yang, Q.; Zhu, Y.; Luo, B.; Lan, F.; Wu, Y.; Gu, Z. pH-Responsive magnetic nanospheres for the reversibly selective capture and release of glycoproteins. *J. Mater. Chem. B* **2017**, *5*, 1236–1245. [CrossRef] [PubMed]
24. Ding, S.; Xing, Y.; Radosz, M.; Shen, Y. Magnetic Nanoparticle Supported Catalyst for Atom Transfer Radical Polymerization. *Macromolecules* **2006**, *39*, 6399–6405. [CrossRef]
25. Ellwanger, A.; Karlsson, L.; Owens, P.K.; Berggren, C.; Crecenzi, C.; Ensing, K.; Bayoudh, S.; Cormack, P.A.; Sherrington, D.; Sellergren, B. Evaluation of methods aimed at complete removal of template from molecularly imprinted polymers. *Analyst* **2001**, *126*, 784–792. [CrossRef]
26. Lorenzo, R.A.; Carro, A.M.; Alvarez-Lorenzo, C.; Concheiro, A. To remove or not to remove? The challenge of extracting the template to make the cavities available in Molecularly Imprinted Polymers (MIPs). *Int. J. Mol. Sci.* **2011**, *12*, 4327–4347. [CrossRef]
27. Gerdon, A.E.; Wright, D.W.; Cliffel, D.E. Quartz Crystal Microbalance Detection of Glutathione-Protected Nanoclusters Using Antibody Recognition. *Anal. Chem.* **2005**, *77*, 304–310. [CrossRef]
28. Diltemiz, S.E.; Hür, D.; Ersöz, A.; Denizli, A.; Say, R. Designing of MIP based QCM sensor having thymine recognition sites based on biomimicking DNA approach. *Biosens. Bioelectron.* **2009**, *25*, 599–603. [CrossRef]
29. Tai, D.-F.; Lin, Y.-F.; Lu, K.-H.; Chen, G.-Y.; Shu, H.-C. A Direct Immersion System for Peptide Enrichment. *J. Chin. Chem. Soc.* **2012**, *59*, 338–344. [CrossRef]
30. Bossi, A.M.; Sharma, P.S.; Montana, L.; Zoccatelli, G.; Laub, O.; Levi, R. Fingerprint-Imprinted Polymer: Rational Selection of Peptide Epitope Templates for the Determination of Proteins by Molecularly Imprinted Polymers. *Anal. Chem.* **2012**, *84*, 4036–4041. [CrossRef]
31. Van Rosmalen, M.; Krom, M.; Merkx, M. Tuning the Flexibility of Glycine-Serine Linkers To Allow Rational Design of Multidomain Proteins. *Biochemistry* **2017**, *56*, 6565–6574. [CrossRef] [PubMed]
32. Dong, H.; Sharma, M.; Zhou, H.-X.; Cross, T.A. Glycines: Role in α-Helical Membrane Protein Structures and a Potential Indicator of Native Conformation. *Biochemistry* **2012**, *51*, 4779–4789. [CrossRef] [PubMed]
33. Transue, T.R.; Krahn, J.M.; Gabel, S.A.; Derose, A.E.F.; London, R.E. X-ray and NMR Characterization of Covalent Complexes of Trypsin, Borate, and Alcohols. *Biochemistry* **2004**, *43*, 2829–2839. [CrossRef] [PubMed]
34. Isidro-Llobet, A.; Kenworthy, M.N.; Mukherjee, S.; Kopach, M.E.; Wegner, K.; Gallou, F.; Smith, A.G.; Roschangar, F. Sustainability Challenges in Peptide Synthesis and Purification: From R&D to Production. *J. Org. Chem.* **2019**, *84*, 4615–4628. [CrossRef] [PubMed]
35. Wei, Y.; Thyparambil, A.A.; Latour, R.A. Protein helical structure determination using CD spectroscopy for solutions with strong background absorbance from 190 to 230 nm. *Biochim. Biophys. Acta (BBA) Proteins Proteom.* **2014**, *1844*, 2331–2337. [CrossRef] [PubMed]
36. Farjadian, F.; Ghasemi, S.; Mohammadi-Samani, S.; Ghasemia, S. Hydroxyl-modified magnetite nanoparticles as novel carrier for delivery of methotrexate. *Int. J. Pharm.* **2016**, *504*, 110–116. [CrossRef] [PubMed]
37. Gao, Z.; Yi, Y.; Zhao, J.; Xia, Y.; Jiang, M.; Cao, F.; Zhou, H.; Wei, P.; Jia, H.; Yong, X. Co-immobilization of laccase and TEMPO onto amino-functionalized magnetic Fe3O4 nanoparticles and its application in acid fuchsin decolorization. *Bioresour. Bioprocess.* **2018**, *5*, 27. [CrossRef]

38. Kurtan, U.; Baykal, A. Fabrication and characterization of Fe_3O_4@APTES@PAMAM-Ag highly active and recyclable magnetic nanocatalyst: Catalytic reduction of 4-nitrophenol. *Mater. Res. Bull.* **2014**, *60*, 79–87. [CrossRef]
39. Farjadian, F.; Hosseini, M.; Ghasemi, S.; Tamami, B. Phosphinite-functionalized silica and hexagonal mesoporous silica containing palladium nanoparticles in Heck coupling reaction: Synthesis, characterization, and catalytic activity. *RSC Adv.* **2015**, *5*, 79976–79987. [CrossRef]
40. Iijima, M.; Kamiya, H. Surface Modification for Improving the Stability of Nanoparticles in Liquid Media. *KONA Powder Part. J.* **2009**, *27*, 119–129. [CrossRef]
41. Boitard, C.; Lamouri, A.; Menager, C.; Griffete, N. Whole Protein Imprinting over Magnetic Nanoparticles Using Photopolymerization. *ACS Appl. Polym. Mater.* **2019**, *1*, 928–932. [CrossRef]
42. Xu, J.; Prost, E.; Haupt, K.; Bui, B.T.S. Direct and sensitive determination of trypsin in human urine using a water-soluble signaling fluorescent molecularly imprinted polymer nanoprobe. *Sens. Actuators B Chem.* **2018**, *258*, 10–17. [CrossRef]
43. Lin, H.; Zhang, C.; Lin, Y.; Chang, Y.; Crommen, J.; Wang, Q.; Jiang, Z.; Guo, J. A strategy for screening trypsin inhibitors from traditional Chinese medicine based on a monolithic capillary immobilized enzyme reactor coupled with offline liquid chromatography and mass spectrometry. *J. Sep. Sci.* **2019**, *42*, 1980–1989. [CrossRef] [PubMed]
44. Bayramoglu, G.; Çelikbıçak, Ö.; Arica, Y.; Salih, B. Trypsin Immobilized on Magnetic Beads via Click Chemistry: Fast Proteolysis of Proteins in a Microbioreactor for MALDI-ToF-MS Peptide Analysis. *Ind. Eng. Chem. Res.* **2014**, *53*, 4554–4564. [CrossRef]
45. Zdarta, J.; Antecka, K.; Jędrzak, A.; Synoradzki, K.; Łuczak, M.; Jesionowski, T. Biopolymers conjugated with magnetite as support materials for trypsin immobilization and protein digestion. *Colloids Surf. B Biointerfaces* **2018**, *169*, 118–125. [CrossRef] [PubMed]
46. Arica, M.Y.; Şenel, S.; Alaeddinoğlu, N.G.; Patir, S.; Denizli, A. Invertase immobilized on spacer-arm attached poly(hydroxyethyl methacrylate) membrane: Preparation and properties. *J. Appl. Polym. Sci.* **2000**, *75*, 1685–1692. [CrossRef]
47. Homaei, A. Enhanced activity and stability of papain immobilized on CNBr-activated sepharose. *Int. J. Biol. Macromol.* **2015**, *75*, 373–377. [CrossRef]
48. Shojaei, F.; Homaei, A.; Taherizadeh, M.R.; Kamrani, E. Characterization of biosynthesized chitosan nanoparticles from Penaeus vannamei for the immobilization of P. vannamei protease: An eco-friendly nanobiocatalyst. *Int. J. Food Prop.* **2017**, *20* (Suppl. 2), 1413–1423.
49. Atacan, K.; Çakıroğlu, B.; Özacar, M. Improvement of the stability and activity of immobilized trypsin on modified Fe_3O_4 magnetic nanoparticles for hydrolysis of bovine serum albumin and its application in the bovine milk. *Food Chem.* **2016**, *212*, 460–468. [CrossRef]
50. Bayramoglu, G.; Yılmaz, M.; Şenel, A. Ülkü; Arıca, M.Y. Preparation of nanofibrous polymer grafted magnetic poly(GMA-MMA)-g-MAA beads for immobilization of trypsin via adsorption. *Biochem. Eng. J.* **2008**, *40*, 262–274. [CrossRef]
51. Zhang, H.; Jiang, J.; Zhang, H.; Zhang, Y. Efficient Synthesis of Molecularly Imprinted Polymers with Enzyme Inhibition Potency by the Controlled Surface Imprinting Approach. *ACS Macro Lett.* **2013**, *2*, 566–570. [CrossRef]
52. Blat, Y. Non-Competitive Inhibition by Active Site Binders. *Chem. Biol. Drug Des.* **2010**, *75*, 535–540. [CrossRef] [PubMed]
53. Bayramoglu, G.; Özalp, V.C.; Arica, M.Y. Magnetic Polymeric Beads Functionalized with Different Mixed-Mode Ligands for Reversible Immobilization of Trypsin. *Ind. Eng. Chem. Res.* **2013**, *53*, 132–140. [CrossRef]

Article

Determination of Cobalt Spin-Diffusion Length in Co/Cu Multilayered Heterojunction Nanocylinders Based on Valet–Fert Model

Saeko Mizoguchi [1], Masamitsu Hayashida [2] and Takeshi Ohgai [2,*]

1. Graduate School of Engineering, Nagasaki University, Bunkyo-machi 1–14, Nagasaki 852-8521, Japan; bb52119647@ms.nagasaki-u.ac.jp
2. Faculty of Engineering, Nagasaki University, Bunkyo-machi 1–14, Nagasaki 852-8521, Japan; hayashida@nagasaki-u.ac.jp
* Correspondence: ohgai@nagasaki-u.ac.jp; Tel.: +81-95-819-2638

Abstract: Anodized aluminum oxide (AAO) nanochannels of diameter, D, of ~50 nm and length, L, of ~60 μm (L/D: approx. 1200 in the aspect ratio), were synthesized and applied as an electrode for the electrochemical growth of Co/Cu multilayered heterojunction nanocylinders. We synthesized numerous Co/Cu multilayered nanocylinders by applying a rectangular pulsed potential deposition method. The Co layer thickness, t_{Co}, ranged from ~8 to 27 nm, and it strongly depended on the pulsed-potential condition for Co layers, E_{Co}. The Cu layer thickness, t_{Cu}, was kept at less than 4 nm regardless of E_{Co}. We applied an electrochemical in situ contact technique to connect a Co/Cu multilayered nanocylinder with a sputter-deposited Au thin layer. Current perpendicular-to-plane giant magnetoresistance (CPP-GMR) effect reached up to ~23% in a Co/Cu multilayered nanocylinder with ~4760 Co/Cu bilayers (t_{Cu}: 4 nm and t_{Co}: 8.6 nm). With a decrease in t_{Co}, $(\Delta R/R_p)^{-1}$ was linearly reduced based on the Valet–Fert equation under the condition of $t_F > l_F^{sf}$ and $t_N < l_N^{sf}$. The cobalt spin-diffusion length, l_{Co}^{sf}, was estimated to be ~12.5 nm.

Keywords: anodization; nanochannel; electrodeposition; nanocylinder; cobalt; copper; heterojunction; multilayer; magnetoresistance; spin-diffusion length

1. Introduction

Fert et al. and Grünberg et al. discovered the current-in-plane giant magnetoresistance (CIP-GMR) effect that the electric current passes through the in-plane direction of Fe/Cr multilayered thin films [1,2]. Schwarzacher et al. demonstrated the CIP-GMR effect by using the electrodeposited Co-Ni/Cu multilayered thin films [3]. After that, several research works have been reported that the electrodeposited ferromagnetic multilayered thin films exhibited the CIP-GMR effect [4–7]. However, considering an industrial application to a magnetic readout head in a hard disk drive (HDD), there are some issues concerning the quality of multilayered structure of an electrodeposited CIP-GMR device because it has a quite larger interface area (~10^{-6} m^2) rather than the square of average crystal size (~10^{-16} m^2).

On the contrary, a nanocylinder-based GMR sensor can realize an ideal sharp interface because the interface area (~10^{-16} m^2) is a similar order to the square of average crystal size (~10^{-16} m^2). These multilayered heterojunction nanocylinders with a large aspect ratio have a potential application to a magnetic readout head in a HDD, a magnetoresistive random access memory (MRAM) and high-sensitive metal-based magnetic field sensor with a small temperature coefficient (alternative to a Hall sensor), and so on. Piraux et al. and Blondel et al. demonstrated the current perpendicular-to-plane giant magnetoresistance (CPP-GMR) effect by using the Co/Cu multilayered nanocylinders which were electrodeposited into ion-track-etched polycarbonate membranes [8,9]. After that, several research works have been reported that the CPP-GMR effect was observed in the ferromagnetic multilayered nanocylinders which were electrodeposited into anodized aluminum oxide (AAO) templates [10–17]. Evans et al. reported that the Co-Ni/Cu multilayered nanocylinders, which were electrodeposited into commercially available AAO membranes (~300 nm in diameter, D and ~60 μm in length, L), exhibited a CPP-GMR effect of ~55% at room temperature [10]. They revealed that the Co-Ni alloy layer thickness, t_{Co} of about 5 nm and Cu layer thickness, t_{Cu} of about 2 nm were optimum values to exhibit a large CPP-GMR effect. Tang et al. also reported that the electrodeposited Co/Cu multilayered nanocylinders in commercial AAO templates showed a CPP-GMR effect of ~13.5% at room temperature [12]. They found that t_{Co} of ~8 nm and t_{Cu} of ~10 nm were optimum values to show a large CPP-GMR effect. Shakya et al. reported that the FeCoNi/Cu multilayered nanocylinders in commercial AAO templates showed a CPP-GMR effect of ~15% at room temperature [14]. Zhang et al. also reported that Ni-Fe/Cu/Co/Cu multilayered nanocylinders, which were electrodeposited into home-made AAO templates (D = 120 nm), exhibited a GMR effect of ~45% at room temperature [15]. Han et al. reported that the Co/Cu multilayered nanocylinders in home-made AAO templates (D = 50 nm) showed a CPP-GMR effect of ~13% at room temperature [16]. They revealed that t_{Co} of ~50 nm and t_{Cu} of ~5 nm were optimum values to demonstrate a large CPP-GMR effect. On the contrary, Xi et al. reported that the Co/Cu multilayered nanocylinders in home-made AAO templates (D = 80 nm) showed a small magnetoresistance effect of ~0.16% at room temperature [17]. The above research works have been conducted using AAO templates with an aspect ratio less than 250. It is estimated that the spin-valve response in the axial direction is improved by decreasing the nanocylinder diameter due to enhancing the magnetic shape anisotropy. Recently, we have demonstrated that Co/Cu multilayered nanocylinders, which were electrodeposited into a home-made AAO template (D = 75 nm and L = 70 μm), exhibited a CPP-GMR effect of ~23.5% at room temperature [18]. Hence, in the present study, to improve the CPP-GMR performance in the axial direction, we created Co/Cu multilayered nanocylinders electrodeposited into nanochannels with the diameter of ~50 nm (the aspect ratio is more than 1000). The spin-diffusion length in the cobalt layers was then determined based on the Valet–Fert equation.

2. Materials and Methods

A commercially available aluminum rod was mechanically and anodically polished in the cross-section (10 mm in diameter) to give a specular surface. During the anodic polishing process, bath voltage was maintained at 50 V for 120 s in an ethyl alcohol solution with 25 vol.% perchloric acid ($HClO_4$) (FUJIFILM Wako Pure Chemical Corpo., Osaka, Japan). Afterward, to make an AAO nanochannel film, the polished cross-section was anodically oxidized in an electrolytic bath (0.3 mol/L oxalic acid) using a power supply (Bipolar DC Power Supply, BP4610, NF Corp., Yokohama, Japan). The nanochannel structure of an AAO film is strongly affected by anodization parameters [19,20]. In this study, the anodization voltage was kept at 50 V for 12 h. The AAO film was separated from an aluminum surface in an ethyl alcohol solution containing 50 vol.% perchloric acid ($HClO_4$). During this separation process, the bath voltage was maintained at 55 V for 3 s. The separated films were employed as nanochannel templates for the electrodeposition of nanocylinders. To cover the nanochannels, a thick gold layer (250 nm) was formed on a surface of an AAO film using a DC magnetron sputter-deposition system (Auto Fine Coater, JFC-1600, JEOL Ltd., Tokyo, Japan). The thick gold layer works as a cathode in the nanochannels. A porous, thin gold layer (60 nm) was also formed on the other side surface of the AAO films without covering the nanochannels. The porous, thin gold layer functions as a floating electrode to make in situ contact with nanocylinders during electrodeposition. A pure gold wire was applied as a counter electrode, while an Ag/AgCl electrode was used as a reference electrode. An aqueous electrolytic solution was prepared using 0.5 mol/L cobalt (II) amido-sulfate (Co $(SO_3NH_2)_2$ $4H_2O$) (Mitsuwa Chemicals Co. Ltd., Osaka, Japan), 0.005 mol/L copper (II) sulfate ($CuSO_4$ $5H_2O$) (FUJIFILM Wako Pure Chemical Corpo., Osaka, Japan), 0.4 mol/L boric acid (H_3BO_3) (FUJIFILM Wako Pure Chemical Corpo., Osaka, Japan). The bath temperature was maintained at 40 °C, and the pH was adjusted to 4.0. To optimize the cathode potential for electrodeposition of Cu and Co layers, the linear sweep voltammetry technique was employed using an automatic polarization system (Electrochemical Measurement System, HZ-7000, Hokuto Denko Corp., Tokyo, Japan). Co/Cu multilayered nanocylinders with Cu layers (from 1.2 to 3.8 nm) and Co layers (from 7.8 to 26.8 nm) were grown into AAO nanochannels with an ultra-large aspect ratio of ~1200 using a rectangular pulsed-potential deposition process.

The bilayer thickness of Cu and Co was estimated from the AAO nanochannel length divided by the filling time. Each layer thickness of Co and Cu was determined from the bilayer thickness and the molar fraction using an energy-dispersive X-ray spectroscopy (EDX, EDX-800HS, Shimadzu Corp., Kyoto, Japan) and a field emission scanning electron microscopy with an energy-dispersive X-ray spectroscopy (FE-SEM-EDS, JSM-7500FA, JEOL Ltd., Tokyo, Japan). The constituent phases of the electrodeposited Co/Cu nanocylinders were investigated using an X-ray diffractometer (XRD, MiniFlex 600-DX, Rigaku Corp., Tokyo, Japan). After the electrodeposition, the nanocylinders were recovered from the AAO template by dissolving them in a sodium hydroxide aqueous solution (5 mol/L). The obtained nanocylinders were observed using a transmission electron microscope (TEM, JEM-2010-UHR, JEOL Ltd., Tokyo, Japan). Using the Co/Cu nanocylinders embedded in an AAO membrane, magnetization and magnetoresistance performance were evaluated using a vibrating-sample-magnetometer (VSM, TM-VSM1014-CRO, Tamakawa Co. Ltd., Sendai, Japan) and a source meter (DC voltage current source monitor, ADCMT6242, ADC Corp., Saitama, Japan). The magnetic field in-plane and perpendicular to the AAO film plane was applied while increasing the field up to 10 kOe. The perpendicular magnetic field corresponds with the axial direction of nanocylinders. The GMR value, G_{MR}, can be defined by the following Equation (1).

$$G_{MR} = \frac{R^{AP} - R^P}{R^P} \tag{1}$$

Here, R^P is the resistance with a maximum magnetic field of 10 kOe, and R^{AP} is the resistance without a magnetic field.

3. Demagnetization Factor and Valet–Fert Model in Multilayered Heterojunction Nanocylinders

The demagnetized field, H_d, can be expressed by the following Equation (2).

$$H_d = \left(\frac{N_d}{\mu_0}\right) \times I \tag{2}$$

Here, N_d is a demagnetization factor, μ_0 represents a magnetic permeability in a vacuum, and I stands for the magnetization strength. N_d can be expressed by Equation (3) as a function of aspect ratio, $k = L/D$ (L: nanocylinder length, D: nanocylinder diameter).

$$N_d = \frac{1}{k^2-1}\left\{\frac{k}{\sqrt{k^2-1}}\ln\left(k+\sqrt{k^2-1}\right)-1\right\} \tag{3}$$

If a nanocylinder has a diameter D of 50 nm and length L of 60 μm, the aspect ratio, $k = L/D$, is 1200. In this case, the demagnetization factor, N_d, can be estimated to be 4.3×10^{-6}, which is almost zero. The spin-valve response in the axial direction will be improved by reducing the demagnetizing field with increased magnetic shape anisotropy.

Based on the Valet–Fert theory, under the conditions of $t_F > l_F^{sf}$ and $t_N < l_N^{sf}$, the spin-valve type GMR value has an inverse proportional relationship with the ferromagnetic layer thickness, t_F, as shown by the following Equations (4)–(6) [21–23].

$$\frac{R^P}{R^{ap}-R^p} = \frac{\left(\frac{\rho_F^*}{\rho_F} - (\beta^e)^2\right)}{2p(\beta^e)^2 l_F^{sf}} t_F \tag{4}$$

$$\rho_F^e = \rho_F^* + \rho_{mix} \qquad (5)$$

$$\beta^e = \frac{\beta}{\left[1 + \frac{\rho_{mix}}{\rho_F^*}\right]} \qquad (6)$$

Here, R^p and R^{ap} are resistance with and without a magnetic field, respectively, while t_F, and l_F^{sf} are the thickness of ferromagnetic layers and spin-diffusion length, respectively. ρ_F^* and ρ_{mix} are the resistivity and spin mixing resistance of ferromagnetic layers, respectively. β is the asymmetric coefficient of bulk scattering spin, and p is the constant ranging from 0.33 to 0.49. Piraux et al. reported that β^e, ρ_F^*, and ρ_F^e were 0.31 ± 0.02, 25 μΩcm, and 29 μΩcm, respectively, in their study on Co/Cu multilayered nanocylinders (D = 90 nm), which were electrodeposited from a sulfuric acid solution at room temperature. The ferromagnetic metal spin-diffusion length, l_F^{sf}, can be obtained from the approximate expression slope using the experimental data of present study.

On the contrary, under the condition of $t_F < l_F^{sf}$ and $t_N < l_N^{sf}$, the GMR value has the following relationship with the non-magnetic layer thickness, t_N, shown in Equation (7).

$$\left(\frac{R^{ap} - R^p}{R^{ap}}\right)^{-1/2} = \frac{\rho_F^* t_F + 2r_b^*}{\beta \rho_F^* t_F + 2\gamma r_b^*} + \frac{\rho_N^* t_N}{\beta \rho_F^* t_F + 2\gamma r_b^*} \qquad (7)$$

Here, ρ_N^* represents the non-magnetic layer resistivity. r_b^* represents interface resistance. In contrast, γ is the asymmetric coefficient of the interface spin. Consequently, Equations (4) and (7) can be simply expressed as the following Equations (8) and (9). Here, a, b, and c mean proportional constants.

$$\frac{R^p}{R^{ap} - R^p} = c \times t_F \qquad (8)$$

$$\left(\frac{R^{ap} - R^p}{R^{ap}}\right)^{-1/2} = a \times t_N + b \qquad (9)$$

In this study, the thickness of ferromagnetic layer, t_F, was varied to determine the spin-diffusion length in the ferromagnetic layer according to Equation (8).

4. Results and Discussion

4.1. Template Synthesis and Electrodeposition Process of Co/Cu Heterojunction Nanocylinders

Figure 1 shows the FE-SEM images of the top-side view (Figure 1a), the cross-sectional view (Figure 1b), and the bottom-side view (Figure 1c) of an AAO nanochannel film that separated from a cross-section of an aluminum rod. The separated AAO film had an ideal nanochannel structure with ~50 nm in diameter. The nanochannel length, which is identical to the AAO film thickness, was ~60 μm.

Figure 2 shows the cathodic (blue line) and anodic (green and red lines) scanned polarization curves (Tafel slope) for Cu and Co electrodeposition from an aqueous solution containing Cu^{2+} and Co^{2+} ions. The Tafel plot was then employed to reveal the reduction behavior of Cu^{2+} ions by magnifying the relatively small current range. According to the Nernst equation, E_{Cu}^{eq} for Cu/Cu^{2+} is estimated to be +0.07 V vs. Ag/AgCl, while E_{Co}^{eq} for Co/Co^{2+} is also calculated to be -0.48 V vs. Ag/AgCl, as follow by Equation (10).

$$E^{eq} = E^0 + \frac{RT}{nF} \ln\left[M^{n+}\right] \qquad (10)$$

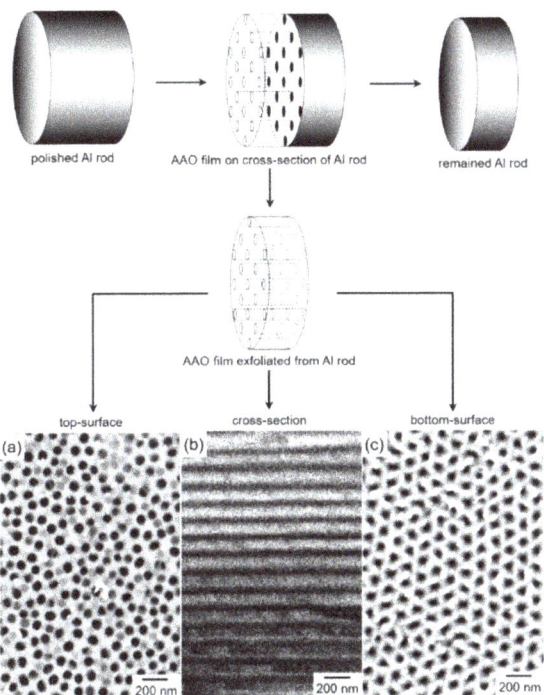

Figure 1. FE-SEM images of top-view (**a**), cross-section (**b**), and bottom-view (**c**) of an anodized aluminum oxide nanochannel template which was exfoliated from the cross-section of an aluminum rod.

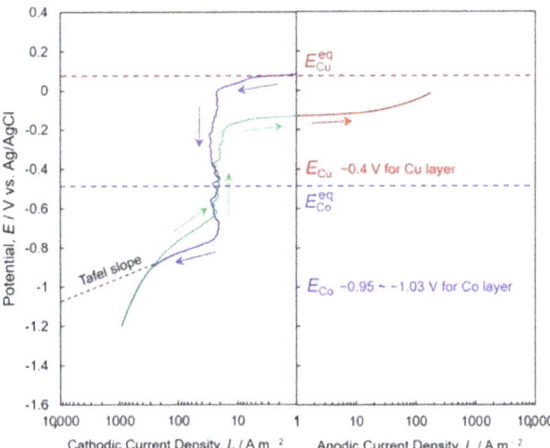

Figure 2. Cathodic (blue line) and anodic (green and red lines) scanned polarization curves (Tafel slope) for Cu and Co electrodeposition from an aqueous solution containing 0.5 M Co (SO$_3$NH$_2$)$_2$·4H$_2$O, 0.005 M CuSO$_4$·5H$_2$O and 0.4 M H$_3$BO$_3$.

Here, E^{eq} and E^0 are the equilibrium potential and standard potential, respectively. R, F, n, and T are gas constant, Faraday constant, ionic valence, and absolute temperature, respectively. [M^{n+}] is the activity of the metal ions. As shown in Figure 2 (cathodic scan: blue line), the cathode current density starts to rise at +0.07 V, which is close to

E_{Cu}^{eq}. It is well known that the normal metal ions, such as Cu^{2+}, Sn^{2+}, Zn^{2+} ions are immediately reduced to the metallic state without substantial overvoltage in an acidic aqueous solution [24]. Hence, this cathode current rising results from Cu^{2+} ions' reduction.

The cathode potential significantly polarizes to -0.80 V at the current density of around 23 A m^{-2}. In the range of current density, Cu^{2+} ions seem to reach a diffusion limit. Moreover, an increase in the cathode current density can be observed at -0.80 V, which is quite less noble than E_{Co}^{eq}. It is well-known that Co^{2+} ions are reduced to a metallic state, accompanying a substantial overvoltage owing to the multi-step reduction process, which was reported by Bockris et al. [25]. Furthermore, in the potential region less noble than -1.2 V, the current density reached over 1000 A m^{-2}, and the cathode potential polarized significantly due to the diffusion limit of Co^{2+} ions [26]. On the contrary, in the anodic scan (green and red lines), the anodic current was observed at -0.13 V. This current seems to be caused by the dissolution of electrodeposited Co. For the pulsed potential deposition of Co/Cu multilayers, the suitable cathode potential for Cu layer, E_{Cu} should be less nobler than E_{Cu}^{eq} (+0.07 V) and initial dissolution potential for Co (-0.13 V). Additionally, E_{Cu} should be nobler than E_{Co}^{eq} (-0.48 V) to avoid Co contamination. Hence, in the present study, E_{Cu} was fixed to -0.4 V, while the suitable cathode potential for Co layer, E_{Co} should be less nobler than E_{Co}^{eq} (-0.48 V) and initial deposition potential for Co (-0.80 V). To prevent Cu contamination, quite less nobler potential than -0.80 V is desirable for Co deposition. Moreover, E_{Co} should be nobler than the diffusion limit potential for Co^{2+} ions (-1.2 V). In this study, Co layer thickness should be controlled within the several tens of nanometer range to investigate the spin-diffusion length based on the Valet–Fert equation. Therefore, E_{Co} was determined to the range from -0.95 V~-1.03 V.

As shown in Figure 3, we synthesized Co/Cu multilayered nanocylinders by switching the cathode potential from -0.4 V (for 1.0 s) to -0.95 V~-1.03 V (for 0.1 s) to adjust the thickness of each layer within several nanometer scale. When the nanocylinders reached the Au thin layer on an AAO template, the reduction current was suddenly enhanced due to the in-situ electric contact with the Au thin layer and formation of hemispheric metal caps as shown in Figure 4. The time for filling AAO nanochannels with Co/Cu multilayered nanocylinders, T_F, was determined from the time-dependence of observed current at the wide range of pulsed-potential deposition time as shown in Figure 4.

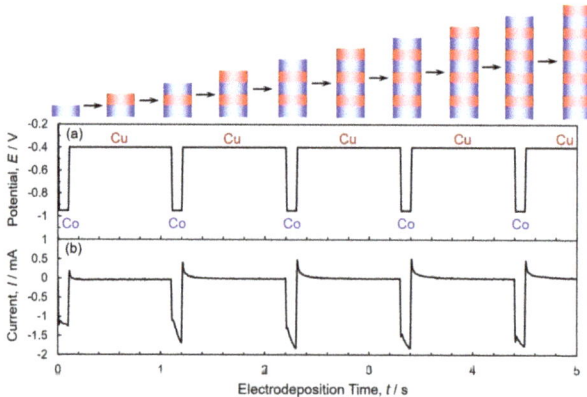

Figure 3. Time-dependence of applied potential (**a**) and observed current (**b**) at the beginning of pulsed-potential deposition time for growing Co/Cu multilayered nanocylinders. The cathode potential was alternatingly changed between -0.4 V (1.0 s) and -0.95 V (0.1 s).

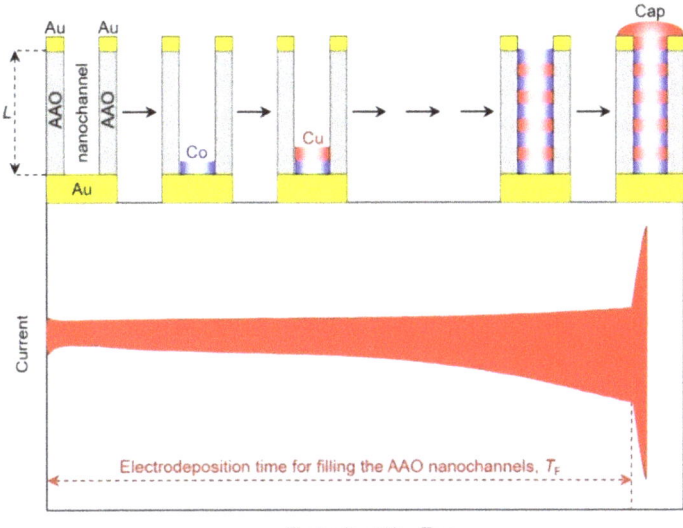

Figure 4. Schematic image for filling the AAO nanochannels with Co/Cu multilayered nanocylinders at the wide range of pulsed-potential deposition time.

The growth rate of Co/Cu multilayered nanocylinders, R_g, can be estimated from dividing the AAO nanochannel's length, L, by the filling time, T_F. Furthermore, Co/Cu bilayer thickness, $t_{Co/Cu}$, can be also estimated from the following Equation (11).

$$t_{Co/Cu} = L \frac{T_{Co} + T_{Cu}}{T_F} \qquad (11)$$

Here, T_{Co} and T_{Cu} are the pulse-deposition time for each Co and Cu layer, respectively. In the present study, T_{Co} and T_{Cu} correspond to 0.1 s and 1.0 s, respectively.

Figure 5a,b show the effect of E_{Co} (pulsed potential for Co layer deposition) on the nanocylinder growth rate, R_g and Co/Cu bilayer thickness, $t_{Co/Cu}$, respectively. When E_{Co} was shifted to the less noble region, R_g and $t_{Co/Cu}$ increased logarithmically up to 27.9 nm s^{-1} and 30.7 nm, respectively. Based on Tafel equation ($\eta = a + b\log i$), the overpotential, η, is proportional to the logarithm of current, $\log i$, when the charge transfer process controls the electrochemical reaction. It is well-known that the nanocylinder growth rate and bilayer thickness are a linear relationship with the electrodeposition current density based on Faraday's laws of electrolysis. Hence, R_g and $t_{Co/Cu}$ should be increased logarithmically with increasing the overpotential. The composition of Co, X_{Co} and that of Cu, X_{Cu} in each sample were also determined from EDX analysis (EDX-800HS, Shimadzu, Kyoto, Japan) as shown in Figure 5c. All over the potential range from −0.95 V to −1.03 V, the average X_{Co} and X_{Cu} were 87.58% and 12.42%, respectively. The compositions were also investigated by FE-SEM-EDS analysis (JSM-7500FA, JEOL, Tokyo, Japan). The average X_{Co} and X_{Cu} were also determined to 87.96% and 12.04%, respectively. If the Cu impurities in Co layers are negligible, each average layer thickness of Co and Cu, t_{Co} and t_{Cu}, can be estimated from the following Equations (12) and (13), respectively.

$$t_{Co} = t_{Co/Cu} \frac{X_{Co}}{100} \qquad (12)$$

$$t_{Cu} = t_{Co/Cu} \frac{X_{Cu}}{100} \qquad (13)$$

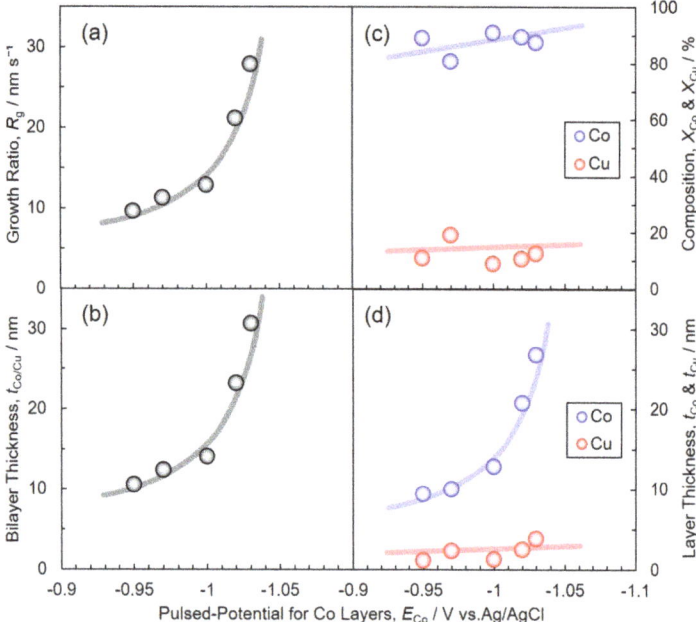

Figure 5. Effects of pulsed-potential for Co layers, E_{Co}, on the growth rate of nanocylinders, R_g (**a**), Co/Cu bilayer thickness, $t_{Co/Cu}$ (**b**), the average composition, X_{Co} and X_{Cu} (**c**), and the average layer thickness, t_{Co} and t_{Cu} (**d**). T_{Co}, E_{Cu} and T_{Cu} were fixed to 0.1 s, −0.40 V and 1.0 s, respectively.

The effect of E_{Co} on t_{Co} and t_{Cu} is shown in Figure 5d. The t_{Cu} was almost constant at less than 4 nm all over the potential range. On the other hand, t_{Co} became thicker as E_{Co} was shifted to a less noble region. According to the above results, it was revealed that t_{Co} can be controlled within the range from 8 to 27 nm by tuning E_{Co}.

4.2. Structure of Co/Cu Heterojunction Nanocylinders

Figure 6 shows TEM bright-field images of Co/Cu multilayered nanocylinders. The samples were prepared by ranging the pulsed-potential for Co layer, E_{Co} as the following: Figure 6a E_{Co} = −0.95 V, Figure 6b E_{Co} = −0.97 V and Figure 6c,c' E_{Co} = −1.00 V. While the other parameters: T_{Co}, E_{Cu} and T_{Cu} were fixed to 0.1 s, −0.40 V and 1.0 s, respectively. The Co/Cu multilayered nanocylinders were separated from AAO templates. As shown in Figure 6, the diameter of Co/Cu multilayered nanocylinder is ~50 nm, which is almost identical to the diameter of AAO nanochannels as shown in Figure 1. The nanocylinder also has a multilayered heterojunction structure. The layer thickness of a dark thick layer is ~10 nm while that of a light thin layer is ~2 nm. The thick and thin layers correspond to the Co and Cu layers, respectively, considering the estimated layer thickness, as shown in Figure 5d.

Figure 7 renders the effect of E_{Co} on the XRD profiles of Co/Cu multilayered nanocylinders. As shown in Figure 7, the observed peaks at 2θ = 41.25°, 44.1°, 44.4°, and 47.25° are derived from hcp-Co (100), fcc-Co (111), hcp-Co (002), and hcp-Co (101), respectively. The diffraction peak, which is derived from fcc-Co, is observed at 2θ = 44.1°. The presence of fcc-Co could be caused by the phase transformation from the hcp to the fcc structure because a part of the Co layer seems to contain Cu as the impurity element. Other researchers have also reported that the fcc-Co phase existed in the X-ray diffraction pattern on their Co/Cu multilayered films [27]. In contrast, the diffraction peak of fcc-Co disappeared when the pulsed potential for Co layer was set to a less nobler region. The peak disappearance results from an increase in the Co layer thickness, as shown in Figure 5d.

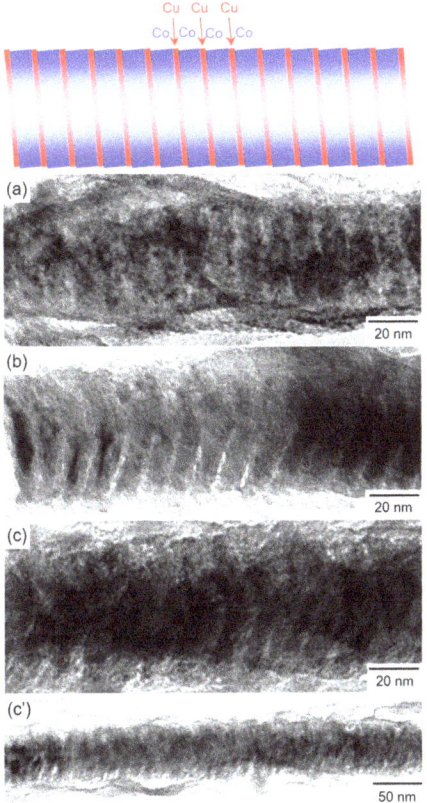

Figure 6. TEM images of Co/Cu multilayered nanocylinders that were separated from an anodized aluminum oxide nanochannel template. (**a**) E_{Co} = −0.95 V, (**b**) E_{Co} = −0.97 V, (**c,c′**) E_{Co} = −1.00 V. T_{Co}, E_{Cu} and T_{Cu} were fixed to 0.1 s, −0.40 V and 1.0 s, respectively.

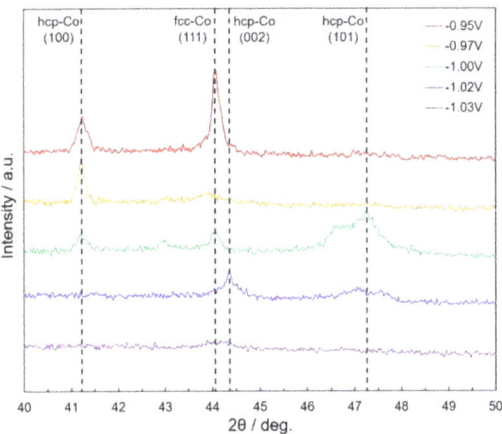

Figure 7. Effect of pulsed-potential for Co layer, E_{Co} on the X-ray diffractograms of Co/Cu multilayered nanocylinders. E_{Co} was set for −0.95 V, −0.97 V, −1.00 V, −1.02 V and −1.03 V. T_{Co}, E_{Cu} and T_{Cu} were fixed to 0.1 s, −0.40 V and 1.0 s, respectively.

4.3. Magnetoresistance Properties of Co/Cu Multilayered Heterojunction Nanocylinders

The effect of E_{Co} on the magnetic and magnetoresistance hysteresis curves of Co/Cu multilayered nanocylinder arrays is shown in Figure 8. The hysteresis curves, which were obtained in the magnetic field perpendicular to the AAO film, are plotted in the solid lines, while the curves that obtained in-plane direction are plotted in the dotted lines. As shown in the dotted lines of Figure 8a–d, it is quite difficult to achieve the saturation magnetization with a magnetic field in-plane direction to the AAO film due to a substantial demagnetizing field, H_d. The demagnetization factor, N_d with in-plane direction can be estimated to ~0.5. On the other hand, as shown by the solid lines, it is relatively easy to achieve the saturation magnetization with a perpendicular magnetic field to the AAO film plane. As shown in Equation (2), H_d will be minimal in a perpendicular direction, which corresponds to the axial direction of a nanocylinder. In this case, the external magnetic field will be effective and not reduced. Hence, the saturation magnetization can be realized by a small external magnetic field (~2 kOe) in the long axis direction of nanocylinders [28].

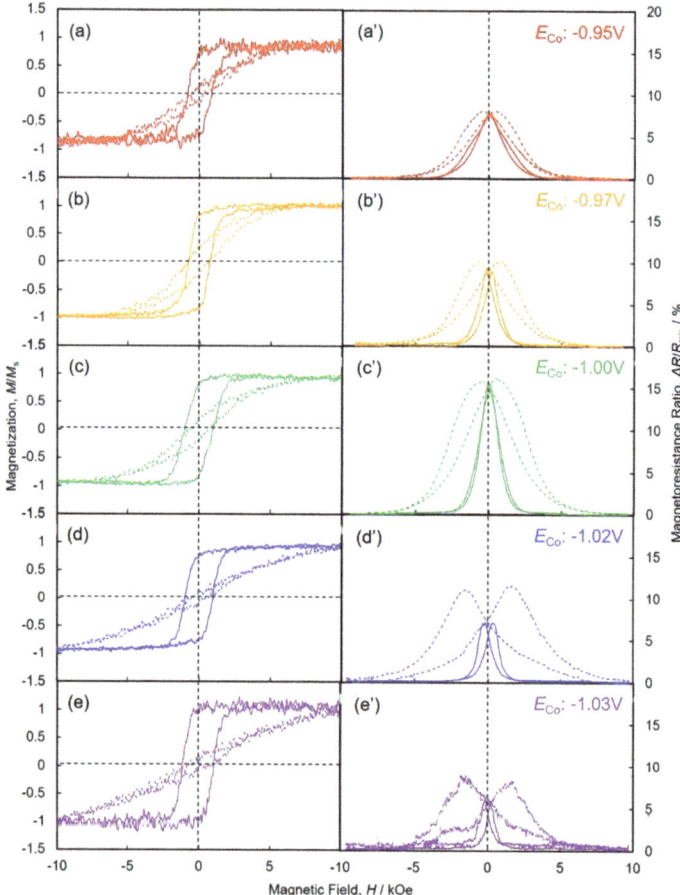

Figure 8. Effect of pulsed-potential for Co layers, E_{Co} on the magnetic and magnetoresistance hysteresis loops of AAO nanochannel films with Co/Cu multilayered nanocylinder arrays. E_{Co} was set for −0.95 V (**a,a′**), −0.97 V (**b,b′**), −1.00 V (**c,c′**), −1.02 V (**d,d′**) and −1.03 V (**e,e′**). T_{Co}, E_{Cu} and T_{Cu} were fixed to 0.1 s, −0.40 V and 1.0 s, respectively. The magnetic field was applied to in-plane (dotted lines) and perpendicular (solid lines) directions to the multilayer interfaces.

If the resistance of a multilayered structure can be expressed by the linear relationship with the composition, the resistance of a Co/Cu multilayered nanocylinder can be defined using the resistivities of a Co layer and a Cu layer as shown in Equation (14).

$$R = \left(\rho_{Co} \frac{X_{Co}}{100} + \rho_{Cu} \frac{X_{Cu}}{100} \right) \frac{L}{S} \tag{14}$$

Here, R is the resistance of a Co/Cu nanocylinder. ρ_{Co} and ρ_{Cu} are the resistivity of a Co layer (64.2 Ω/nm) and a Cu layer (16.8 Ω/nm), respectively. L and S are the length (60 µm) and the cross-section area (~6360 nm^2) of a nanocylinder, respectively. According to Equation (14), R will increase with increasing X_{Co} because ρ_{Co} is larger than ρ_{Cu}. Based on our experimental results, the resistance of a Co/Cu multilayered nanocylinder, which was synthesized by an electrochemical in situ contact process, corresponded to the estimated value for the parallel contacts with only 1~3 nanocylinders regardless of the composition.

As shown by the dotted lines of Figure 8a–d, the magnetoresistance of Co/Cu multilayered nanocylinder arrays decreased like a Gaussian curve. The resistance reached the minimum in the range more than ~7 kOe as the magnetic field increased slowly in the in-plane (parallel) direction. On the other hand, the magnetoresistance ratio decreased quickly and reached zero at ~2 kOe with an increasing magnetic field in the perpendicular (axial) direction, as shown by the solid lines.

The GMR value of Co/Cu multilayered nanocylinder arrays, which were electrodeposited at E_{Co} of -1.03 V, was ~9%, as shown in Figure 8e'. While, the GMR value of the nanocylinder arrays, which were electrodeposited at E_{Co} of -1.00 V, increased up to ~16%, as shown in Figure 8c'. It has been reported that the GMR value increases as the number of interfaces between ferromagnetic and non-magnetic layers increases [29]. As shown in Figure 5d, the Co layer thickness became thinner as the pulsed potential was shifted to a noble region. This decrease in the Co layer thickness increases the number of layer interfaces. Hence, this increase in GMR seems to be caused by decreases in the Co layer thickness. For further improving the CPP-GMR performance, the Co layer thickness, t_{Co} was decreased by shortening the pulse-deposition time for Co layer, T_{Co}. To maintain the throwing power for the pulse-deposition, the pulsed-potential for Co layer, E_{Co} was kept to less nobler than -1.03 V. Figure 9 show the magnetoresistance hysteresis loops of an AAO nanochannel film with Co/Cu multilayered nanocylinder arrays. The nanocylinder arrays were electrodeposited using the pulse parameters of $E_{Co} = -1.05$ V, $T_{Co} = 0.03$ s, $E_{Cu} = -0.4$ V and $T_{Cu} = 1.0$ s. As shown in Figure 9, the CPP-GMR value reached up to ca. 23% in the Co/Cu multilayered nanocylinder with 8.6 nm in t_{Co} and 4 nm in t_{Cu}.

Table 1 shows the summary of CPP-GMR performances (at room temperature) of electrodeposited multilayered nanocylinders that were reported by the other researchers. Most researchers have reported that the CPP-GMR value reached up to ca. 15~20% at room temperature in the t_{Co} range from ca. 5 to 20 nm and the t_{Cu} range from ca. 5 to 10 nm. Those values give good agreement with the value obtained in the present study.

Figure 10 shows the effect of Co layer thickness on the GMR value and $(\Delta R/R_p)^{-1}$ of Co/Cu multilayered nanocylinders. As shown in Figure 10a, the GMR value increases with a decreasing Co layer thickness. In the Co layer thickness of 8.6 nm, the GMR value reached up to ~23%. As shown in Figure 10b, with a decrease in the thickness of the Co layer, $(\Delta R/R_p)^{-1}$ decreases linearly [13]. This tendency corresponds well to Valet–Fert Equation (8). The spin-diffusion length of magnetic metal can also be estimated from the slope of approximate expression in Figure 10b. Consequently, the cobalt spin-diffusion length, l_{Co}^{sf}, was estimated to be ~12.5 nm. As the thickness of the Co layer, t_{Co}, is from 8 to 27 nm, the condition of $t_F > l_F^{sf}$ in the Valet–Fert model seems to be satisfied by the results in the present study.

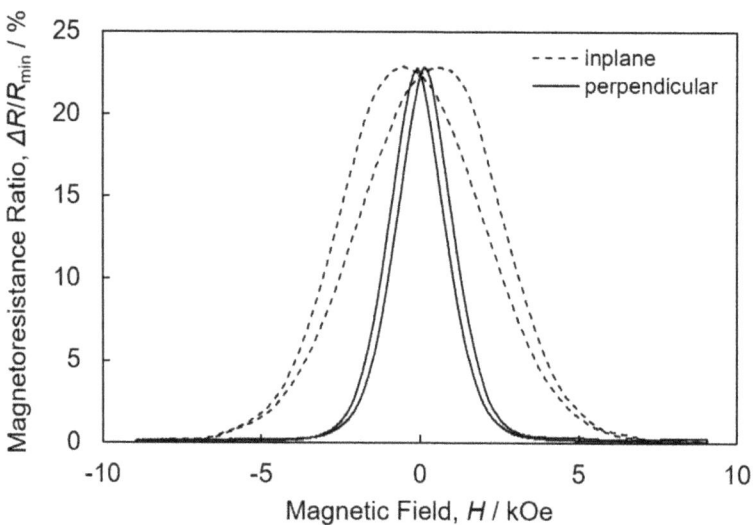

Figure 9. Magnetoresistance hysteresis loops of an AAO nanochannel film with Co/Cu multilayered nanocylinder arrays. The sample was synthesized using the pulse parameters of E_{Co} = −1.05 V, T_{Co} = 0.03 s, E_{Cu} = −0.4 V and T_{Cu} = 1.0 s. The magnetic field was applied to in-plane (dotted lines) and perpendicular (solid lines) directions to the multilayer interfaces.

Table 1. Summary of CPP-GMR performance (at room temperature) of multilayered nanocylinders electrodeposited into AAO that were reported by the other researchers. The nanocylinders in Refs. [8,9] were electrodeposited into ion-track-etched polycarbonate membranes.

Authors	FM/NM	GMR/%	D/nm	L/µm	L/D	t_{Co}/nm	t_{Cu}/nm	Source Title	Year	Ref.
Piraux et al.	Co/Cu	15	40	10	250	10	10	*Appl. Phys. Lett.*	1994	[8]
Blondel et al.	Co/Cu	14	80	6	75	5	5	*Appl. Phys. Lett.*	1994	[9]
Evans et al.	CoNi/Cu	55	300	60	200	5	2	*Appl. Phys. Lett.*	2000	[10]
Ohgai et al.	Co/Cu	15	60	2	33	10	10	*J. Appl. Electrochem.*	2004	[11]
Tang et al.	Co/Cu	14	300	60	200	8	10	*J. Appl. Phys.*	2006	[12]
Tang et al.	CoNi/Cu	23	300	60	200	10	4	*Phys. Rev. B*	2007	[13]
Shakya et al.	FeCoNi/Cu	15	300	60	200	14	10	*J. Magn. Magn. Mater*	2012	[14]
Zhang et al.	FeNi/Cu/Co	45	120	2	17	25	15	*J. Mater. Sci. M. E.*	2015	[15]
Han et al.	Co/Cu	13	50	11	220	50	5	*Adv. Cond. Mat. Phys.*	2016	[16]
Xi et al.	Co/Cu	0.16	80	3	38	200	5	*Physica B*	2017	[17]
Kamimura et al.	Co/Cu	24	75	70	933	19	1.4	*Nanomaterials*	2020	[18]

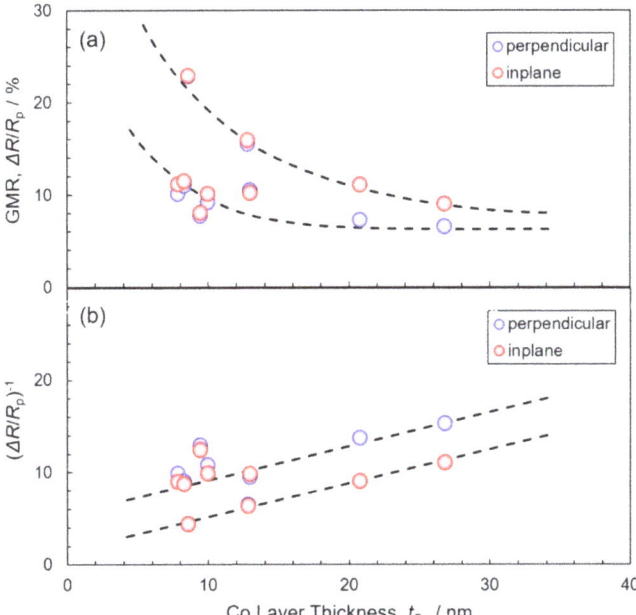

Figure 10. Effect of Co layer thickness on GMR (**a**) and $(\Delta R/R_p)^{-1}$ (**b**) in electrodeposited Co/Cu multilayered nanocylinders.

5. Conclusions

AAO nanochannel films (D: ~50 nm, L: ~60 μm) were fabricated using an anodization and exfoliation technique from a metallic aluminum rod. The Co/Cu multilayered nanocylinders were fabricated by alternating the cathode potentials for Cu and Co deposition to adjust the Co layer thickness within ~30 nm. From the TEM images of the Co/Cu multilayered nanocylinders, it was confirmed that the Co and Cu layers were alternately laminated, and the diameter of the nanocylinders was the same as the pore diameter of the AAO template. The multilayered nanocylinders with alternating Cu and Co layers contained both hcp and fcc phases of cobalt. The multilayered nanocylinders with alternating Cu and Co layers reached saturation magnetization with a small magnetic field (~2 kOe) in the axial direction of nanocylinders due to the substantial aspect ratio. As the Co layer thickness decreased, the GMR reached up to approx. 23%. When decreasing the Co layer thickness, $(\Delta R/R_p)^{-1}$ linearly decreased according to the Valet–Fert equation; this can be explained under the condition of $t_F > l_F^{sf}$ and $t_N < l_N^{sf}$. The cobalt spin-diffusion length, l_{Co}^{sf}, was estimated to be ~12.5 nm by the slope of approximate expression.

Author Contributions: S.M. and M.H. carried out experiments, analyzed data, and wrote the manuscript; T.O. designed the study, supervised the project, and analyzed data. All authors have read and agreed to the published version of the manuscript.

Funding: This research was funded by the Japan Society for the Promotion of Science, grant number 18H01754.

Institutional Review Board Statement: Not applicable.

Informed Consent Statement: Not applicable.

Acknowledgments: The authors acknowledge financial support from the Japan Society for the Promotion of Science.

Conflicts of Interest: The authors declare that they have no competing interests.

References

1. Baibich, M.N.; Broto, J.M.; Fert, A.; Nguyen, V.D.F.; Petroff, F. Giant Magnetoresistance of (001)Fe/(001)Cr Magnetic Super-Lattices. *Phys. Rev. Lett.* **1988**, *61*, 2472–2475. [CrossRef] [PubMed]
2. Binasch, G.; Grünberg, P.; Saurenbach, F.; Zinn, W. Enhanced Magnetoresistance in Fe-Cr Layered Structures with Antiferro-Magnetic Interlayer Exchange. *Phys. Rev. B* **1989**, *39*, 4828. [CrossRef] [PubMed]
3. Alper, M.; Attenborough, K.; Hart, R.; Lane, S.; Lashmore, D.S.; Younes, C.; Schwarzacher, W. Giant Magnetoresistance in Electrodeposited Superlattices. *Appl. Phys. Lett.* **1993**, *63*, 2144–2146. [CrossRef]
4. Lenczowski, S.; Schönenberger, C.; Gijs, M.; De Jonge, W. Giant Magnetoresistance of Electrodeposited Co/Cu Multilayers. *J. Magn. Magn. Mater.* **1995**, *148*, 455–465. [CrossRef]
5. Tóth, B.G.; Peter, L.; Dégi, J.; Revesz, A.; Oszetzky, D.; Molnár, G.; Bakonyi, I. Influence of Cu Deposition Potential on the Giant Magnetoresistance and Surface Roughness of Electrodeposited Ni–Co/Cu Multilayers. *Electrochim. Acta* **2013**, *91*, 122–129. [CrossRef]
6. Sahin, T.; Kockar, H.; Alper, M. Properties of Electrodeposited CoFe/Cu Multilayers: The Effect of Cu Layer Thickness. *J. Magn. Magn. Mater.* **2015**, *373*, 128–131. [CrossRef]
7. Zsurzsa, S.; Peter, L.; Kiss, L.; Bakonyi, I. Magnetic and Magnetoresistance Studies of Nanometric Electrodeposited Co Films and Co/Cu Layered Structures: Influence of Magnetic Layer Thickness. *J. Magn. Magn. Mater.* **2017**, *421*, 194–206. [CrossRef]
8. Piraux, L.; George, J.-M.; Despres, J.F.; Leroy, C.; Ferain, E.; Legras, R.; Ounadjela, K.; Fert, A. Giant Magnetoresistance in Magnetic Multilayered Nanowires. *Appl. Phys. Lett.* **1994**, *65*, 2484–2486. [CrossRef]
9. Blondel, A.; Meier, J.P.; Doudin, B.; Ansermet, J. Giant Magnetoresistance of Nanowires of Multilayers. *Appl. Phys. Lett.* **1994**, *65*, 3019–3021. [CrossRef]
10. Evans, P.R.; Yi, G.; Schwarzacher, W. Current Perpendicular to Plane Giant Magnetoresistance of Multilayered Nanowires Electrodeposited in Anodic Aluminum Oxide Membranes. *Appl. Phys. Lett.* **2000**, *76*, 481–483. [CrossRef]
11. Ohgai, T.; Hoffer, X.; Gravier, L.; Ansermet, J.P. Electrochemical Surface Modification of Aluminium Sheets for Application to Nano-Electronic Devices: Anodization Alminium and Cobalt-Copper Electrodeposition. *J. Appl. Electrochem.* **2004**, *34*, 1007–1012. [CrossRef]
12. Tang, X.-T.; Wang, G.-C.; Shima, M. Perpendicular Giant Magnetoresistance of Electrodeposited Co/Cu-Multilayered Nanowires in Porous Alumina Templates. *J. Appl. Phys.* **2006**, *99*, 033906. [CrossRef]
13. Tang, X.T.; Wang, G.C.; Shima, M. Layer Thickness Dependence of CPP Giant Magnetoresistance in Individual CoNi/Cu mul-tilayer Nanowire Grown by Electrodeposition. *Phys. Rev. B* **2007**, *75*, 134404. [CrossRef]
14. Shakya, P.; Cox, B.; Davis, D. Giant Magnetoresistance and Coercivity of Electrodeposited Multilayered FeCoNi/Cu and CrFeCoNi/Cu. *J. Magn. Magn. Mater.* **2012**, *324*, 453–459. [CrossRef]
15. Zhang, W.; Deng, H.; Li, H.; Su-Wei, Y.; Wang, H. Synthesis and Magnetic Properties of Ni–Fe/Cu/Co/Cu Multilayer Nanowire Arrays. *J. Mater. Sci. Mater. Electron.* **2015**, *26*, 2520–2524. [CrossRef]
16. Han, J.; Qin, X.; Quan, Z.; Wang, L.; Xu, X. Perpendicular Giant Magnetoresistance and Magnetic Properties of Co/Cu Nan-Owire Arrays Affected by Period Number and Copper Layer Thickness. *Adv. Condens. Matter Phys.* **2016**, *2016*, 1–9. [CrossRef]
17. Xi, H.; Gao, Y.; Liu, Z.; Han, G.; Lu, J.; Li, Y. Inter- and Intra-nanowire Magnetic Interaction in Co/Cu Multilayer Nanowires Deposited by Electrochemical Deposition. *Phys. B Condens. Matter* **2017**, *518*, 77–80. [CrossRef]
18. Kamimura, H.; Hayashida, M.; Ohgai, T. CPP-GMR Performance of Electrochemically Synthesized Co/Cu Multilayered Nanowire Arrays with Extremely Large Aspect Ratio. *Nanomaterials* **2020**, *10*, 5. [CrossRef]
19. Lee, W.; Ji, R.; Gösele, U.; Nielsch, K. Fast Fabrication of Long-Range Ordered Porous Alumina Membranes by Hard Anodization. *Nat. Mater.* **2006**, *5*, 741–747. [CrossRef]
20. Ohgai, T.; Mizumoto, M.; Nomura, S.; Kagawa, A. Electrochemical Fabrication of Metallic Nanowires and Metal Oxide na-Nopores. *Mater. Manuf. Process.* **2007**, *22*, 440–443. [CrossRef]
21. Valet, T.; Fert, A. Theory of the Perpendicular Magnetoresistance in Magnetic Multilayers. *Phys. Rev. B* **1993**, *48*, 7099–7113. [CrossRef] [PubMed]
22. Piraux, L.; Dubois, S.; Fert, A.; Belliard, L. The Temperature Dependence of the Perpendicular Giant Magnetoresistance in Co/Cu Multilayered Nanowires. *Eur. Phys. J. B* **1998**, *4*, 413–420. [CrossRef]
23. Fert, A.; Piraux, L. Magnetic nanowires. *J. Magn. Magn. Mater.* **1999**, *200*, 338–358. [CrossRef]
24. Nakano, H.; Ohgai, T.; Fukushima, H.; Akiyama, T.; Kammel, R. Factors Determining the Critical Current Density for Zinc Deposition in Sulfate Solutions. *Metall* **2001**, *55*, 676–681.
25. Bockris, J.O.; Kita, H. Analysis of Galvanostatic Transients and Application to the Iron Electrode Reaction. *J. Electrochem. Soc.* **1961**, *108*, 676–685. [CrossRef]
26. Ohgai, T.; Tanaka, Y.; Fujimaru, T. Soft Magnetic Properties of Ni-Cr and Co-Cr Alloy Thin Films Electrodeposited from Aqueous Solutions Containing Trivalent Chromium Ions and Glycine. *J. Appl. Electrochem.* **2012**, *42*, 893–899. [CrossRef]
27. Rafaja, D.; Schimpf, C.; Klemm, V.; Schreiber, G.; Bakonyi, I.; Péter, L. Formation of Microstructural Defects in Electrodeposited Co/Cu Multilayers. *Acta Mater.* **2009**, *57*, 3211–3222. [CrossRef]
28. Ohgai, T.; Washio, R.; Tanaka, Y. Anisotropic Magnetization Behavior of Electrodeposited Nanocrystalline Ni-Mo Alloy Thin Films and Nanowires Array. *J. Electrochem. Soc.* **2012**, *159*, H800–H804. [CrossRef]
29. Dulal, S.; Charles, E. Effect of Interface Number on Giant Magnetoresistance. *J. Phys. Chem. Solids* **2010**, *71*, 309–313. [CrossRef]

Article

Magnetic Imaging of Encapsulated Superparamagnetic Nanoparticles by Data Fusion of Magnetic Force Microscopy and Atomic Force Microscopy Signals for Correction of Topographic Crosstalk

Marc Fuhrmann [1], Anna Musyanovych [2], Ronald Thoelen [3], Sibylle von Bomhard [2] and Hildegard Möbius [1,*]

[1] Department of Computer Sciences/Micro Systems Technology, University of Applied Sciences Kaiserslautern, Amerika Str. 1, 66482 Zweibrücken, Germany; marc.fuhrmann@hs-kl.de
[2] Nanoparticle Technology Department, Fraunhofer IMM, Carl-Zeiss-Str. 18-20, 55129 Mainz, Germany; anna.musyanovych@imm.fraunhofer.de (A.M.); Sibylle.von.Bomhard@imm.fraunhofer.de (S.v.B.)
[3] Institute for Materials Research, Hasselt University, Martelarenlaan 42, 3500 Hasselt, Belgium; ronald.thoelen@uhasselt.be
* Correspondence: hildegard.moebius@hs-kl.de; Tel.: +49-631-3724-5412

Received: 23 November 2020; Accepted: 9 December 2020; Published: 11 December 2020

Abstract: Encapsulated magnetic nanoparticles are of increasing interest for biomedical applications. However, up to now, it is still not possible to characterize their localized magnetic properties within the capsules. Magnetic Force Microscopy (MFM) has proved to be a suitable technique to image magnetic nanoparticles at ambient conditions revealing information about the spatial distribution and the magnetic properties of the nanoparticles simultaneously. However, MFM measurements on magnetic nanoparticles lead to falsifications of the magnetic MFM signal due to the topographic crosstalk. The origin of the topographic crosstalk in MFM has been proven to be capacitive coupling effects due to distance change between the substrate and tip measuring above the nanoparticle. In this paper, we present data fusion of the topography measurements of Atomic Force Microscopy (AFM) and the phase image of MFM measurements in combination with the theory of capacitive coupling in order to eliminate the topographic crosstalk in the phase image. This method offers a novel approach for the magnetic visualization of encapsulated magnetic nanoparticles.

Keywords: atomic force microscopy; magnetic force microscopy; hybrid nanoparticles; polystyrene; data fusion

1. Introduction

Magnetic nanoparticles encapsulated in a polymer matrix are of increasing importance for medical applications such as magnetic drug delivery, contrast agent for magnetic resonance imaging (MRI) and hyperthermia for cancer treatment [1–5].

Especially superparamagnetic iron oxide nanoparticles (SPIONs) are of high interest due to their unique magnetic properties. However, there is still the need for localized magnetic characterization of the encapsulated SPIONs. Magnetic Force Microscopy (MFM) has proved to be a suitable tool to image SPIONs and to map SPIONs embedded in a polymer film giving information simultaneously about spatial distribution and the magnetic behavior [6–12]. Passeri et al. used MFM measurements for the detection of the magnetic core of magneto ferritin and for the determination of the diameter of agglomerates in niosomes for drug delivery [13]. However, MFM measurements on nanoparticles face the difficulty that the magnetic signals interfere with topographic crosstalk because of the distance change between tip and substrate when measuring the nanoparticles [12,14]. This mirroring of surface structures in MFM phase images is still an issue in MFM research [15–17]. The origin of the crosstalk was experimentally proven and theoretically explained by capacitive coupling between tip and substrate [12]. This effect becomes relevant for structures smaller than the tip radius such as surface roughness or measurements on nanoparticles [18]. In interleave mode, a first scanline measures the topography of the sample, a second scanline following the topography of the first scan at a defined distance, the lift height, measures the phase image. In MFM, the phase image corresponds in principle to long range magnetic forces of the sample. However, the distance change between tip and substrate measuring above nanostructures leads to a positive phase shift indicating a positive force gradient, which might be erroneously interpreted by a repulsive magnetic force. Various methods have been suggested to minimize the topographic crosstalk. Angeloni et al. suggested a change in tip magnetization to distinguish between magnetic and electrostatic forces [19]. Analyzing the parameters relevant for the crosstalk opens several possibilities to reduce this effect [12]. Choosing a substrate with a small contact potential difference between substrate and tip, the crosstalk can be reduced. Introducing a voltage between tip and substrate to cancel the contact potential difference between tip and substrate minimizes the crosstalk as well, but the additional voltage is a further parameter and may influence the measurements [16,17]. Choosing a tip with a small tip radius also reduces the crosstalk having the disadvantage of a smaller tip magnetization. It was shown that introducing a dielectric layer between substrate and nanoparticle the topographic crosstalk can be reduced significantly because, in this case, the distance change following the topography is small compared to the overall distance between substrate and tip [18]. For measuring magnetic nanoparticles one possibility is to embed the nanoparticles in a dielectric layer in order to completely remove the topographic crosstalk [11].

In order to compensate for the topographic crosstalk in general and independently of the sample, a numerical method is needed that calculates the capacitive coupling and the topographic influence on the phase image data for each measuring point. In this paper we present the concept of data fusion of Atomic Force Microscopy (AFM) topography and MFM phase signals to correct the phase signals from topographic crosstalk. This method allows to obtain pure magnetic signals without introducing further measurement parameters such as an additional voltage and without introducing additional process steps such as the embedding of the nanoparticles. As a model system to test the concept of data fusion, unloaded polystyrene nanoparticles and polystyrene (PS) nanoparticles loaded with SPIONs are investigated.

Data fusion on unloaded PS nanoparticles prove the concept of data fusion to compensate the topographic crosstalk completely. Data fusion on single SPIONs reveal the importance of the correction of the topographic crosstalk in order to obtain pure magnetic signals. The measurements confirm the superparamagnetic state of the SPIONs. Data fusion on PS nanoparticles loaded with SPIONs give pure magnetic signals, which show the distribution of the SPIONs in the PS capsules in accordance with Transmission Electron Measurements (TEM). It is for the first time possible to obtain spatially resolved magnetic information of encapsulated SPIONs. Only attractive forces are observed, indicating that the encapsulated SPIONs are still in the superparamagnetic state.

With the help of data fusion of AFM and MFM measurements it is now possible to discuss and interpret magnetic phase images without falsification due to the topographic crosstalk

2. Materials and Methods

The synthesis of polystyrene nanoparticles is based on the method reported by Musyanovych et al. [20]. Polystyrene nanoparticles with encapsulated magnetite were synthesized through free radical miniemulsion polymerization using a ferrofluid of oleic acid-stabilized iron oxide nanoparticles (Webcraft GmbH, Gottmadingen, Germany) without further purification. In order to compare the influence of the production method on the manufactured particles, sufficient amounts of organic and aqueous phases were prepared to have identical starting conditions for the preparation of the particles. The synthesis was performed in a ratio of 1:0 and 1:0.2 polystyrene to magnetite.

For the sample preparation, the polystyrene magnetite nanoparticles in an aqueous solution were diluted with ultrapure water. Single drops of 1–3 mL of the solution were pipetted onto the substrates and allowed to dry.

MFM measurements were performed on a Bruker Dimension Icon atomic force microscope (Bruker AXS, Karslruhe, Germany). The standard methods tapping mode for topography measurements and dynamic lift mode for phase measurements were used. The lift height for measurements was 50 nm. KPFM measurements with identical tips were performed to measure the contact potential difference between substrate and measuring tip. In this work ASYMFM-HM tips were used. The collected measurement data were processed and evaluated in the SPM (scanning probe microscopy) software NanoScope Analysis (version 1.9, Bruker AXS, Karslruhe, Germany) provided by Bruker, as well as the free analysis software Gwyddion (version 2.55, www.gwyddion.net) for Data matrix extraction [21]. For data fusion and graphing, OriginPro (version 2020b, OriginLab Corp., Northampton, MA, USA) was used.

Data mapping is an important step towards eliminating topographic crosstalk and requires accurate measurement evaluation. An AFM generates a measurement image by dividing a predefined area into a discrete number of lines and scanning them one by one. The resolution of the measurement thus depends primarily on the number of lines but also on the number of measuring points per line and therefore is also size-dependent on the predefined area. Measurements were performed at a sample/line rate of 256. In this case, the scanned area is divided into 256 lines with each line having 256 measuring points. Each measurement point thus has the topography information from the first trace and the corresponding phase information from the second trace to the relevant reference point of the topography. As the measurement progresses, a data matrix is formed with X (lines) times Y (measuring points per line) measured values. In addition to the topography values, this matrix also contains phase values. Thus, for each spatially resolved point of the matrix there is topography and phase information in relation to each measuring point. It is of great importance that individual data points are exactly assigned to their measured phase. The measurement data are extracted purely numerically as a 256 × 256 data matrix. With the help of data analysis tools like OriginPro, height values can thus be linked and evaluated with those of the measured and calculated phase. The data matrix cleaned up by the topographical crosstalk can then be transferred and evaluated. A simplified form of data mapping is the linking of data within a cross-section in the area. This reduces the previously measured two-dimensional area into a one-dimensional measurement line and links the topographic measurement points of this line to the measured phase values. Using the Nanoscope Analysis measuring program, the measurement can be analyzed in two separate windows. Topography and phase values can be linked with each other by this procedure.

The morphology was investigated by transmission electron microscopy (TEM) using Zeiss Libra 120 of (Carl Zeiss NTS GmbH, Oberkochen, Germany), operating at an acceleration voltage of 120 kV. The particle dispersions were diluted with demineralized water, dropped on a 300-mesh carbon-coated copper grid and dried at ambient temperature. No additional contrasting was applied.

3. Theory

3.1. Capacitive Coupling

In interleave mode the lift height z between tip and surface structure is kept constant at every measuring point. Measuring small structures, for example, nanoparticles, the distance change between tip and substrate leads to a contribution to the capacity between tip and substrate depending on the topography of the structure t_{topo}.

The distance change between tip and substrate following the topography in the interleave mode leads in total to a positive phase shift in the MFM signal:

$$\Delta\phi_{el} = -\frac{Q}{k}\left(F'(z+t_{topo}) - F'(z)\right) = -\frac{Q}{k}\varepsilon_0\left(\frac{A_{eff}}{(z+t_{topo})^3}(V_{CPD})^2 - \frac{A_{eff}}{(z)^3}(V_{CPD})^2\right) > 0, \quad (1)$$

where F': force gradient acting on the tip during the MFM measurement; Q: cantilever quality factor; k: spring constant; ε_0: vacuum dielectric constant; z: lift height; $t_{topo}(x,y)$: distance parameter of topography; $t_{topo} = 0$ defines the baseline of the substrate for the calculation and is the deepest point of the topography; V_{CPD}: contact potential difference between substrate and tip; A_{eff}: effective area of the capacitor responsible of the capacitive coupling.

A_{eff} increases with increasing tip–substrate distance [12,14]. According to our previous work a parabolic tip shape is used to calculate A_{eff} [14]. The radius of A_{eff} is defined by the value of the force gradient falling below 0.1% ($p = 0.001$) of the value of the force gradient between the top of the tip and the substrate. $A_{eff} = \pi\, r_{eff}^2$ with

$$r_{eff} = \sqrt{\left(\sqrt[3]{\frac{z_1^3}{p}} - z_1\right)\cdot r_{tip}}, \quad (2)$$

$$z_1(x,y) = z + t_{topo}(x,y), \quad (3)$$

where p: percentage factor.

Measuring the topography gives the distance parameter $t_{topo}(x,y)$ and allows the calculation of the positive phase shift due to topographic crosstalk.

3.2. Data Fusion of AFM and MFM

Figure 1 depicts the process of data fusion of AFM and MFM measurements for correction of topographic crosstalk in MFM phase images:

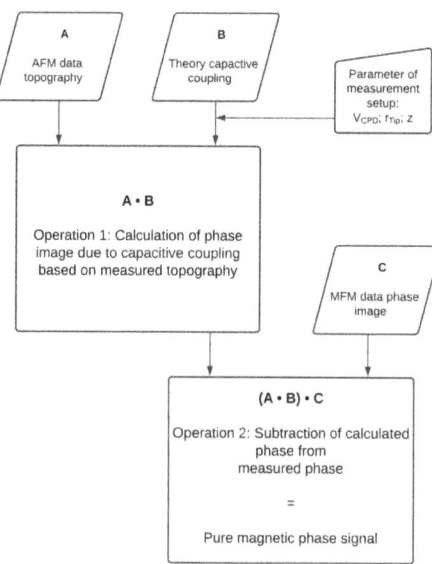

Figure 1. Workflow of data fusion procedure for the elimination of topographic crosstalk.

AFM topographic data are used to calculate the topographic crosstalk in MFM phase images by using the measured AFM data as $t_{topo}(x,y)$ in Equation (1) (Operation 1 in Figure 1). The exact tip radius r_{tip} is determined by Scanning Electron Microscopy and V_{CPD} is determined by KPFM measurements. In all measurements $p = 0.001$ achieved the best agreement with the measured data. The calculated phase image corresponding to the topographic crosstalk is then subtracted from the measured MFM phase image data (Operation 2 in Figure 1) resulting in a phase image depicting the pure magnetic signal.

4. Results and Discussion

4.1. Measurements on Non-Magnetic PS Nanoparticles in Comparison to Measurements on SPIONs

All interleave mode measurements on pure PS nanoparticles as well as on single SPIONs show a topographic crosstalk in the MFM phase image. Figure 2 represents topographic AFM measurements on a single PS nanoparticle with a diameter of 60 nm and the corresponding phase image (a) and the AFM topography and phase image of two single SPIONs with a diameter of 8 and 12 nm, respectively (b):

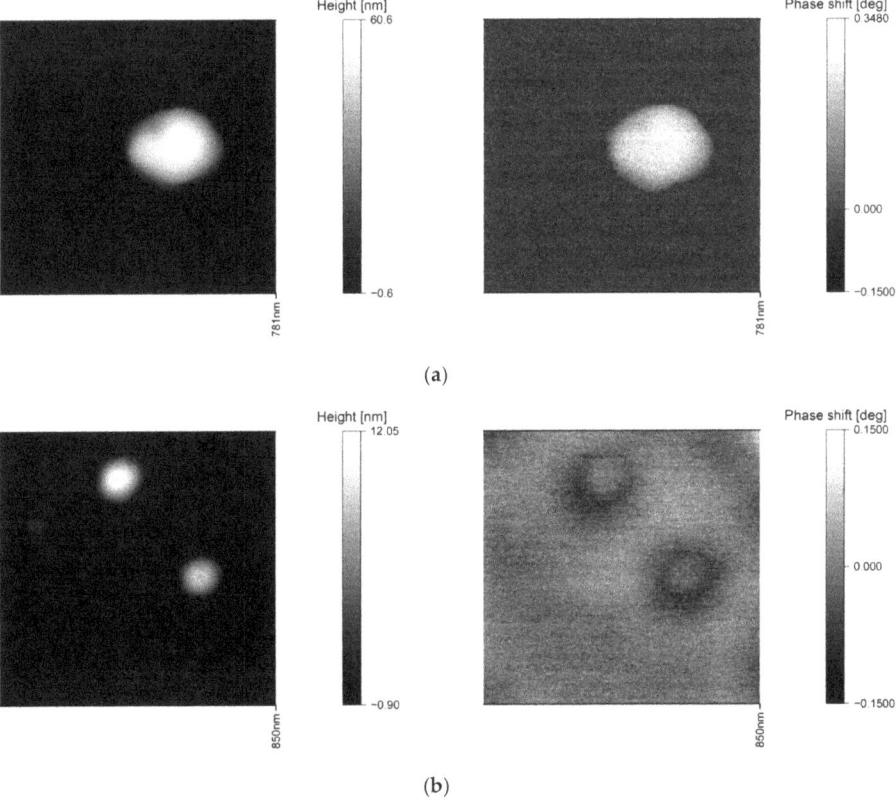

Figure 2. (a) Atomic Force Microscopy (AFM) topography (left) and the corresponding phase image in lift mode (right) of a single polystyrene (PS)-nanoparticle with a diameter of 60 nm (silicon substrate; $z = 50$ nm; $r_{tip} = 80$ nm). (b) AFM topography (left) and the corresponding phase image in lift mode (right) of two single superparamagnetic iron oxide nanoparticles (SPIONs) with a diameter of 12 and 8 nm (silicon substrate; $z = 50$ nm; $r_{tip} = 80$ nm).

The measurements clearly demonstrate that the interpretation of the phase image is not possible without correction of the topographic crosstalk: the non-magnetic PS nanoparticles (Figure 2a) show a positive phase instead of the expected zero phase and the SPIONs (Figure 2b) show a ring of negative phase around a positive phase instead of a completely negative phase due to their superparamagnetic character.

The data fusion process, as depicted in Figure 1, was applied to these two systems, pure PS nanoparticles and SPIONs, as shown in Figure 3. As proved in our previous work the dielectric constant of the nanoparticles has no influence on the capacitive coupling and, therefore, has not been taken into account [14]. The roughness of the substrate (on average around 1 Å) is small compared to the size of the nanoparticles and therefore does not contribute significantly to the topographic crosstalk.

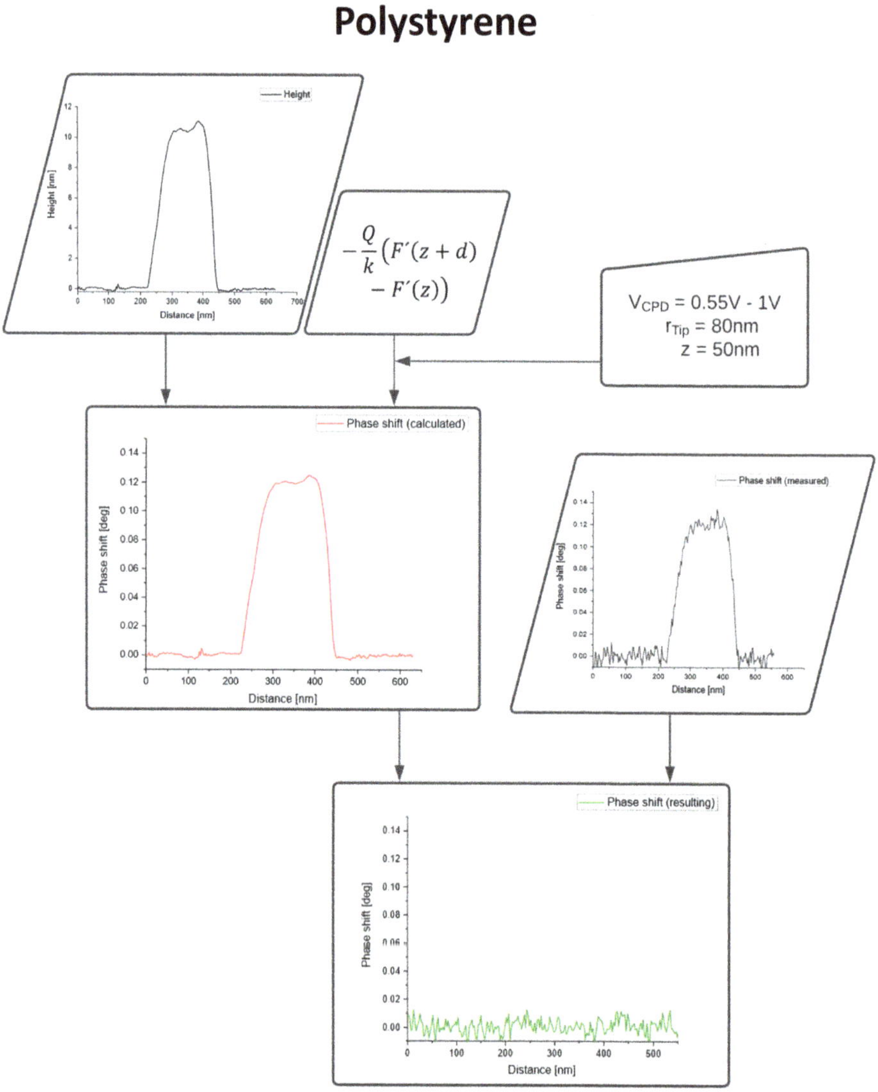

Figure 3. Data fusion of AFM topography and Magnetic Force Microscopy (MFM) phase image for polystyrene nanoparticles.

For pure PS particles the topographic crosstalk is compensated completely, resulting in a phase image of zero as expected for non-magnetic nanoparticles, as shown in Figure 3.

The charts in Figure 4 illustrate the correction of the topographic crosstalk for SPIONs.

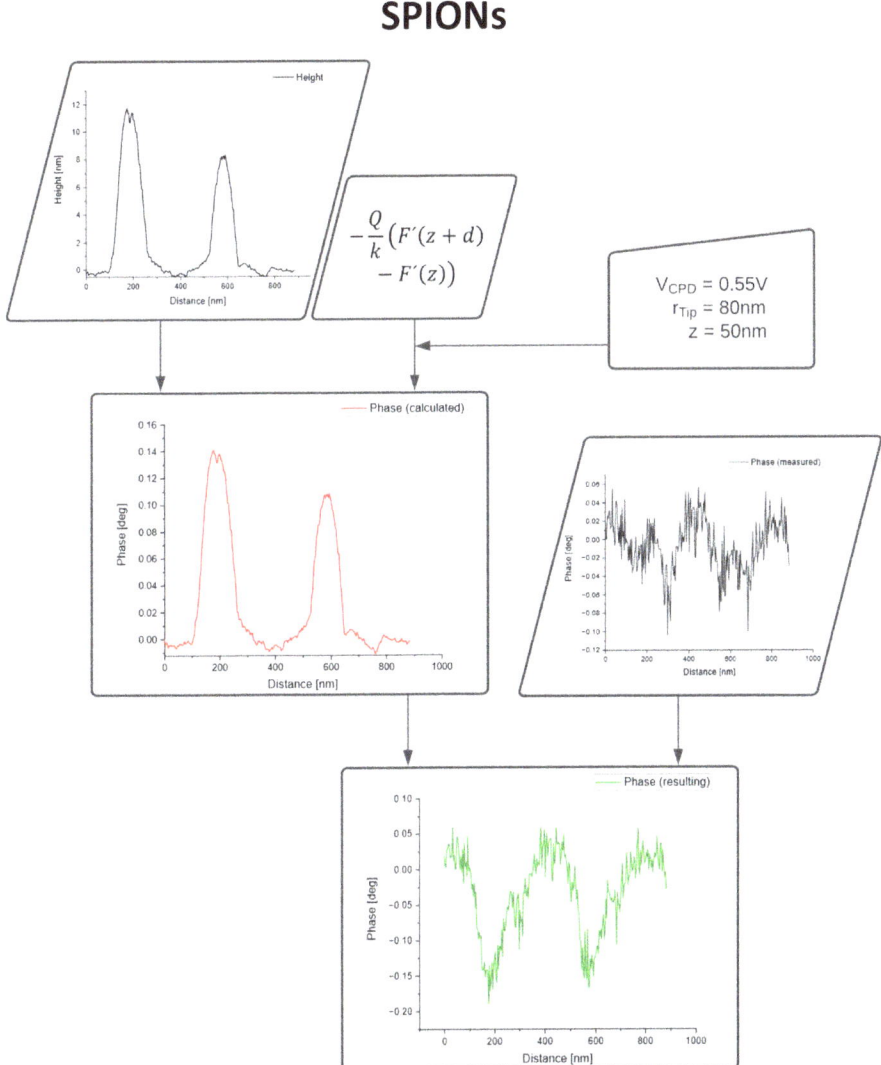

Figure 4. Data fusion of AFM topography and MFM phase image for SPIONs.

The magnetic phase image only shows negative values as expected for superparamagnetic particles. Figure 5 demonstrates the process of data fusion for PS particles with different diameters ranging from 12 to 78 nm.

As the polystyrene nanoparticles are non-magnetic, the measured positive MFM phase image clearly indicates significant contributions due to capacitive coupling (column 1). The second column shows cross-sections through the phase images of the measured data. The third column contains the calculated phase shifts (Operation 1 in Figure 1) due to capacitive coupling and based on the topography AFM measurements. The fourth column presents the results of Operation 2, the subtraction of measured and calculated phase in Figure 1, in form of a cross-section. It is clearly seen that the topographic crosstalk is eliminated and the phase signal approaches the measurement noise. The fifth segment visualizes the elimination of the crosstalk in the two-dimensional phase images. For all particles, the topographic crosstalk is almost completely removed.

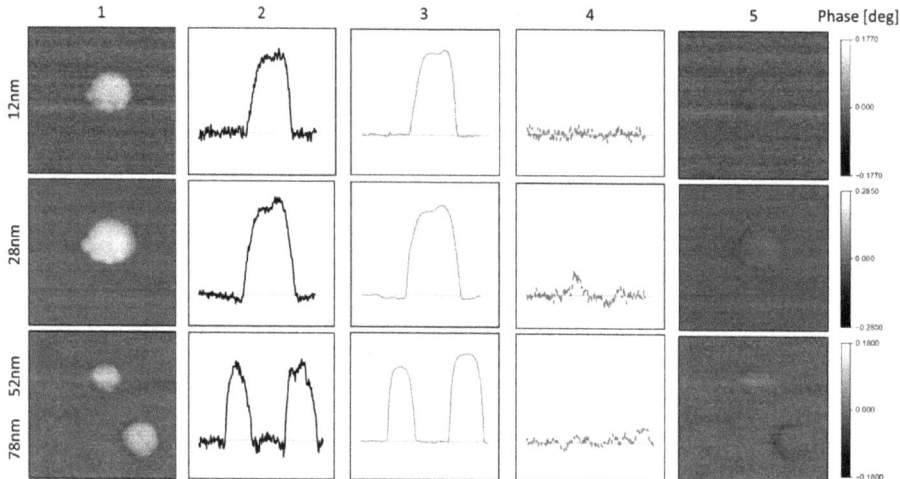

Figure 5. Data Fusion of different polystyrene capsules (top to bottom): (**1**) measured phase shift, (**2**) cross-section of measured phase shift, (**3**) calculated phase shift based on topographic cross-section, (**4**) cross-section of resulting phase shift by data fusion, (**5**) resulting phase image. (Silicon substrate; $V_{CPD} = 0.55$ V $- 1$ V; $z = 50$ nm; $r_{tip} = 80$ nm).

These measurements prove that the process of data fusion is an appropriate method for elimination of topographic crosstalk for a wide range of nanoparticle sizes.

4.2. Measurements on SPIONs Encapsulated in PS

Data fusion of AFM and MFM measurements now allow to investigate the localized magnetic behavior of encapsulated SPIONs. SPIONs with a diameter ranging from 7 to 10 nm diameter are encapsulated in PS nanoparticles with a diameter from 18 to 100 nm. The measured phase images (column 1) show a positive phase shift (white color) surrounded by a ring of negative phase shift (black color). The calculation of the topographic crosstalk based on the topographic AFM-measurements demonstrates that the measured positive phase shift is not due to repulsive magnetic forces but only due to topographic crosstalk (column 3 in Figure 6). Removing the topographic crosstalk, the corrected phase images only show negative values indicating the superparamagnetic character of the SPIONS.

The corrected phase images of PS nanoparticles show that the SPIONs are located at the outer edge of the PS nanoparticles, which is in good agreement with TEM measurements shown in Figure 7. The pure magnetic signals (columns 4 and 5 in Figure 6) only show attractive forces indicating that the encapsulated SPIONS are still in their superparamagnetic state in accordance to VSM measurements on similar nanoparticles [21].

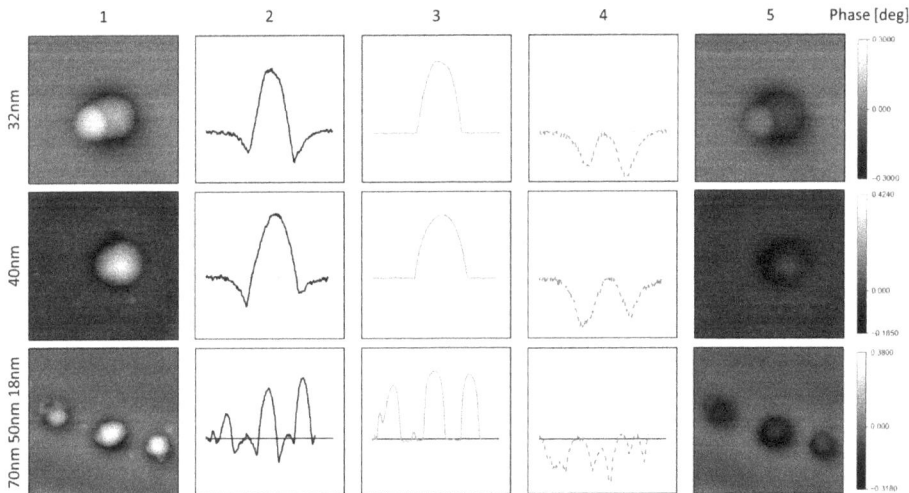

Figure 6. Data fusion for polystyrene particles filled with magnetite (ratio 1:0.2) with different diameters (top to bottom): (**1**) measured phase shift, (**2**) cross-section of measured phase shift, (**3**) calculated phase shift based on topographic cross-section, (**4**) cross-section of resulting phase shift by data fusion, (**5**) resulting phase image. (Silicon substrate; V_{CPD} = 0.55 V; z = 50 nm; r_{tip} = 80 nm).

Figure 7. Transmission Electron Measurements (TEM) image of PS nanoparticles with encapsulated SPIONs.

5. Conclusions

In summary, a numerical method was developed, which in general allows the correction of topographic crosstalk in MFM measurements. This method is based on data fusion of the AFM topography and the MFM phase image in combination with the theory of capacitive coupling.

The success of the data fusion was demonstrated by measurements on pure polystyrene nanoparticles of different sizes serving as a non-magnetic model system. With this method it was then possible to magnetically characterize in SPIONs encapsulated in polystyrene. The measurements demonstrate the superparamagnetic behavior of the SPIONs. It is now possible to magnetically image encapsulated SPIONs without falsifications due to the topographic crosstalk. The correction by data fusion presented in this paper thus offers a solution not only for nanoparticles but also for various applications that are affected by topographic crosstalk in lift mode measurements. An implementation of the measurement software is conceivable and could be directly incorporated into the analysis by an automated background process.

Author Contributions: Conceptualization, M.F. and H.M.; methodology, M.F. and H.M.; software, M.F.; validation, M.F. and H.M.; formal analysis, M.F. and H.M.; investigation, M.F.; resources, A.M., S.v.B. and H.M.; data curation, M.F.; writing—original draft preparation, M.F. and H.M.; writing—review and editing, M.F., A.M., R.T. and H.M.; visualization, M.F., A.M. and H.M.; supervision, R.T. and H.M.; project administration, H.M.; funding acquisition, H.M. All authors have read and agreed to the published version of the manuscript.

Funding: This research was funded by the European Funds of Regional Development (EFRE), grant number 84004058, co-funded by the German state Rhineland-Palatinate and Karl Otto Braun GmbH and Co. Kg.

Acknowledgments: The presented results are part of the ongoing InnoProm-project "TRAPP—Nanocarrier in carrier matrix for transdermal applications" co-funded by the German state Rhineland-Palatinate, the European Funds for Regional Development (EFRE), and Karl Otto Braun GmbH and Co. Kg.

Conflicts of Interest: The authors declare no conflict of interest.

References

1. Yadav, H.K.S.; Almokdad, A.A.; Shaluf, S.I.M.; Debe, M.S. Polymer-based nanomaterials for drug-delivery carriers. In *Nanocarriers for Drug Delivery*; Elsevier: Amsterdam, The Netherlands, 2019; pp. 531–556.
2. Din, F.U.; Aman, W.; Ullah, I.; Qureshi, O.S.; Mustapha, O.; Shafique, S.; Zeb, A. Effective use of nanocarriers as drug delivery systems for the treatment of selected tumors. *Int. J. Nanomed.* **2017**, *12*, 7291–7309. [CrossRef] [PubMed]
3. Jose, J.; Kumar, R.; Harilal, S.; Mathew, G.E.; Parambi, D.G.T.; Prabhu, A.; Uddin, M.S.; Aleya, L.; Kim, H.; Mathew, B. Magnetic nanoparticles for hyperthermia in cancer treatment: An emerging tool. *Environ. Sci. Pollut. Res.* **2019**, *27*, 19214–19225. [CrossRef] [PubMed]
4. Paysen, H.; Loewa, N.; Stach, A.; Wells, J.; Kosch, O.; Twamley, S.; Makowski, M.R.; Schaeffter, T.; Ludwig, A.; Wiekhorst, F. Cellular uptake of magnetic nanoparticles imaged and quantified by magnetic particle imaging. *Sci. Rep.* **2020**, *10*, 1–8. [CrossRef] [PubMed]
5. Liu, J.; Yu, W.; Han, M.; Liu, W.; Zhang, Z.; Zhang, K.; Shi, J. A specific "switch-on" type magnetic resonance nanoprobe with distance-dominate property for high-resolution imaging of tumors. *Chem. Eng. J.* **2021**, *404*, 126496. [CrossRef]
6. Schwarz, A.; Wiesendanger, R. Magnetic sensitive force microscopy. *Nano Today* **2008**, *3*, 28–39. [CrossRef]
7. Collins, L.; Belianinov, A.; Proksch, R.; Zuo, T.; Zhang, Y.; Liaw, P.K.; Kalinin, S.V.; Jesse, S. G-mode magnetic force microscopy: Separating magnetic and electrostatic interactions using big data analytics. *Appl. Phys. Lett.* **2016**, *108*, 193103. [CrossRef]
8. Raşa, M.; Philipse, A.P. Scanning probe microscopy on magnetic colloidal particles. *J. Magn. Magn. Mater.* **2002**, *252*, 101–103. [CrossRef]
9. Paolino, P.; Tiribilli, B.; Bellon, L. Direct measurement of spatial modes of a microcantilever from thermal noise. *J. Appl. Phys.* **2009**, *106*, 094313. [CrossRef]
10. Li, L.H.; Chen, Y. Electric contributions to magnetic force microscopy response from graphene and MoS2 nanosheets. *J. Appl. Phys.* **2014**, *116*, 213904. [CrossRef]
11. Krivcov, A.; Schneider, J.; Junkers, T.; Möbius, H. Magnetic force microscopy of in a polymer matrix embedded single magnetic nanoparticles. *Phys. Status Solidi* **2019**, *216*, 1800753. [CrossRef]
12. Krivcov, A.; Junkers, T.; Möbius, H. Understanding electrostatic and magnetic forces in magnetic force microscopy: Towards single superparamagnetic nanoparticle resolution. *J. Phys. Commun.* **2018**, *2*, 075019. [CrossRef]
13. Passeri, D.; Dong, C.; Reggente, M.; Angeloni, L.; Barteri, M.; Scaramuzzo, F.A.; De Angelis, F.; Marinelli, F.; Antonelli, F.; Rinaldi, F.; et al. Magnetic force microscopy. *Biomatter* **2014**, *4*, e29507. [CrossRef] [PubMed]
14. Fuhrmann, M.; Krivcov, A.; Musyanovych, A.; Thoelen, R.; Möbius, H. The role of nanoparticles on topographic cross-talk in electric force microscopy and magnetic force microscopy. *Phys. Status Solidi* **2020**, *217*, 1900828. [CrossRef]

15. Angeloni, L.; Passeri, D.; Reggente, M.; Rossi, M.; Mantovani, D.; Lazzaro, L.; Nepi, F.; De Angelis, F.; Barteri, M. Experimental issues in magnetic force microscopy of nanoparticles. In Proceedings of the AIP Conference Proceedings; AIP Publishing LLC: Melville, NY, USA, 2015; Volume 1667, p. 020010.
16. Das Neves, J.; Amiji, M.M.; Bahia, M.F.; Sarmento, B. Nanotechnology-based systems for the treatment and prevention of HIV/AIDS. *Adv. Drug Deliv. Rev.* **2010**, *62*, 458–477. [CrossRef] [PubMed]
17. Yu, J.; Ahner, J.; Weller, D. Magnetic force microscopy with work function compensation. *J. Appl. Phys.* **2004**, *96*, 494–497. [CrossRef]
18. Krivcov, A.; Ehrler, J.; Fuhrmann, M.; Junkers, T.; Möbius, H. Influence of dielectric layer thickness and roughness on topographic effects in magnetic force microscopy. *Beilstein J. Nanotechnol.* **2019**, *10*, 1056–1064. [CrossRef] [PubMed]
19. Angeloni, L.; Passeri, D.; Reggente, M.; Mantovani, D.; Rossi, M. Removal of electrostatic artifacts in magnetic force microscopy by controlled magnetization of the tip: Application to superparamagnetic nanoparticles. *Sci. Rep.* **2016**, *6*, 26293. [CrossRef] [PubMed]
20. Musyanovych, A.; Rossmanith, R.; Tontsch, C.; Landfester, K. Effect of hydrophilic comonomer and surfactant type on the colloidal stability and size distribution of carboxyl-and amino-functionalized polystyrene particles prepared by miniemulsion polymerization. *Langmuir* **2007**, *23*, 5367–5376. [CrossRef] [PubMed]
21. Ramírez, L.P.; Landfester, K. Magnetic polystyrene nanoparticles with a high magnetite content obtained by miniemulsion processes. *Macromol. Chem. Phys.* **2003**, *204*, 22–31. [CrossRef]

Publisher's Note: MDPI stays neutral with regard to jurisdictional claims in published maps and institutional affiliations.

© 2020 by the authors. Licensee MDPI, Basel, Switzerland. This article is an open access article distributed under the terms and conditions of the Creative Commons Attribution (CC BY) license (http://creativecommons.org/licenses/by/4.0/).

Article

Coating Effect on the ^{1}H—NMR Relaxation Properties of Iron Oxide Magnetic Nanoparticles

Francesca Brero [1,*], Martina Basini [2], Matteo Avolio [1], Francesco Orsini [2], Paolo Arosio [2], Claudio Sangregorio [3,4], Claudia Innocenti [3,4], Andrea Guerrini [4], Joanna Boucard [5], Eléna Ishow [5], Marc Lecouvey [6], Jérome Fresnais [7], Lenaic Lartigue [5] and Alessandro Lascialfari [1,2]

[1] Dipartimento di Fisica and INFN, Università degli Studi di Pavia, Via Bassi 6, 27100 Pavia, Italy; matteo.avolio01@universitadipavia.it (M.A.); alessandro.lascialfari@unipv.it (A.L.)
[2] Dipartimento di Fisica and INFN, Università degli Studi di Milano, Via Celoria 16, 20133 Milano, Italy; martina.basini@gmail.com (M.B.); francesco.orsini@unimi.it (F.O.); paolo.arosio@unimi.it (P.A.)
[3] ICCOM-CNR, via Madonna del Piano 10, 50019 Sesto Fiorentino (FI), Italy; csangregorio@iccom.cnr.it (C.S.); claudia.innocenti@unifi.it (C.I.)
[4] Dipartimento di Chimica "U. Schiff" and INSTM, Università degli Studi di Firenze, Via della Lastruccia 3-13, 50019 Sesto Fiorentino (FI), Italy; andrea.guerrini@sns.it
[5] CNRS, CEISAM UMR 6230, Université de Nantes, F-44000 Nantes, France; joanna.boucard@laposte.net (J.B.); elena.ishow@univ-nantes.fr (E.I.); lenaic.lartigue@univ-nantes.fr (L.L.)
[6] CSPBAT-UMR CNRS 7244, Université Sorbonne Paris Nord, 74 rue Marcel Cachin, 93017 Bobigny, France; marc.lecouvey@univ-paris13.fr
[7] CNRS, Laboratoire de Physico-chimie des Electrolytes et Nanosystèmes Interfaciaux, Sorbonne Université, PHENIX—UMR 8234, CEDEX 05 F-75252 Paris, France; jerome.fresnais@sorbonne-universite.fr
* Correspondence: francesca.brero01@universitadipavia.it; Tel.: +39-0382-987-483

Received: 24 June 2020; Accepted: 19 August 2020; Published: 24 August 2020

Abstract: We present a ^{1}H Nuclear Magnetic Resonance (NMR) relaxometry experimental investigation of two series of magnetic nanoparticles, constituted of a maghemite core with a mean diameter d_{TEM} = 17 ± 2.5 nm and 8 ± 0.4 nm, respectively, and coated with four different negative polyelectrolytes. A full structural, morpho-dimensional and magnetic characterization was performed by means of Transmission Electron Microscopy, Atomic Force Microscopy and DC magnetometry. The magnetization curves showed that the investigated nanoparticles displayed a different approach to the saturation depending on the coatings, the less steep ones being those of the two samples coated with P(MAA-*stat*-MAPEG), suggesting the possibility of slightly different local magnetic disorders induced by the presence of the various polyelectrolytes on the particles' surface. For each series, ^{1}H NMR relaxivities were found to depend very slightly on the surface coating. We observed a higher transverse nuclear relaxivity, r_2, at all investigated frequencies (10 kHz ≤ ν_L ≤ 60 MHz) for the larger diameter series, and a very different frequency behavior for the longitudinal nuclear relaxivity, r_1, between the two series. In particular, the first one (d_{TEM} = 17 nm) displayed an anomalous increase of r_1 toward the lowest frequencies, possibly due to high magnetic anisotropy together with spin disorder effects. The other series (d_{TEM} = 8 nm) displayed a r_1 vs. ν_L behavior that can be described by the Roch's heuristic model. The fitting procedure provided the distance of the minimum approach and the value of the Néel reversal time ($\tau \approx 3.5 \div 3.9 \cdot 10^{-9}$ s) at room temperature, confirming the superparamagnetic nature of these compounds.

Keywords: magnetic nanoparticles; Superparamagnetism; Nuclear Magnetic Resonance; Magnetic Resonance Imaging; coating; polyelectrolytes

1. Introduction

Imaging techniques play a fundamental role in every branch of medicine [1–6]. Among these, Magnetic Resonance Imaging (MRI) has played a leading role, as it combines the possibility of obtaining 3D images with a spatial resolution down to a few micrometers, the absence of limits for the penetration depth and the use of non-ionizing electromagnetic radiation [7]. The reconstruction of MRI acquisitions is mainly based on the analysis of the Nuclear Magnetic Resonance (NMR) signal coming from the water protons of different liquids/organs/tissues, on which appropriate magnetic field gradients are applied. The search for a higher sensitivity and the continuous optimization of methods and tools in MRI requires the development of efficient contrast agents (CAs) (i.e., biocompatible and biodegradable materials properly designed in terms of geometry, interactions with water and magnetic properties) that can be injected into the body to produce an optimized image contrast [8–14]. The presence of CAs, in fact, induces a decrease in the nuclear relaxation times T_1 and T_2 of protons, producing a local increase (T_1-relaxing CAs) or decrease (T_2-relaxing CAs) of the NMR signal in areas of the body that contain the agent, making them appear with unequal brightness/darkness in the MRI image [15]. Approximately 10 years after the first application on humans of the paramagnetic MRI CAs (Young et al. in 1981 [16]), more complex magnetic nanostructures based on iron oxide particles, typically magnetite Fe_3O_4 or maghemite γ-Fe_2O_3, were introduced into the market (Endorem®/Feridex®, Feraheme®, Combidex®, Clariscan® and Resovist®) [17–19]. Most of them were withdrawn from the market; nevertheless, Resovist® is still sold in a few countries, and Feraheme® is approved for the treatment of iron deficiency in adult chronic kidney disease patients. However, superparamagnetic properties (which mainly lead to a reduction of the T_2 of the solvent nuclei), low toxicity and the improved synthesis control on the size, shape and surface of the magnetic nanoparticles (MNPs) (due to recently developed synthesis procedures) make them very versatile from an applicative point of view [20].

The efficiency of such particles for diagnostics has indeed been demonstrated to depend on several magnetic (nature of metal ion, spin topology, magnetic anisotropy), morphological/structural (core diameter, shape, crystallinity degree, coating thickness) and chemical (water exchange dynamics, principally due to coating hydrophilicity, permeability and thickness) parameters [21–26]. Moreover, a great potential of MNPs relies on their reactive surface, which can be exploited for the anchorage of several molecules with different functionalities. Indeed, over the last two decades, several research groups have tried to functionalize the nanoparticle surface with specific targeting agents, such as antigens and antibodies, or load them with cargo such as drugs, fluorescent dyes, radiotracers, etc. [27]. Ideally, these nanosystems could selectively reach the targeted tissues and organs, increase the image contrast and, at the same time, release a drug or heat that region (through Magnetic Fluid Hyperthermia [28–31]) to induce cell death. Thus, these nanoparticles can combine properties useful in diagnostics with other properties compatible with therapy, thus becoming potential *theranostic* agents.

This paper studies the morpho-dimensional, magnetic and relaxometric properties of aqueous dispersions of two series of γ-Fe_2O_3 superparamagnetic nanoparticles (with mean diameters $d_{TEM} \approx 17 \pm 2.5$ nm and $\approx 8 \pm 0.4$ nm) coated with four different types of biocompatible negative polyelectrolytes. The purpose of our study is to investigate how the different kinds of polymer coatings can influence the behavior of the longitudinal (T_1) and transverse (T_2) ^1H NMR relaxation times of MNP suspensions, which have also been shown to depend on the size of the magnetic core.

2. Materials and Methods

2.1. Tuning the Coating of Maghemite Nanoparticles

2.1.1. Polyelectrolytes Serving as MNP Coatings

We used four different MNP polyelectrolyte coatings: (i) Poly(acrylic acid), named PAA-A (average M_n = 1800 g·mol^{-1}), purchased from Sigma–Aldrich (St. Louis, MO, USA) and used as received; (ii) a copolymer issued from the random esterification of poly(methacrylic acid) (PMAA) chains with polyethylene glycol (PEG$_{2000}$), PMAA-g-PEG$_{2000}$, named PEG-B (M$_n$ = 5.86 × 10^4 g·mol^{-1}); (iii) and (iv), two comb-like polymers fabricated by reversible addition-fragmentation chain transfer (RAFT) based on PMMA and poly(ethylene glycol) methyl ether methacrylate (MAPEG$_{2000}$) with two different chain transfer agents, P(MAA-*stat*-MAPEG$_{2000}$), named respectively PEG-C for the hydrophobic transfer agent (M$_n$ = 3.99 × 10^4 g·mol^{-1}) and PEG-D for the hydrophilic transfer agent (M$_n$ = 2.87 × 10^4 g·mol^{-1}) [32,33].

2.1.2. Fabrication Procedure of Magnetic Nanoparticles

Low diameter (average magnetic core size d$_{TEM}$ ≈ 8 ± 0.4 nm) maghemite-based MNPs (samples A-8, B-8, C-8 and D-8, with PAA-A, PEG-B, PEG-C and PEG-D coatings, respectively) were synthesized following Massart's protocol relying on the coprecipitation of iron(II) and iron(III) chloride salts in the presence of ammonium hydroxide [34]. High-diameter (average magnetic core d$_{TEM}$ ≈ 17 ± 2.5 nm) maghemite-based nanoparticles (samples A-17, B-17, C-17 and D-17, with PAA-A, PEG-B, PEG-C and PEG-D coatings, respectively) were prepared by a modified Massart's method [35]. Briefly, iron chloride salt was dissolved in HCl acidic solutions (2 mol·L^{-1}) and deoxygenated. Subsequently, 6.6 mL of FeCl$_3$·6H$_2$O solution (1 mol·L^{-1}) and 1.7 mL of FeCl$_2$·4H$_2$O solution (2 mol·L^{-1}) were mixed together and heated up to 70 °C under an argon atmosphere. Under vigorous stirring, a tetrapropylammonium hydroxide solution (1 mol·L^{-1}, 64.4 mL) was injected at a 0.7 mL·min^{-1} rate using a syringe pump and then mixed for an additional 20 min. The two suspensions were oxidized to maghemite by an acidic solution of iron nitrate and redispersed in nitric acid [34]. Purification of the dispersion was performed by successive magnetic decantation steps. A size sorting by selective precipitation was conducted to obtain a narrow polydispersity [36]. Coating with the different polyelectrolytes, chosen for their biocompatibility from the perspective of *in cellulo* MRI, was achieved after a protocol already described in the literature [33]. The polymer powder was added to the targeted acidic dispersion of maghemite MNPs (0.06 wt.%) (for example, for 2.5 mL of iron oxide suspension, 5 mg of PAA-A or 15 mg of PEG-B, C, D were added). A 1.4 mol·L^{-1} solution of ammonium hydroxide was added dropwise under stirring to reach a final pH above 8. Dialysis against Millipore water using Spectra/Por^{TM} membrane (regenerated cellulose) with an 8–10 kDa or 300 kDa cut-off was performed for 48 h to remove the excess of polyelectrolytes, while the solution pH reached around 7 after neutralization.

2.2. Characterization Methods

Experimental Details

The nanoparticle morphology was investigated by transmission electron microscopy (TEM). Images were recorded using a MO-Jeol 123S0 (80 kV) TEM equipped with a GATAN Orius 11 Megapixel Camera. A few drops of suspensions of the nanosystems were deposited onto holey carbon-coated copper grids (300 mesh) purchased from Agar Oxford Instruments.

Atomic Force Microscopy (AFM), performed by means of a Bruker Nanoscope Multimode IIId AFM system operating in tapping mode in air, was used to estimate the total size of the MNPs (core plus organic coating). The measurements were performed using a silicon rectangular cantilever (NSG01, NT_MDT, length of 120 μm, spring constant of 2.5 N/m and a resonance frequency of about 130 kHz). The samples were prepared by drying a drop of very diluted aqueous solution of MNPs on a mica substrate.

The hydrodynamic diameters of the nanosystems were measured with a Zetasizer Nano ZS ZEN 3600 (Malvern Instruments, Worcestershire, UK). Measurements were collected at 25 °C and averaged over three acquisitions, and correlograms were fitted with a Cumulant algorithm. The results presented are given after a lognormal fitting of the mean size volume histogram.

The electrophoretic mobility of the nanosytems was determined with a Zetasizer Nano ZS ZEN 3600 (Malvern Instruments, Worcestershire, UK). From the electrophoretic mobility, the zeta potential, ζ, was deduced using Smoluchowski's approximation. All the measurements were performed at 25 °C in disposable folded capillary cells (DTS1070) and repeated three times.

The DC magnetic measurements were carried out by a VSM magnetometer (PPMS Quantum Design Ltd., San Diego, CA, USA) and a SQUID magnetometer (MPMS by Quantum Design Ltd., San Diego, CA, USA) operating in the 2–300 K temperature range and $-5 \leq \mu_0 H \leq +5$ Tesla magnetic field range. Zero Field Cooled/Field Cooled magnetizations were acquired in a 5 milliTesla probe magnetic field after cooling the sample without (ZFC) and with (FC) the applied field. Due to the small quantity of synthetized products, the magnetic material content in the samples could not be estimated with accuracy. This inaccuracy and the large experimental error in the sample weight only allowed us to assume a rough estimate of the saturation magnetization.

The NMR-dispersion profiles were collected at room temperature by measuring the T_1 and the T_2 relaxation times, varying the Larmor frequency of the investigated nuclei ($2\pi\nu_L = \gamma B_0$, where $\gamma = 2.67513 \times 10^8$ rad s^{-1} T^{-1} is the gyromagnetic factor of ^{1}H, from 10 kHz up to 60 MHz). For low-frequency relaxation measurements (from 0.01 MHz to 7.2 MHz), the Fast-Field-Cycling technique was used by means of a Smartracer Stelar NMR relaxometer. High-frequency relaxation measurements (up to 60 MHz) were performed using a Stelar Spinmaster Fourier transform nuclear magnetic resonance spectrometer. For $\nu_L < 7.2$ MHz, pre-polarized Saturation Recovery (for T_1) and spin-echo (for T_2) sequences were adopted. For frequencies $\nu_L > 7.2$ MHz, non-pre-polarized Saturation Recovery (SR) and Carr Purcell Meiboom Gill (CPMG) pulse sequences were used for the T_1 and T_2 measurements, respectively.

3. Results and Discussion

3.1. Nanoparticles Synthesis

The coatings of maghemite MNPs were chosen so that they met three main objectives: easy synthetic access, biocompatibility and hydrophilicity. The structures of the four used polyelectrolytes (PAA-A, PEG-B, PEG-C and PEG-D) are sketched in Figure 1. To confer stealth properties to the coated NPs, PEGylated chains, well known to help nanoparticles evade the mononuclear phagocytic system after in vivo injection, were added to the polymer backbone by using post-esterification (PEG-B) or reversible addition-fragmentation chain transfer polymerization (PEG-C and PEG-D). Two series of coated MNPs, differing by the size of the inorganic core, were actually generated by adding an excess of each polyelectrolyte to an acidic solution of maghemite nanoparticles. After dialysis against Millipore water, alkalinization using ammonium hydroxide was performed so as to favor the anchoring of the polyelectrolyte carboxylate units to the naked surface of the iron oxide nanoparticles.

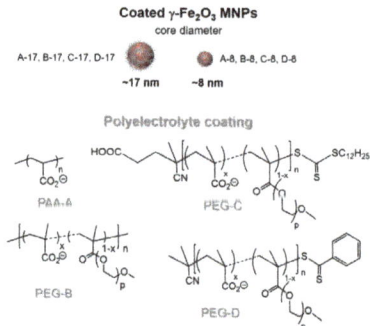

Figure 1. Structures of the investigated MNPs as a function of their core diameter and polyelectrolyte coating, depicted as PAA-A, PEG-B, PEG-C and PEG-D.

3.2. Morphological Characterization

Both series of samples consist of spherical MNPs, as deduced from the TEM and AFM images, which are presented in Figure 2a,b for A-17. Representative histograms of the core size for A-17 (first series) and A-8 (second series) are reported in Figure 3, while the average and standard deviation of the core diameter distribution are reported in Table 1. After statistical counting of more than 300 nanoparticles, performed using the ImageJ software, the average diameter and the standard deviation were determined by fitting the data to a log-normal distribution:

$$p_x(x, \mu_y, \sigma_y) = \frac{1}{\sqrt{2\pi}\sigma_y x} \exp\left[-\frac{1}{2}\left(\frac{\ln x - \mu_y}{\sigma_y}\right)^2\right] \quad (1)$$

where x represents the different values of the diameter, $\mu_y = \ln(d_{TEM})$, where d_{TEM} is the mean diameter, and σ_y is the standard deviation. The size distribution for each sample is within 16% around the mean value for both series (Table 1), i.e., large enough to include the mean values of the others samples of the same series.

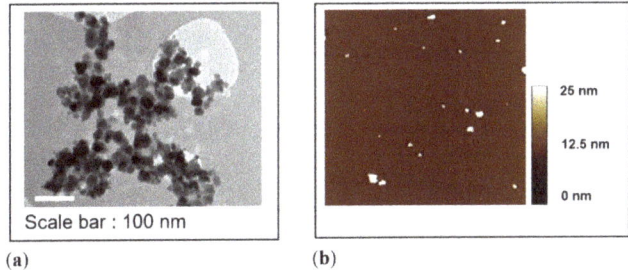

Figure 2. Representative images of sample A-17 obtained by means of: (**a**) bright field TEM and (**b**) AFM over an area of 3 × 3 µm².

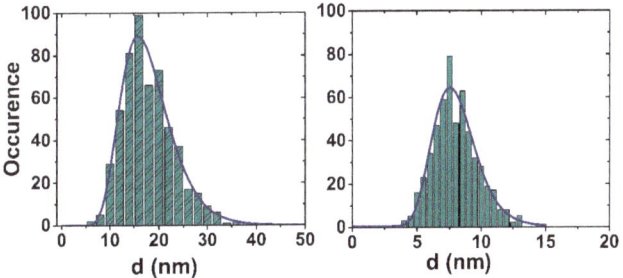

Figure 3. Histograms reporting the distribution of the core sizes for the two series of MNPs, as obtained from TEM analysis (A-17, left and A-8, right). The distributions are fitted to a log-normal function; the mean value and the standard deviation are reported in Table 1.

Table 1. Mean diameters ± standard deviation of the inorganic cores for the first and second series, obtained by TEM.

Sample	d_{TEM} nm
first series samples	17.0 ± 2.5
second series samples	8.0 ± 0.4

For the samples of the first series (high diameter), the MNPs' morphology was also investigated by Tapping Mode Atomic Force Microscopy, which allowed for the evaluation of the overall size of the MNPs, i.e., the diameter of the magnetic core together with its coating. Besides, AFM can distinguish the presence of MNP agglomerates and single MNPs, as shown in the topographic image of Figure 2b. As expected, the MNPs' average diameter d_{AFM} obtained by AFM is greater than the diameter estimated from the TEM data, due to the presence of the polymeric coating, whose thickness, as calculated from the difference of the diameters obtained through the two different techniques [($d_{AFM} - d_{TEM}$)/2 (data not reported)], is in the order of 1 ÷ 1.5 nm, depending on the sample.

All of the samples have a negative zeta potential due to the presence of acrylate units on the various polyelectrolytes. Each sample has a zeta potential in the −28 to −48 mV range, indicating a good colloidal stability. The hydrodynamic diameters of the magnetic nanoparticles with a core of 17 nm vary little with the nature of the stabilizing polyelectrolyte and remain within a range of 71 to 85 nm. The increase in diameter, when compared with that obtained by TEM, is consistent with the presence of the polyelectrolyte and a solvation layer on the surface of the nanoparticles. Likewise, the value of 21 nm for the hydrodynamic diameter of the 8-nm nanoparticles samples stabilized by the PAA is coherent with the presence of the polyelectrolyte and of the solvation layer.

3.3. Magnetic Measurements

The ZFC/FC magnetization curves are reported in Figure 4a for the samples A-17 and A-8, which were measured in the form of powders. The temperature of the maximum in the ZFC curve, commonly identified as the blocking temperature of the system, occurs at T_B ~45 K for the smaller diameter MNPs. For the larger diameter MNPs, the maximum is broadened around the end of the measuring temperature range ($T_B \geq$ 260–300 K), suggesting that this latter series of samples is in a sort of transition between "blocked/unblocked" (superparamagnetic) regimes at room temperature. As a reminder, T_B is proportional to the competition between the magnetic energy barrier (Ea ≈ K_{eff}V) and the magnetization reversal process, which, in turn, increases with the effective anisotropy constant (K_{eff}) and the volume (V) of the MNPs. Thus, the large difference observed in the T_B values of the two series reflects this dependence.

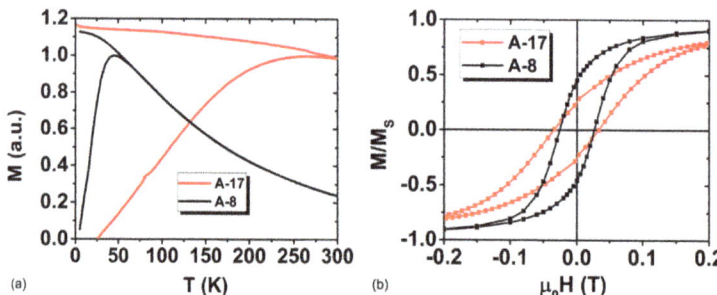

Figure 4. (a) ZFC/FC magnetization curves collected with a magnetic field $\mu_0 H = 5 \times 10^{-3}$ Tesla and (b) low field hysteresis loops at 2.5 K for A-17 and A-8.

The magnetization curves acquired at low (2.5 K) and high (300 K) temperatures are shown in Figure 5 for the 17-nm series. The curves are normalized to the corresponding saturation magnetization (M_s) value to better compare their shape features. Samples of the 17-nm series present a similar coercivity ($\mu_0 H_C$ = 35 milliTesla) at a low temperature, a similar magnetic remanence, M_R/M_s = 0.3 at 2.5 K, and a similar susceptibility, χ, at 300 K, with the exception of D-17, which displays slightly higher M_R and χ values. On the contrary, a different approach to saturation (high field region), particularly evident at a low temperature, is observed among the samples; these can be ordered, from slowest to fastest to reach saturation, in the following sequence: C-17, B-17, A-17 and D-17.

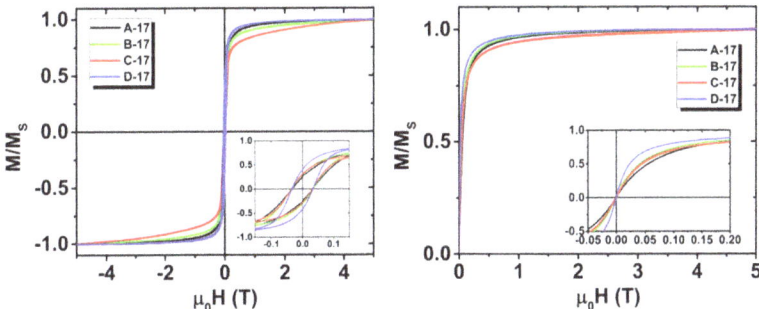

Figure 5. Magnetization curves at 2.5 K (left panel) and 300 K (right panel) for the first series. In the insets, details of the curves at low magnetic fields are shown.

A linear approach to saturation is commonly reported in nanoparticle systems and is related to the spin disorder on the particle surface, which affects the magnetization alignment upon increasing the field. In the present case, the different approach that was observed could thus be ascribed to the different modifications of the particle surface induced by: (i) slight variations in the synthesis procedure and/or (ii) the presence of different coatings. Due to their interconnection, distinguishing between these two contributions is not an easy task and will require a more detailed analysis, which is beyond the scope of this work. Interestingly, we can note that a negligible coercivity is recorded in the magnetization curves at room temperature (300 K) for all the samples, suggesting that the transition to the "unblocked" state mentioned above occurred for most of the particles of the 17 nm series at this temperature. The magnetization curves for the second series (not shown) present roughly similar features among the samples at low fields, with $\mu_0 H_C$ = 25 milliTesla and M_R/M_s = 0.4 at 2.5 K. A comparison of the low field hysteresis at 2.5 K for representative samples of the two series is shown in Figure 4b.

3.4. ^{1}H NMR Relaxation

Proton NMR relaxation in superparamagnetic colloids occurs because of the fluctuations of the dipolar magnetic coupling between nanoparticle magnetization and proton spins. The relaxation rate is described by an outer sphere model that includes the Curie relaxation, where the dipolar interaction fluctuates because of both the translational diffusion process and the Néel reversal (i.e., the flip of the magnetization vector from one direction of the easy magnetization axis to the opposite one). The sensitivity of MNPs as contrast agents was evaluated through the nuclear relaxivities r_1 (longitudinal relaxivity) and r_2 (transverse relaxivity), which were calculated by means of the following equation:

$$r_i = [(1/T_i)_{meas} - (1/T_i)_{dia}]/C \qquad i = 1, 2 \qquad (2)$$

where $(1/T_i)_{meas}$ is the value measured on the samples, $(1/T_i)_{dia}$ is the relaxation rate of the dispersant in the absence of superparamagnetic nanoparticles and C is the iron molar concentration within the sample.

3.4.1. Experimental Data

17 nm MNPs (1st Series)

The experimental longitudinal relaxivity profile (r_1) of the 17-nm series of MNPs is represented in Figure 6a. All samples show a continuous increase of the longitudinal relaxivity lowering the Larmor frequencies, with no detectable maximum. This behavior can be explained qualitatively by taking into account the energy related to the crystal's internal magnetic anisotropy at a low frequency.

The absence of a maximum is common for spherical maghemite-based particles with diameters d_{TEM} above approximately 15 nm. However, here the flattening of the $r_1(\nu)$ curves at a low frequency, which is expected for high anisotropy systems, is not observed [37].

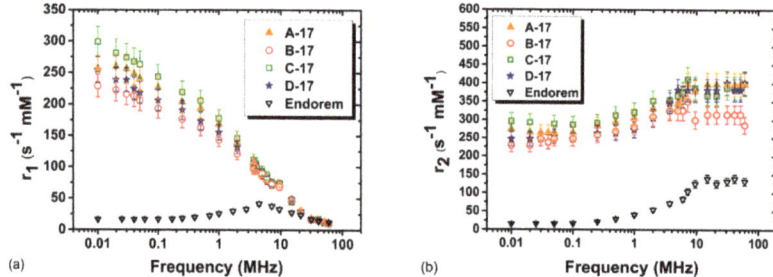

Figure 6. (a) Longitudinal r_1 and (b) transverse r_2 NMRD profiles collected at room temperature in the Larmor frequency range $0.01 \leq \nu_L \leq 60$ MHz for the first series of polymer-coated MNPs. For comparison, the relaxivity values of Endorem, as reported by Basini et al., are shown [22].

The transverse relaxivity vs. frequency behavior (Figure 6b) is similar for all samples, and at a high magnetic field $\mu_0 H \sim 1.41$ Tesla (close to the clinical one), r_2 reaches the value ~ 285 mM^{-1}s^{-1} for sample B-17 and ~ 400 mM^{-1}s^{-1} for samples A-17, C-17 and D-17. At Larmor frequencies $\nu_L > 5$–10 MHz, sample B-17 shows r_2 values smaller than those of the other samples, also slightly reflecting differences in r_1. The spin disorder, induced by the different polymer and/or the agglomeration effects, possibly generate a lower magnetization value in the case of sample B-17. Table 2 summarizes the r_1, r_2 and r_2/r_1 values at the two frequencies, namely 60 MHz and 15 MHz. The values are compared to those of Endorem, a commercial T_2 contrast agent, no longer used since 2012, but which still remains a good reference for assessing the relaxation efficiency of relaxing T_2 superparamagnetic nanoparticles. The r_2/r_1 value is greater than 2, indicating that all the ferrofluids act as negative contrast agents, this value being the threshold conventionally used to distinguish T_1-relaxing agents and T_2-relaxing agents [38].

Table 2. Longitudinal and transverse relaxivity values and their ratio at 15 and 60 MHz, for the 1st series of MNPs aqueous dispersions at room temperature.

Sample	Frequency	r_1 (s^{-1}mM^{-1})	r_2 (s^{-1}mM^{-1})	r_2/r_1
A-17	60 MHz	12.4 (1.0)	396.8 (31.7)	32
	15 MHz	48.4 (3.9)	396.8 (31.7)	8.2
B-17	60 MHz	11.3 (0.9)	285.7 (22.8)	25.3
	15 MHz	47.1 (3.8)	314.9 (25.2)	6.7
C-17	60 MHz	10.9 (0.9)	398.7 (31.9)	36.6
	15 MHz	44.5 (3.6)	365.5 (29.2)	8.2
D-17	60 MHz	12.6 (1.0)	401.8 (32.1)	31.9
	15 MHz	49.9 (4.0)	381.7 (30.5)	7.6
Endorem	60 MHz	12.3 (1.0)	131.6 (10.5)	10.7
	15 MHz	27.5 (2.2)	138.9 (11.1)	5

8 nm MNPs (2nd Series)

The relaxivity profiles of the 8-nm series of MNPs are presented in Figure 7. From Figure 7a, it is possible to observer that the longitudinal relaxivity r_1 of this series of MNPs behaves as expected for ultrasmall superparamagnetic particles. The maximum of the longitudinal relaxivity is located between 1 and 20 MHz, and diminishes just by a factor of ~4 at 60 MHz. For sample A-8, the maximum is shifted towards a higher frequency, suggesting slightly smaller sizes. At low fields, all samples present a slight dispersion at 300–400 kHz, indicating that, at room temperature, their magnetization is not completely locked along the magnetic 'easy' axis.

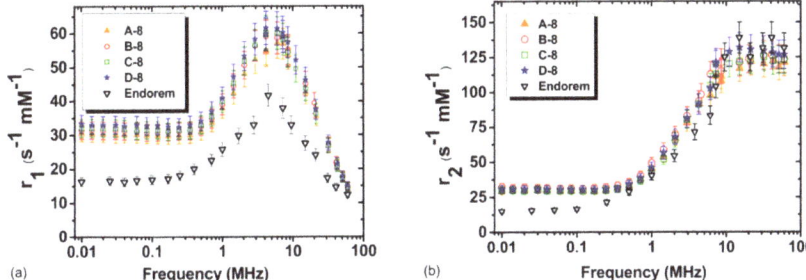

Figure 7. (**a**) Longitudinal r_1 and (**b**) transverse r_2 NMRD profiles collected at room temperature in the Larmor frequency range $0.01 \leq \nu_L \leq 60$ MHz for the second series of polymer-coated MNPs. For comparison, the relaxivity values of Endorem, as reported by Basini et al., are shown [22].

In Table 3, the relaxivities of γ-Fe$_2$O$_3$ nanoparticles are compared to those of Endorem.

Table 3. Longitudinal and transverse relaxivity values and their ratio at 15 and 60 MHz, for the 2nd series of MNPs aqueous dispersions at room temperature.

Sample	Frequency	r_1 (s^{-1}mM^{-1})	r_2 (s^{-1}mM^{-1})	r_2/r_1
A-8	60 MHz	14.7 (1.2)	121.0 (9.7)	8.2
	15.1 MHz	42.5 (3.4)	116.7 (9.3)	2.7
B-8	60 MHz	14.7 (1.2)	123.7 (9.9)	8.4
	15.1 MHz	45.1 (3.6)	121.1 (9.7)	2.7
C-8	60 MHz	14.5 (1.2)	123.1 (9.9)	8.5
	15.1 MHz	43.6 (3.5)	122.13 (9.8)	2.8
D-8	60 MHz	14.9 (1.2)	126.7 (10.1)	8.5
	15.1 MHz	46.2 (3.7)	131.9 (10.6)	2.9
Endorem	60 MHz	12.3 (1.0)	131.6 (10.5)	10.7
	15 MHz	27.5 (2.2)	138.9 (11.1)	5

The transverse relaxivity vs. frequency behavior (Figure 7b) is similar for all samples, and at a high magnetic field $\mu_0 H$ ~1.41 Tesla, r_2 reaches the value ~125 mM^{-1}s^{-1}. Then, the transverse relaxometric performance of these samples at the typical clinical frequency ~60 MHz (crucial for darkening the MRI images and thus increasing the sensitivity) is comparable to that of the commercial CA. Additionally, one can note that the frequency behavior of the r_2 relaxation curve is also similar to that of the commercial compound, although the latter shows lower values for $\nu < 7$ MHz.

It is interesting to note that the r_2/r_1 ratio at 60 MHz assumes a value of ~8 for our MNPs and ~11 for Endorem®. The r_2/r_1 values at 60 and 15.1 MHz are comparable to those of Endorem, indicating that the two substances have a very similar efficiency as T_2 contrast agents.

3.4.2. Analysis of NMR Results

To analyze the NMR longitudinal relaxivity profiles (i.e., r_1 vs. frequency) at room temperature, the heuristic model of Roch et al. [37] was employed. We were able to use this model (valid for an ensemble of single nanoparticles) because the coincidence of r_1 and r_2 values for each sample at low frequencies, approximately ν < 0.1 MHz, ensured the absence of particle aggregation at the dilution used for the NMR measurements. In more detail, the longitudinal NMRD profiles were fitted using the following expression:

$$\frac{1}{T_1} = \frac{32\pi}{135{,}000} \mu_{SP}^2 \gamma_I^2 \frac{N_A C}{RD} 7P\frac{L(x)}{x} J^F \left[\Omega(\omega_S,\omega_0),\tau_D,\tau_N\right] + \left[7Q\frac{L(x)}{x} + 3\left(1 - L^2(x) - 2\frac{L(x)}{x}\right)\right] \cdot J^F(\omega_L,\tau_D,\tau_N) + 3L^2(x)\cdot J^A\left(\sqrt{2\omega_I\tau_D}\right) \quad (3)$$

where μ_{SP} is the effective magnetic moment of the MNPs experienced by the ^{1}H nuclei, γ_I is the proton gyromagnetic ratio, N_A is the Avogadro's number, C is the molar concentration of iron in the MNPs, R is the minimum approach distance between the protons and MNPs, $L(x)$ is the Langevin's function ($L(x) = \coth x - 1$, where $x = \frac{\mu_{SP} B_0}{k_B T}$), D is the diffusion coefficient of the medium, $\tau_D = R^2/D$ is the diffusion time that characterizes the fluctuation of the hyperfine interaction between the nuclear magnetic moments of the ^{1}H nuclei of the solvent (here water) and the nanoparticle magnetic moment, $\tau_N = \tau_0 e^{\frac{KV}{k_B T}}$ is the Néel relaxation time at room temperature, and ω_S and ω_I are the electron and proton transition frequencies, respectively. The parameters P and Q are related to the degree of magnetic anisotropy of the system, being the weight of the spectral density functions J^A (Ayant, high fields) and J^F (Freed, low fields), respectively. In particular, $P = 0$ and $Q = 1$ for highly anisotropic systems, while $P = 1$ and $Q = 0$ for weakly anisotropic systems.

For the 17-nm samples, we were not able to fit the experimental data because their size was at the limit of validity of the Roch's heuristic model (for which nanoparticles should have a mean diameter < 20 nm). Moreover, the fitting process was made difficult by the broad size distribution for all samples (see Figure 3). The non-applicability of the Roch's model to this series of MNPs was confirmed by the fact that it did not predict any r_1 increase at the lowest frequencies, as displayed by our experimental data.

Conversely, for the smaller MNPs (second series of samples), we were able to fit the experimental r_1 data. In Figure 8, the r_1 fitting curves obtained by means of the Roch model for samples A-8, B-8, C-8 and D-8 are shown. The parameters of physical interest, obtained by the fit of the experimental data of Figure 8 to Equation (3) (deduced from the Roch's model), are the saturation magnetization M_s, the magnetic core radius r, the distance R of minimum approach (of the bulk water protons to the MNP magnetic center) and the Néel relaxation time τ_N.

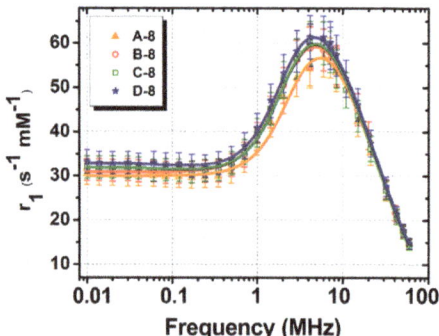

Figure 8. Longitudinal r_1 NMRD profiles (symbols) collected at room temperature in the Larmor frequency range $0.01 \leq \nu_L \leq 60$ MHz for the 8-nm series of polymer-coated MNPs. The solid lines represent the best fit obtained by applying the Roch's model (see text).

In the fitting procedure, we let the saturation magnetization M_S parameter vary between 60 and 70 Am^2/kg$_{\gamma\text{-Fe2O3}}$, as expected from the literature data for particles with a similar size and substantially confirmed by our magnetic measurements for our samples. For the particle radius, by considering the size distribution width, we fixed an upper limit of r ≈ 5 nm. For the water diffusion coefficient, we used the theoretical value D = 2.3 × 10^{-9} m^2s^{-1} at 293 K.

For all samples, we observed the following: (i) The saturation values, constrained in the range specified above, allowed for the fitting procedure convergence, obtaining M_s = 70 ± 4 Am^2/$kg_{\gamma-Fe2O3}$; (ii) The core radius r was slightly higher than the one estimated by TEM (r = 5.0 ± 0.4 nm), a result possibly consistent with the width of the size distributions (Figure 3); (iii) The distance of the minimum approach was 1 ÷ 2 nm greater than the core radius. This latter result points out the tendency of the coating to prevent the diffusion of water inside itself; (iv) The values of τ_N were consistent with the ones typical for nanoparticles of this size, in particular τ_N ≈ 3.5 ÷ 3.9 × 10^{-9} s.

4. Conclusions

In this work, we employed NMR relaxometry to investigate the dependence of the MRI contrast efficiency (i.e., the nuclear relaxivities) on the organic coating of maghemite-based MNPs. In particular, we studied MNPs dispersed in water with two different diameters (d_{TEM} ≈ 8 ± 0.4 nm and ≈ 17 ± 2.5 nm) and four different coatings, i.e., PAA, PMAA-g-PEG and two P(MAA-*stat*-MAPEG) with different transfer agents. A structural, morpho-dimensional and magnetic characterization of the nanoparticles was performed by means of Transmission Electron Microscopy (TEM), Atomic Force Microscopy (AFM) and DC magnetometry. The magnetization curves, particularly at a low temperature, displayed a different approach to saturation depending on the coating, the M_s approaching rate of the sample coated with hydrophobic P(MAA-*stat*-MAPEG) being the slowest one, followed by that coated with PMAA-g-PEG. These results seem to suggest that the hydrophobic P(MAA-*stat*-MAPEG) and PMAA-g-PEG coatings favor a higher spin disorder at the particle surface. The r_1-NMRD profiles show the same behavior for samples with the same core size but with different coatings, indicating that the type of coating used in this work does not evidently influence the longitudinal relaxometric properties. For the transverse relaxivity, we observed a similar trend, except for the sample of the 17-nm series coated with PMAA-g-PEG, which had a lower r_2, in particular for ν_L > 5–10 MHz. Remarkably, all samples showed high r_2 values at 60 MHz (~120 $mM^{-1}s^{-1}$ for d_{TEM} ≈ 8 nm and 300–400 $mM^{-1}s^{-1}$ for d_{TEM} ≈ 17 nm), which were comparable to or higher than the transverse relaxivity of the commercial compound Endorem®. Thus, our samples are promising superparamagnetic T_2 contrast agents for MRI, especially in the case of d_{TEM} ≈ 17 nm. This conclusion is supported by the values of the r_2/r_1 ratio, which generally provides an indication as to how magnetic nanoparticles may behave in their application as contrast-enhancing agents. In our case, for the larger diameter series and at the most used clinical frequency (ν_L ~60 MHz), the r_2/r_1 values were three times larger than the one for Endorem, allowing us to envision a possible superparamagnetic CA dose reduction in clinical use.

Author Contributions: Investigation, F.B., M.B., M.A., F.O., P.A., C.S., C.I., A.G., J.B., E.I., M.L., J.F. and L.L.; Resources, E.I., M.L., J.F. and L.L.; Supervision, A.L.; Writing—original draft, F.B.; Writing—review & editing, F.O., P.A., C.S., C.I., E.I., L.L. and A.L. Acquisition and/or analysis, interpretation of NMR data, P.A., M.A., M.B., F.B.; synthesis and/or characterization of MNPs, J.B., E.I., M.L., J.F., L.L., C.S., C.I., A.G.; AFM measurements, F.O.; supervision, A.L. All authors have read and agreed to the published version of the manuscript.

Funding: This research received no external funding.

Acknowledgments: Région des Pays de la Loire is gratefully acknowledged for its strong support through the PhD program of LUMOMAT RFI (ONASSIS project) as well as CNRS through the MITI program "Nano Challenge: Health and Welfare" (ETHICAM project) and grant number PICS07354. The INSTM-Regione Lombardia projects MAGNANO and MOTORSPORT, the EU-COST projects RADIOMAG TD-1402 and EURELAX CA-15209, and the INFN project HADROMAG are gratefully acknowledged. Julien Poly (Université de Haute-Alsace, IS2M UMR CNRS 7361) is warmly acknowledged for the synthesis of polyelectrolytes PEG-B to PEG-D and his very rich and extremely stimulating discussions. The authors strongly acknowledge Carole La and Marion Rivoal (LPGN-CNRS 6112, University of Nantes) for the iron content titrations using ICP-OES. TEM experiments were performed in the BIBS microscopy platform (UR1268 BIA, INRA).

Conflicts of Interest: The authors declare no conflict of interest.

References

1. Bremer, C.; Ntziachristos, V.; Weissleder, R. Optical-based molecular imaging: Contrast agents and potential medical applications. *Eur. Radiol.* **2003**, *13*, 231–243. [CrossRef] [PubMed]
2. Britz-Cunningham, S.H.; James Adelstein, S. Molecular targeting with radionuclides: State of the science. *J. Nucl. Med.* **2003**, *44*, 1945–1961. [PubMed]
3. Wu, D.; Huang, L.; Jiang, M.S.; Jiang, H. Contrast agents for photoacoustic and thermoacoustic imaging: A review. *Int. J. Mol. Sci.* **2014**, *15*, 23616–23693. [CrossRef] [PubMed]
4. Gambhir, S.S. Molecular imaging of cancer with positron emission tomography. *Nat. Rev. Cancer* **2002**, *2*, 683–693. [CrossRef]
5. Ring, E.F.J.; Ammer, K. Infrared thermal imaging in medicine. *Physiol. Meas.* **2012**, *33*, R33. [CrossRef]
6. Mettler, F.; Guiberteau, M. *Essentials of Nuclear Medicine Imaging*; Elsevier Saunders: Philadelphia, PA, USA, 2012; ISBN 9781455701049.
7. Chan, R.W.; Lau, J.Y.C.; Lam, W.W.; Lau, A.Z. Magnetic resonance imaging. In *Encyclopedia of Biomedical Engineering*; Elsevier Inc.: Amsterdam, The Netherlands, 2018; ISBN 9780128051443.
8. Xiao, Y.D.; Paudel, R.; Liu, J.; Ma, C.; Zhang, Z.S.; Zhou, S.K. MRI contrast agents: Classification and application (Review). *Int. J. Mol. Med.* **2016**, *38*, 1319–1326. [CrossRef]
9. Hao, D.; Ai, T.; Goerner, F.; Hu, X.; Runge, V.M.; Tweedle, M. MRI contrast agents: Basic chemistry and safety. *J. Magn. Reson. Imaging* **2012**, *36*, 1060–1071. [CrossRef]
10. Na, H.B.; Hyeon, T. Nanostructured T1 MRI contrast agents. *J. Mater. Chem.* **2009**, *19*, 6267–6273. [CrossRef]
11. Caravan, P.; Ellison, J.J.; McMurry, T.J.; Lauffer, R.B. Gadolinium(III) chelates as MRI contrast agents: Structure, dynamics, and applications. *Chem. Rev.* **1999**, *99*, 2293–2352. [CrossRef]
12. Shokrollahi, H. Contrast agents for MRI. *Mater. Sci. Eng. C* **2013**, *33*, 4485–4497. [CrossRef]
13. Taylor, P.M. Contrast agents. In *Imaging and Technology in Urology: Principles and Clinical Applications*; Springer: London, UK, 2012; ISBN 9781447124221.
14. Na, H.B.; Song, I.C.; Hyeon, T. Inorganic nanoparticles for MRI contrast agents. *Adv. Mater.* **2009**, *21*, 2133–2148. [CrossRef]
15. Merbach, A.; Helm, L.; Tóth, É. *The Chemistry of Contrast Agents in Medical Magnetic Resonance Imaging*, 2nd ed.; John Wiley & Sons: Chichester, UK, 2013; ISBN 9781119991762.
16. Young, I.R.; Clarke, G.J.; Baffles, D.R.; Pennock, J.M.; Doyle, F.H.; Bydder, G.M. Enhancement of relaxation rate with paramagnetic contrast agents in NMR imaging. *J. Comput. Tomogr.* **1981**, *5*, 543–547. [CrossRef]
17. Wang, Y.X.J. Current status of superparamagnetic iron oxide contrast agents for liver magnetic resonance imaging. *World J. Gastroenterol.* **2015**, *21*, 13400. [CrossRef] [PubMed]
18. Bobo, D.; Robinson, K.J.; Islam, J.; Thurecht, K.J.; Corrie, S.R. Nanoparticle-Based Medicines: A Review of FDA-Approved Materials and Clinical Trials to Date. *Pharm. Res.* **2016**, *33*, 2373–2387. [CrossRef] [PubMed]
19. Reimer, P.; Balzer, T. Ferucarbotran (Resovist): A new clinically approved RES-specific contrast agent for contrast-enhanced MRI of the liver: Properties, clinical development, and applications. *Eur. Radiol.* **2003**, *13*, 1266–1276. [CrossRef]
20. González-Gómez, M.A.; Belderbos, S.; Yañez-Vilar, S.; Piñeiro, Y.; Cleeren, F.; Bormans, G.; Deroose, C.M.; Gsell, W.; Himmelreich, U.; Rivas, J. Development of superparamagnetic nanoparticles coated with polyacrylic acid and aluminum hydroxide as an efficient contrast agent for multimodal imaging. *Nanomaterials* **2019**, *9*, 1626. [CrossRef]
21. Casula, M.F.; Floris, P.; Innocenti, C.; Lascialfari, A.; Marinone, M.; Corti, M.; Sperling, R.A.; Parak, W.J.; Sangregorio, C. Magnetic resonance imaging contrast agents based on iron oxide superparamagnetic ferrofluids. *Chem. Mater.* **2010**, *22*, 1739–1748. [CrossRef]
22. Basini, M.; Guerrini, A.; Cobianchi, M.; Orsini, F.; Bettega, D.; Avolio, M.; Innocenti, C.; Sangregorio, C.; Lascialfari, A.; Arosio, P. Tailoring the magnetic core of organic-coated iron oxides nanoparticles to influence their contrast efficiency for Magnetic Resonance Imaging. *J. Alloys Compd.* **2019**, *770*, 58–66. [CrossRef]
23. Orlando, T.; Albino, M.; Orsini, F.; Innocenti, C.; Basini, M.; Arosio, P.; Sangregorio, C.; Corti, M.; Lascialfari, A. On the magnetic anisotropy and nuclear relaxivity effects of Co and Ni doping in iron oxide nanoparticles. *J. Appl. Phys.* **2016**, *119*, 134301. [CrossRef]

24. Johnson, N.J.J.; He, S.; Nguyen Huu, V.A.; Almutairi, A. Compact Micellization: A Strategy for Ultrahigh T1 Magnetic Resonance Contrast with Gadolinium-Based Nanocrystals. *ACS Nano* **2016**, *10*, 8299–8307. [CrossRef]
25. Fresnais, J.; Ma, Q.Q.; Thai, L.; Porion, P.; Levitz, P.; Rollet, A.L. NMR relaxivity of coated and non-coated size-sorted maghemite nanoparticles. *Mol. Phys.* **2019**, *117*, 990–999. [CrossRef]
26. Vangijzegem, T.; Stanicki, D.; Panepinto, A.; Socoliuc, V.; Vekas, L.; Muller, R.N.; Laurent, S. Influence of experimental parameters of a continuous flow process on the properties of very small iron oxide nanoparticles (VSION) designed for T1-weighted magnetic resonance imaging (MRI). *Nanomaterials* **2020**, *10*, 757. [CrossRef] [PubMed]
27. Magro, M.; Vianello, F. Bare iron oxide nanoparticles: Surface tunability for biomedical, sensing and environmental applications. *Nanomaterials* **2019**, *9*, 1608. [CrossRef] [PubMed]
28. Dutz, S.; Buske, N.; Landers, J.; Gräfe, C.; Wende, H.; Clement, J.H. Biocompatible magnetic fluids of co-doped iron oxide nanoparticles with tunable magnetic properties. *Nanomaterials* **2020**, *10*, 1019. [CrossRef] [PubMed]
29. Murgulescu, I.; Ababei, G.; Stoian, G.; Danceanu, C.; Lupu, N.; Chiriac, H. Fe-Cr-Nb-B magnetic nanoparticles prepared by arc discharge for hyperthermia. *J. Optoelectron. Adv. Mater.* **2019**, *21*, 733–739.
30. Ortega, D.; Pankhurst, Q.A. Magnetic hyperthermia. *Nanoscience* **2013**, *1*, e88. [CrossRef]
31. Bonvin, D.; Arakcheeva, A.; Millán, A.; Piñol, R.; Hofmann, H.; Mionić Ebersold, M. Controlling structural and magnetic properties of IONPs by aqueous synthesis for improved hyperthermia. *RSC Adv.* **2017**, *7*, 13159–13170. [CrossRef]
32. Linot, C.; Poly, J.; Boucard, J.; Pouliquen, D.; Nedellec, S.; Hulin, P.; Marec, N.; Arosio, P.; Lascialfari, A.; Guerrini, A.; et al. PEGylated Anionic Magnetofluorescent Nanoassemblies: Impact of Their Interface Structure on Magnetic Resonance Imaging Contrast and Cellular Uptake. *ACS Appl. Mater. Interfaces* **2017**, *9*, 14242–14257. [CrossRef]
33. Faucon, A.; Maldiney, T.; Clément, O.; Hulin, P.; Nedellec, S.; Robard, M.; Gautier, N.; De Meulenaere, E.; Clays, K.; Orlando, T.; et al. Highly cohesive dual nanoassemblies for complementary multiscale bioimaging. *J. Mater. Chem. B* **2014**, *2*, 7747–7755. [CrossRef]
34. Ménager, C.; Sandre, O.; Mangili, J.; Cabuil, V. Preparation and swelling of hydrophilic magnetic microgels. *Polymer* **2004**, *45*, 2475–2481. [CrossRef]
35. Santoyo Salazar, J.; Perez, L.; De Abril, O.; Truong Phuoc, L.; Ihiawakrim, D.; Vazquez, M.; Greneche, J.M.; Begin-Colin, S.; Pourroy, G. Magnetic iron oxide nanoparticles in 10-40 nm range: Composition in terms of magnetite/maghemite ratio and effect on the magnetic properties. *Chem. Mater.* **2011**, *23*, 1379–1386. [CrossRef]
36. Lefebure, S.; Dubois, E.; Cabuil, V.; Neveu, S.; Massart, R. Monodisperse magnetic nanoparticles: Preparation and dispersion in water and oils. *J. Mater. Res.* **1998**, *13*, 2975–2981. [CrossRef]
37. Roch, A.; Muller, R.N.; Gillis, P. Theory of proton relaxation induced by superparamagnetic particles. *J. Chem. Phys.* **1999**, *110*, 5403–5411. [CrossRef]
38. Umut, E. Surface Modification of Nanoparticles Used in Biomedical Applications. In *Modern Surface Engineering Treatments*; InTech: Rijeka, Croatia, 2013; pp. 185–208.

© 2020 by the authors. Licensee MDPI, Basel, Switzerland. This article is an open access article distributed under the terms and conditions of the Creative Commons Attribution (CC BY) license (http://creativecommons.org/licenses/by/4.0/).

Article

Application of Magnetosomes in Magnetic Hyperthermia

Nikolai A. Usov [1,2,3,*] and Elizaveta M. Gubanova [3]

[1] National University of Science and Technology «MISiS», 119049 Moscow, Russia
[2] Pushkov Institute of Terrestrial Magnetism, Ionosphere and Radio Wave Propagation, Russian Academy of Sciences, IZMIRAN, Troitsk, 108480 Moscow, Russia
[3] National Research Nuclear University "MEPhI", 115409 Moscow, Russia; elizaveta.gubanova.1995@mail.ru
* Correspondence: usov@obninsk.ru

Received: 26 May 2020; Accepted: 2 July 2020; Published: 5 July 2020

Abstract: Nanoparticles, specifically magnetosomes, synthesized in nature by magnetotactic bacteria, are very promising to be usedin magnetic hyperthermia in cancer treatment. In this work, using the solution of the stochastic Landau–Lifshitz equation, we calculate the specific absorption rate (SAR) in an alternating (AC) magnetic field of assemblies of magnetosome chains depending on the particle size D, the distance between particles in a chain a, and the angle of the applied magnetic field with respect to the chain axis. The dependence of SAR on the a/D ratio is shown to have a bell-shaped form with a pronounced maximum. For a dilute oriented chain assembly with optimally chosen a/D ratio, a strong magneto-dipole interaction between the chain particles leads to an almost rectangular hysteresis loop, and to large SAR values in the order of 400–450 W/g at moderate frequencies f = 300 kHz and small magnetic field amplitudes H_0 = 50–100 Oe. The maximum SAR value only weakly depends on the diameter of the nanoparticles and the length of the chain. However, a significant decrease in SAR occurs in a dense chain assembly due to the strong magneto-dipole interaction of nanoparticles of different chains.

Keywords: magnetotactic bacteria; magnetosome chain; magnetic hyperthermia; low frequency hysteresis loops; numerical simulation

1. Introduction

Magnetic hyperthermia [1–3] is currently considered as a very promising method of cancer treatment. Laboratory tests [4–7] and clinical studies [8–10] show that local, dosed heating of tumors leads to a delay in their growth and even complete decay and disappearance. For this, it is necessary in several sessions to maintain the temperature of the tumor at the level of 43–45 °C for 30 min [1–3]. In magnetic hyperthermia local heating of biological tissues is achieved by introducing magnetic nanoparticles into a tumor and applying to them a low frequency alternating (AC) magnetic field [1–3,11,12]. The fundamental advantage of magnetic hyperthermia lies in the possibility of local heating of tumors located deep in a human body, since the low-frequency magnetic field is relatively poorly shielded by the biological medium, in contrast to high-frequency and laser heating methods [13,14], which are suitable for the treatment of tumors localized near body surface.

However, for the successful implementation of the magnetic hyperthermia in practice it is necessary to use assemblies of magnetic nanoparticles with a sufficiently high specific absorption rate (SAR) in AC magnetic field of moderate frequency, f < 1 MHz, and amplitude H_0 of the order of 100–200 Oe. It was proved [15] that the effect of an AC magnetic field on living organisms is relatively safe under the condition $H_0 f \le 5 \times 10^8$ A(ms)$^{-1}$, though recently safety limit was increased [16] up to $H_0 f \le 5 \times 10^9$ A(ms)$^{-1}$. In addition, only mildly toxic and biodegradable nanoparticles can be used in magnetic hyperthermia to reduce the risk of possible side effects. Based on these requirements, magnetic nanoparticles of iron oxides are considered to be the most suitable for application in magnetic hyperthermia [1–3,11,12,16–19].

Unfortunately, the popular chemical methods for the synthesis of magnetic nanoparticles of iron oxides [1,20–22] in most cases give assemblies with a wide distribution of nanoparticles in size and shape. Moreover, the obtained nanoparticles turn out to be polycrystalline, which leads to a low saturation magnetization of particles in comparison with the corresponding single-crystal three-dimensional samples [23]. For such assemblies it is hardly possible to obtain sufficiently high SARs under the above restrictions on the frequency and amplitude of AC magnetic field [24,25]. Furthermore, the subsequent uncontrolled agglomeration of nanoparticles in the biological medium leads to a further decrease of the assembly SAR [26,27].

The results of numerical simulations [28–30], confirmed by a number of experimental data [25,31–36] show that in order to achieve sufficiently high SARs at moderate amplitudes and frequencies of AC magnetic field, one has to use single-crystal magnetic nanoparticles with high saturation magnetization. It is necessary also to ensure a narrow particle size distribution near the optimally chosen nanoparticle diameter. Due to difficulties with chemical synthesis, lately, much attention has been paid to experimental and theoretical studies of assemblies of magnetosomes, which are synthesized in nature by magnetotactic bacteria [5,25,37–45]. Magnetosomes grow inside bacteria under optimal physiological conditions. Therefore, they have a perfect crystalline structure, a quasi-spherical shape, and a fairly narrow particle size distribution. Most magnetotactic bacteria synthesize nanoparticles that are close in chemical composition to high-purity maghemite, γ-Fe_2O_3 [42] or magnetite, Fe_3O_4 [45].

It was shown in the pioneer work of Hergt, et al. [37] that the SAR in the oriented assembly of magnetosomes reaches a very high value of 960 W/g at a frequency f = 410 kHz and magnetic field amplitude H_0 = 126 Oe. High SAR values in magnetosome assemblies were also obtained in a number of subsequent experimental studies [5,25,39,45]. It is known [37,38,43] that various types of magnetotactic bacteria synthesize nanoparticles with different characteristic sizes, from 20 to 50 nm. In the bacteria the nanoparticles are arranged in the form of long chains consisting of 6–30 particles of approximately the same diameter. Existing experimental techniques make it possible to isolate magnetosomes from bacteria both in the form of single particles and in the form of chains with different numbers of nanoparticles in the chain [25,37–43].

The behavior of the assembly of magnetosome chains in an AC magnetic field is of great interest from a theoretical point of view. It was already emphasized earlier [46–48] that the orientation of individual nanoparticles and their chains along the applied AC magnetic field is important to obtain appreciable SARs for magnetosome assembly. In this paper, based on the solution of the stochastic Landau–Lifshitz equation [49–52], the low-frequency hysteresis loops of assemblies of magnetosome chains extracted from magnetotactic bacteria are calculated. This theoretical approach makes it possible to take into account the complicated magnetic anisotropy of particles, the influence of thermal fluctuations of magnetic moments, and strong magnetic dipole interaction between chain particles. In this work the dependence of the assembly SAR on the particle diameter, on the length of the chains, and the distance between particles in the chain is studied in detail. We also investigate the dependence of SAR on the orientation of applied magnetic field with respect to the chain axis, as well as on the average density of the chain assembly. It is shown that with an optimal choice of magnetosome chain geometry, sufficiently high SAR values, of the order of 400–450 W/g, can be obtained at a frequency f = 300 kHz and at small and moderate field amplitudes, H_0 = 50–100 Oe. However, it is found that the SAR of the chain assembly decreases significantly with an increase in its average density. It seems that the results obtained will be useful for the optimal choice of the geometric parameters of magnetosome chains to further increase their heating ability for successful application in magnetic hyperthermia.

2. Model and Methods

Calculations of SAR of magnetosome chains were performed in this work by numerical simulation using stochastic Landau–Lifshitz equation [49–52]. The latter governs the dynamics of the unit magnetization vector $\vec{\alpha}_i$ of i-th single-domain nanoparticle of the magnetosome chain

$$\frac{\partial \vec{\alpha}_i}{\partial t} = -\gamma_1 \vec{\alpha}_i \times \left(\vec{H}_{ef,i} + \vec{H}_{th,i}\right) - \kappa\gamma_1 \vec{\alpha}_i \times \left(\vec{\alpha}_i \times \left(\vec{H}_{ef,i} + \vec{H}_{th,i}\right)\right), \ i = 1, 2, \ldots, N_p. \tag{1}$$

where γ is the gyromagnetic ratio, κ is phenomenological damping parameter, $\gamma_1 = \gamma/(1 + \kappa^2)$, $\vec{H}_{ef,i}$ is the effective magnetic field, $\vec{H}_{th,i}$ is the thermal field, and N_p is the number of nanoparticles in the chain. The effective magnetic field acting on a separate nanoparticle can be calculated as a derivative of the total chain energy

$$\vec{H}_{ef,i} = -\frac{\partial W}{M_s V \partial \vec{\alpha}_i}. \tag{2}$$

The total magnetic energy of the chain $W = W_a + W_Z + W_m$ is a sum of the magnetic anisotropy energy W_a, Zeeman energy W_Z of the particles in applied magnetic field, and the energy of mutual magneto-dipole interaction of the particles W_m. Since magnetosomes released from bacteria are coated with thin non-magnetic shells, there is no direct contact between the magnetic cores of the nanoparticles. Therefore, the exchange interaction of neighboring nanoparticles in the chain can be neglected.

The magneto-crystalline anisotropy energy of magetosomes with a perfect crystalline structure is that of cubic type [52,53]

$$W_a = K_c V \sum_{i=1}^{N_p} \left(\left(\vec{\alpha}_i \vec{e}_{1i}\right)^2 \left(\vec{\alpha}_i \vec{e}_{2i}\right)^2 + \left(\vec{\alpha}_i \vec{e}_{1i}\right)^2 \left(\vec{\alpha}_i \vec{e}_{3i}\right)^2 + \left(\vec{\alpha}_i \vec{e}_{2i}\right)^2 \left(\vec{\alpha}_i \vec{e}_{3i}\right)^2 \right) \tag{3}$$

where $(\vec{e}_{1i}, \vec{e}_{2i}, \vec{e}_{3i})$ is a set of orthogonal unit vectors that determine an orientation of i-th nanoparticle of the chain. It is assumed that the easy anisotropy axes of various nanoparticles in the chain are randomly oriented with respect to each other.

The Zeeman energy of the chain in applied AC magnetic field $\vec{H}_0 \sin(\omega t)$ is given by

$$W_Z = -M_s V \sum_{i=1}^{N_p} \left(\vec{\alpha}_i \vec{H}_0 \sin(\omega t)\right) \tag{4}$$

For nearly spherical uniformly magnetized nanoparticles the magnetostatic energy of the chain can be represented as the energy of the point interacting dipoles located at the particle centers $\vec{r}_i$ within the chain. Then, the energy of magneto-dipole interaction is

$$W_m = \frac{M_s^2 V^2}{2} \sum_{i \neq j} \frac{\vec{\alpha}_i \vec{\alpha}_j - 3(\vec{\alpha}_i \vec{n}_{ij})(\vec{\alpha}_j \vec{n}_{ij})}{\left|\vec{r}_i - \vec{r}_j\right|^3} \tag{5}$$

where $\vec{n}_{ij}$ is the unit vector along the line connecting the centers of i-th and j-th particles, respectively.

The thermal fields $\vec{H}_{th,i}$ acting on various nanoparticles of the chain are statistically independent, with the following statistical properties [49] of their components

$$\left\langle H_{th,i}^{(\alpha)}(t) \right\rangle = 0; \ \left\langle H_{th,i}^{(\alpha)}(t) H_{th,i}^{(\beta)}(t_1) \right\rangle = \frac{2k_B T \kappa}{\gamma M_s V_i} \delta_{\alpha\beta} \delta(t - t_1), \ \alpha, \beta = (x, y, z) \tag{6}$$

where k_B is the Boltzmann constant, $\delta_{\alpha\beta}$ is the Kroneker symbol, and $\delta(t)$ is the delta function. The numerical simulation procedure is described in details in [54,55].

The SAR value of an assembly of magnetic nanoparticles in an AC magnetic field with a frequency f is determined by the integral [28]

$$SAR = \frac{fM_s}{\rho} \oint \langle \vec{m} \rangle d\vec{H} \qquad (7)$$

where ρ is the density of magnetic material and $\langle \vec{m} \rangle$ is the reduced magnetic moment of the assembly.

3. Results

In this work the calculations of SAR of magnetosome chain assemblies are performed in the frequency range f = 250–350 kHz, at small and moderate amplitudes of an AC magnetic field, H_0 = 50–150 Oe, since the use of AC magnetic fields of small amplitude is preferable in a medical clinic.

3.1. Dilute Chain Assembly

Let us consider first the properties of dilute oriented assemblies of magnetosome chains neglecting the magnetic dipole interaction between various chains. It is of considerable interest to study the dependence of the SAR of such an assembly on the nanoparticle diameter D, on the number of particles in the chain N_p, and on the orientation of the external magnetic field with respect to the chain axis. For completeness, the chains with different average distances a between the centers of the nanoparticles are considered. Note that all geometric parameters mentioned can be adopted properly during the experimental design of a chain from individual magnetosomes of approximately the same diameter [5,7,40–42]. In particular, the distance a between the centers of successive nanoparticles in a chain is determined by the thickness of non-magnetic shells on their surfaces. The latter protect the nanoparticles from the aggressive action of the medium. It is worth noting that, since the cubic magnetic anisotropy constant of magnetite is negative, $K_c = -1 \times 10^5$ erg/cm^3, the quasi-spherical magnetite nanoparticles have eight equivalent directions of easy anisotropy axes [53]. In the calculations performed it is assumed that the easy anisotropy axes of individual magnetosomes are randomly oriented, since it is hardly possible to make multiple easy anisotropy axes of various magnetosomes parallel when a chain is created. Saturation magnetization of the magnetosomes is assumed to be M_s = 450 emu/cm^3 [25,45]. It is also supposed that the magnetosome chain cannot rotate as a whole, being distributed in a medium with a sufficiently high viscosity or being tightly bound to surrounding tissue.

Figure 1a shows the results of SAR calculation for dilute oriented assemblies of magnetosome chains with particles of various diameters, D = 20–50 nm. The number of particles in the chains is fixed at N_p = 30, AC magnetic field frequency f = 300 kHz, field amplitude H_0 = 50 Oe. The AC magnetic field is applied parallel to the chain axis, the easy anisotropy axes of various particles in the chain are randomly oriented. The calculation results are averaged over a sufficiently large number of independent chain realizations, N_{exp} = 40–60. The temperature of the system is T = 300 K.

As Figure 1a shows, in all cases considered there is a significant dependence of SAR on the reduced distance a/D between the centers of the particles of the chain. The assembly SAR reaches a maximum value at the ratios a/D = 1.5, 1.85, 2.2, and 3.6 for particles with diameters D = 20, 30, 40, and 50 nm, respectively.

The SAR decreases sharply after reaching the maximum. It is interesting to note that the SAR at the maximum only weakly depends on the particle diameter. Indeed, according to Figure 1a for particles with diameters D = 20, 30, 40, and 50 nm, the maximum SAR values are given by 440, 444, 454, and 409 W/g, respectively.

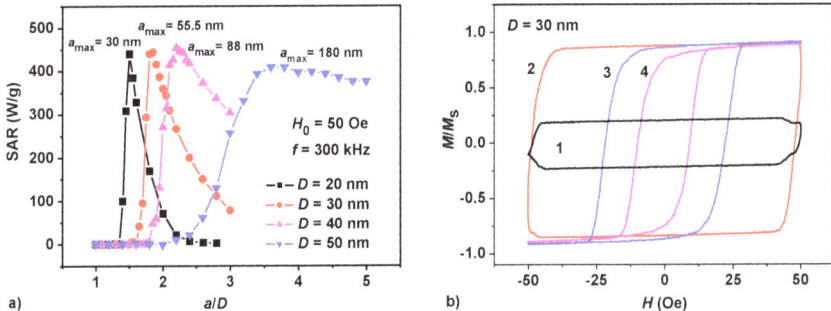

Figure 1. (a) Dependence of assembly SAR on the reduced distance a/D between the particle centers for chains with various particle diameters. (b) Evolution of the shape of the low-frequency hysteresis loop for chains of particles with diameter $D = 30$ nm for various reduced distances: (1) $a/D = 1.7$, (2) $a/D = 1.85$, (3) $a/D = 2.4$, (4) $a/D = 3.0$.

Figure 1b shows the evolution of the low frequency hysteresis loops of a dilute chain assembly with nanoparticle diameter $D = 30$ nm as a function of the reduced distance a/D between the nanoparticle centers. It is known [28,29] that the SAR of an assembly of magnetic nanoparticles is proportional to the area of the low frequency hysteresis loop. In Figure 1b, it can be seen that in accordance with Figure 1a the maximum area of the hysteresis loop for particles with diameter $D = 30$ nm corresponds to the reduced distance $a/D = 1.85$.

It is interesting to note that the position of the SAR maximum is practically independent on the frequency. For example, according to Figure 2a for the case of particles with diameter $D = 30$ nm the SAR maximum corresponds to the ratio $a/D \approx 1.8$ in the frequency range $f = 250$–350 kHz. At the same time, as Figure 2b shows, the position of the maximum and the SAR value at the maximum substantially depend on the amplitude of the AC magnetic field. Therefore, the choice of the optimal a/D ratio for a magnetosome chain should be consistent with the given value of H_0. With increasing H_0 the maximum of the assembly SAR grows, and the position of the maximum falls at a shorter distance between the particles of the chain. Indeed, in Figure 2b the maximum values of SAR = 440.1, 854.6, and 1281.0 W/g are observed at $a/D = 1.5$, 1.35, and 1.2 for the magnetic field amplitudes $H_0 = 50$, 100, and 150 Oe, respectively. In general, as Figures 1 and 2 show, for a dilute oriented assembly of magnetosome chains with an optimal choice of the a/D ratio one can obtain rather high SAR values, of the order of 400–450 W/g already in an AC magnetic field with a relatively small amplitude $H_0 = 50$ Oe.

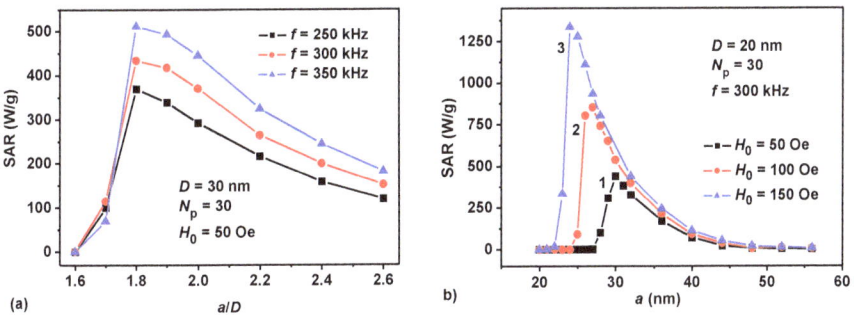

Figure 2. (a) Frequency dependence of SAR for assembly of magnetosome chains with particle diameter $D = 30$ nm in an AC magnetic field with an amplitude $H_0 = 50$ Oe. (b) Dependence of SAR on the distance a between the particle centers for particles with diameter $D = 20$ nm for various amplitudes of AC magnetic field: (1) $H_0 = 50$ Oe, (2) $H_0 = 100$ Oe, (3) $H_0 = 150$ Oe.

The significant dependence of the assembly SAR on the a/D ratio, and on the amplitude of the AC magnetic field H_0, shown in Figures 1a and 2b, is explained by the influence of a strong interacting field H_d acting between closely located particles of the chain. Figure 3a shows the instantaneous distribution of interacting fields for an individual chain consisting of $N_p = 30$ nanoparticles of diameter $D = 20$ nm at a time when the AC magnetic field is close to zero. The z axis is assumed to be parallel to the axis of the chain, the amplitude and frequency of the AC magnetic field are $H_0 = 50$ Oe, and $f = 300$ kHz, respectively. As Figure 3a shows, in the central part of the chain, due to the summation of the magnetic fields of individual nanoparticles, the longitudinal component of the interacting field reaches sufficiently large values, $H_{dz} = 250$ Oe, significantly exceeding the amplitude of the AC magnetic field, $H_0 = 50$ Oe.

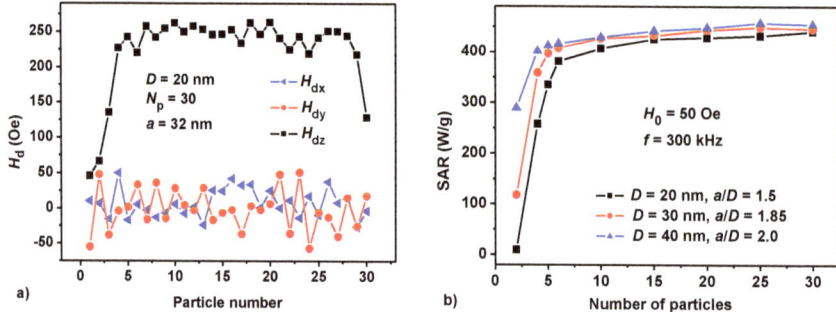

Figure 3. (**a**) Distribution of the components of the interaction field depending on the position of the nanoparticle in the chain. (**b**) Dependence of the SAR of a dilute chain assembly on the number of particles N_p in the chains with particles of different diameters, $D = 20$–40 nm, assuming the optimal ratios a/D for various chains.

On the contrary, the transverse field components turn out to be relatively small $|H_{dx}|$, $|H_{dy}| < 50$ Oe. Obviously, the irregular distribution of the interacting field on individual particles, shown in Figure 3a is associated with a random orientation of the easy anisotropy axes of individual nanoparticles.

As Figure 3a shows the interacting field is greatly reduced at the chain ends. Therefore, the magnetization switching field H_{sf} of the chain is determined by the conditions near its ends. Calculations show that for $H_{sf} < H_0$ the chain magnetization reverses as a whole in a process similar to the giant Barkhausen jump in iron-rich amorphous ferromagnetic microwires [56]. The switching field H_{sf} of the chain increases with decreasing average distance between the particles in the chain, since the intensity of the magnetic dipole interaction increases if neighboring nanoparticles are located closer to each other. When the distance between the particle centers decreases, the switching field of the chain can reach values exceeding the amplitude of AC magnetic field, $H_{sf} > H_0$. Under this condition, the magnetization reversal of the chain is impossible. As Figure 1b shows, when the reduced distance between the particle centers decreases from $a/D = 1.85$ to $a/D = 1.7$, the area of the assembly hysteresis loop reduces sharply. This leads to a sharp drop in the SAR of the corresponding assembly in Figure 1a. However, the drop in the hysteresis loop area in Figure 1b does not occur immediately to zero, due to fluctuations in the switching fields H_{sf} of individual chains of the assembly. As a result, even at $a/D = 1.7$ some of the assembly chains are still able to reverse their magnetizations in applied magnetic field. This, however, leads to a sharp narrowing of the vertical size of the low frequency hysteresis loop of the whole assembly. With a further decrease of the a/D ratio, the fraction of chains capable of magnetization reversal at given magnetic field amplitude H_0 tends to zero. As a result, the ability of the assembly to absorb the energy of an AC magnetic field disappears.

Similar considerations explain the behavior of the assembly SAR as a function of the AC magnetic field amplitude. As Figure 2b shows, with an increase in H_0 the magnetization reversal of the chain is possible at smaller reduced distances a/D. This leads to a shift in the positions of the SAR maximum to smaller reduced distances and to increase in the maximum SAR values.

It is important to note the characteristic change of the shape of the low frequency hysteresis loop of a dilute magnetosome assembly as a function of a/D ratio. As Figure 1b shows, the hysteresis loop is nearly rectangular and has maximal area for optimal ratio $(a/D)_0$. For ratios smaller than optimal, $a/D < (a/D)_0$, the vertical size of the hysteresis loop is considerably decreased, $M/M_s < 1$. On the other hand, for ratios larger than optimal, $a/D > (a/D)_0$, the width of the assembly hysteresis loop is reduced, $2H_{max} < 2H_0$.

For an infinite periodic chain of single-domain nanoparticles with uniaxial magnetic anisotropy, the switching field for magnetization reversal was analytically determined by Aharoni [57]. However, in the present case an analytical consideration is hardly possible due to the randomness in the orientation of the easy anisotropy axes of individual nanoparticles and the influence of thermal fluctuations of the particle magnetic moments at a finite temperature.

It is interesting to note that the effect of sharp variation of the SAR of a dilute assembly of magnetosome chains as a function of the a/D ratio only slightly depends on the number of particles in the chains, provided that it exceeds the value of the order of $N_p = 4$–6. For example, Figure 3b shows the dependences of SAR on the number of particles, $N_p = 2$–30, in the chain assemblies with particles of various diameters, $D = 20$–40 nm. The calculations presented in Figure 3b are performed at optimal ratios $a/D = 1.5$, 1.85, and 2.2 for particles with diameters $D = 20$, 30, and 40 nm, respectively. These optimal a/D values were determined previously at frequency $f = 300$ kHz and amplitude $H_0 = 50$ Oe. According to Figure 3b, regardless of the nanoparticle diameter, a sharp increase in SAR as a function of the number of particles in the chains occurs in the range $N_p \leq 4$–5. However, with a further increase in the chain length, the SAR values of the assemblies change slowly.

Let us consider now the dependence of the SAR of a dilute oriented chain assembly on the angle θ of AC magnetic field with respect to the chain axis. The calculations presented in Figure 4 are carried out at frequency $f = 300$ kHz and magnetic field amplitude $H_0 = 50$ Oe, the number of particles in the chains being $N_p = 30$. The reduced distances between the centers of neighboring nanoparticles in the chains are chosen optimal, so that $a/D = 1.5$, 1.85, and 2.2 for particles with diameters $D = 20$, 30, and 40 nm, respectively.

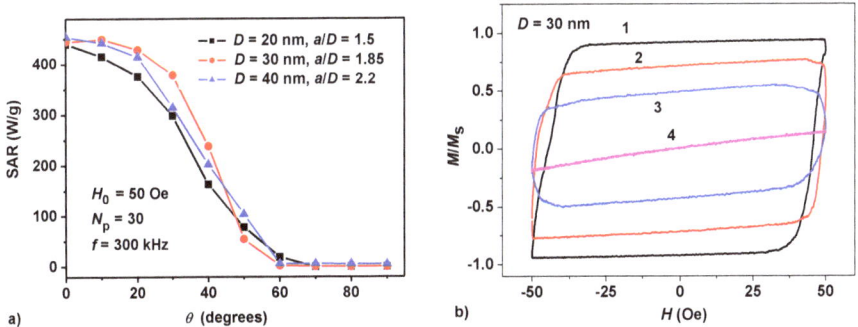

Figure 4. (a) Dependence of the assembly SAR on the angle of AC magnetic field with respect to the chain axis for assemblies of particles of different diameters. (b) The evolution of low frequency hysteresis loops for assembly of particles with diameter $D = 30$ nm for different directions of applied magnetic field with respect to the common chain axis: (1) $\theta = 0$, (2) $\theta = 30°$, (3) $\theta = 40°$, (4) $\theta = 60°$.

As Figure 4a shows, the maximum SAR value is achieved when the AC magnetic field is parallel to the chain axis, $\theta = 0$. It is remarkable, however, that in a rather wide range of angles, $\theta \leq 30$–40°, the SAR of the assembly varies slightly. Only for angles $\theta > 50°$ the assembly SAR drops sharply, so that for magnetic field directed perpendicular to the chain axis the assembly SAR is close to zero.

Figure 4b shows the evolution of the low frequency hysteresis loops for an assembly of nanoparticles with a diameter of $D = 30$ nm and ratio $a/D = 1.85$ for various values of the angle θ. In accordance with Figure 4a, the hysteresis loop of the maximum area corresponds to the angle $\theta = 0$, and for angles $\theta \geq 60°$ the hysteresis loop area is tends to zero.

The fact that the angular dependence of assembly SAR turns out to be rather slow in the range of angles $\theta \leq 30$–$40°$ means that the results presented in Figures 1–3 are approximately true for dilute partially oriented assemblies of magnetosome chains. Our calculations also show that a small variation in particle diameters within the same chain, random variations in the distances between the centers of particles in the chain, and small random deviations of the particle positions from the straight line do not significantly affect the low frequency hysteresis loops averaged over a representative assembly of magnetosome chains.

3.2. Interaction of Magnetosome Chains

In this section the results obtained above are generalized to the case of dilute assembly of clusters of magnetosome chains of various densities. Inside the cluster due to the close arrangement of the chains the magnetic dipole interaction of nanoparticles of various chains should be taken into account. Figure 5a shows the results of calculating the low frequency hysteresis loops of an assembly of oriented cylindrical clusters with an overall diameter $D_{cl} = 280$ nm, and a height $L_{cl} = 480$ nm. The axis of the chains is parallel to the axis of the cylinder, but the positions of the chains in the cluster are distributed randomly, as shown schematically in Figure 5b. The AC magnetic field is applied along the chain axis. In the calculations performed it is assumed that the centers of the nanoparticles in the chains are located at an optimal distance, $a = 2.2D$, for nanoparticles with diameter $D = 40$ nm. This allows one to compare the numerical results with the data shown in Figure 1a for a dilute assembly of individual chains with the same nanoparticle diameter. The frequency and amplitude of the AC magnetic field are $f = 300$ kHz and $H_0 = 50$ Oe, respectively.

In the illustrative calculations performed, the number of particles in the chains is fixed at $N_p = 6$, but the number of chains N_{ch} in a cluster of given diameter $D_{cl} = 280$ nm varied from 4 to 10. In this manner we can change the total number of nanoparticles in the cluster, $N_p N_{ch}$, as well as the cluster filling density, $\eta = N_p N_{ch} V / V_{cl}$, where $V = \pi D^3/6$ is the volume of the nanoparticle, $V_{cl} = \pi D_{cl}^2 L_{cl}/4$ being the volume of the cylindrical cluster. SAR calculations of dilute assemblies of clusters with different numbers of chains N_{ch} are averaged over a sufficiently large number of independent realizations of random clusters, $N_{exp} = 40$–60.

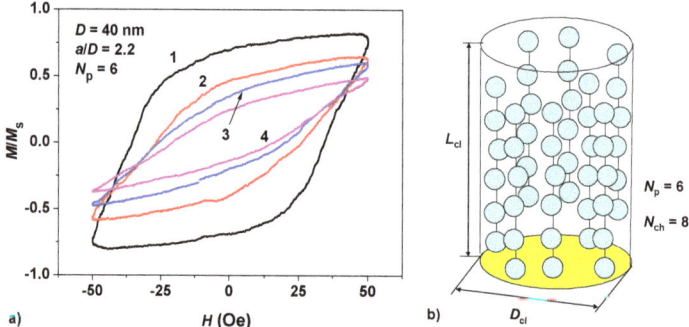

Figure 5. (a) Low-frequency hysteresis loops of dilute assemblies of cylindrical clusters of interacting chains of nanoparticles depending on the number of chains located inside the cluster: (1) $N_{ch} = 4$, (2) $N_{ch} = 6$, (3) $N_{ch} = 8$, (4) $N_{ch} = 10$. (b) The model of random oriented cluster of magnetosome chains used in the calculations.

As Figure 5a shows, due to an increase in the intensity of the magnetic dipole interaction inside the clusters, the area of the assembly hysteresis loop rapidly decreases with an increase in the cluster filling density. The cluster filling density increases as η = 0.027, 0.041, 0.054, and 0.068 for the clusters of the given size with the number of chains within the cluster N_{ch} = 4, 6, 8, and 10, respectively. It is found that with increase in the cluster filling density, the SAR of the assembly of clusters rapidly decreases as follows: SAR = 270.9, 145.5, 98.2, and 62.5 W/g. It should be noted that for a dilute assembly of non-interacting chains (see Figure 1a) with the same chain geometry, frequency and amplitude of the AC magnetic field, the SAR of the assembly at maximum reaches the value 454 W/g for the particles with optimal ratio a/D = 2.2. Thus, a significant drop in the SAR assembly due to the magnetic dipole interaction of individual chains must be taken into account when analyzing experimental data.

4. Discussion

As emphasized above, the properties of magnetosome assemblies are currently being actively studied for application in biomedicine [7,25,37–48]. In particular, SARs of magnetosome assemblies are measured [25,37,41,42,45] in an AC magnetic field under various conditions. In addition, assemblies of magnetosome chains are used in laboratory experiments [7,38,40,42] for the successful treatment of glioblastoma in mice using magnetic hyperthermia.

Low-frequency hysteresis loops of dilute assemblies of isolated magnetosomes with an average diameter D = 45 ± 6 nm were measured [25] at selected frequencies of 75 kHz, 149 kHz, 302 kHz and 532 kHz, in the range of magnetic fields up to 570 Oe. The measurements were carried out for magnetosome assemblies distributed in water, and in agarose gel with an increased viscosity. It was shown that in both cases the SAR of the assemblies almost linearly depend on the AC field frequency. For the assembly of magnetosomes dispersed in water, a very high value of SAR = 880 W/g was obtained at a frequency f = 200 kHz and magnetic field amplitude H_0 = 310 Oe. In agarose gel the SAR of the assembly at the same frequency and magnetic field amplitude turns out to be somewhat lower, SAR = 635 W/g, since in a medium with increased viscosity, the Brownian contribution to SAR is significantly reduced. These data are in agreement with the high SAR values obtained previously for magnetosome assemblies by Hergt et al. [37].

Recently very large SARs were obtained by the same authors [45] for assemblies of "whole" magnetotactic bacteria distributed in water. The importance of the orientation of the chains of magnetosomes located inside the bacteria in the direction of the AC magnetic field is emphasized again. Since the chains of magnetosomes are located inside bacteria with sizes 2–5 × 0.5 μm, they turn out to be efficiently separated by organic material. This reduces the intensity of the magnetic dipole interaction of nanoparticles belonging to different chains. It is shown [45] that the heating efficiency of oriented assembly of magnetotactic bacteria is appreciably higher than the one obtained [25] for the assembly of isolated magnetosomes. Actually, the SAR values of the assembly of magnetotactic bacteria measured by means of AC magnetometry reaches the huge value of 2400 W/g at frequency f = 300 kHz and magnetic field amplitude H_0 = 380 Oe. Nevertheless, it seems hardly possible to uniformly distribute and correctly orient within the tumor large bacteria of 2–5 μm in length. This can probably be done much easier using short chains of magnetosomes having 4–6 nanoparticles in length. For example, the optimal length of a chain consisting of five nanoparticles with diameter D = 20 nm is given by only 140 nm. In addition, when working with chains created from individual magnetosomes, it becomes possible to optimize the geometry of the chains, that is, to ensure the optimal a/D ratio, matching it with the applied value of H_0.

In this regard, a technique for working with magnetosomes developed by Alphandery with co-workers [7,38,40–42] seems to be promising. They removed most of the organic materials, including endotoxins, from magnetosomes extracted from magnetotactic bacteria. The nanoparticles are then stabilized with different bio-degradable and biocompatible coating agents of various thickness, such as poly-L-lysine, oleic acid, citric acid, or carboxy-methyl-dextran [40]. Then, individual magnetosomes coated with these shells are assembled into chains with a small number of particles, N_p = 4–6, and introduced into the tumor for magnetic hyperthermia. A typical concentration of magnetic nanoparticles introduced into a tumor is given by 25 µg in iron of nanoparticles per mm^3 of tumor [7,41,42]. Taking into account the density of magnetosomes around 5 g/cm^3, and assuming nearly uniform distribution of the magnetosome chains within the tumor, one obtains the average density of the assembly $\eta = 0.005$. Such chain assembly can be considered quite diluted, so that the results obtained in Section 3.1 above can be applied for the theoretical estimation of the assembly SAR.

However, Le Fèvre et al. [42] obtained rather small SAR values for the assembly of the magnetosome chains studied, only 40 W/g$_{Fe}$ at frequency $f = 198$ kHz. As a result, in order to obtain the required tumor temperature of 43–46 °C, it was necessary to use rather large amplitudes of the AC magnetic field, $H_0 = 110$–310 Oe. Similarly, only relatively small values of SAR = 89–196 W/g$_{Fe}$ were obtained in [40] at frequency $f = 198$ kHz at sufficiently large magnetic field amplitudes, $H_0 = 340$–470 Oe. This may indicate that the geometric structure of the magnetosome chains, in particular, the a/D ratio, was not optimal in these experiments. Indeed, the individual magnetosomes used to construct the magnetosome chains [42] were covered by rather thin poly-L-lysine shells with a thickness of $t = 4$–17 nm. Therefore, for a significant fraction of the magnetosome chains with an average particle diameter $D = 40.5 \pm 8.5$ nm, the ratio $a/D = 1 + 2t/D \approx 1.2$–1.6 may turn out to be far from the optimal value for the amplitudes of the AC magnetic field used. Similarly, magnetosomes with an average diameter $D = 40$–50 nm were covered with shells of various chemical compositions [41], but of sufficiently small thickness, $t = 2$–6 nm.

Another reason for the small SAR values observed in experiments [41,42] may be associated with the formation of dense clusters of magnetosome chains inside the tumor. As shown in Section 3.2, in a dense cluster of magnetosome chains, a significant decrease in SAR value occurs due to the strong magneto-dipole interaction of nanoparticles of different chains. Finally, as shown in Section 3.1, the unfavorable orientation of the chains with respect to the direction of applied magnetic field can also cause a significant drop in the assembly SAR value. In this regard, it would be interesting to check the possible effect of a strong magnetic field of a Neodymium permanent magnet on the orientation of short magnetosome chains ($N_p = 4$–6) within a tumor. A constant magnetic field can be temporarily applied during the procedure of introducing magnetosome chains into a tumor to provide correct chain orientation.

5. Conclusions

It is now widely accepted [1–3] that, for use in magnetic hyperthermia, it is extremely important to ensure a high quality of magnetic nanoparticles. For dilute assemblies of single-domain magnetic nanoparticles with perfect crystalline structure, sufficiently high saturation magnetization and narrow particle size distribution, sufficiently high SAR can be obtained [25,31–35] at moderate frequencies and amplitudes of an AC magnetic field satisfying the criteria of Brezovich [15] or Hergt et al. [16]. These experimental results are in satisfactory agreement with the earlier theoretical estimates [28,29]. However, in many experimental studies, it was found [26,27,58,59] that when an assembly of magnetic nanoparticles is introduced into a tumor, uncontrolled agglomeration of nanoparticles occurs with the formation of dense clusters of nanoparticles of various geometric structures. As a result, the SAR value of the assembly significantly decreases [26,27,60] due to the influence of a strong magnetic dipole interaction between the nanoparticles of the cluster. Recent theoretical calculations show [54,55] that with an increase in the average density of a nanoparticle assembly in the range $\eta = 0.01$–0.35, the maximum SAR value of the assembly of nanoparticles with different types of magnetic anisotropy decreases by about 5–6 times, compared with the SAR of dilute assembly of the same nanoparticles. Thus, the strong magneto-dipole interaction between the nanoparticles, generally speaking, negatively affects the ability of the assembly to absorb the energy of AC magnetic field [1,26,27,54,55].

An important exception to this rule is a dilute oriented assembly of magnetosome chains in an AC magnetic field applied parallel to the chain axis [45–48]. In this case the intense magnetic dipole interaction between the nanoparticles of the chain plays a positive role. Due to the influence of mutual magneto-dipole interaction within the chain, the magnetic moments of magnetically soft iron oxide nanoparticles are oriented approximately along the chain axis. In addition, the magnetization reversal of the chain in a magnetic field parallel to its axis occurs in a process similar to the giant Barkhausen jump [56]. As a result, the shape of the low-frequency hysteresis loop becomes nearly rectangular, which leads to a significant increase in the SAR value of a dilute assembly of chains in comparison with the corresponding assembly of isolated magnetosomes [45].

In this paper, this effect is studied in detail using numerical modeling by solving the stochastic Landau–Lifshitz equation for assemblies of magnetosome chains of various geometries. In addition to the results of previous studies [46–48], attention is paid to the correct selection of the geometric parameters of the chain, namely, the reduced distance a/D between the centers of the particles of the chain, consistent with the used value of the magnetic field amplitude H_0. It is shown that assemblies of magnetosome chains of different lengths have comparable SAR values, provided that the number of particles in the chain exceeds N_p = 4–5. However, the SAR of an oriented chain assembly significantly decreases for large angles $\theta > 50°$ of the magnetic field direction with respect to chain axis. In addition, the SAR value of the oriented assembly of magnetosome chains also decreases rapidly with an increasing average density of the assembly due to increase in the intensity of a magneto-dipole interaction between the nanoparticles belonging to various chains.

Author Contributions: Conceptualization, N.A.U.; software, N.A.U. and E.M.G.; numerical simulation, E.M.G.; writing-original draft preparation, N.A.U. and E.M.G. All authors have read and agreed to the published version of the manuscript.

Funding: This research was funded by the Ministry of Science and Higher Education of the Russian Federation in the framework of Increase Competitiveness Program of NUST «MISiS», contract № K2-2019-012.

Conflicts of Interest: The authors declare no conflict of interest.

References

1. Périgo, E.A.; Hemery, G.; Sandre, O.; Ortega, D.; Garaio, E.; Plazaola, F.; Teran, F.J. Fundamentals and Advances in Magnetic Hyperthermia. *Appl. Phys. Rev.* **2015**, *2*, 041302. [CrossRef]
2. Blanco-Andujar, C.; Teran, F.J.; Ortega, D. Current Outlook and Perspectives on Nanoparticle-Mediated Magnetic Hyperthermia. In *Iron Oxide Nanoparticles for Biomedical Applications*; Mahmoudi, M., Laurent, S., Eds.; Elsevier: Amsterdam, The Netherlands, 2018; pp. 197–245. [CrossRef]
3. Mahmoudi, K.; Bouras, A.; Bozec, D.; Ivkov, R.; Hadjipanayis, C. Magnetic hyperthermia therapy for the treatment of glioblastoma: A review of the therapy's history, efficacy and application in humans. *Int. J. Hyperth.* **2018**, *34*, 1316–1328. [CrossRef] [PubMed]
4. Ito, A.; Tanaka, K.; Kondo, K.; Shinkai, M.; Honda, H.; Matsumoto, K.; Saida, T.; Kobayashi, T. Tumor regression by combined immunotherapy and hyperthermia using magnetic nanoparticles in an experimental subcutaneous murine melanoma. *Cancer Sci.* **2003**, *94*, 308–313. [CrossRef] [PubMed]
5. Alphandéry, E.; Faure, S.; Seksek, O.; Guyot, F.; Chebbi, I. Chains of magnetosomes extracted from AMB 1 magnetotactic bacteria for application in alternative magnetic field cancer therapy. *ACS Nano* **2011**, *5*, 6279–6296. [CrossRef]
6. Kobayashi, T. Cancer hyperthermia using magnetic nanoparticles. *Biotechnol. J.* **2011**, *6*, 1342–1347. [CrossRef]
7. Alphandéry, E.; Idbaih, A.; Adam, C.; Delattre, J.-Y.; Schmitt, C.; Guyot, F.; Chebbi, I. Development of non-pyrogenic magnetosome minerals coated with poly-l-lysine leading to full disappearance of intracranial U87-Luc glioblastoma in 100% of treated mice using magnetic hyperthermia. *Biomaterials* **2017**, *141*, 210–222. [CrossRef]
8. Johannsen, M.; Gneveckow, U.; Eckelt, L.; Feussner, A.; Waldofner, N.; Scholz, R.; Deger, S.; Wust, P.; Loening, S.A.; Jordan, A. Clinical hyperthermia of prostate cancer using magnetic nanoparticles: Presentation of a new interstitial technique. *Int. J. Hyperth.* **2005**, *21*, 637–647. [CrossRef]
9. van Landeghem, F.K.; Maier-Hauff, K.; Jordan, A.; Hoffmann, K.T.; Gneveckow, U.; Scholz, R.; Thiesen, B.; Brück, W.; von Deimling, A. Post-mortem studies in glioblastoma patients treated with thermotherapy using magnetic nanoparticles. *Biomaterials* **2009**, *30*, 52–57. [CrossRef]
10. Matsumine, A.; Takegami, K.; Asanuma, K.; Matsubara, T.; Nakamura, T.; Uchida, A.; Sudo, A. A novel hyperthermia treatment for bone metastases using magnetic materials. *Int. J. Clin. Oncol.* **2011**, *16*, 101–108. [CrossRef]

11. Pankhurst, Q.A.; Thanh, N.T.K.; Jones, S.K.; Dobson, J. Progress in applications of magnetic nanoparticles in biomedicine. *J. Phys. D Appl. Phys.* **2009**, *42*, 224001. [CrossRef]
12. Dutz, S.; Hergt, R. Magnetic nanoparticle heating and heat transfer on a microscale: Basic principles, realities and physical limitations of hyperthermia for tumour therapy. *Int. J. Hyperth.* **2013**, *29*, 790–800. [CrossRef] [PubMed]
13. El-Sayed, I.H.; Huang, X.; El-Sayed, M.A. Selective laser photo-thermal therapy of epithelial carcinoma using anti-EGFR antibody conjugated gold nanoparticles. *Cancer Lett.* **2006**, *239*, 129–135. [CrossRef] [PubMed]
14. Espinosa, A.; Kolosnjaj-Tabi, J.; Abou-Hassan, A.; Sangnier, A.P.; Curcio, A.; Silva, A.K.A.; Di Corato, R.; Neveu, S.; Pellegrino, T.; Liz-Marzán, L.M.; et al. Magnetic (hyper)thermia or photo-thermia? Progressive comparison of iron oxide and gold nanoparticles heating in water, in cells, and in vivo. *Adv. Funct. Mater.* **2018**, *28*, 1803660. [CrossRef]
15. Brezovich, I.A. Low frequency hyperthermia: Capacitive and ferromagnetic thermoseed methods. *Med. Phys. Monogr.* **1988**, *16*, 82–111.
16. Hergt, R.; Dutz, S.; Müller, R.; Zeisberger, M. Magnetic Particle Hyperthermia: Nanoparticle Magnetism and Materials Development for Cancer Therapy. *J. Phys. Condens. Matter* **2006**, *18*, S2919–S2934. [CrossRef]
17. Reddy, L.H.; Arias, J.L.; Nicolas, J.; Couvreur, P. Magnetic Nanoparticles: Design and Characterization, Toxicity and Biocompatibility, Pharmaceutical and Biomedical Applications. *Chem. Rev.* **2012**, *112*, 5818–5878. [CrossRef]
18. Silva, A.K.A.; Espinosa, A.; Kolosnjaj-Tabi, J.; Wilhelm, C.; Gazeau, F. Medical applications of iron oxide nanoparticles. In *Iron Oxides: From Nature to Applications*; Faivre, D., Ed.; Wiley-VCH: Weinheim, Germany, 2016; pp. 425–471. [CrossRef]
19. Kolosnjaj-Tabi, J.; Lartigue, L.; Javed, Y.; Luciani, N.; Pellegrino, T.; Wilhelm, C.; Alloyeau, D.; Gazeau, F. Biotransformations of magnetic nanoparticles in the body. *Nano Today* **2016**, *11*, 280–284. [CrossRef]
20. Sun, S.; Zeng, H. Size-controlled synthesis of magnetite nanoparticles. *J. Am. Chem. Soc.* **2002**, *124*, 8204–8205. [CrossRef]
21. Daou, T.J.; Pourroy, G.; Bégin-Colin, S.; Grenèche, J.M.; Ulhaq-Bouillet, C.; Legaré, P.; Bernhardt, P.; Leuvrey, C.; Rogez, G. Hydrothermal Synthesis of Monodisperse Magnetite Nanoparticles. *Chem. Mater.* **2006**, *18*, 4399–4404. [CrossRef]
22. Hui, C.; Shen, C.; Yang, T.; Bao, L.; Tian, J.; Ding, H.; Li, C.; Gao, H.-J. Large-Scale Fe_3O_4 Nanoparticles Soluble in Water Synthesized by a Facile Method. *J. Phys. Chem. C* **2008**, *112*, 11336–11339. [CrossRef]
23. Bautin, V.A.; Seferyan, A.G.; Nesmeyanov, M.S.; Usov, N.A. Properties of polycrystalline nanoparticles with uniaxial and cubic types of magnetic anisotropy of individual grains. *J. Magn. Magn. Mater.* **2018**, *460*, 278–284. [CrossRef]
24. Kallumadil, M.; Tada, M.; Nakagawa, T.; Abe, M.; Southern, P.; Pankhurst, Q.A. Suitability of commercial colloids for magnetic hyperthermia. *J. Magn. Magn. Mater.* **2009**, *321*, 1509–1513. [CrossRef]
25. Muela, A.; Munoz, D.; Martin-Rodriguez, R.; Orue, I.; Garaio, E.; Abad Diaz de Cerio, A.; Alonso, J.; Garcia, J.A.; Fdez-Gubieda, M.L. Optimal Parameters for Hyperthermia Treatment Using Biomineralized Magnetite Nanoparticles: A Theoretical and Experimental Approach. *J. Phys. Chem. C* **2016**, *120*, 24437–24448. [CrossRef]
26. Etheridge, M.L.; Hurley, K.R.; Zhang, J.; Jeon, S.; Ring, H.L.; Hogan, C.; Haynes, C.L.; Garwood, M.; Bischof, J.C. Accounting for biological aggregation in heating and imaging of magnetic nanoparticles. *Technology* **2014**, *2*, 214–228. [CrossRef] [PubMed]
27. Sanz, B.; Calatayud, M.P.; Biasi, E.D.; Lima, E., Jr.; Mansilla, M.V.; Zysler, R.D.; Ibarra, M.R.; Goya, G.F. In silico before in vivo: How to predict the heating efficiency of magnetic nanoparticles within the intracellular space. *Sci. Rep.* **2016**, *6*, 38733. [CrossRef] [PubMed]
28. Usov, N.A. Low frequency hysteresis loops of superparamagnetic nanoparticles with uniaxial anisotropy. *J. Appl. Phys.* **2010**, *107*, 123909. [CrossRef]
29. Carrey, J.; Mehdaoui, B.; Respaud, M. Simple models for dynamic hysteresis loop calculations of magnetic single-domain nanoparticles: Application to magnetic hyperthermia optimization. *J. Appl. Phys.* **2011**, *109*, 83921. [CrossRef]
30. Jonasson, C.; Schaller, V.; Zeng, L.; Olsson, E.; Frandsen, C.; Castro, A.; Nilsson, L.; Bogart, L.K.; Southern, P.; Pankhurst, Q.A.; et al. Modelling the effect of different core sizes and magnetic interactions inside magnetic nanoparticles on hyperthermia performance. *J. Magn. Magn. Mater.* **2019**, *477*, 198–202. [CrossRef]

31. Mehdaoui, B.; Meffre, A.; Carrey, J.; Lachaize, S.; Lacroix, L.-M.; Gougeon, M.; Chaudret, B.; Respaud, M. Optimal Size of Nanoparticles for Magnetic Hyperthermia: A Combined Theoretical and Experimental Study. *Adv. Funct. Mater.* **2011**, *21*, 4573–4581. [CrossRef]
32. Guardia, P.; Di Corato, R.; Lartigue, L.; Wilhelm, C.; Espinosa, A.; Garcia-Hernandez, M.; Gazeau, F.; Manna, L.; Pellegrino, T. Water-soluble iron oxide nanocubes with high values of specific absorption rate for cancer cell hyperthermia treatment. *ACS Nano* **2012**, *6*, 3080–3091. [CrossRef]
33. Di Corato, R.; Espinosa, A.; Lartigue, L.; Tharaud, M.; Chat, S.; Pellegrino, T.; Ménager, C.; Gazeau, F.; Wilhelm, C. Magnetic hyperthermia efficiency in the cellular environment for different nanoparticle designs. *Biomaterials* **2014**, *35*, 6400–6411. [CrossRef]
34. Unni, M.; Uhl, A.M.; Savliwala, S.; Savitzky, B.H.; Dhavalikar, R.; Garraud, N.; Arnold, D.P.; Kourkoutis, L.F.; Andrew, J.S.; Rinaldi, C. Thermal decomposition synthesis of iron oxide nanoparticles with diminished magnetic dead layer by controlled addition of oxygen. *ACS Nano* **2017**, *11*, 2284–2303. [CrossRef] [PubMed]
35. Nemati, Z.; Alonso, J.; Rodrigo, I.; Das, R.; Garaio, E.; Garcia, J.A.; Orue, I.; Phan, M.-H.; Srikanth, H. Improving the heating efficiency of iron oxide nanoparticles by tuning their shape and size. *J. Phys. Chem. C* **2018**, *122*, 2367–2381. [CrossRef]
36. Navarro, E.; Luengo, Y.; Veintemillas, S.; Morales, M.P.; Palomares, F.J.; Urdiroz, U.; Cebollada, F.; González, J.M. Slow magnetic relaxation in well crystallized, monodispersed, octahedral and spherical magnetite nanoparticles. *AIP Adv.* **2019**, *9*, 125143. [CrossRef]
37. Hergt, R.; Hiergeist, R.; Zeisberger, M.; Schüler, D.; Heyen, U.; Hilger, I.; Kaiser, W.A. Magnetic properties of bacterial magnetosomes as potential diagnostic and therapeutic tools. *J. Magn. Magn. Mater.* **2005**, *293*, 80–86. [CrossRef]
38. Alphandéry, E.; Chebbi, I.; Guyot, F.; Durand-Dubief, M. Use of bacterial magnetosomes in the magnetic hyperthermia treatment of tumours: A review. *Int. J. Hyperth.* **2013**, *29*, 801–809. [CrossRef] [PubMed]
39. Timko, M.; Molcan, M.; Hashim, A.; Skumiel, A.; Müller, M.; Gojzewski, H.; Jozefczak, A.; Kovac, J.; Rajnak, M.; Makowski, M.; et al. Hyperthermic effect in suspension of magnetosomes prepared by various methods. *IEEE Trans. Magn.* **2013**, *49*, 250–254. [CrossRef]
40. Alphandéry, E. Application of magnetosomes synthesized by magnetotactic bacteria in medicine. *Front. Bioeng. Biotechnol.* **2014**, *2*, 5. [CrossRef]
41. Mandawala, C.; Chebbi, I.; Durand-Dubief, M.; Le Fèvre, R.; Hamdous, Y.; Guyot, F.; Alphandery, E. Biocompatible and stable magnetosome minerals coated with poly-l-lysine, citric acid, and carboxy-methyl-dextran for application in the magnetic hyperthermia treatment of tumors. *J. Mater. Chem. B* **2017**, *5*, 7644–7660. [CrossRef]
42. Le Fèvre, R.; Durand-Dubief, M.; Chebbi, I.; Mandawala, C.; Lagroix, F.; Valet, J.-P.; Idbaih, A.; Adam, C.; Delattre, J.-Y.; Schmitt, C.; et al. Enhanced antitumor efficacy of biocompatible magnetosomes for the magnetic hyperthermia treatment of glioblastoma. *Theranostics* **2017**, *7*, 4618–4631. [CrossRef]
43. Marcano, L.; Muñoz, D.; Martin-Rodriguez, R.; Orue, I.; Alonso, J.; Garcia-Prieto, A.; Serrano, A.; Valencia, S.; Abrudan, R.; FernándezBarquin, L.; et al. Magnetic Study of Co-Doped Magnetosome Chains. *J. Phys. Chem. C* **2018**, *122*, 7541–7550. [CrossRef]
44. Orue, I.; Marcano, L.; Bender, P.; García-Prieto, A.; Valencia, S.; Mawass, M.A.; Gil-Cartón, D.; Alba Venero, D.; Honecker, D.; Garcıa-Arribas, A.; et al. Configuration of the magnetosome chain: A natural magnetic nanoarchitecture. *Nanoscale* **2018**, *10*, 7407–7419. [CrossRef] [PubMed]
45. Gandia, D.; Gandarias, L.; Rodrigo, I.; Robles-Garcia, J.; Das, R.; Garaio, E.; Garcia, J.A.; Phan, M.-H.; Srikanth, H.; Orue, I.; et al. Unlocking the Potential of Magnetotactic Bacteria as Magnetic Hyperthermia Agents. *Small* **2019**, *15*, 1902626. [CrossRef] [PubMed]
46. Martinez-Boubeta, C.; Simeonidis, K.; Makridis, A.; Angelakeris, M.; Iglesias, O.; Guardia, P.; Cabot, A.; Yedra, L.; Estrade, S.; Peiro, F.; et al. Learning from nature to improve the heat generation of iron–oxide nanoparticles for magnetic hyperthermia applications. *Sci. Rep.* **2013**, *3*, 1652. [CrossRef] [PubMed]
47. Serantes, D.; Simeonidis, K.; Angelakeris, M.; Chubykalo-Fesenko, O.; Marciello, M.; Morales, M.P.; Baldomir, D.; Martinez-Boubeta, C. Multiplying Magnetic Hyperthermia Response by Nanoparticle Assembling. *J. Phys. Chem. C* **2014**, *118*, 5927–5934. [CrossRef]
48. Simeonidis, K.; Morales, M.P.; Marciello, M.; Angelakeris, M.; de la Presa, P.; Lazaro-Carrillo, A.; Tabero, A.; Villanueva, A.; Chubykalo-Fesenko, O.; Serantes, D. In-situ particles reorientation during magnetic hyperthermia application: Shape matters twice. *Sci. Rep.* **2016**, *6*, 38382. [CrossRef]

49. Brown, W.F., Jr. Thermal fluctuations of a single-domain particle. *Phys. Rev.* **1963**, *130*, 1677–1686. [CrossRef]
50. Garcia-Palacios, J.L.; Lazaro, F.J. Langevin-dynamics study of the dynamical properties of small magnetic particles. *Phys. Rev. B* **1998**, *58*, 14937–14958. [CrossRef]
51. Scholz, W.; Schrefl, T.; Fidler, J. Micromagnetic simulation of thermally activated switching in fine particles. *J. Magn. Magn. Mater.* **2001**, *233*, 296–304. [CrossRef]
52. Coffey, W.T.; Kalmykov, Y.P.; Waldron, J.T. *The Langevin Equation: With Applications to Stochastic Problems in Physics, Chemistry and Electrical Engineering*, 2nd ed.; World Scientific: Singapore, 2004; p. 704, ISBN 978-981-238-462-1.
53. Usov, N.A.; Peschany, S.E. Theoretical hysteresis loops of single-domain particles with cubic anisotropy. *J. Magn. Magn. Mater.* **1997**, *174*, 247–260. [CrossRef]
54. Usov, N.A.; Serebryakova, O.N.; Tarasov, V.P. Interaction effects in assembly of magnetic nanoparticles. *Nanoscale Res. Lett.* **2017**, *12*, 489. [CrossRef] [PubMed]
55. Usov, N.A.; Nesmeyanov, M.S.; Gubanova, E.M.; Epshtein, N.B. Heating ability of magnetic nanoparticles with cubic and combined anisotropy. *Beilstein J. Nanotechnol.* **2019**, *10*, 305–314. [CrossRef] [PubMed]
56. Gudoshnikov, S.A.; Grebenshchikov, Y.B.; Ljubimov, B.Y.; Palvanov, P.S.; Usov, N.A.; Ipatov, M.; Zhukov, A.; Gonzalez, J. Ground state magnetization distribution and characteristic width of head to head domain wall in Fe-rich amorphous microwire. *Phys. Stat. Sol.* **2009**, *206*, 613–617. [CrossRef]
57. Aharoni, A. Nucleation of magnetization reversal in ESD magnets. *IEEE Trans. Magn.* **1969**, *5*, 207–210. [CrossRef]
58. Branquinho, L.C.; Carriao, M.S.; Costa, A.S.; Zufelato, N.; Sousa, M.H.; Miotto, R.; Ivkov, R.; Bakuzis, A.F. Effect of magnetic dipolar interactions on nanoparticle heating efficiency: Implications for cancer hyperthermia. *Sci. Rep.* **2013**, *3*, 2887. [CrossRef]
59. Conde-Leboran, I.; Baldomir, D.; Martinez-Boubeta, C.; Chubykalo-Fesenko, O.; Morales, M.P.; Salas, G.; Cabrera, D.; Camarero, J.; Teran, F.J.; Serantes, D. A single picture explains diversity of hyperthermia response of magnetic nanoparticles. *J. Phys. Chem. C* **2015**, *119*, 15698–15706. [CrossRef]
60. Gudoshnikov, S.A.; Liubimov, B.Y.; Popova, A.V.; Usov, N.A. The influence of a demagnetizing field on hysteresis losses in a dense assembly of superparamagnetic nanoparticles. *J. Magn. Magn. Mater.* **2012**, *324*, 3690–3695. [CrossRef]

© 2020 by the authors. Licensee MDPI, Basel, Switzerland. This article is an open access article distributed under the terms and conditions of the Creative Commons Attribution (CC BY) license (http://creativecommons.org/licenses/by/4.0/).

Article

Room Temperature Magnetic Memory Effect in Cluster-Glassy Fe-Doped NiO Nanoparticles

Ashish Chhaganlal Gandhi [1], Tai-Yue Li [1], B. Vijaya Kumar [2], P. Muralidhar Reddy [3], Jen-Chih Peng [4], Chun-Ming Wu [4] and Sheng Yun Wu [1],*

[1] Department of Physics, National Dong Hwa University, Hualien 97401, Taiwan; acg.gandhi@gmail.com (A.C.G.); tim312508@gmail.com (T.-Y.L.)
[2] Department of Chemistry, Nizam College, Osmania University, Hyderabad 500001, India; vijaychemou@gmail.com
[3] Department of Chemistry, University College of Science, Osmania University, Hyderabad 500007, Telangana, India; pmdreddy@gmail.com
[4] SIKA, National Synchrotron Radiation Research Center, Hsinchu 30076, Taiwan; peng.hanz@nsrrc.org.tw (J.-C.P.); cmw@ansto.gov.au (C.-M.W.)
* Correspondence: sywu@mail.ndhu.edu.tw; Tel.: +886-3-890-3717

Received: 11 June 2020; Accepted: 2 July 2020; Published: 4 July 2020

Abstract: The Fe-doped NiO nanoparticles that were synthesized using a co-precipitation method are characterized by enhanced room-temperature ferromagnetic property evident from magnetic measurements. Neutron powder diffraction experiments suggested an increment of the magnetic moment of $3d$ ions in the nanoparticles as a function of Fe-concentration. The temperature, time, and field-dependent magnetization measurements show that the effect of Fe-doping in NiO has enhanced the intraparticle interactions due to formed defect clusters. The intraparticle interactions are proposed to bring additional magnetic anisotropy energy barriers that affect the overall magnetic moment relaxation process and emerging as room temperature magnetic memory. The outcome of this study is attractive for the future development of the room temperature ferromagnetic oxide system to facilitate the integration of spintronic devices and understanding of their fundamental physics.

Keywords: room temperature; magnetic memory effect; intraparticle interactions; 4:1 defect cluster; Fe-doped NiO

1. Introduction

The antiferromagnetic (AF) metal oxide nanoparticles (NPs) have attracted enormous attention because of their promising technological applications and fundamental physics. The effect of finite size leads to the accumulation of frustrating surface spins and various point defects at the surface of the AF NPs, which results in an interesting magnetic and optical properties that differ significantly from their bulk counterparts [1]. Among the various AF materials, nickel oxide (NiO) is one of the few p-type semiconductors (acceptor state induced by the nickel vacancy (V_{Ni}) with a wide-bandgap E_g = 4 eV) having face-centered-cubic (fcc) crystal symmetry. In the bulk form, NiO possesses AF ordering with the Neel transition temperature T_N of 523 K [2]. However, NiO nanostructure exhibits anomalous magnetic properties that are very sensitive to size, Ni vacancy defects, morphology, and crystal structure, thus showing a wide variety of intriguing phenomena. It has been reported that below a particle size of d = 30 nm, the long-range ordered Ni^{2+}–O^{2-}–Ni^{2+} superexchange interaction breakdown, due to an enhanced V_{Ni} defect resulting in a weak ferromagnetic (FM) like properties [3,4]. Furthermore, the effect of frustrating surface spins become more dominant below d = 10 nm [5–7]. The NiO nanostructures have been used in rechargeable batteries [8], magnetic recording media [9], the next-generation resistive switching memory devices [10], and so on. Recently, functionalized NiO nanostructures have attracted great research interest from both fundamental and application point of view [11,12]. The further development in functionalized NiO nanostructure is focused on obtaining a room temperature (RT) ferromagnetism without comprising the structure in order to facilitate the possible integration of spintronic devices.

According to recent findings, RT ferromagnetism in NiO NPs can be achieved through transition metal (TM)-doping, for example, Fe (either due to substitution or the formed defect clusters), which opens up their potential applications in the future advanced spintronic devices [13–15]. The properties of such a system can be tailored by controlling both the particle size and Fe-dopant concentration, which results in a complex magnetic property [16–22]. For instance, the doping of Fe^{3+} ions in NiO either could replace the Ni ions or occupy an interstitial site. The substituted Fe^{3+} ions can alter the Ni^{2+}–O^{2-}–Fe^{3+} superexchange interaction and so the AF properties. Whereas, Fe ions at the interstitial site could form 4:1 defect cluster consisting of tetravalent interstitial Fe_i^{4+} and four V_{Ni}, in total being four times negatively charged [16,23–27]. Such a complex structure could result in interesting magnetic properties. According to previous reports, bare and the Fe-doped NiO NPs both exhibit low-temperature magnetic memory effect [16,28,29]. Such a type of nanoscale system can be used as a "thermal assistant memory cell" in digital information storage [30–32].

However, the magnetic memory effect is mostly observed in the low-temperature region far below the RT, and this is the major obstruct precluding its application in nanotechnology. In the past, the above obstacle has been foiled through introducing exchange-coupling [30,33] and particle size distribution [34]. In this study, we have reported RT magnetic memory effect from Fe-doped NiO NPs synthesized using a co-precipitation method followed by thermal treatment in the air. A thorough investigation of structural and magnetic properties was carried out using synchrotron radiation powder X-ray diffraction (PXRD), neutron powder diffraction (NPD), and dc magnetometer. Our findings suggest that the RT magnetic memory effect in Fe-doped NiO NPs is mediated through intrinsic intraparticle interactions.

2. Materials and Methods

All of the analytical grade Nickel (II) nitrate hexahydrate ($Ni(NO_3)_2 \cdot 6H_2O$), iron (III) nitrate nonahydrate ($Fe(NO_3)_3 \cdot 9H_2O$), and ammonium bicarbonate (NH_4HCO_3) were procured from S.D. Fine-Chemicals Ltd., India, and used as received without further purifications. NiO and Fe-doped $Ni_{1-x}Fe_xO$ with x(%) varying from 0 to 10% compositions were prepared. A schematic Scheme 1 shows the process of $Ni_{1-x}Fe_xO$ NPs preparation while using the co-precipitation method. The stoichiometric amounts of $Ni(NO_3)_2$ and $Fe(NO_3)_3$ (molar ratio of Ni to Fe is 0, 0.5, 1, 5, and 10%) were dissolved in double-distilled water (DDW) separately (Scheme 1a). Subsequently, ferric nitrate solution was added dropwise to nickel nitrate solution under continuous stirring for 1 h. The pH of the solution was maintained at 8 by adding the NH_4HCO_3 solution (Scheme 1b). The light green precipitate was formed by adding NH_4HCO_3. The resultant precipitate was washed several times with DDW (Scheme 1c) and then dried at 100 °C for 12 h in a hot air oven (Scheme 1d). The obtained powder was grounded in an agate mortar and calcined at 600 °C for 4 h in a muffle furnace (Scheme 1e). The obtained pure NiO powder was gray-green. The effect of Fe doping up to 1% leads to a slight change in color to light brownish gray-green. However, a drastic change in color to dark brown was noted on 5%, and on 10% powder becomes red-brown (Scheme 1d).

A morphological analysis of powdered samples was carried out using the field-emission scanning electron microscopy (FE-SEM, JEOL JSM-6500F microscope, Tokyo, Japan). Synchrotron radiation PXRD measurements were carried out at the National Synchrotron Radiation Research Center in Hsinchu, Taiwan (beamline BL01C2, λ = 0.7749 Å). For the investigation of magnetic structural properties, the unpolarised NPD spectra were collected on 2 g sample at 10 K and 300 K. The NPD experiments were carried out on SIKA (the high flux cold neutron triple-axis spectrometer) at OPAL reactor, ANSTO (λ = 2.35 Å). With a wide range of incident energy from 2.5 to 30 meV and three detector options provide great flexibility on neutron study. SIKA was configured as diffraction mode, which used diffraction detector to gain a strong diffraction signal. Incident energy was configured at 14.87 meV with open-open-60 collimation. Sample pre-slit and post-slit were adjusted to the sample size in order to improve the signal quality. Sample environment CF11 was used to control the sample temperature. The measurements were collected based on the counts of the beam monitor to ensure the precise neutron flux on the sample. The measurements of magnetic properties were carried out while using a superconducting quantum interference device (SQUID) magnetometer (Quantum Design, SQUID-VSM Ever Cool, San Diego, CA, USA).

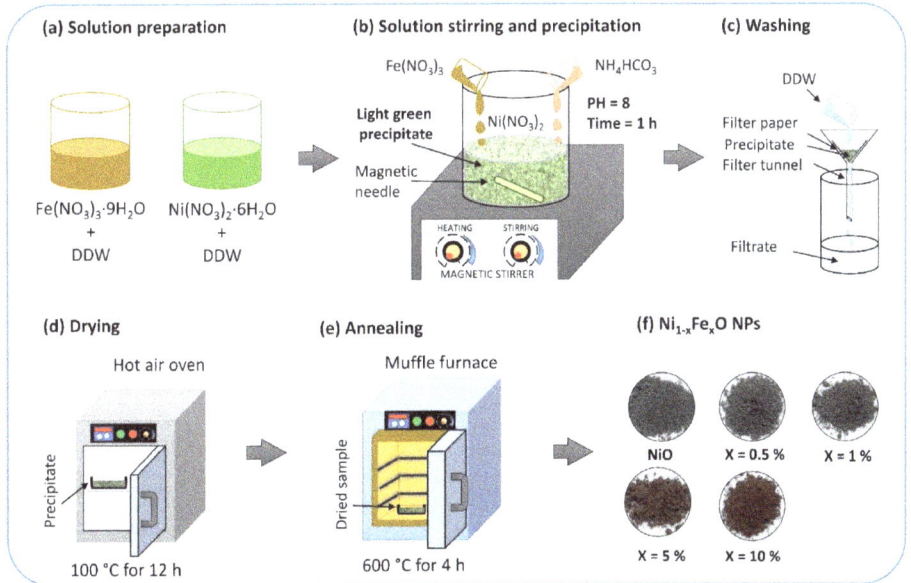

Scheme 1. (a–f) Schematic representation of $Ni_{1-x}Fe_xO$ NPs preparation process using the co-precipitation method.

3. Results

3.1. Morphological and Elemental Analysis

The facetted NPs with a broad shape distribution were observed from the SEM images of NiO, 0.5 to 10% samples that are shown in Figure 1a–e, respectively. From SEM images, an increase of aggregation of particles with the increase of doping concentration can be seen. The mean diameter $<d>$ of NPs is estimated by fitting a log-normal distribution function: $f(d) = \frac{1}{\sqrt{2\pi}d\sigma} \exp\left[-\frac{(\ln d - \ln\langle d \rangle)}{2\sigma^2}\right]$ to the histogram that was obtained from SEM images of NiO (Figure 1f) and 0.5 to 10% sample (Figure 1g, top to bottom), where the value of σ represents a standard deviation of the fitted function. The effect of Fe-doping from 0 to 10% in NiO leads to a decrease of particle size (i.e., mean diameter $<d>$) from 63(1) nm to 44(2) nm. Furthermore, along with the big size particles, small size NPs with a diameter below <5 nm were also visible in SEM images of 5 and 10% samples (Figure 1d,e). The small size particles could be related to the Fe_3O_4 impurity phase (supported by an observed drastic color change and further confirmed from the synchrotron radiation PXRD experiment Figure S1 in the supporting information and the magnetic measurements) [32]. Fe_3O_4 is a ferrimagnetic material with a high Curie temperature (T_C = 850 K). The presence of such a strong magnetic impurity phase overshadows the intrinsic magnetic properties of Fe-doped NiO materials. The focus of the present study is to study the RT magnetic and memory effect from Fe-doped NiO NPs without compromising the structural properties. Hence, we will not discuss the 5% and 10% samples having the Fe_3O_4

impurity phase. The crystalline size that was obtained from the most intense (200) PXRD peak of 0 to 1% samples varies between 68 nm to 64 nm (Table S1). The decrease of particle size with the increase of Fe-concentration and appearance of small size NPs above 1% Fe-concentration are consistent with the previous findings [24].

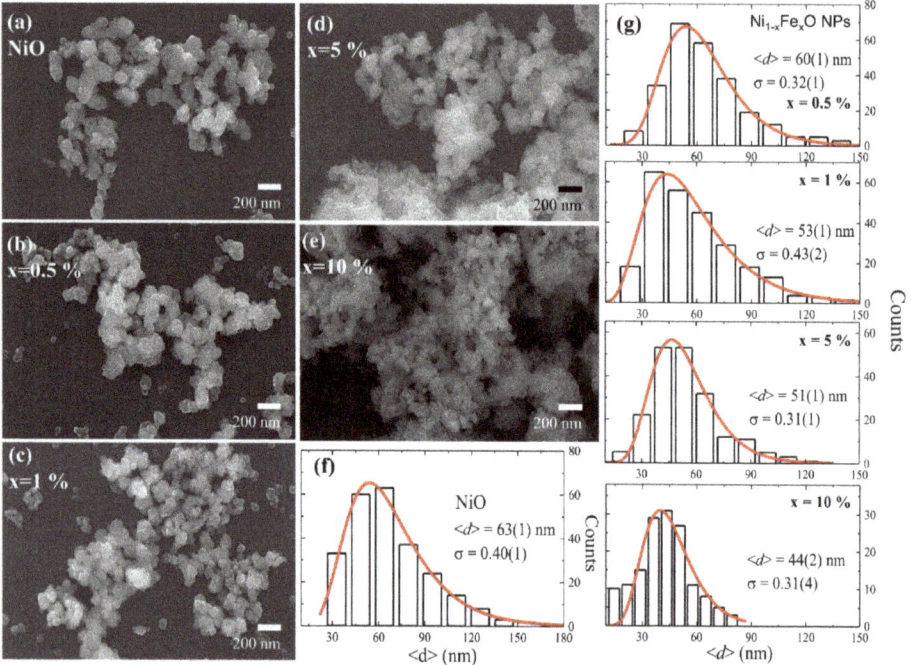

Figure 1. (a–e) SEM images of NiO, 0.5, 1, 5 and 10% samples, respectively. Histogram obtained from (f) NiO, (g) 0.5, 1, 5, and 10% samples (top to bottom). The solid red lines represent a fit using the Log-normal distribution function.

3.2. Structural Properties

In the paramagnetic phase (T > T_N = 523 K) NiO possesses cubic $Fm\bar{3}m$ symmetry. Below T_N, the structure of NiO undergoes a weak cubic-to-rhombohedral distortion (space group $R\bar{3}m$) due to the magnetostriction effect. The rhombohedral distortion can simply be noticed from the splitting of (220) reflection in diffraction peak profile, and the effect increases with the decrease of temperature [35,36]. The NPD spectra of particles that were recorded at 300 K contain a peak originating from both the nuclear and magnetic origin (Figure 2). A slight enhancement in the intensity of the magnetic peak is noted on lowering the sample temperature to 10 K. However, we did not observe any splitting of (220) reflection in pure and Fe-doped NiO NPs. Therefore, initially, we have analyzed the diffraction in the cubic $Fm\bar{3}m$ symmetry with a propagation vector k = $\left(\frac{1}{2}, \frac{1}{2}, \frac{1}{2}\right)$ and a lattice parameter of 4.1783(2) Å. The Rietveld refined NPD spectra of NiO, 0.5%, and 1% samples taken at 300 K and 10 K are shown in Figure 2a,b (bottom to top), and Table S2 summarizes the corresponding fitting parameters. The refinement of the NPD spectra of NiO NPs corroborates the conventional two magnetic sublattices of NiO with antiparallel orientation within (111) plane (inset of Figure 2a, in the bottom panel). The large AF exchange interaction between the next-nearest-neighbor is at the origin of the Neel temperature of NiO [37]. At 300 K, the obtained value of the Ni ordered magnetic moment from NiO NPs is 1.232(18) μ_B and at 10 K, it enhances to 1.258(18) μ_B [37,38]. Rietveld refinement of NPD spectra was further carried out using rhombohedral $R\bar{3}m$ symmetry to examine whether pure and Fe-doped

NiO NPs undergo cubic-to-rhombohedral distortion, and the corresponding fitting parameters are summarized in Table S3. Interestingly, an improved magnetic moment of 1.636(30) μ_B is obtained from NiO NPs at 300 K and 10 K, it enhances to 1.831(23) μ_B, which is 8.45% lower than the spin-only value of 2.0 μ_B. The obtained low value could be related to the presence of magnetically disordered shell at the surface of the NPs [39,40]. Note that, here, the overall goal of the NPD experiment is to understand the effect of Fe-doping on the magnetic moment of $3d$ ions in the AF NiO NPs. Hence, even though refinement using rhombohedral symmetry return results near to spin-only value than that of cubic symmetry, it will not affect the magnetic memory results as long as both symmetries yield a similar increase in magnetic moment trend with Fe-doping concentration. Furthermore, the observed difference between the results of the refinement is worth discussion; however, it is outside of the scope of current work and, therefore, can be discussed further in the near future. The subsequent analysis of NPD spectra from 0.5 and 1% samples using both cubic and rhombohedral symmetry yields a similar magnetic structure to that of undoped NiO NPs, but with an enhanced magnetic moment (see Tables S2 and S3). Therefore, no obvious change in the chemical structure was observed up to 1% Fe-doping, although the effect has led to an increment of the magnetic moment in the NPs.

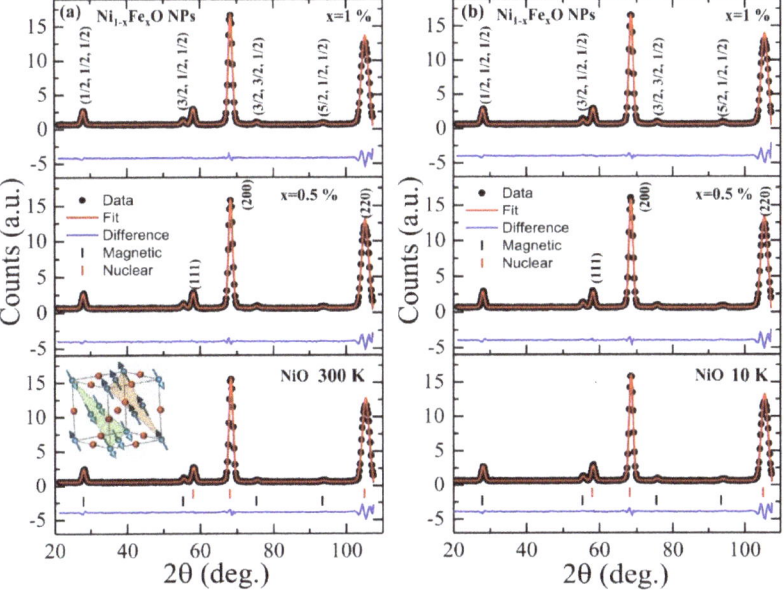

Figure 2. The Rietveld refined (red line) neutron powder diffraction (NPD) spectra (dots) from NiO, 0.5, and 1% samples (bottom to top) taken at (**a**) 300 K and (**b**) 10 K are calculated with $Fm\bar{3}m$ symmetry. The blue line represents the difference between experimental and fitted spectra. The vertical black and short red lines represent the magnetic and nuclear brags reflection positions. Inset in the bottom panel of (**a**): a unit cell of NiO with two magnetic sublattices.

3.3. Exchange Bias

ZFC and FC magnetic hysteresis M(H_a) loop measurements were both carried out at 300 K to study the effect of Fe-doping on the magnetic properties of NiO NPs. Initially, during ZFC measurement, the SQUID magnet was reset to remove any stray field at 400 K, followed by cooling the sample to the desired temperature. During FC measurement, the sample was cooled down from 400 K in an external magnetic field of 10 kOe to the desired temperature. Figure 3a shows the magnified ZFC M(H_a) loops near zero-field, where inset gives full M(H_a) loops measured over ±30 kOe from Ni$_{1-x}$Fe$_x$O NPs. The observed non-zero coercivity and the linear increasing behavior of the magnetization in the high-field region correspond to FM and AF two-component behavior, respectively. The value of coercivity (H_C) and the magnetization increase with the increase of Fe-concentration from 0 to 1%. On the other hand, a saturation-like magnetization behavior with an enhanced magnetization is obtained from 5 and 10% samples attributed to the Fe_3O_4 impurity phase (Figure S2) [22]. The net enhancement in the magnetization because of Fe-doping ($M_{Fe} = M - \chi_{NiO}H$) can be quantified by subtracting the contribution from NiO. Figure 3b depicts M_{Fe} vs. H_a curves with a tendency towards saturation from 0.5 and 1% samples giving a maximum value of $M_{Fe} = 0.11$ emu/g and 0.24 emu/g at $H_a = 30$ kOe, respectively.

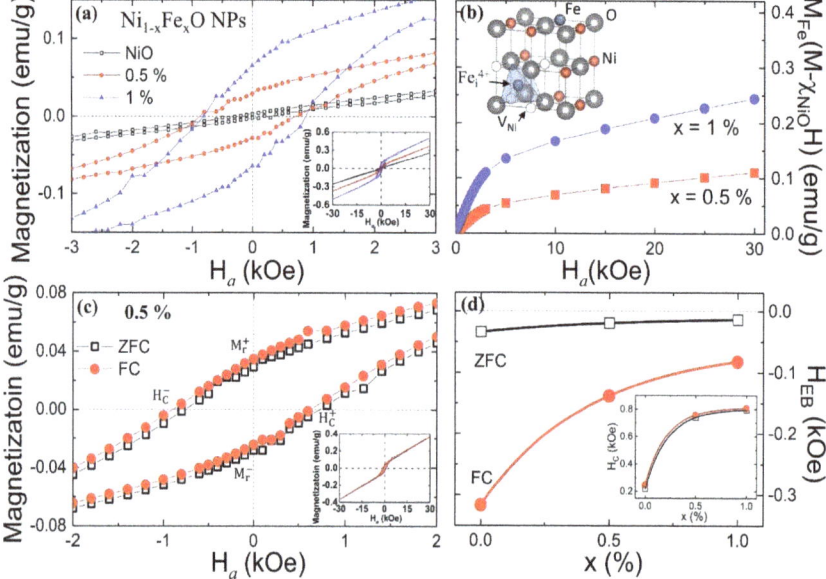

Figure 3. (a) Magnified ZFC-M(H_a) loops near zero-field from $Ni_{1-x}Fe_xO$ NPs at 300 K. (b) Field dependent magnetization M_{Fe}(M-χ_{NiO}H) from 0.5% and 1% samples. (c) Magnified ZFC-FC M(H_a) loop measured at 300 K from 0.5% sample. (d) Composition x dependency of exchange-bias field H_{EB}. Inset: (a) full ZFC-M(H_a), (b) 4:1 defect cluster consisting of tetravalent interstitial Fe_i^{4+}, four nickel vacancies V_{Ni}, substituted Fe at Ni site (blue sphere), (c) full ZFC-FC M(H_a) loops over ±30 kOe field, and (d) composition x-dependency of coercivity H_C.

The ZFC M(H_a) loops also revealed first-field-induced magnetic anisotropy, which can be quantified as a spontaneous exchange bias (EB) field $H_{EB} = \left([H_C^+] - [H_C^-]\right)/2$, where H_C^+ and H_C^- corresponds to coercivity in the first- and second-curve of the M(H_a) loop at which the magnetization is zero [3,41]. Figure 3c compares the ZFC and FC M(H_a) loops near zero-field, where the inset gives full M(H_a) loops measured over ±30 kOe from 0.5% sample. The effect of the cooling-field leads to further enhancement in the magnetic anisotropy. Figure 3d gives a plot of H_{EB} obtained from both ZFC and FC M(H_a) loops vs. Fe-concentration x, where the inset of the figure depicts the coercivity $H_C = \left(H_C^+ - H_C^-\right)/2$ vs. x. Almost similar values of H_C were obtained from both ZFC and FC M(H_a) loops, whereas an enhanced value of the EB field from FC M(H_a) loops. A sudden enhancement in the values of coercivity H_C can be seen from a 0.5% Fe-doped NiO sample, reaching a maximum value of 809 Oe at 1% (inset of Figure 3d). A maximum value of $H_{EB} = -316$ Oe is obtained after FC M(H_a) loop from NiO NPs, and its value decreases with the increase of x. Note that the obtained values from M(H_a) loops are not intrinsic since saturated hysteresis is not achievable even with a maximum field of 50 kOe. Moreover, the obtained conventional and spontaneous EB field from NiO NPs is consistent with previous findings and it can be attributed to the formed uncompensated core and disordered shell-type structure [3,4]. The observed reduction in the EB field from Fe-doped NiO NPs and its further reduction with the increase of Fe-concentration could be understood by assuming the presence of a 4:1 defect cluster in the core of the NPs (inset of Figure 3b) [16]. The 4:1 defect cluster consists of tetravalent interstitial iron Fe_i^{4+} and four V_{Ni}, in total, being four times negatively charged. The formation of such defect clusters could result in the enhancement of V_{Ni} in the core of NPs and, consequently, suppression in AF anisotropy of host NiO [27]. Therefore, for a very weak AF anisotropy, one could only see an enhancement in the H_C without any EB field [42].

3.4. Temperature Dependence of Magnetization

The temperature-dependent magnetization measurements were carried out using ZFC and FC protocols. During ZFC measurement, the applied magnetic field was set to zero while using oscillator mode, and then the SQUID magnet was reset to remove any stray field at 400 K. Figure 4a,b show the $M(H)/H_a$ vs. T plots for NiO, 0.5%, and 1% samples at external fields H_a of 500 Oe and 5 kOe, respectively. An increase in the magnetization with Fe-concentration can be seen consistent with the NPD and $M(H_a)$. At 500 Oe, the ZFC-FC curves of all samples remain bifurcated, even up to 400 K (defined as irreversible temperature T_{irr} where $(M_{FC} - M_{ZFC}) = 0$), suggesting blocking temperature T_B (ZFC maximum) lying above the measured temperature range. With the increase of the external magnetic field, a relatively broadened ZFC curve appears from Fe-doped NiO NPs as compared to pure NiO, and its maximum shifts towards lower temperatures. The obtained T_B at 5 kOe field from NiO, 0.5%, and 1% sample is around 350, 334, and 316 K, respectively. Whereas, the T_{irr}, which can be considered as the onset temperature of the freezing process for NiO, 0.5%, and 1% sample, lies above 400 K. Assuming the assemblies of non-interacting NPs, the relaxation of magnetization with a uniaxial magnetic anisotropy can be described by the Néel–Arrhenius law: $K(x) = 25k_B T_B(x)/V(x)$, where $K(x)$ is the magnetocrystalline anisotropy of NPs, k_B is Boltzmann constant, and $V(x)$ is the median of the particle volume distribution. The calculated value of K from NiO, 0.5% and 1% samples at 5 kOe is 0.9232×10^4 erg/cm^3, 1.020×10^4 erg/cm^3, and 1.400×10^4 erg/cm^3, respectively. The obtained value of K from NiO NPs is much smaller than that of value from bulk $K_{AF}(0) = 4.96 \times 10^6$ erg/cm^3 [43]. The increase in the value of K with the Fe-concentration could be a consequence of the decrease of particle size.

Irrespective of the external magnetic field, a sudden increase in the magnetization can be seen in the low-temperature region from ZFC-FC curves of NiO NPs, which could be related to the collective freezing of disordered surface spins. On the other hand, the response of ZFC-FC curves from 0.5% and 1% samples varies with an external magnetic field. At 500 Oe, a plateau is observed below 100 K from the FC curve of 0.5%, and 1% samples. Such FC curve shape in the low-temperature region is usually observed for super-spin-glasses (SSG) and super-ferromagnetic (SFM) materials, suggesting collective spin behavior [44]. At a finite interparticle distance in the magnetic system, more complex systems can be encountered due to magnetic interactions, especially of dipolar origin. Such a collective spin behavior in the magnetic system evolves with the increase of the interaction strength, modified superparamagnetic (SPM) behavior, SSG, and SFM states [45]. Therefore, in the present system, due to the presence of a 4:1 defect cluster in the core of the NPs and the enhanced magnetization, the interparticle and intraparticle interactions at a finite distance could have enhanced further with the increase of Fe-doping concentration. At 5 kOe, a saturation like behavior can be seen from the FC curve, whereas a huge broadening in the ZFC curve associated with the large distribution of energy barriers. The above findings suggest the existence of two almost independent contributions to the measured $M(H_a)$ and $M(T)$ curves: one reflecting the antiferromagnetism of NiO (linear in H and weakly temperature-dependent). Another reflecting the NP size distribution and weak ferromagnetic (excess) magnetic moment with particle anisotropies yielding relaxation time according to Arrhenius dynamics and M vs. H_a curves governed by Langevin functions.

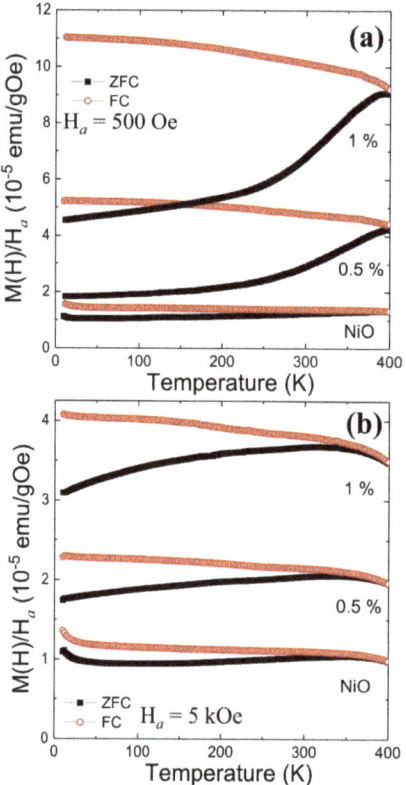

Figure 4. (**a**,**b**) M(H)/H_a vs. T plots for NiO, 0.5, and 1% samples measured using ZFC and FC protocols at external fields H_a of 500 Oe and 5 kOe, respectively.

3.5. Time Dependence of Magnetization

Time dependency of the magnetization relaxation M(t) measurement was carried out both with and without the external magnetic field in order to investigate the effect of Fe-doping concentration on the magnetic anisotropy energy barriers. Typically, the sample was initially cool down from 400 K to 300 K in an applied field of 500 Oe, and the M(t) curves were recorded for a duration of 2 h at zero and 500 Oe fields. The magnetic field was set to zero while using oscillator mode, and then the SQUID magnet was reset to remove any stray field at 300 K. Figure 5 compares the normalized time-dependent magnetic moment relaxation obtained in 0 and 500 Oe from NiO, 0.5% and 1% samples. Very weak relaxation in the magnetic moment was observed from the M(t) curves measured at 500 Oe field. Whereas, a broad relaxation with about 6.56% drop in the magnetization was obtained from NiO NPs at 0 Oe field. The observed broad relaxation is governed by distribution in the particle size as well as exchange coupling anisotropy. Interestingly, with the increase of Fe-doping concentration from 0.5% to 1%, relaxation broadened, and the amount of magnetization dropped further from 4.58% to 4.21%. The above findings suggest the effect of 1% Fe-doping has slow down the magnetic moment relaxation by around 35.8% concerning NiO at 300 K over 2 h. Because, firstly, the mean diameter distribution obtained from SEM images has shown a very similar size distribution for NiO, 0.5, and 1% samples, and, secondly, the exchange coupling anisotropy has reduced further with the increase of Fe-doping concentration. Hence, apart from size distribution, the intraparticle interactions due to formed 4:1 defect clusters may have created additional magnetic anisotropy energy barriers affecting the overall magnetic moment relaxation process. In this context, a well-known stretched exponential function was successfully used elsewhere to describe the effect of distribution in anisotropy energy barriers on the process of magnetic moment relaxation [46]. The solid lines in Figure 5 represent a satisfactory fit using a stretched exponential function $M(t) = m_o - m_e exp\left(-(t/\tau)^\beta\right)$ and the fitted value of τ and β are depicted. m_o is an intrinsic magnetic component, m_e glassy component, τ characteristic relaxation time, and β is stretching parameter. m_e and τ are the function of measuring temperature and time, whereas β ($0 < \beta \leq 1$) is a function of the measuring temperature only. In the above expression, depending on the value of β, the system either relaxes with a single time constant ($\beta = 1$), or it involves activation against multi magnetic anisotropy energy barriers ($\beta < 1$). The fitted value of β obtained from NiO, 0.5, and 1% samples is 0.47, 0.51, and 0.60, respectively, which suggests activation against multiple anisotropy energy barriers from both undoped and Fe-doped NiO NPs. Therefore, along with particle size distribution existence of multiple anisotropy energy barriers possibly bears a correlation to the signature of the presence of intraparticle interactions in Fe-doped NiO NPs.

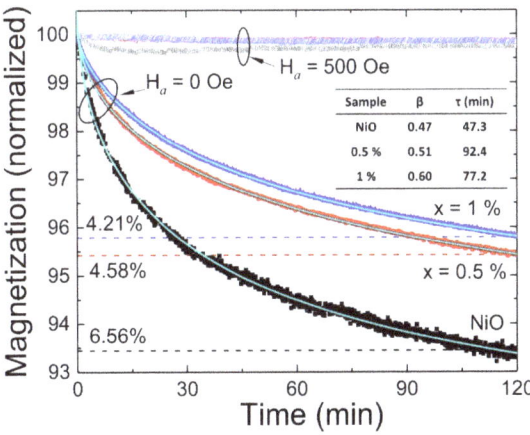

Figure 5. A plot of normalized magnetic moment relaxation measured concerning time M(t) at 300 K for 2 h from NiO (black), 0.5% (Red) and 1% (blue) samples in an applied field H_a = 500 Oe (light color) and 0 Oe (dark color). The solid line represents a satisfactory linear fit to M(t) curves measured in 0 Oe using stretched exponential function with fitted parameters, as depicted in the inset. The horizontal dashed lines correspond to the amount of relaxation for 2 h in zero applied field for each composition.

3.6. Magnetic Memory Effect

Both FC and ZFC protocols suggested by Sun et al. were used to study the magnetic memory effect for possible future applications [47]. FC magnetic memory test: sample was first cooled down in an applied field H_a = 500 Oe from 400 K, where sporadic stops for t_{wt} = 1 h in 0 Oe were given at various stopping temperatures T_S (300 K, 200 K, and 100 K, below T_B); above curve is designated as M_{FC}^{Cool}. Subsequently, magnetization was recorded while warming in the same applied field and the curve is designated as M_{FC}^{Mem}. Figure 6a–c depict the obtained FC memory effect results from NiO, 0.5%, and 1% samples, respectively. The amount of recovery of spins depends upon how fast the NPs realign to the applied magnetic field and, therefore, it can be quantified as $\Delta M(T_S) = M_{FC}^{cool} - M_{FC}^{mem}$ (Figure 6d). The value of $\Delta M(T_S)$ at each intermittent stopping temperature, T_S reaches to its maximum and shows an increasing trend with the increase of temperature. An enhanced value of $\Delta M(T_S)$ is obtained from a 0.5% sample, and its value increases further with the increase of Fe-concentration to 1% (inset of Figure 6d). The above findings suggest that the step-like time dynamic magnetization measurement is reproduced upon warming from Fe-doped NiO NPs at RT. On the other hand, strongly exchange-coupled NiO NPs do not retain any sign of memory effect. Furthermore, an increasing behavior of M_{FC}^{Cool} with the decrease of temperature can be seen from Fe-doped NiO NPs. Such type of behavior is commonly assigned to a non-interacting SPM system [48]. Contrary to it, an interacting spin-glass (SG) system shows a decrease of M_{FC}^{Cool} with a decrease in temperature [49]. The SG system can be identified by measuring the ZFC memory effect. ZFC magnetic memory test: the sample was first ZFC from 400 K to 10 K, and then magnetization was recorded during heating in an applied field of 500 Oe; this curve is designated as a reference curve. The sample was again ZFC from 400 K, but now with a stop-and-wait protocol at 200 K and 100 K, where the sample is aged for 1 h duration, and then again cooled down to 10 K. Subsequently, magnetization was recorded, as done in the reference curve; this curve is designated as wait curve. The field was set to zero using the inbuild oscillatory mode, and then the stray field was removed by resetting the SQUID magnet. Usually, the difference between the wait and reference curve is characterized by a dip at the waiting temperature [48,49]. However, no dip was seen from both NiO and Fe-doped NiO samples (Figure S3).

The RT memory effect was further investigated by studying the effect of both the temperature-cooling and -heating cycle and the field switching with ZFC and FC magnetization relaxation protocols in a 1% sample (Figure 7) [47]. In ZFC (FC) relaxation magnetization measurement, sample was initially cooled from 400 to 300 K under zero-field (500 Oe) and the magnetization M(t) was measured for t_1 = 4000 s at 500 Oe field (zero fields); after that, the sample was cooled to 280 K in the same magnetic field and magnetization was measured over time t_2 = 4000 s. Finally, the sample was warmed back to 300 K and magnetization was measured for t_3 = 4000 s. The relaxation in magnetization returns to the previous level, even after a temporary period that the sample was cooled to 280 K (Figure 7a). A similar phenomenon is also observed from ZFC and FC magnetization relaxation when the magnetic field was switched to 0 Oe and 500 Oe during t_2 with the temporary cooling at 280 K, respectively (Figure 7b). The above measurement shows that magnetization returns to the previous level when the temperature and magnetic field are returned to the previous condition at 300 K and 500 Oe. However, both ZFC and FC relaxation magnetization does not restore its previous state before the temporary heating to 320 K, demonstrating no memory effect (Figure 7c).

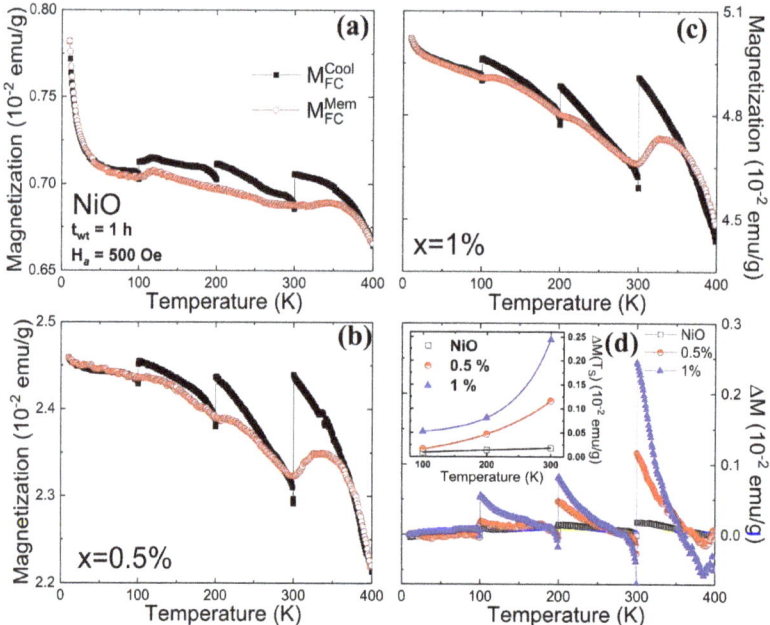

Figure 6. The FC magnetic memory effect from (**a**) NiO, (**b**) 0.5%, and (**c**) 1% samples. (**d**) The difference in magnetization $\Delta M = M_{FC}^{cool} - M_{FC}^{mem}$ concerning temperature. The value of maximum $\Delta M(T_S)$ at each stopping temperature, T_S is given in the inset of a figure (**d**).

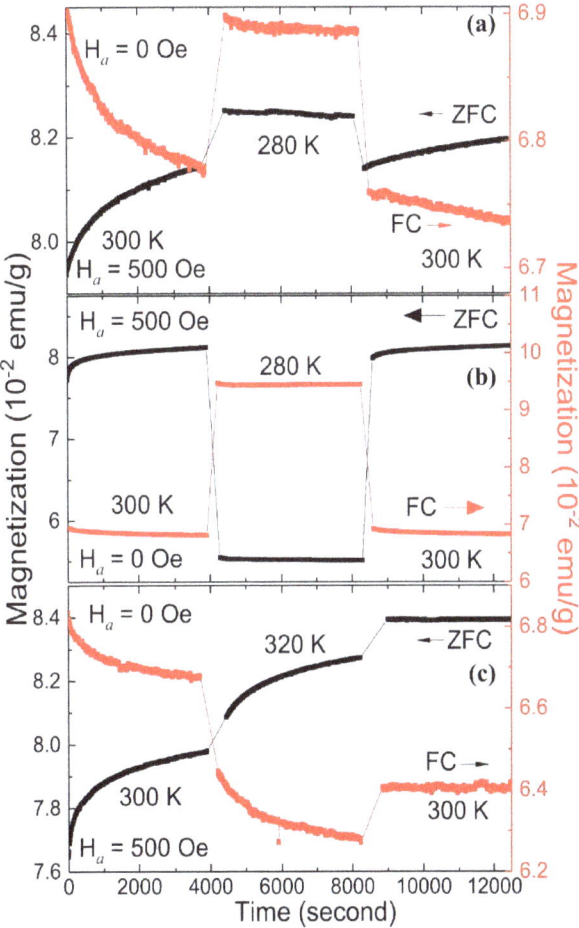

Figure 7. The effect of temperature-cooling (**a**) without (**b**) with field switching and (**c**) temperature-heating cycle without field switching during ZFC and FC magnetization relaxation from 1% sample.

4. Discussion and Conclusions

The observed asymmetric response concerning positive and negative temperature changes is following a hierarchical model for an interacting particle system [50,51]. The above model is also applicable to the non-interacting SPM system [32] and the exchange-coupled system [33], according to recent findings. In the former case, the distribution in energy barriers originated from the particle size distribution, whereas, in the latter case, from interface exchange-coupling. However, in the studied system, one could see that the effect of the cooling field has led to an enhancement in the exchange bias field both in the pure NiO and Fe-doped NiO NPs. The observed decrease in the EB field with the increase of Fe-doping concentration and the absence of memory effect in strongly exchange-coupled NiO NPs (having size distribution, i.e., anisotropy distribution) suggested RT memory effect and EB phenomenon are independent. Hence, it appears that, along with particles uniaxial anisotropy [17], the intraparticle interactions due to formed 4:1 defect clusters [16,23] may have paved the way for creating an additional anisotropy energy barrier, resulting in an appearance of RT memory effect from Fe-doped NiO NPs. In conclusion, $Ni_{1-x}Fe_xO$ NPs with x(%) varying from 0 to 1% and having a similar structure as that of NiO without any impurity phase was successfully synthesized while using the co-precipitation method. The effect of Fe-doping from 0 to 1% leads to a decrease of particle size from 63 nm to 53 nm and an increment of the magnetic moment of 3d ions (evident from NPD experiment). The RT ferromagnetic properties and magnetic memory effect is accomplished in Fe-doped NiO NPs studied by different protocols of ZFC and FC magnetization relaxation measurements. Our findings suggested that, as compared to non-interacting or interacting SPM NPs and exchange-coupled systems, the intrinsic intraparticle interaction in the Fe-doped NiO system has created an additional anisotropy energy barrier that can be tailored simply by controlling Fe-dopant concentration. The outcome of this is technologically attractive for the future development of RT ferromagnetism in AF NiO in order to facilitate the possible integration of spintronic devices.

Supplementary Materials: The following are available online at http://www.mdpi.com/2079-4991/10/7/1318/s1, Figure S1: Rietveld refined (red line) PXRD spectra (dots) from $Ni_{1-x}Fe_xO$ NPs, Figure S2: M(H_a) loops measured from 5% and 10% samples at 300 K, Figure S3: ZFC magnetic memory effect: Difference between the wait and reference curve from (**a**) NiO, (**b**) 0.5%, and 1% samples. Table S1: Rietveld refined parameters obtained from PXRD spectra in cubic $Fm\bar{3}m$ phase. Table S2: Rietveld refined parameters obtained from NPD spectra in cubic $Fm\bar{3}m$ phase. Table S3: Rietveld refined parameters obtained from NPD spectra in rhombohedral $R\bar{3}m$ phase.

Author Contributions: Conceptualization, methodology, validation, formal analysis, investigation, writing—original draft preparation, visualization, project administration, A.C.G.; investigation, T.-Y.L.; investigation, resources, B.V.K.; investigation, resources, P.M.R.; resources, J.-C.P.; resources, C.-M.W.; resources, data curation, supervision, writing—review and editing, funding acquisition, S.Y.W. All authors have read and agreed to the published version of the manuscript.

Funding: This research was funded by the ministry of science and technology (MOST) of the Republic of China, grant numbers MOST-107-2112-M-259005-MY3 and MOST-107-2811-M-259-005. The article processing charge (APC) was funded by MOST.

Conflicts of Interest: The authors declare no conflict of interest.

References

1. Kodama, R.H.; Makhlouf, S.A.; Berkowitz, A.E. Finite Size Effects in Antiferromagnetic NiO Nanoparticles. *Phys. Rev. Lett.* **1997**, *79*, 1393–1396. [CrossRef]
2. Gandhi, A.C.; Wu, S.Y. Strong Deep-Level-Emission Photoluminescence in NiO Nanoparticles. *Nanomaterials* **2017**, *7*, 231. [CrossRef] [PubMed]
3. Gandhi, A.C.; Pant, J.; Wu, S.Y. Dense inter-particle interaction mediated spontaneous exchange bias in NiO nanoparticles. *RSC Adv.* **2016**, *6*, 2079–2086. [CrossRef]
4. Mandal, S.; Banerjee, S.; Menon, K.S.R. Core-shell model of the vacancy concentration and magnetic behaviour for antiferromagnetic nanoparticle. *Phys. Rev. B* **2009**, *80*, 214420. [CrossRef]
5. Gandhi, A.C.; Pant, J.; Pandit, S.D.; Dalimbkar, S.K.; Chan, T.S.; Cheng, C.L.; Ma, Y.R.; Wu, S.Y. Short-Range Magnon Excitation in NiO Nanoparticles. *J. Phys. Chem. C* **2013**, *117*, 18666–18674. [CrossRef]
6. Rinaldi-Montes, N.; Gorria, P.; Martínez-Blanco, D.; Fuertes, A.B.; Fernández Barquín, L.; Rodríguez Fernández, J.; de Pedro, I.; Fdez-Gubieda, M.L.; Alonso, J.; Olivi, L.; et al. Interplay between microstructure and magnetism in NiO nanoparticles: Breakdown of the antiferromagnetic order. *Nanoscale* **2014**, *6*, 457–465. [CrossRef]
7. Tiwari, S.D.; Rajeev, K.P. Signatures of spin-glass freezing in NiO nanoparticles. *Phys. Rev. B* **2005**, *72*, 104433. [CrossRef]
8. Lee, D.U.; Fu, J.; Park, M.G.; Liu, H.; Kashkooli, A.G.; Chen, Z. Self-Assembled NiO/Ni(OH)$_2$ Nanoflakes as Active Material for High-Power and High-Energy Hybrid Rechargeable Battery. *Nano Lett.* **2016**, *16*, 1794–1802. [CrossRef]
9. Rinaldi-Montes, N.; Gorria, P.; Martínez-Blanco, D.; Fuertes, A.B.; Puente-Orench, I.; Olivi, L.; Blanco, J.A. Size effects on the Néel temperature of antiferromagnetic NiO nanoparticles. *AIP Adv.* **2016**, *6*, 056104. [CrossRef]
10. Oka, K.; Yanagida, T.; Nagashima, K.; Kawai, T.; Kim, J.S.; Park, B.H. Resistive-Switching Memory Effects of NiO Nanowire/Metal Junctions. *J. Am. Chem. Soc.* **2010**, *132*, 6634–6635. [CrossRef]
11. Sasaki, T.; Devred, F.; Eloy, P.; Gaigneaux, E.M.; Hara, T.; Shimazu, S.; Ichikuni, N. Development of supported NiO nanocluster for aerobic oxidation of 1-Phenylethanol and elucidation of reaction mechanism via X-ray analysis. *Bull. Chem. Soc. Jpn.* **2009**, *92*, 840–846. [CrossRef]

12. Ouyang, Y.; Huang, R.; Xia, X.; Ye, H.; Jiao, X.; Wang, L.; Lei, W.; Hao, Q. Hierarchical structure electrodes of NiO ultrathin nanosheets anchored to $NiCo_2O_4$ on carbon cloth with excellent cycle stability for asymmetric supercapacitors. *Chem. Eng.* **2018**, *355*, 416–427. [CrossRef]
13. Wang, J.; Cai, J.; Lin, Y.-H.; Nan, C.-W. Room-temperature ferromagnetism observed in Fe-doped NiO. *Appl. Phys. Lett.* **2005**, *87*, 202501. [CrossRef]
14. Shim, J.H.; Hwang, T.; Lee, S.; Park, J.H.; Han, S.J.; Jeong, Y.H. Origin of ferromagnetism in Fe- and Cu-codoped ZnO. *Appl. Phys. Lett.* **2005**, *86*, 082503. [CrossRef]
15. Lee, H.-J.; Jeong, S.-Y.; Cho, C.R.; Park, C.H. Study of diluted magnetic semiconductor: Co-doped ZnO. *Appl. Phys. Lett.* **2002**, *81*, 4020–4022. [CrossRef]
16. Gandhi, A.C.; Pradeep, R.; Yeh, Y.C.; Li, T.Y.; Wang, C.Y.; Hayakawa, Y.; Wu, S.Y. Understanding the Magnetic Memory Effect in Fe-Doped NiO Nanoparticles for the Development of Spintronic Devices. *ACS Appl. Nano Mater.* **2019**, *2*, 278–290. [CrossRef]
17. Moura, K.O.; Lima, R.J.S.; Coelho, A.A.; Souza-Junior, E.A.; Duque, J.G.S.; Meneses, C.T. Tuning the surface anisotropy in Fe-doped NiO nanoparticles. *Nanoscale* **2014**, *6*, 352–357. [CrossRef]
18. Kurokawa, A.; Sakai, N.; Zhu, L.; Takeuchi, H.; Yano, S.; Yanoh, T.; Onuma, K.; Kondo, T.; Miike, K.; Miyasaka, T.; et al. Magnetic properties of Fe-doped NiO nanoparticles. *J. Korean Phys. Soc.* **2013**, *63*, 716. [CrossRef]
19. Mishra, A.K.; Bandyopadhyay, S.; Das, D. Structural and magnetic properties of pristine and Fe-doped NiO nanoparticles synthesized by the co-precipitation method. *Mater. Res. Bull.* **2012**, *47*, 2288–2293. [CrossRef]
20. Mallick, P.; Rath, C.; Biswal, R.; Mishra, N.C. Structural and magnetic properties of Fe doped NiO. *Indian J. Phys.* **2009**, *83*, 517–523. [CrossRef]
21. Manna, S.; Deb, A.K.; Jagannath, J.; De, S.K. Synthesis and Room Temperature Ferromagnetism in Fe Doped NiO Nanorods. *J. Phys. Chem. C* **2008**, *112*, 10659–10662. [CrossRef]
22. He, J.H.; Yuan, S.L.; Yin, Y.S.; Tian, Z.M.; Li, P.; Wang, Y.Q.; Liu, K.L.; Wang, C.H. Exchange bias and the origin of room-temperature ferromagnetism in Fe-doped NiO bulk samples. *J. Appl. Phys.* **2008**, *103*, 023906. [CrossRef]
23. Grimes, R.W.; Anderson, A.B.; Heuer, A.H. Interaction of dopant cations with 4:1 defect clusters in non-stoichiometric 3d transition metal monoxides: A theoretical study. *J. Phys. Chem. Solids* **1987**, *48*, 45–50. [CrossRef]
24. Krafft, K.N.; Martin, M. A Thermogravimetric Study of the Non-Stoichiometry of Iron-Doped Nickel Oxide $(Ni_{1-x}Fe_x)_{1-\delta}O$. *Korean J. Ceram.* **1998**, *4*, 156–161.
25. Hoser, A.; Martin, M.; Schweika, W.; Carlsson, A.E.; Caudron, R.; Pyka, N. Diffuse neutron scattering of iron-doped nickel oxide. *Solid State Ion.* **1994**, *72*, 72–75. [CrossRef]
26. Haaß, F.; Buhrmester, T.; Martin, M. High-temperature in situ X-ray absorption studies on the iron valence in iron-doped nickel oxide $(Ni_{1-x}Fe_x)_{1-\delta}O$. *Solid State Ion.* **2001**, *141*, 289–293. [CrossRef]
27. Haaß, F.; Buhrmester, T.; Martin, M. Quantitative elaboration of the defect structure of iron doped nickel oxide $(Ni_{0.955}Fe_{0.045})_{1-\delta}O$ by in situ X-ray absorption spectroscopy. *Phys. Chem. Chem. Phys.* **2001**, *3*, 4806–4810.
28. Gandhi, A.C.; Chan, T.S.; Pant, J.; Wu, S.Y. Strong Pinned-Spin-Mediated Memory Effect in NiO Nanoparticles. *Nanoscale Res. Lett.* **2017**, *12*, 1–8.
29. Bisht, V.; Rajeev, K.P. Memory and aging effects in NiO nanoparticles. *J. Phys. Condens. Matt.* **2009**, *22*, 016003. [CrossRef]
30. Xu, L.; Gao, Y.; Malik, A.; Liu, Y.; Gong, G.; Wang, Y.; Tian, Z.; Yuan, S. Field pulse induced magnetic memory effect at room temperature in exchange coupled $NiFe_2O_4$/NiO nanocomposites. *J. Mag. Mag. Mater.* **2019**, *469*, 504–509. [CrossRef]
31. Chakraverty, S.; Ghosh, B.; Kumar, S.; Frydman, A. Magnetic coding in systems of nanomagnetic particles. *Appl. Phys. Lett.* **2006**, *88*, 042501. [CrossRef]
32. Tsoi, G.M.; Senaratne, U.; Tackett, R.J.; Buc, E.C.; Naik, R. Memory effects and magnetic interactions in a γ-Fe_2O_3 nanoparticle system. *J. Appl. Phys.* **2005**, *97*, 10J507. [CrossRef]
33. Tian, Z.; Xu, L.; Gao, Y.; Yuan, S.; Xia, Z. Magnetic memory effect at room temperature in exchange coupled $NiFe_2O_4$-NiO nanogranular system. *Appl. Phys. Lett.* **2017**, *111*, 182406. [CrossRef]
34. Dhara, S.; Chowdhury, R.R.; Bandyopadhyay, B. Strong memory effect at room temperature in nanostructured granular alloy $Co_{0.3}Cu_{0.7}$. *RSC Adv.* **2015**, *5*, 95695–95702. [CrossRef]
35. Balagurov, A.M.; Bobrikov, I.A.; Sumnikov, S.V.; Yushankhai, V.Y.; Grabis, J.; Kuzmin, A.; Mironova-Ulmane, N.; Sildos, I. Neutron diffraction study of microstructural and magnetic effects in fine particle NiO powders. *Phys. Status Solidi* **2016**, *253*, 1529–1536. [CrossRef]

36. Brok, E.; Lefmann, K.; Deen, P.P.; Lebech, B.; Jacobsen, H.; Nilsen, G.J.; Keller, L.; Frandsen, C. Polarized neutron powder diffraction studies of antiferromagnetic order in bulk and nanoparticle NiO. *Phys. Rev. B* **2015**, *91*, 014431. [CrossRef]
37. Roth, W.L. Magnetic Structures of MnO, FeO, CoO, and NiO. *Phys. Rev.* **1958**, *110*, 1333–1341. [CrossRef]
38. Lisa, T.; Matthew, G.T.; David, A.K.; Peter, M.M.T.; Paul, J.S.; Andrew, L.G. Exploration of antiferromagnetic CoO and NiO using reverse Monte Carlo total neutron scattering refinements. *Phys. Scr.* **2016**, *91*, 114004.
39. Rinaldi-Montes, N.; Gorria, P.; Martínez-Blanco, D.; Fuertes, A.B.; Fernández Barquín, L.; Rodríguez Fernández, J.; de Pedro, I.; Fdez-Gubieda, M.L.; Alonso, J.; Olivi, L.; et al. On the exchange bias effect in NiO nanoparticles with a core(antiferromagnetic)/shell (spin glass) morphology. *J. Phys. Conf. Ser.* **2015**, *663*, 012001. [CrossRef]
40. Cooper, J.F.K.; Ionescu, A.; Langford, R.M.; Ziebeck, K.R.A.; Barnes, C.H.W.; Gruar, R.; Tighe, C.; Darr, J.A.; Thanh, N.T.K.; Ouladdiaf, B. Core/shell magnetism in NiO nanoparticles. *J. Appl. Phys.* **2013**, *114*, 083906. [CrossRef]
41. Magnan, H.; Bezencenet, O.; Stanescu, D.; Belkhou, R.; Barbier, A. Beyond the magnetic domain matching in magnetic exchange coupling. *Phys. Rev. Lett.* **2010**, *105*, 097204. [CrossRef]
42. Nogués, J.; Sort, J.; Langlais, V.; Skumryev, V.; Suriñach, S.; Muñoz, J.S.; Baró, M.D. Exchange bias in nanostructures. *Phys. Rep.* **2005**, *422*, 65–117. [CrossRef]
43. Kondoh, H. Antiferromagnetic resonance in NiO in far-infrared region. *J. Phys. Soc. Jpn.* **1960**, *15*, 1970–1975. [CrossRef]
44. Perzanowski, M.; Marszalek, M.; Zarzycki, A.; Krupinski, M.; Dziedzic, A.; Zabila, Y. Influence of Superparamagnetism on Exchange Anisotropy at CoO/[Co/Pd] Interfaces. *ACS Appl. Mater. Interfaces* **2016**, *8*, 28159–28165. [CrossRef] [PubMed]
45. Chen, X.; Bedanta, S.; Petracic, O.; Kleemann, W.; Sahoo, S.; Cardoso, S.; Freitas, P.P. Superparamagnetism versus superspin glass behaviour in dilute magnetic nanoparticle systems. *Phys. Rev. B* **2005**, *72*, 214436. [CrossRef]
46. Khan, N.; Mandal, P.; Prabhakaran, D. Memory effects and magnetic relaxation in single-crystalline $La_{0.9}Sr_{0.1}CoO_3$. *Phys. Rev. B* **2014**, *90*, 024421. [CrossRef]
47. Sun, Y.; Salamon, M.B.; Garnier, K.; Averback, R.S. Memory effects in an interacting magnetic nanoparticle system. *Phys. Rev. Lett.* **2003**, *91*, 167206. [CrossRef]
48. Sasaki, M.; Jönsson, P.E.; Takayama, H.; Mamiya, H. Aging and memory effects in superparamagnets and superspin glasses. *Phys. Rev. B* **2005**, *71*, 104405. [CrossRef]
49. Bandyopadhyay, M.; Dattagupta, S. Memory in nanomagnetic systems: Superparamagnetism versus spin-glass behaviour. *Phys. Rev. B* **2006**, *74*, 214410. [CrossRef]
50. Parisi, G. Infinite number of order parameters for spin-glasses. *Phys. Rev. Lett.* **1979**, *43*, 1754–1756. [CrossRef]
51. Parisi, G. Order Parameter for Spin-Glasses. *Phys. Rev. Lett.* **1983**, *50*, 1946–1948. [CrossRef]

© 2020 by the authors. Licensee MDPI, Basel, Switzerland. This article is an open access article distributed under the terms and conditions of the Creative Commons Attribution (CC BY) license (http://creativecommons.org/licenses/by/4.0/).

Article

Magnetocrystalline and Surface Anisotropy in CoFe$_2$O$_4$ Nanoparticles

Alexander Omelyanchik [1,2], María Salvador [1,3], Franco D'Orazio [4], Valentina Mameli [5,6], Carla Cannas [5,6], Dino Fiorani [1], Anna Musinu [5,6], Montserrat Rivas [3], Valeria Rodionova [2], Gaspare Varvaro [1] and Davide Peddis [1,6,7,*]

1. Institute of Structure of Matter–CNR, Monterotondo Stazione, 00016 Rome, Italy; asomelyanchik@kantiana.ru (A.O.); salvadormaria@uniovi.es (M.S.); Dino.Fiorani@ism.cnr.it (D.F.); gaspare.varvaro@ism.cnr.it (G.V.)
2. Institute of Physics, Mathematics and Information Technology, Immanuel Kant Baltic Federal University, 236041 Kaliningrad, Russia; vvrodionova@kantiana.ru
3. Department of Physics, University of Oviedo, 33204 Gijón, Spain; mrivasardisana@gmail.com
4. The Department of Physical and Chemical Science, University of L'Aquila, Via Vetoio, Coppito, 67100 L'Aquila, Italy; franco.dorazio@aquila.infn.it
5. Department of Geological and Chemical Sciences, University of Cagliari, Cittadella Universitaria, 09042 Monserrato, Italy; valentina.mameli@unica.it (V.M.); ccannas@unica.it (C.C.); musinu@unica.it (A.M.)
6. National Interuniversity Consortium of Materials Science and Technology (INSTM), Via Giuseppe Giusti 9, 50121 Firenze, Italy
7. Department of Chemistry and Industrial Chemistry (DCIC), University of Genova, 16146 Genova, Italy
* Correspondence: davide.peddis@unige.it; Tel.: +39-010-776-7974

Received: 19 May 2020; Accepted: 26 June 2020; Published: 30 June 2020

Abstract: The effect of the annealing temperature T_{ann} on the magnetic properties of cobalt ferrite nanoparticles embedded in an amorphous silica matrix (CoFe$_2$O$_4$/SiO$_2$), synthesized by a sol-gel auto-combustion method, was investigated by magnetization and AC susceptibility measurements. For samples with 15% w/w nanoparticle concentration, the particle size increases from ~2.5 to ~7 nm, increasing T_{ann} from 700 to 900 °C. The effective magnetic anisotropy constant (K_{eff}) increases with decreasing T_{ann}, due to the increase in the surface contribution. For a 5% w/w sample annealed at 900 °C, K_{eff} is much larger (1.7×10^6 J/m^3) than that of the 15% w/w sample (7.5×10^5 J/m^3) annealed at 700 °C and showing comparable particle size. This indicates that the effect of the annealing temperature on the anisotropy is not only the control of the particle size but also on the core structure (i.e., cation distribution between the two spinel sublattices and degree of spin canting), strongly affecting the magnetocrystalline anisotropy. The results provide evidence that the magnetic anisotropy comes from a complex balance between core and surface contributions that can be controlled by thermal treatments.

Keywords: magnetic nanoparticles; cobalt ferrite; magnetic anisotropy

1. Introduction

Within the last few years, magnetic nanoparticles have contributed to the development of a variety of cutting edge technologies in fields such as ferrofluids [1], microwave devices [2], biomedicine [3,4], or catalysis [5,6]. The growing interest that magnetic nanoparticles attract demands a fundamental understanding of their properties, which are very different from their bulk counterparts. In this context, spinel ferrites are excellent candidates thanks to their tunable physico-chemical properties [7]. Their general chemical formula is MFe$_2$O$_4$, where M^{2+} can be any divalent metal (e.g., M^{2+} = Fe^{2+}, Co^{2+}, Zn^{2+}, Ni^{2+}, Mn^{2+}, etc.). The atomic arrangement corresponds to a face-centered cubic structure of the oxygen atoms, with Fe^{3+} and M^{2+} occupying the tetrahedral (T_d) and octahedral (O_h) sites [7]. Such a structure makes magnetic spinel nanoparticles particularly attractive. It provides a tool to tailor their magnetic properties (e.g., magnetic crystalline anisotropy and saturation magnetization) by the variation of the cation distribution between the two sublattices. This can be done by changing the chemical composition, the preparation method, and thermal treatments [8–10].

Magnetic properties of spinel ferrite nanoparticles are also strongly affected by the presence of a non-collinear spin structure (i.e., spin canting). The spin-canting is due to competing interactions between sublattices [11,12], as confirmed by polarized neutron scattering [13] and ^{57}Fe Mössbauer experiments [14,15]. This symmetry breaking induces changes in the topology of the surface magnetic moments and, consequently, in the exchange integrals (through super-exchange angles and/or distances between moments), thus leading to a change in the surface anisotropy [15]. Therefore, the magnetic properties of ferrite nanoparticles with a spinel structure are due to a complex interplay of several effects, among which surface disorder, cationic distribution, and spin canting are dominant [14,16].

The present work is aimed at investigating the effect of the annealing temperature on the magnetic properties of nanocomposites consisting of $CoFe_2O_4$ nanoparticles dispersed in a silica matrix ($CoFe_2O_4/SiO_2$). The results show that the thermal treatment plays an important role, along with the particle size, in controlling the surface and core contributions to the magnetic anisotropy and saturation magnetization.

2. Materials and Methods

A set of $CoFe_2O_4$ nanoparticles uniformly embedded in a silica matrix with 15% (w/w) concentration of the magnetic phase were synthesized by a sol-gel auto-combustion method and treated afterward at three different annealing temperatures (T_{ann} = 700, 800 and 900 °C). Synthesis and morpho-structural characterization of all the samples was already described in detail elsewhere [8,14,17,18]

The $Fe(NO_3)_3 \cdot 9H_2O$ (Sigma Aldrich 98%, Darmstadt, Germany), $Co(NO_3)_2 \cdot 6H_2O$ (Sigma Aldrich 98%, Darmstadt, Germany), citric acid (Sigma Aldrich 99.5%, Darmstadt, Germany) and of 25% ammonia solution (Carlo Erba Reagenti SpA, Cornaredo, Italy) were used without further purification. In this process, 1-molar iron and cobalt nitrate aqueous solutions in a 2:1 ratio, respectively, and citric acid (CA) with 1:1 molar ratio of metals to CA were prepared, and pH-adjusted to ~2 by aqueous ammonia addition. Tetraethoxysilane (TEOS, Sigma Aldrich 98%, Darmstadt, Germany) in ethanol was used as a silica precursor and, after its addition and vigorous stirring for 30 min, the sols were placed in an oven to gel in static air at 40 °C for 24 h. The gels underwent successively a thermal treatment at 300 °C for 15 min, where the auto-combustion reaction took place.

The temperature was then raised to 900 °C in steps of 100 °C and kept for 1 h at the treatment temperature. The X-ray diffraction (XRD) patterns [18] (Figure S1, reported in supporting materials) show a big halo due to the amorphous silica; the main reflections due to the cubic cobalt ferrite phase start to appear at 700 °C, and they become more and more evident at 800 and 900 °C [18]. For this reason, investigation of the magnetic properties was focused on samples treated at 700, 800 and 900 °C, hereafter named N15T700, N15T800, and N15T900. Transmission electron microscopy (TEM) (Figure S2, reported in supporting materials) shows the presence of crystalline particles for all the samples. The heating process led to the progressive growth of the particles and their structural ordering. The high resolution TEM images performed on N15T900 (Figure S2, Supplementary Materials) confirm the particles' spherical morphology. The observed set of fringes corresponds to the (311) lattice planes of the cobalt ferrite phase with a distance of 2.4 Å.

The particle size distribution obtained by TEM image analysis can be fitted by a log-normal function [19]:

$$P = \frac{A}{D\sigma\sqrt{2\pi}} exp - \left[\frac{ln^2(D/\langle D_{TEM}\rangle)}{2\sigma^2}\right] \quad (1)$$

where $<D_{TEM}>$ is the median of the variable "*diameter*" (Table 1) and σ is the standard deviation. An increase in particle size with the increase in annealing temperature is observed.

Table 1. Structural and magnetic properties.

Sample	d_{TEM}	T_{max}	T_{irr}	$<T_B>_{CH.}$ [2]	$PD_{CH.}$	$<T_B>_{H.M.}$ [2]	$PD_{H.M.}$
	(nm)	(K)	(K)	(K)	(%)	(K)	(%)
N15T700	2.5(2) [1]	29(1)	57(5)	18(1)	3.26	16(1)	4.56
N15T800	5.3(5)	43(1)	70(5)	22(2)	2.86	25(2)	2.44
N15T900	6.6(5)	53(1)	82(3)	29(1)	2.41	31(2)	2.45

[1] Uncertainties in the last digits are given in parenthesis; [2] Average blocking temperature extracted from thermoremanent magnetization (TRM) ($<T_B>_{CH.}$) and ($<T_B>_{H.M.}$) from Hansen and Mørup method are reported with their corresponding percentual polydispersity index.

DC-magnetization measurements were performed using a SQUID magnetometer (Quantum Design Inc., San Diego, CA, USA) equipped with a superconducting magnet producing fields up to 5 T. AC-susceptibility measurements were performed at different frequencies (20–800 Hz) as a function of the temperature using a susceptometer (Model ACS 7000, Lake Shore Cryotronics Inc., Weterville, OH, USA).

3. Results and Discussions

The temperature dependence of the zero-field-cooled/field-cooled (ZFC/FC) magnetizations is shown in Figure 1a. The sample was cooled down to 4.2 K from room temperature in the absence of an applied field. Then, the ZFC curve was recorded in a field of 5 mT while warming up to 325 K. In contrast, the FC curve was recorded after having cooled the sample down (from 325 to 4.2 K) with the same field applied. The shape of the FC-curves suggests that interparticle interactions are negligible [19–21]. The temperature corresponding to the maximum in the ZFC curve, T_{max}, (Table 1) increases with the annealing temperature. According to Gittleman et al. [22], T_{max} is related to the average blocking temperature $<T_B>$ through the equation:

$$T_{max} \approx \beta <T_B> \qquad (2)$$

where β is a constant that, for a log-normal distribution of particle sizes, is in the range of 1.5–2.5. The temperature at which the ZFC and FC curves merge is the irreversibility temperature (T_{irr}), and it corresponds to the blocking temperature of the particles with the maximum anisotropy. As expected, both T_{irr} and T_{max} grow with increasing size (i.e., increasing temperature). The difference between T_{irr} and T_{max} reflects the width of the blocking temperature distribution in the absence of magnetic interparticle interactions and it is correlated to the volume distribution. In our samples, such difference is weakly dependent on the annealing temperature, indicating that the thermal treatment does not significantly affect the distribution of the blocking temperatures. This is confirmed by the thermoremanent magnetization (TRM) curves [21] (Figure 2b, see supplementary information for details). Indeed, the shape of the energy barrier distribution is similar for the three samples, confirming that the sources of anisotropy are basically the same and that the interparticle interactions are weak. Two different models have been proposed to determine the blocking temperature distribution, yielding to its mean value and standard deviation. Starting from the model proposed by Chantrell and co-workers [23], the distribution of the anisotropy energy barriers was fitted by a log-normal function to determine the mean value of the blocking temperature ($<T_B>_{CH}$), reported in Table 1 [24–26]. We give details of the fit and values of the standard deviation (σ_{TRM}) in the supporting information (Figures S3 and S4).

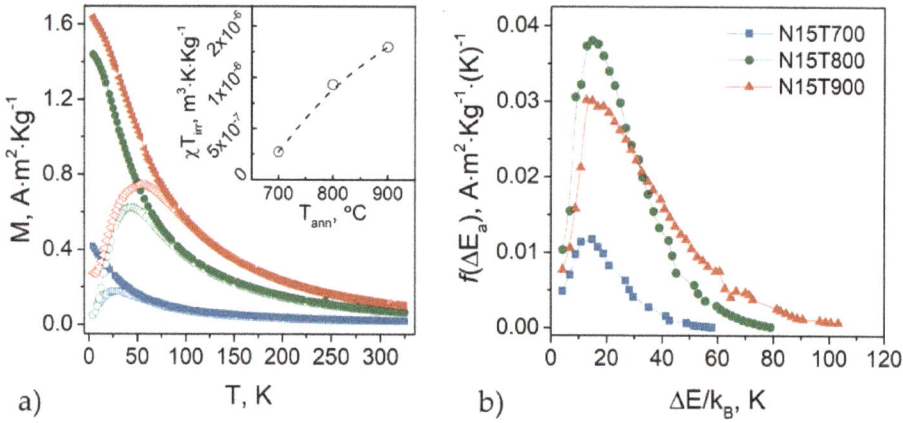

Figure 1. (a) Zero-field-cooled (ZFC) (empty symbols) and field-cooled (FC) (solid symbols) magnetization curves. Inset: product of the magnetic susceptibility times the irreversibility temperature as a function of the annealing temperature. (b) Energy barrier distribution obtained from the first derivative of the thermoremanent magnetization $M_{TRM}(T)$ versus temperature.

Hansen and Mørup proposed a phenomenological approach to calculate the mean blocking temperature ($<T_B>_{H.M.}$) and its standard deviation ($\sigma_{H.M.}$) [27] for a log-normal distribution of the particles volume, and negligible interparticle interactions. They found that $<T_B>_{H.M.}$ and $\sigma_{H.M}$ can be expressed with known values of T_{irr} and T_{max} from $<T_B>_{H.M.} = T_{max} [1.792 + 0.186 \cdot \ln(T_{irr}/T_{max} - 0.918)]^{-1} + 0.0039 \cdot T_{irr}$ and $\sigma_{H.M.} = 0.624 + 0.397 \ln(T_{irr}/T_{max} - 0.665)$. $<T_B>_{H.M.}$ values are given in 16(1), 25(2) and 31(2) K for samples (Table 1), their standard deviation values being 0.73, 0.61, and 0.57 for N15T700, N15T800 and N15T900, respectively.

The values of the mean blocking temperatures extracted by the two models are equal within the experimental errors (Table 1). The percentual polydispersity of the blocking temperatures is defined as:

$$PD\% = 100 \times \frac{\sigma}{\langle T_B \rangle} \tag{3}$$

The $PD\%$ value obtained for Chantrell and Hansen–Mørup models ($PD_{CH.}$ and $PD_{H.M.}$) decreases with increasing particle size, although this trend is more evident for the Chantrell model.

The inset in Figure 1a shows the product of the susceptibility times the irreversibility temperature (χT_{irr}) as a function of the annealing temperature (T_{ann}). These results indicate a strong increase in the ferrimagnetic phase between 700 and 900 °C, which can be ascribed to the rise in the particle volume [28].

The dynamic magnetic properties were investigated by AC-susceptibility measurements in a field of 2.5 mT at frequencies υ from 5 Hz to 10 kHz, in the temperature interval 18–310 K. According to the Néel–Arrhenius model, the relaxation process of the particle moments is driven by thermal activation and described, in the absence of interparticle interactions, by the Arrhenius law $\tau_N = \tau_0 \exp(K_{eff}V/k_BT)$. Since $T = T_B$ when $\tau_m = 1/\upsilon_m$, a linear relation between $\ln(\tau m)$ and $1/T_B$ can be derived:

$$\ln(\tau_m) = \ln(\tau_0) + \frac{K_{eff}V}{k_B T_B}. \tag{4}$$

In Figure 2a, the linear relationship between $\ln(\tau m)$ versus $1/T_B$ is reported for the three samples. The values of the effective magnetic anisotropy constant, K_{eff}, and the characteristic relaxation times, τ_0, obtained from the linear fitting of Equation (4), are given in Table 2.

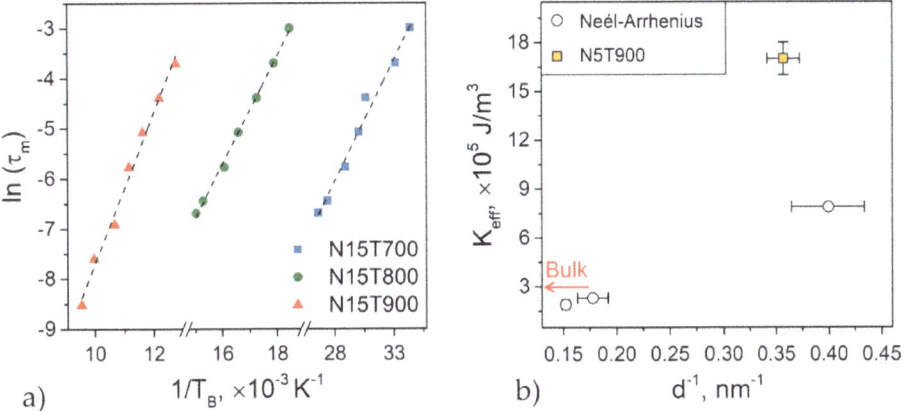

Figure 2. (a) Logarithm of the measurement time τ_m versus $1/T_B$ and its linear fit (dashed line); (b) effective anisotropy constant K_{eff} of N15T700, N15T800 and N15T900 obtained from fitting $ln(\tau_m)$ versus $1/T_B$ by Neél–Arrhenius model (empty circles). The value from N5T900 was taken from reference [14] and that of bulk cobalt ferrite from reference [7].

Table 2. Magnetic parameters obtained from AC magnetic susceptibility measurements.

Sample	Neél–Arrhenius [1]		Vogel–Fulcher [2]	
	K_{eff} (J m^{-3})	τ_0 (s)	K_{eff} (J m^{-3})	T_0 (K)
N15T700	$7.9(4) \times 10^5$	1.9×10^{-9}	$11(1) \times 10^5$	$-1(3)$
N15T800	$2.3(2) \times 10^5$	8.2×10^{-14}	$1.3(1) \times 10^5$	$14(2)$
N15T900	$1.9(2) \times 10^5$	1.5×10^{-14}	$0.92(1) \times 10^5$	$32(3)$

[1] Effective magnetic anisotropy constant (K_{eff}) and characteristic relaxation time (τ_0) obtained from the fitting to Neél–Arrhenius law (Equation (4)); [2] K_{eff} and the interaction temperature term, T_0, assuming $\tau_0 = 10^{-10}$ s, from Vogel–Fulcher law (Equation (5)).

For the sample annealed at the lowest temperature (N15T700), the τ_0 value has a coherent physical meaning (1.9×10^{-9} s), confirming the absence of interparticle interactions. On the other hand, for samples N15T800 and N15T900, the τ_0 value is much smaller. This fact indicates that the Neél–Arrhenius model is not appropriate to describe the dynamical behavior of these samples, suggesting that weak interparticle interactions are present.

According to the Vogel–Fulcher law, weak interparticle interactions are accounted for by a temperature term T_0 [29–31]:

$$ln(\tau_m) = ln(\tau_0) + \frac{K_{eff}V}{k_B(T_B + T_0)}. \qquad (5)$$

The values of T_0 and K_{eff} (Table 2) have been obtained from the fitting of Equation (5) by fixing the specific relaxation time τ_0 equal to 10^{-10} s for all the samples [14,26]. In sample N15T700, T_0 is almost zero, consistent with the absence of interparticle interactions. Then, T_0 rises with the annealing temperature, indicating an increase in the dipolar interactions due to the enhancement of the particle magnetic moment. It is worth underlining that the value of K_{eff} obtained by Neél–Arrhenius and Vogel–Fulcher models are similar for sample N15T700 where the interactions can be considered negligible. A difference in the K_{eff} values deduced from the two models is observed for N15T800 and N15T900 due to magnetic interactions.

On the other hand, in both models, K_{eff} increases with a decreasing particle size (i.e., decreasing annealing temperature). We measured a rise of ~30% when the diameter goes from 6.6 (N15T900) to 5.6 nm (N15T800), while a much higher growth of ~70% is observed when it goes from 6.6 (N15T800) to 2.5 nm (N15T700). This result indicates that the surface anisotropy increases with a decrease in the particle size, but its role becomes dominant in tiny particles (e.g., N15T700). This idea is also confirmed by the fact that the K_{eff} values of N15T800 and N15T900, which are lower than the value of the bulk material (3×10^5 J/m^3 [14,32]), indicating that the magnetic structure also plays a crucial role. The smaller anisotropy in CoFe$_2$O$_4$ nanoparticles compared to the bulk value can be related to a change in the cation distribution with the size, induced by the annealing treatment. This phenomenon was already observed in CoFe$_2$O$_4$ particles [14,33] and explained by a modification of the cation distribution leading to a change in the magneto-crystalline anisotropy mainly determined by the distribution of the Co^{2+} ions between O_h and T_d sites. Indeed, here the cause can be a lower fraction of Co^{2+} ions in the octahedral sites, having larger anisotropy ($+850 \times 10^{-24}$ J/ion) (due to the orbital contribution in the crystal field $^4T_1$ ground energy term) than Co^{2+} ions in a tetrahedral site (-79×10^{-24} J/ion; $^4A_2$ term) [8].

To highlight these results, Figure 2b reports the K_{eff} value of an additional sample consisting of CoFe$_2$O$_4$ nanoparticles embedded in a silica matrix with a 5% (w/w) concentration of magnetic phase annealed at 900 °C (hereafter named N5T900). For this sample, the average particle size (2.8 ± 0.3 nm [14]) is very close to that of the N15T700 (2.5 ± 0.5 nm), with the same percentual polydispersity (see Figure S5 and Table S1 in the supporting information). It is important to underline that the interparticle interactions in both N5T900 and N15T700 samples are negligible, as indicated by their corresponding ZFC-FC and δM-plots [34] (Figures S6 and S7, respectively, in the supporting information).

Despite the two samples having the same morphological features, K_{eff} is much larger for N5T900, which could be related to the cation distribution change caused by the annealing. The highest temperature produces a larger occupancy of O_h sites by the Co^{2+} ions in sample N5T900 [14].

Low-temperature (5 K) magnetization loops of the samples N15T700 and N5T900 are reported in Figure 3. They are not saturated due to their high anisotropy (the same as samples N15T800 and N15T900, reported in Figure S8 of the supporting information). The saturation magnetization (M_S) has been estimated by fitting the high field range of the curves to the equation [35]:

$$M(H) = M_S \cdot \left(1 - \frac{a}{H} - \frac{b}{H^2}\right) + H \cdot \chi_{SAT} \qquad (6)$$

where a and b are the fitting parameters, and χ_{SAT} is the "non-saturated" magnetic susceptibility (for high applied fields). The latter is strongly related to the non-collinear spin structure due to competing interactions between sublattices, and to the symmetry breaking at the particle surface [36,37].

Figure 4a shows that M_S increases with particle size (i.e., annealing temperature), as expected. In the same figure, we plot M_S for sample N5T900 (2.8 ± 0.3 nm particle size). Despite N5T900 and N15T700 having the same particle size, M_S for N5T900 is almost twice than for N15T700. Considering that the magnetic interparticle interactions are negligible in both samples, this difference can be ascribed to the combined effect of cation distribution, spin-canting, and surface anisotropy [14,16]. The non-saturated susceptibility (Figure 4b) increases with decreasing particle size (i.e., decreasing annealing temperature) [31]. The trend of χ_{SAT} indicates, as expected, the more substantial contribution of the surface magnetic anisotropy for smaller particles. It is worth emphasizing that N15T700 has a higher value of χ_{SAT}, indicating that the surface contribution to the effective magnetic anisotropy is higher in N15T700 than in N15T900. The energy barrier distribution can confirm this. In fact, despite N5T900 and N15T700 having the same $PD\%$ of the TEM diameter, the $PD\%$ for T_B calculated by I.I.M. model is much higher for N15T700 ($PD\%$ T_B 4.56) than for N5T900 ($PD\%$ T_B 2.45).

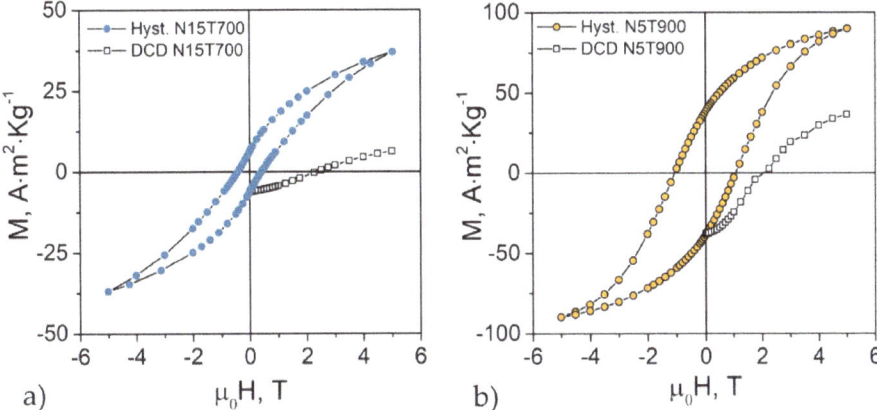

Figure 3. Field-dependence of magnetization and direct current demagnetization (DCD) curves measured at 5 K for (**a**) sample N15T700 and (**b**) reference sample N5T900 with the same particle size [14].

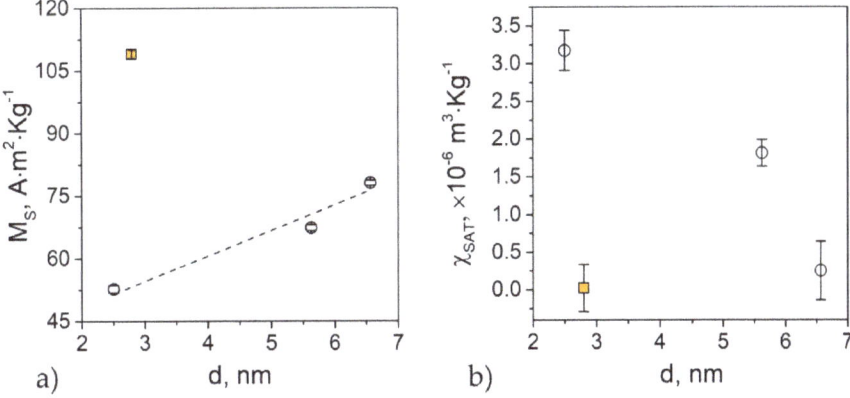

Figure 4. (**a**) Saturation magnetization, M_S, and (**b**) non-saturated susceptibility, χ_{SAT}, obtained by fitting Equation (6). The dashed line represents a guide for the eye. The square in both graphs corresponds to the reference sample N5T900.

Given that the particle volume is comparable in the two samples, this fact can be explained by the mentioned surface anisotropy contribution. Labarta and co-workers have shown that, for spinel ferrite nanoparticles, even when the size distribution is narrow, surface anisotropy can produce a substantial broadening of the anisotropy energy distribution. This effect is an obvious consequence of the different size dependence of the energy contributions from the core and the surface [38]. Because the volume content of the surface spin layer increases with a decrease in size, and it becomes more significant for ultra-small particles (<10 nm).

Then, although χ_{SAT} and the anisotropy energy barrier distribution indicate a more significant contribution of the surface component to the anisotropy in N15T700, the value K_{eff} obtained by AC measurements is higher in N5T900. This could be associated with an increase in the magneto-crystalline component of the anisotropy.

The squares in Figure 3 represent the low temperature (5 K) direct current demagnetization (DCD) remanent curves measured following the protocol in ref. [39]: (1) application of a quasi-saturating field of 5 T; (2) application of a gradually increasing magnetic field in the reversal direction; (3) switching off the magnetic field and collection of the remanent magnetization value after each iteration. Since each measurement is performed at zero field, M_{DCD} is only sensitive to the irreversible component of the magnetization and only the blocked particles contribute to the remanent magnetization. The curve shape is linked to the switching field distribution, which, in turn, is related to the energy barrier distribution; the value of the field at which the remanent magnetization is equal to zero (called remanent coercivity, H_{Cr}) corresponds to the mean switching field. Although the two samples have different coercivity, the remanent coercivities are close (N15T700: ~2.4 T, N5T900: ~2.1 T). This result is in line with the similar anisotropy fields (N15T700: 5.8(5) T, N5T900: 5.9(6) T) estimated by the Stoner–Wohlfarth model ($H_K = 2K_{eff}/M_S$).

Even though H_{Cr} and H_K are equal within the experimental error, the coercivities of the two samples are different. Since both systems are non-interacting, such differences can be associated with a larger fraction of very small particles that probably are not well crystallized due to the low treatment temperature. We have confirmed this by the trend of χT_{irr} (inset of Figure 1a) and the lower value of the remanent and saturation magnetizations [40].

4. Conclusions

The results provide evidence that the relative surface and core contributions to the effective magnetic anisotropy and saturation magnetization of $CoFe_2O_4$ nanoparticles embedded in a silica matrix can be controlled by the annealing temperature T_{ann}. For samples with 15% w/w of nanoparticles, the value of the effective magnetic anisotropy constant K_{eff} increases and the saturation magnetization decreases by decreasing T_{ann} from 900 to 700 °C, with a decrease in particle size, showing a dominant role of the disordered surface. On the other hand, the comparison between the 15% w/w sample annealed at 700 °C and a 5% w/w sample annealed at 900 °C, with comparable particle size, (2.5 and 2.8 nm with the same size distribution) shows a much larger saturation magnetization and K_{eff} values for the latter one, for which the χ_{SAT} value, related to non-collinear core spin structure and surface disorder, is much lower. The comparison indicates that for the 5% w/w with T_{ann} = 900 °C sample the major contribution to the anisotropy comes from the core, despite its very small particle size. This should be due to a better crystallinity and change in a core structure (e.g., different cation distribution and degree of spin canting) with larger magnetocrystalline anisotropy induced by the higher T_{ann}. In conclusion, the results indicate that the effect of the annealing temperature on the anisotropy and saturation magnetization is not limited to the change in the particle size, increasing with T_{ann}. Besides the decrease in surface disorder, the core structure is also affected by the thermal treatment, which can significantly modify the magnetocrystalline anisotropy and the saturation magnetization.

Supplementary Materials: The following are available online at http://www.mdpi.com/2079-4991/10/7/1288/s1, Figure S1: XRD patterns of CoFe2O4 nanoparticles embedded (15% w/w) in an amorphous silica matrix and annealed at different temperatures (from ref. [18]); Figure S2: TEM image of as-prepared sample (a); dark field images of N15T700 (b) and N15T800 (c); bright field (d) and high resolution image of N15T900 (from ref. [18]); Figure S3: TRM measurements (full symbols) and corresponding distribution of magnetic anisotropy (empty symbols) energies with cooling field of 2.5 mT for the samples N15T700 (a), N15T800 (b), N15T900 (c); Figure S4: distribution of magnetic anisotropy energies (empty symbols) fitted by a log normal function (lines) for the samples N15T700 (squares), N15T800 (triangles), N15T900 (circles). Values of mean Blocking temperature (<T_B>) and standard deviation (σ) are reported in the graph; Figure S5: Particle size distribution extracted by TEM images of the sample N15T700 (left side) and N5T900 (right side); Figure S6: ZFC-FC curves for the N15T700 (a) and N5T900 (b); Figure S7: δM-plot at 5 K for N15T700 sample and reference sample N5T900 [14]; Figure S8: Hysteresis loop recorded at 5 K for N15T900 (a), N15T800 (b) and N15T700 (c); Table S1: Mean particle size (D_{TEM}) σ the standard deviation of the natural logarithm of the variable D and percentual polydispersity determined by Equation (7).

Author Contributions: Conceptualization, D.P.; methodology, D.P. and D.F.; validation, D.P., M.R. and V.R.; investigation, D.P., F.D., C.C., G.V., V.M. and A.M.; data curation, A.O. and M.S.; writing—Original draft preparation, A.O. and M.S.; writing—Review and editing, D.P. and D.F.; visualization, A.O.; supervision, D.P., D.F., V.R. and M.R.; project administration, D.P.; funding acquisition, D.P. and V.R. All authors have read and agreed to the published version of the manuscript.

Funding: The reported study was funded by RFBR according to the research project No. 18-32-01016. This work was supported in part by the 5 top 100 Russian Academic Excellence Project at the Immanuel Kant Baltic Federal University.. MS thanks the University of Oviedo, the Spanish Ministry of Education, Culture and Sport and Banco Santander for a grant (CEI15-24), and Spanish Ministry of Economy and Competitiveness for grant MAT2017-84959-C2-1-R. RAS-Piano Sulcis (CESA Project) and PON AIM (PON Ricerca e Innovazione 2014–2020-Azione I.2-D.D. n.407 del 27 febbraio 2018 "Attraction and International Mobility", Cult-GeoChim project AIM1890410-3) are gratefully acknowledged for financing the past and actual fellowships of V. Mameli.

Conflicts of Interest: The authors declare no conflict of interest.

References

1. Gomes, J.D.A.; Sousa, M.H.; Tourinho, F.A.; Aquino, R.; Da Silva, G.J.; Depeyrot, J.; Dubois, E.; Perzynski, R. Synthesis of core-shell ferrite nanoparticles for ferrofluids: Chemical and magnetic analysis. *J. Phys. Chem. C* **2008**, *112*, 6220–6227. [CrossRef]
2. Pardavi-Horvath, M. Microwave applications of soft ferrites. *J. Magn. Magn. Mater.* **2000**, *216*, 171–183. [CrossRef]
3. Song, G.; Kenney, M.; Chen, Y.S.; Zheng, X.; Deng, Y.; Chen, Z.; Wang, S.X.; Gambhir, S.S.; Dai, H.; Rao, J. Carbon-coated FeCo nanoparticles as sensitive magnetic-particle-imaging tracers with photothermal and magnetothermal properties. *Nat. Biomed. Eng.* **2020**, *4*, 325–334. [CrossRef]
4. Cardoso, V.F.; Francesko, A.; Ribeiro, C.; Bañobre-López, M.; Martins, P.; Lanceros-Mendez, S. Advances in Magnetic Nanoparticles for Biomedical Applications. *Adv. Healthc. Mater.* **2018**, *7*, 1700845. [CrossRef] [PubMed]
5. Waag, F.; Gökce, B.; Kalapu, C.; Bendt, G.; Salamon, S.; Landers, J.; Hagemann, U.; Heidelmann, M.; Schulz, S.; Wende, H.; et al. Adjusting the catalytic properties of cobalt ferrite nanoparticles by pulsed laser fragmentation in water with defined energy dose. *Sci. Rep.* **2017**, *7*, 1–13. [CrossRef] [PubMed]
6. Rizzuti, A.; Dassisti, M.; Mastrorilli, P.; Sportelli, M.C.; Cioffi, N.; Picca, R.A.; Agostinelli, E.; Varvaro, G.; Caliandro, R. Shape-control by microwave-assisted hydrothermal method for the synthesis of magnetite nanoparticles using organic additives. *J. Nanoparticle Res.* **2015**, *17*, 1–16. [CrossRef]
7. da Silva, F.G.; Depeyrot, J.; Campos, A.F.C.; Aquino, R.; Fiorani, D.; Peddis, D. Structural and Magnetic Properties of Spinel Ferrite Nanoparticles. *J. Nanosci. Nanotechnol.* **2019**, *19*, 4888–4902. [CrossRef]
8. Cannas, C.; Musinu, A.; Piccaluga, G.; Fiorani, D.; Peddis, D.; Rasmussen, H.K.; Mørup, S. Magnetic properties of cobalt ferrite-silica nanocomposites prepared by a sol-gel autocombustion technique. *J. Chem. Phys.* **2006**, *125*, 1–11. [CrossRef] [PubMed]
9. Jovanović, S.; Spreitzer, M.; Otoničar, M.; Jeon, J.-H.; Suvorov, D. pH control of magnetic properties in precipitation-hydrothermal-derived $CoFe_2O_4$. *J. Alloys Compd.* **2014**, *589*, 271–277. [CrossRef]
10. Albino, M.; Fantechi, E.; Innocenti, C.; López-Ortega, A.; Bonanni, V.; Campo, G.; Pineider, F.; Gurioli, M.; Arosio, P.; Orlando, T.; et al. Role of Zn^{2+} Substitution on the Magnetic, Hyperthermic, and Relaxometric Properties of Cobalt Ferrite Nanoparticles. *J. Phys. Chem. C* **2019**, *123*, 6148–6157. [CrossRef]
11. Martínez, B.; Obradors, X.; Balcells, L.; Rouanet, A.; Monty, C. Low Temperature Surface Spin-Glass Transition in -gFe_2O_3 Nanoparticles. *Phys. Rev. Lett.* **1998**, *80*, 181–184. [CrossRef]
12. Coey, J.M.D. Non-collinear Spin Arrangement in Ultrafine Ferrimagnetic Crystallites. *Phys. Rev. Lett.* **1971**, *27*, 1140. [CrossRef]
13. Lin, D.; Nunes, A.C.; Majkrzak, C.F.; Berkowitz, A.E. Polarized neutron study of the magnetization density distribution within a $CoFe_2O_4$ colloidal particle II. *J. Magn. Magn. Mater.* **1995**, *145*, 343–348. [CrossRef]
14. Peddis, D.; Mansilla, M.V.; Mørup, S.; Cannas, C.; Musinu, A.; Piccaluga, G.; D'Orazio, F.; Lucari, F.; Fiorani, D. Spin-canting and magnetic anisotropy in ultrasmall $CoFe_2O_4$ nanoparticles. *J. Phys. Chem. B* **2008**, *112*, 8507–8513. [CrossRef]
15. Peddis, D.; Yaacoub, N.; Ferretti, M.; Martinelli, A.; Piccaluga, G.; Musinu, A.; Cannas, C.; Navarra, G.; Greneche, J.M.; Fiorani, D. Cationic distribution and spin canting in $CoFe_2O_4$ nanoparticles. *J. Phys. Condens. Matter* **2011**, *23*, 426004. [CrossRef]
16. Peddis, D. Magnetic Properties of Spinel Ferrite Nanoparticles: Influence of the Magnetic Structure. In *Magnetic Nanoparticle Assemblies*; Trohidou, K.N., Ed.; Pan Stanford Publishing: Singapore, 2014; Volume 7, pp. 978–981, ISBN 9789814411967.

17. Cannas, C.; Musinu, A.; Peddis, D.; Piccaluga, G. Synthesis and Characterization of $CoFe_2O_4$ Nanoparticles Dispersed in a Silica Matrix by a Sol–Gel Autocombustion Method. *Chem. Mater.* **2006**, *18*, 3835–3842. [CrossRef]
18. Muscas, G.; Singh, G.; Glomm, W.R.; Mathieu, R.; Kumar, P.A.; Concas, G.; Agostinelli, E.; Peddis, D. Tuning the size and shape of oxide nanoparticles by controlling oxygen content in the reaction environment: Morphological analysis by aspect maps. *Chem. Mater.* **2015**, *27*, 1982–1990. [CrossRef]
19. Peddis, D.; Jönsson, P.E.; Laureti, S.; Varvaro, G. *Magnetic Interactions: A Tool to Modify the Magnetic Properties of Materials Based on Nanoparticles*; Elsevier: Amsterdam, The Netherlands, 2014; Volume 6.
20. Knobel, M.; Nunes, W.C.; Socolovsky, L.M.; De Biasi, E.; Vargas, J.M.; Denardin, J.C. Superparamagnetism and other magnetic features in granular materials: A review on ideal and real systems. *J. Nanosci. Nanotechnol.* **2008**, *8*, 2836–2857. [CrossRef]
21. Dormann, J.L.; Fiorani, D.; Tronc, E. Magnetic Relaxation in Fine-Particle Systems. *Adv. Chem. Phys.* **1997**, *98*, 283–494.
22. Gittleman, J.I.; Abeles, B.; Bozowski, S. Superparamagnetism and relaxation effects in granular $Ni-SiO_2$ and $Ni-Al_2O_3$ films. *Phys. Rev. B* **1974**, *9*, 3891–3897. [CrossRef]
23. Chantrell, R.W.; El-Hilo, M.; O'Grady, K. Spin-Glass behaviour in fine particle system. *IEEE Trans. Magn.* **1991**, *27*, 3570–3578. [CrossRef]
24. Lavorato, G.C.; Peddis, D.; Lima, E.; Troiani, H.E.; Agostinelli, E.; Fiorani, D.; Zysler, R.D.; Winkler, E.L. Magnetic Interactions and Energy Barrier Enhancement in Core/Shell Bimagnetic Nanoparticles. *J. Phys. Chem. C* **2015**, *119*, 15755–15762. [CrossRef]
25. Liu, C.; Zou, B.; Rondinone, A.J.; Zhang, Z.J. Chemical Control of Superparamagnetic Properties of Magnesium and Cobalt Spinel Ferrite Nanoparticles through Atomic Level Magnetic Couplings. *J. Am. Chem. Soc.* **2000**, *122*, 6263–6267. [CrossRef]
26. Rondinone, A.J.; Liu, C.; Zhang, Z.J. Determination of Magnetic Anisotropy Distribution and Anisotropy Constant of Manganese Spinel Ferrite Nanoparticles. *J. Phys. Chem. B* **2001**, *105*, 7967–7971. [CrossRef]
27. Hansen, M.F.; Mørup, S. Estimation of blocking temperatures from ZFC/FC curves. *J. Magn. Magn. Mater.* **1999**, *203*, 214–216. [CrossRef]
28. Cannas, C.; Gatteschi, D.; Musinu, A.; Piccaluga, G.; Sangregorio, C. Structural and Magnetic Properties of Fe_2O_3 Nanoparticles Dispersed over a Silica Matrix. *J. Phys. Chem. B* **1998**, *102*, 7721–7726. [CrossRef]
29. Pacakova, B.; Kubickova, S.; Reznickova, A.; Niznansky, D.; Vejpravova, J. Spinel Ferrite Nanoparticles: Correlation of Structure and Magnetism. In *Magnetic Spinels—Synthesis, Properties and Applications*; InTech: London, UK, 2017; ISBN 9789537619824.
30. Dormann, J.L.; Bessais, L.; Fiorani, D. A dynamic study of small interacting particles: Superparamagnetic model and spin-glass laws. *J. Phys. C Solid State Phys.* **1988**, *21*, 2015. [CrossRef]
31. del Castillo, V.L.C.D.; Rinaldi, C. Effect of sample concentration on the determination of the anisotropy constant of magnetic nanoparticles. *IEEE Trans. Magn.* **2010**, *46*, 852–859. [CrossRef]
32. Sharifi, I.; Shokrollahi, H.; Amiri, S. Ferrite-based magnetic nanofluids used in hyperthermia applications. *J. Magn. Magn. Mater.* **2012**, *324*, 903–915. [CrossRef]
33. Sharifi, I. Magnetic and structural studies on $CoFe_2O_4$ nanoparticles synthesized by co-precipitation, normal micelles and reverse micelles methods. *J. Magn. Magn. Mater.* **2012**, *324*, 1854–1861. [CrossRef]
34. Omelyanchik, A.; Knezevic, N.; Rodionova, V.; Salvador, M.; Peddis, D.; Varvaro, G.; Laureti, S.; Mrakovic, A.; Kusigerski, V.; Illes, E. Experimental Protocols for Measuring Properties of Nanoparticles Dispersed in Fluids. In Proceedings of the 2018 IEEE 8th International Conference Nanomaterials: Application & Properties (NAP), Zatoka, Ukraine, 9–14 September 2018; pp. 1–5.
35. Morrish, A.H. *The Physical Principles of Magnetism*; Wiley: Hoboken, NJ, USA, 1965; Volume 1, ISBN 0-7803-6029-X.
36. Muscas, G.; Concas, G.; Cannas, C.; Musinu, A.; Ardu, A.; Orru, F.; Fiorani, D.; Laureti, S.; Rinaldi, D.; Piccaluga, G.; et al. Magnetic Properties of Small Magnetite Nanocrystals. *J. Phisical Chem. C* **2013**, *114*, 23378–23384. [CrossRef]
37. Peddis, D.; Cannas, C.; Piccaluga, G.; Agostinelli, E.; Fiorani, D. Spin-glass-like freezing and enhanced magnetization in ultra-small $CoFe_2O_4$ nanoparticles. *Nanotechnology* **2010**, *21*, 125705. [CrossRef] [PubMed]

38. Pérez, N.; Guardia, P.; Roca, A.G.; Morales, M.P.; Serna, C.J.; Iglesias, O.; Bartolomé, F.; García, L.M.; Batlle, X.; Labarta, A. Surface anisotropy broadening of the energy barrier distribution in magnetic nanoparticles. *Nanotechnology* **2008**, *19*, 475704. [CrossRef] [PubMed]
39. Chantrell, R.W.; O'Grady, K. The Magnetic Properties of Fine Particles. In *Applied Magnetism*; Springer: Dordrecht, The Netherlands, 1994; pp. 113–164.
40. El-Hilo, M.; Bsoul, I. Interaction effects on the coercivity and fluctuation field in granular powder magnetic systems. *Phys. B Condens. Matter* **2007**, *389*, 311–316. [CrossRef]

© 2020 by the authors. Licensee MDPI, Basel, Switzerland. This article is an open access article distributed under the terms and conditions of the Creative Commons Attribution (CC BY) license (http://creativecommons.org/licenses/by/4.0/).

Article

Biocompatible Magnetic Fluids of Co-Doped Iron Oxide Nanoparticles with Tunable Magnetic Properties

Silvio Dutz [1,2,*], **Norbert Buske** [3], **Joachim Landers** [4], **Christine Gräfe** [5], **Heiko Wende** [4] and **Joachim H. Clement** [5]

1. Institute of Biomedical Engineering and Informatics (BMTI), Technische Universität Ilmenau, D-98693 Ilmenau, Germany
2. Department of Nano Biophotonics, Leibniz Institute of Photonic Technology (IPHT), D-07745 Jena, Germany
3. MagneticFluids, Köpenicker Landstraße 203, D-12437 Berlin, Germany; ngb.buske@gmail.com
4. Faculty of Physics and Center for Nanointegration Duisburg-Essen (CENIDE), University of Duisburg-Essen, D-47057 Duisburg, Germany; joachim.landers@uni-due.de (J.L.); heiko.wende@uni-due.de (H.W.)
5. Department Hematology and Oncology, Jena University Hospital, D-07747 Jena, Germany; christine.graefe@med.uni-jena.de (C.G.); joachim.clement@med.uni-jena.de (J.H.C.)
* Correspondence: silvio.dutz@tu-ilmenau.de; Tel.: +49-3677-691-309

Received: 29 April 2020; Accepted: 21 May 2020; Published: 27 May 2020

Abstract: Magnetite (Fe_3O_4) particles with a diameter around 10 nm have a very low coercivity (H_c) and relative remnant magnetization (M_r/M_s), which is unfavorable for magnetic fluid hyperthermia. In contrast, cobalt ferrite ($CoFe_2O_4$) particles of the same size have a very high H_c and M_r/M_s, which is magnetically too hard to obtain suitable specific heating power (SHP) in hyperthermia. For the optimization of the magnetic properties, the Fe^{2+} ions of magnetite were substituted by Co^{2+} step by step, which results in a Co doped iron oxide inverse spinel with an adjustable Fe^{2+} substitution degree in the full range of pure iron oxide up to pure cobalt ferrite. The obtained magnetic nanoparticles were characterized regarding their structural and magnetic properties as well as their cell toxicity. The pure iron oxide particles showed an average size of 8 nm, which increased up to 12 nm for the cobalt ferrite. For ferrofluids containing the prepared particles, only a limited dependence of H_c and M_r/M_s on the Co content in the particles was found, which confirms a stable dispersion of the particles within the ferrofluid. For dry particles, a strong correlation between the Co content and the resulting H_c and M_r/M_s was detected. For small substitution degrees, only a slight increase in H_c was found for the increasing Co content, whereas for a substitution of more than 10% of the Fe atoms by Co, a strong linear increase in H_c and M_r/M_s was obtained. Mössbauer spectroscopy revealed predominantly Fe^{3+} in all samples, while also verifying an ordered magnetic structure with a low to moderate surface spin canting. Relative spectral areas of Mössbauer subspectra indicated a mainly random distribution of Co^{2+} ions rather than the more pronounced octahedral site-preference of bulk $CoFe_2O_4$. Cell vitality studies confirmed no increased toxicity of the Co-doped iron oxide nanoparticles compared to the pure iron oxide ones. Magnetic heating performance was confirmed to be a function of coercivity as well. The here presented non-toxic magnetic nanoparticle system enables the tuning of the magnetic properties of the particles without a remarkable change in particles size. The found heating performance is suitable for magnetic hyperthermia application.

Keywords: cobalt ferrite; coercivity; ferrimagnetism; magnetic fluid hyperthermia; magnetic nanoparticles; magnetite

1. Introduction

Magnetic nanoparticles (MNPs) and their biocompatible suspensions (ferrofluids) are very promising materials for biomedical applications [1,2]. Because of their small size and their high surface-to-volume ratio, MNPs show properties which differ from those of the bulk material. Due to their magnetic behavior, MNPs enable a mechanical manipulation in an external field or field gradient, a magnetic determination of their location, and a heating of the particles by an external alternating magnetic field, so that they cannot only be used in therapy, but also in diagnostics and as tracer materials. A broad range of applications like in hyperthermia [3,4], in drug targeting [5] or as contrast and tracer agent for medical imaging like magnetic resonance imaging (MRI) [6] or magnetic particle imaging (MPI) [7,8] of the MNPs are described in the literature.

A lot of different MNP systems with specific cores and coatings have been developed and tested for medical application. Because of their high biocompatibility, iron oxide MNPs are of particular interest for application in medicine [9,10]. Due to a good stability against agglomeration and sedimentation as well as a high cellular uptake, a lot of research is focused on the application of superparamagnetic nanoparticles in the size range of 10–15 nm.

Unfortunately, such particles are not favorable for aimed hyperthermia heating applications, since for maximum hyperthermia heating performance, particles with a slight ferrimagnetism are needed [11,12]. An increase in particles size leads to a transition to ferrimagnetism [13] but also an increased tendency to form agglomerates. A compromise is the application of so-called multicore magnetic nanoparticles (MCNP) with superferrimagnetic behavior, which means ferrimagnetic behavior in the presence of an external magnetic field and superparamagnetic behavior in its absence, which results in a good stability against agglomeration. These particles are very promising for medical applications [14], especially for magnetic fluid hyperthermia because of their high heating performance [15,16]. Nevertheless, these cores are in the size range of several 10 nm and exceed the here biologically favored size from 10 to 15 nm.

An approach to achieve ferrimagnetism for small MNPs is the use of cobalt ferrite as magnetic core material. Unfortunately, this material shows a very pronounced hard magnetic behavior and for particles of about 10 nm, coercivities in the order of 60 kA/m occur [17]. Such coercivities are too high to obtain a therapeutically promising heating performance when using a magnetic field strength harmless for the patient. Thus, pure cobalt ferrite is not a suitable material for the preparation of MNPs in the size from 10 to 15 nm with magnetic properties promising for hyperthermia.

A possible solution to prepare MNPs in the range of 10–15 nm with a mild ferrimagnetic behavior might be Co doping the magnetic iron oxides, expecting a magnetic behavior between those of superparamagnetic iron oxide and hard magnetic cobalt ferrite in dependence of the dotation degree. In the literature, a non-monotonic dependency of coercivity on Co content for non-stoichiometric compositions due to magnetically induced anisotropy is described [18]. This was also observed for nanoparticles [19] but is not expected for MNPs obtained from the here used preparation procedure. The magnetic iron oxide magnetite (Fe_3O_4=FeO · Fe_2O_3) as well as the cobalt ferrite ($CoFe_2O_4$=CoO · Fe_2O_3) show the crystal lattices of an inverse spinel with a cubic crystal system (Fd3m), see Figure 1. The unit cell of the inverse spinel of magnetite $Fe^{2+}(Fe^{3+})_2O_4$ consists of 32 O^{2-} ions (O), eight Fe^{2+} and eight Fe^{3+} ions on the octahedral sites (Fe–O) as well as eight Fe^{3+} ions on the tetrahedral sites (Fe–T). For cobalt ferrite $Co^{2+}(Fe^{3+})_2O_4$, the tetrahedral sites are occupied again by eight Fe^{3+} ions and the octahedral sites preferably by eight Co^{2+} and eight Fe^{3+} ions. This means that magnetite and cobalt ferrite have the same crystal lattice but for cobalt ferrite the eight Fe^{2+} ions on the octahedral sites are replaced by eight Co^{2+} ions, theoretically. Since the lattice constant of magnetite (a = 8.3985 Å) is very similar to that of cobalt ferrite (a = 8.3940 Å) and the ionic radii of Fe^{2+} (78 pm) and Co^{2+} (75 pm) are comparable, it should be possible to replace the Fe^{2+} step-by-step by Co^{2+} in non-stoichiometric ratios during the preparation, to tune the magnetic properties of the obtained MNP. Similar attempts were made before to improve the magnetic properties of tapes for magnetic recording [20].

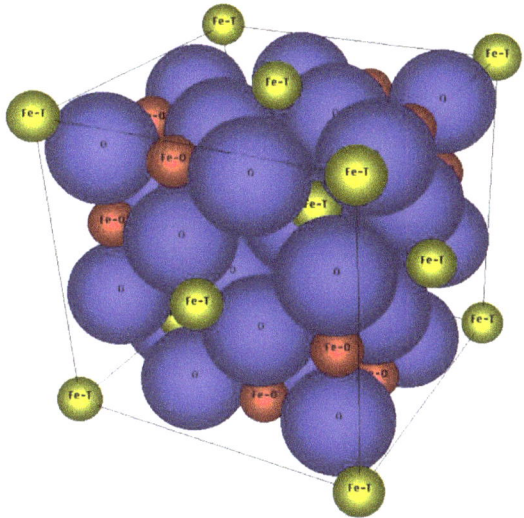

Figure 1. Unit cell of the spinel lattice of magnetite, consisting of 32 O^{2-} ions (O), 8 Fe^{2+} and 8 Fe^{3+} ions on the octahedral sites (Fe–O) as well as 8 Fe^{3+} ions on the tetrahedral sites (Fe–T); created with CrystalDesigner®.

For the experimental investigation of this hypothesis, magnetite was prepared by a co-precipitation process and during the preparation the Fe^{2+} content of the reagents was replaced by Co^{2+} in different ratios. The obtained MNPs were characterized structurally and magnetically. Furthermore, the biocompatibility of the MNPs after citric acid coating was investigated as well as their performance for magnetic fluid hyperthermia.

2. Results and Discussion

2.1. Structural Properties

In this study, magnetic nanoparticles (iron oxide/cobalt ferrite) with tunable magnetic properties were prepared. To adjust the magnetic properties (mainly the coercivity), during the synthesis of magnetite (Fe_3O_4=FeO · Fe_2O_3), the Fe^{2+} ions of magnetite were substituted with Co^{2+} step by step in a stoichiometric range from 0% to 33.3%. Theoretically, a substitution of 0% leads to pure magnetite, whereas 33.3% (one third of all Fe atoms are replaced by Co) leads to the formation of hard magnetic cobalt ferrite ($CoFe_2O_4$=CoO · Fe_2O_3) and for substitution degrees in between, Co-doped magnetite nanoparticles should be obtained. Since the preparation of the particles was performed without an exclusion of oxygen, during and after the preparation a proportion of the particle material will turn to maghemite (γ-Fe_2O_3) by further oxidation. For better understanding, the iron oxide particles will be rigorously denoted as "magnetite" in this paper.

The substitution was confirmed by means of inductively coupled plasma optical emission spectrometry (ICP–OES). For these measurements it was found that the measured Co content of the particle samples was in good accordance with the expected Co content based on the added amount of Co^{2+} ions during the preparation of the particles, see Table 1. ICP–OES determines the total proportion of elements within the particles. Thus, for a substitution of 33.3% of the Fe atoms by Co, a total amount of 25.1 wt% of Co within the MNPs was theoretically obtained. The results show that a defined amount of Co was embedded into the magnetic iron oxide particles. However, these measurements provide only information about the amount of Co within the iron oxide. A detailed investigation of the crystal lattice is needed to determine the position of the Co^{2+} within the lattice and thus, the resulting magnetic phase, see below in this section.

Table 1. Overview on the structural properties of the prepared particles: the Co substitution degree (a) during preparation, the expected Co content (theoretically) as well as the measured resulting Co content (inductively coupled plasma optical emission spectrometry (ICP–OES)), the diameter from TEM (TEM) and XRD (XRD) as well as the hydrodynamic diameter (D_h) from dynamic light scattering (DLS); SD = standard deviation.

a	Co Content		D			D_h	
	Theroretid [%]	CP-OES [%]	TEM [nm]	SD [nm]	XRD [nm]	DLS [nm]	SD [nm]
0.00	0.00	0.0	8.6	±2.1	7.6	106.2	±2.1
0.10	2.51	2.5	8.5	±2.4	7.5	94.6	±1.8
0.25	6.28	6.4	8.6	±2.0	8.2	102.1	±2.2
0.33	8.29	8.6	9.2	±2.5	8.9	97.8	±1.9
0.50	12.56	12.6	9.7	±2.8	9.0	93.4	±1.6
0.75	18.83	19.3	10.9	±2.9	10.8	106.5	±2.1
1.00	25.11	25.1	12.1	±3.1	12.4	108.3	±2.2

The investigation of the dried ferrofluid samples (citric acid-coated particles dispersed in water) by means of transmission electron microscopy (TEM) revealed an overall spherical structure for the single particles, a relatively narrow particle size distribution and a low tendency to form agglomerates for all particles with different Co contents, see Figure 2. The mean particle size for all samples was in the order of 10 nm, see Table 1. For an increasing amount of Co in the particles, a slight increase in the particle core diameter was observed. For the magnetite particles (a = 0), a mean size of 8.6 nm was determined, which increased to 9.7 nm for a Co content of 12.5% (a = 0.5) until it reached 12.1 nm for a Co content of 25% (a = 1).

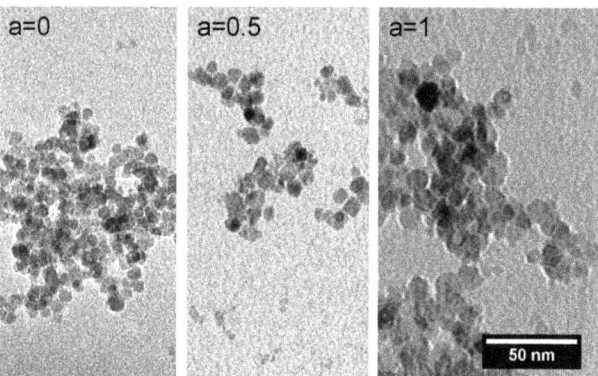

Figure 2. Typical TEM images of pure magnetite (**a** = 0) and the particles with a Co content of 12.5% (**a** = 0.5) as well as 25% (**a** = 1) confirm a slight increase in the particle size for increasing the Co content of the particles. The aggregation of the particles occurs during the sample preparation (drying of the fluid) for the TEM imaging.

The investigation of the hydrodynamic diameter of the citric acid-coated magnetic cores dispersed in water by means of dynamic light scattering (DLS) revealed for all the samples a hydrodynamic diameter of 101.3 ± 6.1 nm. No correlation between the hydrodynamic diameter and Co content was observed, see Table 1. This confirmed a weak agglomeration tendency and a good stabilization of the cores against sedimentation due to the citric acid coating.

By means of X-ray diffraction (XRD), the phase composition and the particle size was investigated. For the plain magnetite (a = 0) and for the cobalt ferrite (a = 1), the found diffraction patterns of an inverse spinel with peak patterns typical for magnetite and cobalt ferrite confirmed the formation of these magnetic phases. For all substitution degrees between 0 and 1, no remarkable changes in the diffractograms were observed, which excludes the formation of impurity phases for non-stoichiometric particles, see Figure 3A. For a = 0, the lattice constant was determined to be 8.3537 Å, which was very close to that of maghemite (8.3515 Å). In combination with a diffraction angle of 62.85° for the 440-peak of this sample, these results confirmed the magnetic phase to be maghemite for the pure iron oxide (a = 0) [21]. When a increases, the lattices constant increases up to 8.3791 Å for a = 1, which is in the order of what it is for magnetite (8.3985 Å) and cobalt ferrite (a = 8.3940 Å). Unfortunately, XRD did not allow us to distinguish between the magnetite and cobalt ferrite.

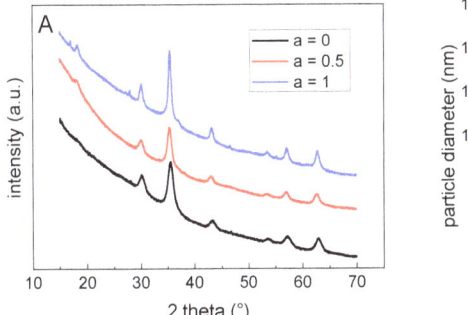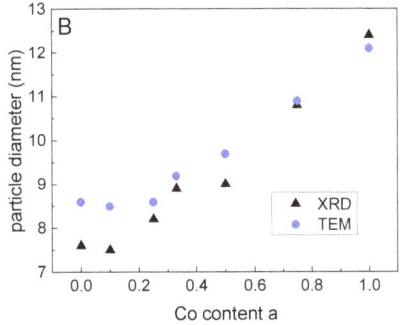

Figure 3. (**A**) X-ray diffractograms of the particles confirm a spinel structure of the particles; for better visibility the curves were shifted manually. (**B**) Particles size derived from XRD and TEM increase upon rising Co content.

By the analysis of the peak broadening of the 440-peak using the Scherrer method, the mean particle sizes were derived from the scattering volume of the particles. The here obtained values confirm the results from the TEM—an increasing Co content leads to a nearly linear slight increase in the particle diameter, see Figure 3B and Table 1.

From these results it can be concluded that non-stoichiometric Fe^{2+}/Fe^{3+} ratios during the particle preparation did not lead to a solid solution of different magnetic phases but rather to an inverse spinel structure with varying ratios of Fe^{2+} to Co^{2+} within the crystal lattice, probably on the octahedral gaps. This means a linear transition from magnetite to cobalt ferrite when changing the amount of Co^{2+} during the precipitation without remarkable changes in particle size. Since from the XRD data solely this hypothesis cannot be proved, a detailed investigation of the lattice structure of the particles, their magnetic alignment behavior and Fe-site occupation was performed via Mössbauer spectroscopy, see this section below.

Figure 4A displays the Mössbauer spectra of the $Co_aFe_{3-a}O_4$ nanoparticles (a = 0–1) recorded at 4.3 K. To resolve the individual subspectra of Fe-ions on the tetrahedral A-(blue) and on the octahedral B-sites (green), a magnetic field of 5 T was applied along the propagation direction of the incident γ-ray. These two observed contributions correspond to Fe^{3+} on the mentioned lattice sites, while for Fe^{2+} a further spectral component of higher isomeric shift and lower hyperfine magnetic field would be expected, which is not visible here [22,23]. In agreement with lattice constants from the XRD, this indicates at least the partial oxidation of the relatively small Fe_3O_4 (a = 0) nanoparticles to maghemite (γ-Fe_2O_3), although minor contents of Fe^{2+} could be overlooked due to partial superposition with the larger B-site Fe^{3+} component. To reproduce these sextet structures, narrow distributions of effective magnetic fields were used to account for the minor variations in the local Fe-surroundings and in spin canting.

As the subspectral intensities are in good approximation proportional to the number of Fe ions on the respective lattice positions, we can obtain information on the Co site occupation by studying the ratio of A- to B-site intensity R_{AB}. For a = 0, we obtained $R_{AB} \approx 0.59 \pm 0.03$, in agreement with the assumption of partially oxidized magnetite, as for magnetite twice as much B- than A-site positions are occupied by Fe ($R_{AB} = 0.5$), while in maghemite the ratio increases due to B-site vacancies ($R_{AB} = 0.6$). Unlike in magnetite and maghemite, in Co-bearing samples R_{AB} will also reflect the Co-site occupation. This is often described in terms of the so-called inversion parameter S [24], which defines the occupation of the additional metal ion rather on the tetrahedral A-site in a regular spinel ($S \approx 0$) or completely on the B-site in an inverse spinel ($S = 1$), as one would deem more likely due to the B-site preference of Co^{2+} [25]. Still, even upon increasing the Co-content, there was no significant variation in relative A- to B-site intensity, which would display a preferred B-site occupation. Instead, for a higher Co^{2+} fraction, R_{AB} slightly decreased. For a = 1.0 ($CoFe_2O_4$) we observed $R_{AB} \approx 0.53 \pm 0.05$, rather indicating a random placement of Co^{2+} on both available lattice positions instead of the strongly preferred occupation of the octahedral B-site. However, in the context of this study, this can be considered as favorable, as the random placement of the Co^{2+} ions has been repeatedly reported to lead to increased saturation magnetization as compared to the inverse spinel structure due to the higher uncompensated B-site sublattice magnetization [26,27].

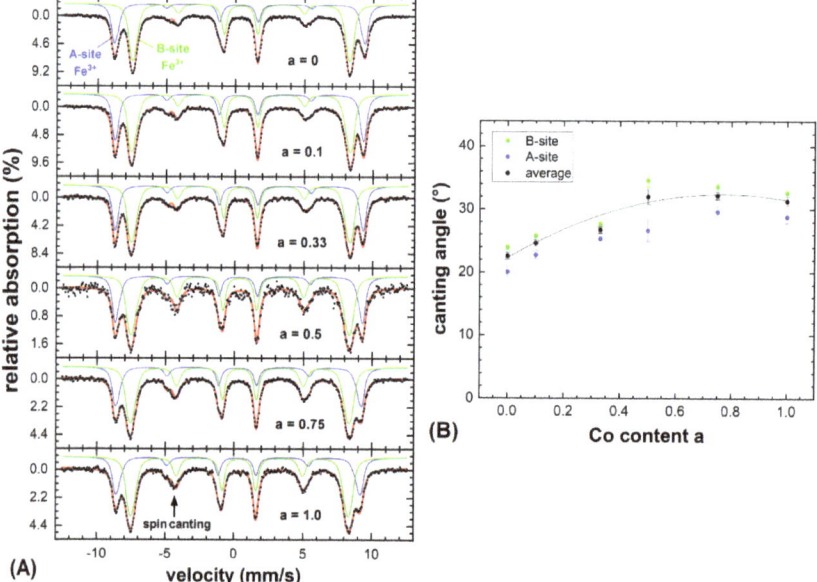

Figure 4. (**A**) Mössbauer spectra of the $Co_aFe_{3-a}O_4$ nanoparticles recorded at 4.3 K in a magnetic field of 5 T applied along the γ-ray incidence direction, composed of Fe^{3+} ions on the tetrahedral A-(blue) and the octahedral B-sites (green). Narrow hyperfine field distributions were used to reproduce the individual subspectra. (**B**) Spin canting angles extracted for the A- and B-site iron and averaged canting angles; the line serves as a guide to the eye only.

In addition to the site occupation probabilities, the Mössbauer spectroscopy provided valuable insight into the site-selective orientation of magnetic moments, as the average angle (canting angle θ) between the γ-ray—identical here to the direction of the applied magnetic field—and the Fe spins can be determined from the intensity ratio of lines 2 and 3 (respectively 5 to 4) of each spectral component following Fermi's golden rule [28]. B-site canting dominates here as often observed in spinel systems, as is expected due to antiferromagnetic B–B superexchange in addition to the superior ferrimagnetic structure [29,30]. More surprisingly, Figure 4B displays an increase in θ upon rising Co content, with a maximum ca. at $a = 0.6$–0.8. While on the one hand smaller nanoparticles exhibit an increasingly stronger spin frustration due to their higher specific surface, on the other hand the higher magnetocrystalline anisotropy of cobalt ferrite as compared to the magnetite often results in stronger spin canting in samples comparable in size and crystallinity. Therefore, this observed trend in θ could be explained by more pronounced surface spin canting in smaller particles for $a \approx 0$ due to their higher specific surface [31] and the higher magnetocrystalline anisotropy of $CoFe_2O_4$ for a approaching 1.0, with both contributions in superposition resulting in a maximum spin frustration at intermediate stoichiometries.

All in all, the Mössbauer spectroscopy indicated primarily Fe^{3+} ions in the six studied particle systems, showing a moderate spin canting and a minor B-site preference of the Co^{2+} ions.

2.2. Magnetic Properties

The quasistatic magnetic properties of the powders of the dried uncoated particles as well as ferrofluids of the citric acid-coated magnetic cores were determined by means of vibrating sample magnetometry (VSM).

The saturation magnetization (M_S) of the uncoated cores (measured at $H = 1275$ kA/m) was determined to be in the range of 42 to 62 Am^2/kg, which represents typical values for nanoparticles of magnetite and cobalt ferrite in this size range [32]. An increasing M_S for increasing Co content was observed, see Figure 5 and Table 2. Since magnetite and cobalt ferrite have a similar bulk saturation magnetization of about 80 Am^2/kg [33], the transition of magnetite to cobalt ferrite might not be the reason for the increasing M_S for higher Co contents. Possible reasons for this behavior might be the occurrence of maghemite (with a bulk saturation magnetization of about 60 Am^2/kg) for lower Co content and the experimentally found increasing particles size for higher Co content, see Table 1. Thus, for the latter, for a lower Co content, smaller particles with a higher surface-to-volume ratio result. Assuming a higher ratio of non-magnetic dead layer on the surface of the particles to the total particles volume for smaller particles, smaller particles show a lower net saturation magnetization. While Mössbauer spectroscopy indicates higher spin frustration for intermediate to high Co content, these findings cannot be directly compared due to different magnetic field amplitude they were obtained at.

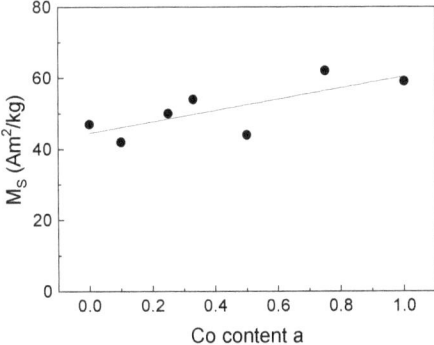

Figure 5. Saturation magnetization (M_S) of powders of bare magnetic particles as function of Co content of the partices, measured by means of VSM at $H = 1275$ kA/m, the line serves as a guide to the eye only.

Table 2. Overview on the magnetic properties of the prepared particles with varying Co content: the saturation magnetization (M_S), the coercivity (H_C), the relative remanence (M_R/M_S) as well as the specific heating power (SHP) for different magnetic fields strengths during hyperthermia experiments at 235 kHz on particles immobilized in gelatine.

a	Co [%]	M_S [Am²/kg]	H_C [kA/m]	M_R/M_S	SHP		
					10 kA/m [W/g]	20 kA/m [W/g]	30 kA/m [W/g]
0.00	0.0	47	0.2	0.00	7	30	33
0.10	2.5	42	2.1	0.03	16	56	64
0.25	6.4	50	2.2	0.03	38	125	172
0.33	8.6	54	3.0	0.04	33	149	231
0.50	12.6	44	11.9	0.10	8	53	355
0.75	19.3	62	49.9	0.27	8	37	151
1.00	25.1	59	84.9	0.42	9	35	72

Figure 6 shows minor loops of hysteresis curves of pure magnetite (a = 0) and particles with a Co content of 12.5% (a = 0.5) as well as 25% (a = 1), normalized to the saturation magnetization of each curve. It becomes obvious that an increasing Co content led to a more pronounced ferrimagnetic behavior (starting from a superparamagnetic behavior for the pure magnetite particles) and the coercivity H_C varied in a wide range. From Figure 7A,B it can be seen, that for the powder samples, a continuous increase in H_C took place for the increasing Co content of the particles. This means that H_C can be tuned in a wide range for this magnetic nanoparticle system by varying the Co content of the particles, see Table 2. This is especially true, as seen in Figure 7B, as where H_C is plotted on a logarithmic scale. It can be seen that even a very small amount (2.5%) of Co within the particles leads to a significant increase in H_C and a resulting blocked magnetic behavior at room temperature.

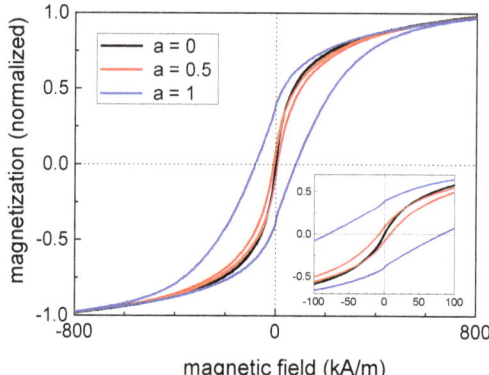

Figure 6. Normalized minor loops of pure magnetite (**a = 0**) and particles with a Co content of 12.5% (**a = 0.5**) as well as 25% (**a = 1**) recorded at room temperature confirm a significant increase in the coercivity for increasing Co content of the particles. The inset depicts a higher magnification of the hysteresis curves at low magnetic field strengths.

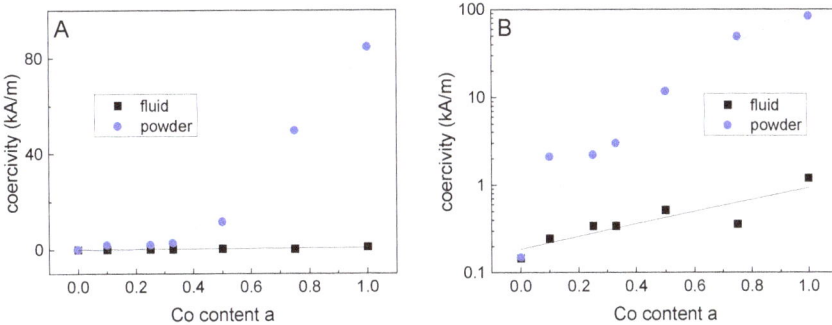

Figure 7. Coercivity of the powder samples (blue spheres) and the fluidic samples (black squares) as a function of the Co content, plotted on (**A**) linear and (**B**) logarithmic scales; the lines serve as a guide to the eye only.

For the VSM measurements on liquid ferrofluid samples it was found that H_C shows only a very weak increase for increasing Co content, see the black squares in Figure 7A,B. This confirms the good colloidal stability of the fluids without particle agglomerates, as already found by DLS. The slight increase in H_C for the fluidic samples with increasing Co content might be attributed to the increasing particles size for higher Co contents. This size increase led to longer Brown relaxation times [34], resulting in a higher H_C for the quasistatic measurements of hysteresis curves by means of VSM. The relative remanence M_R/M_S showed a dependency on the Co content similar to that found for the H_C (see Table 2), as described already before for other particles systems in the transition range from superparamagnetic to ferrimagnetic behavior [13].

The magnetic heating performance of the particles for hyperthermia application was investigated by a magnetic field calorimeter for field strengths of 10, 20 and 30 kA/m for a field frequency of 235 kHz. Since in animal investigations it was found that the particles were immobilized to the tumor tissue immediately after injection [15], measurements were performed on particles immobilized in gelatine.

In good accordance with the theory [35–37], the specific heating power (SHP) increased with increasing field strength in our measurements, see Figure 8. For a lower field strength (10 kA/m), a maximum SHP of 38 W/g was obtained for the sample with a H_C of 2.2 kA/m (a = 0.25/Co = 6.4%). In the medium field (20 kA/m), the maximum SHP is 149 W/g for the sample with a H_C of 3.0 kA/m (a = 0.33 (Co = 8.6%) and for the highest investigated field strength of 30 kA/m, the maximum SHP of 355 W/g was determined for the sample with a H_C of 11.9 kA/m (a = 0.5/Co = 12.6%), see Table 2.

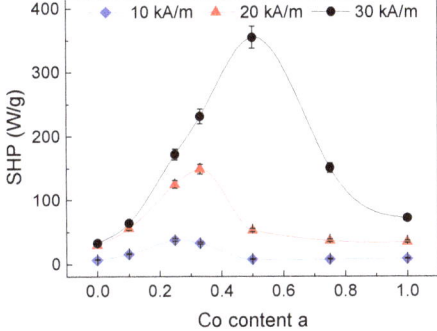

Figure 8. SHP of the immobilized particles for three different field strengths (10, 20, and 30 kA/m); the lines serve as a guide to the eye only, for some data points error bars are within the symbols.

This behavior can be explained by the pairing of hysteresis losses of the magnetization curve (for a saturation magnetic field strength) and the applied field to reverse the magnetization of the particles within the alternating magnetic field generator. As shown before [13], the H_C can serve as a measure to estimate the hysteresis losses of a particle ensemble. For a certain magnetic field strength, the reversal losses increase with increasing H_C due to a higher area of the hysteresis curve until the losses reach a maximum. Exceeding this maximum, a higher H_C led to lower losses, since the magnetic field strength was too low to reverse the magnetization of the particle ensemble completely. Due to a (possible) particle size distribution, only a fraction of the particles ensemble will show reversal losses [11,12] and the effective net losses decrease for a further increasing H_C. This means, for each applied field strength during hyperthermia, particles with optimum losses (H_C) have to be chosen to obtain maximum heating.

For the here presented particle system, the heating performance of the immobilized particles was in a range which was suitable for magnetic hyperthermia treatment. The magnetic heating performance of the particles could be optimized/adapted for the used magnetic field parameters by controlling the Co content (and thus the H_C) of the particles. This means that the heating performance of the particles can be tuned without a considerable change in particle size.

2.3. Cell Viability Analysis

Cell viability analysis enables a statement on the potential cytotoxic effects of the tested MNPs. Human brain microvascular endothelial cells (HBMECs) are a prevalent cell line representing the human blood–brain barrier [38]. In previous studies we investigated their interactions with MNPs where the cells showed a pronounced sensitivity for e.g., polyethylene imine-coated magnetic nanoparticles [39,40]. To study the effect of cobalt ferrite MNPs on human cells, HBMEC cultures were incubated with MNPs in a range of 5 µg/cm^2 –100 µg/cm^2 for 24 h (Figure 9). Cell viability testing was performed with the PrestoBlue reagent, which was demonstrated to allow a sensitive and robust analysis [39,41]. The fluorescence intensity of the reaction product resorufin of the MNP-treated samples was related to the untreated control (U). All MNPs independent of cobalt ion content exhibited a moderate concentration-dependent effect on HBMECs with at least 71.0% ± 4.1% cell viability up to a concentration of 50 µg/cm^2. However, the highest MNP concentration (100 µg/cm^2 = 388.8 µg/mL) severely affected the cells and reduced viability to 41.2% ± 6.9% for a = 1.

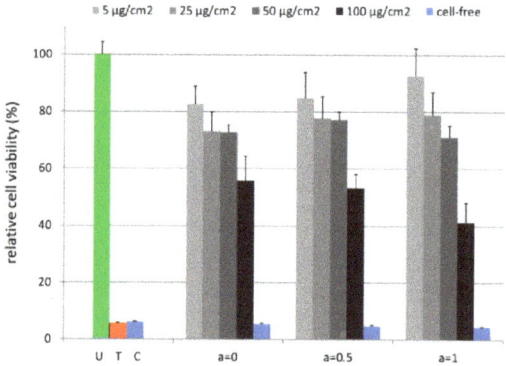

Figure 9. Magnetic nanoparticles (MNPs) with varying Co contents have a minor effect on the cell viability up to a concentration of 50 µg/cm^2. The PrestoBlue Cell Viability Reagent was added after a 24 h incubation of the HBMEC samples. Cell-free samples contain 100 µg/cm^2 MNPs. The 0.1% (v/v) Triton X 100 serves as a toxic control (T). U: untreated HBMEC control; C: cell-free medium control. n = 4–8 technical replicates.

A reason for this observation might be the interference of the MNPs with the PrestoBlue reagent-based assay. It is reported for several cell viability assays that they might interfere with the investigated nanomaterials themselves, with assay components or with reaction products [39,41–44]. Therefore, we carried along cell-free controls supplemented with 100 µg/cm^2 of MNPs. These samples did not exhibit enhanced fluorescence signals compared to the cell-free controls without MNPs (cell-free medium control = C) (Figure 9). To prove whether MNPs interfered with the PrestoBlue reagent product resorufin we measured the cell viability of the untreated cell cultures before and after supplementation with 100 µg/cm^2 of MNPs. These analyses showed a marked decrease in the fluorescence intensity after the addition of MNPs. This pointed clearly to an interaction of the fluorescent reaction product resorufin with the MNPs and thus a masking of the fluorescence signal (Figure S1). In consequence, the cell viability may be underestimated at least for a MNP concentration of 100 µg/cm^2 and especially for a = 1.

Similar drawbacks are reported for the 3-(4,5-Dimethylthiazol-2-yl)-2,5-diphenyltetrazolium bromide (MTT) assay which is frequently used to determine the cytotoxicity of nanomaterials, e.g., cell lines release the MTT formazane by exocytosis under nanoparticle incubation [45] or some nanoparticles absorb fluorescence like carbon black [46]. In general, cell viability evaluation in vitro depends on various parameters. The characteristics of the nanomaterial examined including its shape, composition and charge have to be considered as well as the concentration applied and the duration of the treatment. Furthermore, the in vitro test system like the cell line used or the incubation medium composition has to be taken into account. Moreover, the assay characteristics like the ingredients and the measuring principle play a role [47]. Therefore, it is advisable to evaluate the effects of nanomaterials on cell viability with independent assay systems.

It has to be emphasized that for the medical application of MNPs, typically a concentration of up to 25 µg/cm^2 (97.2 µg/mL) was used. In this range, the herein tested MNPs showed only mild cytotoxic effects. Comparing for each single concentration the cell viability for different a-values, it can be seen that a higher Co content did not lead to increasing cytotoxic effects for concentrations up to 50 µg/cm^2. Only when the concentration exceeded 50 µg/cm^2 a higher Co content lead to stronger toxic effects. However, for such concentrations, even the pure iron oxide (a = 0) shows cytotoxicity.

In order to verify the observed effects of MNPs on cell viability with an independent approach, real-time cell analysis (RTCA) was performed. This label-free and impedance-based technique monitors the biological condition of cells, e.g., the cell number, viability and the adhesion degree [48]. RTCA avoids the interference of MNPs with dyes and allows the monitoring of the dynamic responses of the cells in real-time [49,50].

In our experiments, the MNPs were added 24 h after seeding the HBMECs into E16 plates. The development of the cell culture under treatment was monitored up to 84 h which means an observation period of 60 h. Figure 10 shows the results of RTCA for a = 0 (**A**), a = 0.5 (**B**) and a = 1 (**C**). The data were normalized to the time point of the MNP addition (24 h) and denoted as the normalized cell index. The untreated control showed a steady increase during the observation period. The addition of 25 µg/cm^2 of the cobalt ferrite MNPs did not affect the progress of the cell culture. Nevertheless, it has to be mentioned that the normalized cell index was below the untreated control. Cell cultures incubated with 100 µg/cm^2 MNPs showed a tendency towards a reduction of the normalized cell index after 60 h of treatment. The MNPs per se had no effect on the measurement over the whole period of time (see cell-free samples). Thus, the RTCA supported the idea that cobalt ferrite MNPs do not sustainably affect cell viability up to a concentration of 50 µg/cm^2.

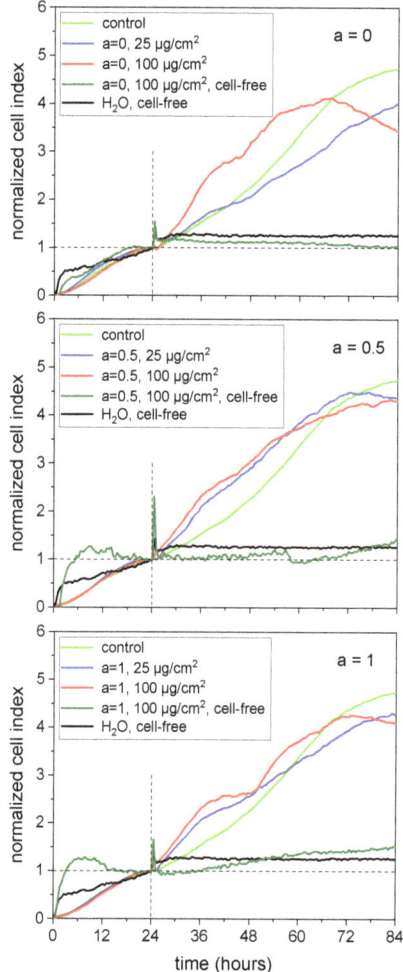

Figure 10. Real-time cell analysis confirmed no reduced viability of human brain microvascular endothelial cells (HBMECs) within 60 h for particles with an increasing Co content in comparison to the untreated control (green line); MNPs were added after 24 h which were indicated by the vertical dashed line. The basis of the normalized cell index was the 24 h value of the samples, respectively (horizontal dashed line). $n = 2$ technical replicates.

The results obtained from the PrestoBlue assay and RTCA were confirmed by vital fluorescent staining microscopy. For that, the HBMECs were incubated with 25, 50 and 100 µg/cm² MNPs for 24 h. In vital cells, the non-fluorescent calcein AM is converted into green-fluorescent calcein by intracellular esterases. Ethidium-homodimer-1 is only able to enter cells when the cell membrane is corrupted, thus labelling dead or damaged cells which are prone to die. As demonstrated in Figure 11, neither iron oxide MNPs (a = 0) nor Co-doped MNPs (a = 0.5 and a = 1) affect the viability of HBMEC. Only a few dead cells are detectable. The ratio of these cells is comparable to the untreated control. Thus, no elevated toxicity can be observed when the iron oxide MNPs are doped with cobalt ions.

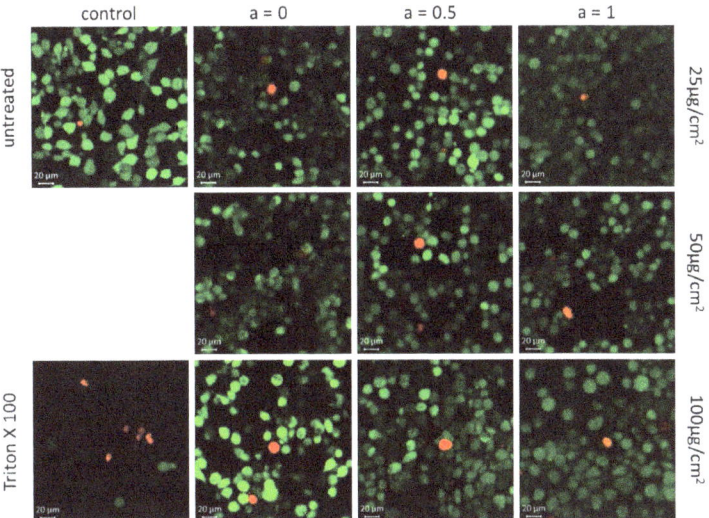

Figure 11. Vital fluorescent staining confirmed no reduced cell viability within 24 h for particles with an increasing Co content; the living cells and dead cells are colored in green and red, respectively. Magnification × 400.

Co–ferrite MNPs were investigated from several groups for their potential cytotoxicity with diverse results [47]. Horev-Azaria and colleagues could demonstrate using several cell lines that Co–ferrite MNPs exhibited a concentration-, cell-line- and duration-dependent cytotoxic effect [51]. They calculated an overall threshold of 200 µM for all cell lines, which was about 13 µg/cm² in our system. Ansari et al. used a modified MTT assay and reported that nanosized cobalt ferrites exhibited only a minor toxic effect in a breast cancer cell line up to a concentration of 300 µg/mL [52].

Resulting from the cell viability experiments with citric acid-coated MNPs it can be concluded, that for concentrations suitable for medical applications, no increased toxicity in our in vitro model was observed for increasing the Co content of the MNPs. The cobalt ferrite particles show almost the same non-toxic behavior as the iron oxide particles, which are approved for medical application.

3. Materials and Methods

3.1. Particle and Ferrofluid Preparation

The magnetic nanoparticles were prepared by co-precipitation in an alkaline media applying a modified protocol of the standard preparation procedure for such particles [53]. For tuning the magnetic properties, the Fe^{2+} ions of magnetite were substituted by Co^{2+} step by step in our study which resulted in a Co-doped inverse spinel lattice with an adjustable Fe^{2+} substitution degree, see Formula (1):

$$(Co^{2+})_a + (Fe^{2+})_{1-a} + (Fe^{3+})_2 + (OH^-)_8 \rightarrow Co_a^{2+} Fe_{1-a}^{2+} Fe_2^{3+} O_4 + 4H_2O \text{ with } a = 0\ldots 1 \quad (1)$$

Formula (1). Reaction equation for the preparation of Co-doped inverse spinels.

A stoichiometric substitution of Fe^{2+} by Co^{2+} of 0% (a = 0) leads to the formation of pure magnetite particles and at a substitution degree of 33.3% (a = 1), one third of all Fe^{2+} ions are replaced by Co^{2+} and pure $CoFe_2O_4$ results.

The particles were prepared from Co^{2+}, Fe^{2+}, and Fe^{3+} chloride mixtures (0.02 molar) at different "a" values by co-precipitation with sodium hydroxide under stirring at 100 °C for a 90 min reaction time. The obtained particles were washed with distilled water using a magnetic separation technique and stabilized in water with hydrochloric acid (intermediate with positively charged particles).

Subsequently, the particles were coated with citric acid, which resulted in negatively charged MNPs at pH7. Finally, the citric acid-coated MNPs were filtrated by a 0.8 µm filter to remove potentially existing agglomerates.

3.2. Structural Characterization

ICP–OES: The elemental composition of the prepared nanoparticles was determined by means of inductively coupled plasma optical emission spectrometry (ICP–OES; Agilent 725 ICP–OES Spectrometer, Agilent Technologies). For the ICP–OES measurement, the nanoparticles were precipitated and dissolved in aqua regia.

TEM: For TEM from aqueous solutions, copper grids were rendered hydrophilic by argon plasma cleaning for 30 s. Then, 10 µL of the respective sample solution was applied to the grid, and excess sample was blotted with a filter paper. TEM images were acquired with a JEM 2010FEF (JEOL, Japan).

XRD: The inner structure of the magnetic nanoparticles was investigated by means of X-ray diffraction (XRD, Panalytical X'pert Pro, Malvern Panalytical, Almelo/The Netherlands). The results of the XRD investigations provided information about the magnetic phase composition and the mean size of the magnetic cores. The size of the cores was calculated from measurements of the XRD line width by using the Scherrer formula. The lattice constant was determined by means of Rietveld analysis.

DLS: The hydrodynamic diameters (dh) of the coated cores were determined by using dynamic light scattering (DLS, Zetasizer nano ZS, Malvern Instruments, Malvern, UK). Before the measurement, samples were diluted in the ratio 1:30 with distilled water and treated in an ultrasonic bath. For size measurements, the z-average of the intensity weighted normalization was used. All measurements were performed in 3 consecutive runs and the obtained values were averaged.

Mössbauer spectroscopy: Mössbauer spectra of the nanoparticle samples were recorded in transmission geometry and constant acceleration mode with a ^{57}Co(Rh) source. Spectra at 4.3 K were measured in a custom-built Thor Cryogenics l-He cryostat containing a setup in split-coil geometry, providing a field of 5 T along the incidence direction of the γ-ray. Individual subspectra were reproduced by using a narrow distribution of effective magnetic fields for the A- and B-sites using the "Pi" program's Mössbauer fitting routine [54].

3.3. Magnetic Characterization

VSM: The magnetic properties were measured at room temperature by using vibrating sample magnetometry (VSM; Micromag 3900, Princeton Measurement Systems, USA). The measurements were performed on liquid samples or dried powders. The overall magnetic behavior of the samples was derived from magnetization, coercivity and relative remanence measured at H = 1275 kA/m. The concentration of MNPs in the fluid was calculated from the obtained saturation magnetization of the fluid samples.

SHP: The specific heating power (SHP) was measured by means of magnetic field calorimetry at field amplitudes of 10, 20 and 30 kA/m and at a frequency of 235 kHz as described before [55]. The SHP of the immobilized particles was determined for particles dispersed in a gelatin gel. It was shown in previous investigations that this method yields a strong immobilization of the particles. For the measurement of the heating curves, the particle suspensions (1 mL in a 2 mL Eppendorf tube) were thermally isolated in a PUR foam block and the temperature measurements were performed by a fiber optic device (Fotemp, OPTOcon, Dresden, Germany). The MNP concentration of the samples for SHP measurements was determined by measuring the specific saturation magnetization using the VSM and assuming the specific magnetization values which were obtained from the powder samples.

3.4. Cell Viability Analysis

Cell culture: The biocompatibility of the prepared samples was studied utilizing human brain microvascular endothelial cells (HBMECs). The HBMECs were cultivated at 37 °C and 5% CO_2 in RPMI 1640 + GlutaMAX™ I (Invitrogen, Karlsruhe, Germany) supplied with 10% (*v/v*) fetal bovine serum (FBS; Biochrom, Berlin, Germany), 100 U/mL penicillin and 0.1 mg/mL streptomycin (Life Technologies, Carlsbad, CA, USA).

PrestoBlue Assay: The cell viability was investigated by using the PrestoBlue Cell Viability Reagent (Invitrogen, Karlsruhe, Germany). The assay was based on the reduction of a non-fluorescent resazurin-based reagent to fluorescent resorufin by metabolically active cells. First, the cells were seeded into black-walled 96-well plates (μ-Clear, F-bottom, Greiner Bio-One, Frickenhausen, Germany) and cultivated for 24 h. The subconfluent cell cultures were incubated with particle concentrations between 5 and 100 μg/cm^2 (which equaled 19.4 to 388.9 μg (MNPs)/mL) in at least quadruplicate. These values had a mean total Co concentration from 2.4 to 48.6 μg/mL in the case of a = 0.5 and from 4.9 to 97.2 μg/mL in the case of a = 1.0. For the control, sterile water was added (untreated control) or the detergent Triton X 100 (0.1%(v/v)). After 24 h of nanoparticle incubation the PrestoBlue reagent was added and the incubation was continued for a further 30 min at 37 °C. After magnetic separation of the MNP towards the outer walls of the wells and an excitation of the cells with light of 550 nm (9 nm bandwidth), a fluorescence signal at 600 nm (20 nm bandwidth) was detected by using the microplate reader Infinite M200 Pro (Tecan, Crailsheim, Germany). By measuring cell-free wells with added particles, the particle-associated auto fluorescence effect was measured and used for the correction of the cell measurements. Similarly, quenching effects were considered by fluorescence measurements immediately before and after the addition of particles to the cells in the wells. From these investigations, the cell survival rate (which was an indicator for the cell toxicity of the particles) was derived for the different particle concentrations at 24 h after the start of particle–cell incubation.

RTCA: The long-term viability of incubated HBMECs was investigated by means of real-time cell analysis (RTCA) using the xCELLigence system by ACEA Biosciences (San Diego, CA, USA) as described previously [40]. For this, 10,000 cells/cm^2 were seeded in each well of a 16 well E plate. The cells were monitored in real time at 37 °C in a humidified atmosphere with 5% CO_2 after a sedimentation waiting step of 30 min. After 24 h cultivation, the particles (25 μg/cm^2 and 100 μg/cm^2) were added to the cells and mixed carefully. RTCA measurements were continued 60 h after adding the MNPs. Plain incubation media served as untreated controls.

Vital fluorescent staining microscopy: The influence of the here prepared MNPs on the viability of the cells was examined by determining the ratio of living cells to dead cells using LIVE/DEAD viability/cytotoxicity fluorescent staining microscopy as described previously [40]. The HBMECs were seeded on glass cover slips (12 mm diameter, Menzel, Braunschweig, Germany) in 24-well tissue culture plates (76,000 cells/cm^2, Greiner Bio-One). After overnight cultivation, the cell cultures were incubated with 25, 50 or 100 μg/cm^2 of the MNP for 24 h, respectively. The cells were washed with PBS and 100 μL of an ethidium homodimer 1/calcein AM mix (0.3 μM calcein AM and 0.7 μM ethidium homodimer 1 in PBS) (Thermo Scientific, Waltham, MA, USA) was added for 3 h. Calcein (excitation: 494 nm, emission: 517 nm) and ethidium homodimer-1 (excitation: 528 nm, emission 617 nm) fluorescence was measured using the confocal laser scanning microscope LSM 510 META (Carl Zeiss Microscopy GmbH, Jena, Germany).

4. Conclusions

In the here presented study, magnetic Co-doped iron oxide nanoparticles with tunable magnetic properties were prepared. For the Co-doping, Fe^{2+} ions were replaced by Co^{2+} in the educts in non-stoichiometric ratios during the preparation, leading to the formation of pure iron oxide, pure cobalt ferrite and non-stoichiometric intermediates between both. The Co content of the obtained particles was in good accordance with the Co amount in the educts. The determination of the core size of the particles revealed a mean particle size in the range of 8–12 nm with the slight tendency of increasing particle sizes for higher Co contents. The measurement of the hydrodynamic diameter of the particles after the citric acid coating showed nearly the same behavior for all the investigated particles ($D_h \approx$ 100 nm), which confirmed that the particles showed a low agglomeration tendency which was not influenced by the Co content. The Mössbauer spectroscopy revealed predominantly

Fe^{3+} in all the samples, while also verifying an ordered magnetic structure with low to moderate surface spin canting. Relative spectral areas of the Mössbauer subspectra indicated a mainly random distribution of Co^{2+} ions rather than the more pronounced octahedral site-preference of bulk cobalt ferrite. The magnetic properties of the obtained particles at room temperature are determined by the Co content. An increasing Co content leads to a linear increase in saturation magnetization in the range of 42–62 Am^2/kg. The coercivity and relative remanence increased in a non-linear manner with increasing Co content. The higher coercivity for the particles with a higher Co content cannot be explained by the larger diameter of those particles but rather by the increasing crystal anisotropy due to the doping. A clear correlation between the coercivity and the magnetic heating performance for the different magnetic field strengths was found. The SHP values for the most promising particle–field combination seem suitable for application in magnetic fluid hyperthermia. By means of cell viability analysis utilizing different independent assays and procedures, no cytotoxic effects were observed for all prepared coated MNPs (all different Co content cores with citric acid coating) up to 50 $\mu g/cm^2$. It is noteworthy that no increased toxicity was found for the Co-doped particles compared to the pure iron oxide ones, which are approved for medical applications.

In summary, we prepared biocompatible ferrofluids consisting of magnetic nanoparticles suitable for application in magnetic hyperthermia. The magnetic properties can be tuned by doping with Co in a wide range without a significant influence on the particles size. This enables an adaption of the magnetic properties of the used particles to the field parameters of the hyperthermia setup, resulting in the optimized heating performance without any change in particles size. In ongoing work, we focus on the investigation of the doping range where the transition from superparamagnetism to the blocked ferrimagnetic nanoparticle state occurs.

Supplementary Materials: The following are available online at http://www.mdpi.com/2079-4991/10/6/1019/s1, Figure S1: Co–ferrite MNPs interact with the fluorescent resorufin.

Author Contributions: S.D. and N.B. conceived and designed the experiments; S.D.; N.B.; J.L.; H.W.; C.G.; and J.H.C. performed the experiments; all authors analyzed the data and contributed equally to the writing of the manuscript. All authors have read and agreed to the published version of the manuscript.

Funding: This research was funded by DEUTSCHE FORSCHUNGSGEMEINSCHAFT (DFG) via SPP 1681, grant numbers CL202/3-2, DU1293/7-3, and WE2623/7.

Acknowledgments: The authors thank C. Schmidt (IPHT Jena) for the XRD measurements, I. Appel (KIT Karlsruhe) for ICP–OES measurements, and C. Jörke for the expert technical assistance. We acknowledge support for the Article Processing Charge by the Thuringian Ministry for Economic Affairs, Science and Digital Society and the Open Access Publication Fund of the Technische Universität Ilmenau.

Conflicts of Interest: The authors declare no conflict of interest.

References

1. Cardoso, V.F.; Francesko, A.; Ribeiro, C.; Banobre-Lopez, M.; Martins, P.; Lanceros-Mendez, S. Advances in Magnetic Nanoparticles for Biomedical Applications. *Adv. Healthc. Mater.* **2018**, *7*. [CrossRef] [PubMed]
2. Krishnan, K.M. Biomedical Nanomagnetics: A Spin Through Possibilities in Imaging, Diagnostics, and Therapy. *IEEE Trans. Magn.* **2010**, *46*, 2523–2558. [CrossRef]
3. Dutz, S.; Hergt, R. Magnetic particle hyperthermia—a promising tumour therapy? *Nanotechnology* **2014**, *25*, 452001. [CrossRef] [PubMed]
4. Maier-Hauff, K.; Ulrich, F.; Nestler, D.; Niehoff, H.; Wust, P.; Thiesen, B.; Orawa, H.; Budach, V.; Jordan, A. Efficacy and safety of intratumoral thermotherapy using magnetic iron-oxide nanoparticles combined with external beam radiotherapy on patients with recurrent glioblastoma multiforme. *J. Neurooncol.* **2011**, *103*, 317–324. [CrossRef] [PubMed]
5. Lubbe, A.S.; Bergemann, C.; Riess, H.; Schriever, F.; Reichardt, P.; Possinger, K.; Matthias, M.; Dorken, B.; Herrmann, F.; Gurtler, R.; et al. Clinical experiences with magnetic drag targeting: A phase I study with 4'-epidoxorubicin in 14 patients with advanced solid tumors. *Cancer Res.* **1996**, *56*, 4686–4693. [PubMed]
6. Estelrich, J.; Sanchez-Martin, M.J.; Busquets, M.A. Nanoparticles in magnetic resonance imaging: From simple to dual contrast agents. *Int. J. Nanomed.* **2015**, *10*, 1727–1741. [CrossRef]

7. Gleich, B.; Weizenecker, R. Tomographic imaging using the nonlinear response of magnetic particles. *Nature* **2005**, *435*, 1214–1217. [CrossRef]
8. Panagiotopoulos, N.; Duschka, R.L.; Ahlborg, M.; Bringout, G.; Debbeler, C.; Graeser, M.; Kaethner, C.; Ludtke-Buzug, K.; Medimagh, H.; Stelzner, J.; et al. Magnetic particle imaging: Current developments and future directions. *Int. J. Nanomed.* **2015**, *10*, 3097–3114. [CrossRef]
9. Dadfar, S.M.; Roemhild, K.; Drude, N.I.; von Stillfried, S.; Knuchel, R.; Kiessling, F.; Lammers, T. Iron oxide nanoparticles: Diagnostic, therapeutic and theranostic applications. *Adv. Drug Deliv. Rev.* **2019**, *138*, 302–325. [CrossRef]
10. Wu, W.; Wu, Z.H.; Yu, T.; Jiang, C.Z.; Kim, W.S. Recent progress on magnetic iron oxide nanoparticles: Synthesis, surface functional strategies and biomedical applications. *Sci. Technol. Adv. Mater.* **2015**, *16*, 023501. [CrossRef]
11. Hergt, R.; Dutz, S.; Roeder, M. Effects of size distribution on hysteresis losses of magnetic nanoparticles for hyperthermia. *J. Phys. Condens. Matter* **2008**, *20*, 385214. [CrossRef]
12. Hergt, R.; Dutz, S.; Zeisberger, M. Validity limits of the Néel relaxation model of magnetic nanoparticles for hyperthermia. *Nanotechnology* **2010**, *21*, 015706. [CrossRef] [PubMed]
13. Dutz, S.; Hergt, R.; Muerbe, J.; Mueller, R.; Zeisberger, M.; Andrae, W.; Toepfer, J.; Bellemann, M.E. Hysteresis losses of magnetic nanoparticle powders in the single domain size range. *J. Magn. Magn. Mater.* **2007**, *308*, 305–312. [CrossRef]
14. Dutz, S. Are Magnetic Multicore Nanoparticles Promising Candidates for Biomedical Applications? *IEEE Trans. Magn.* **2016**, *52*, 1–3. [CrossRef]
15. Dutz, S.; Kettering, M.; Hilger, I.; Muller, R.; Zeisberger, M. Magnetic multicore nanoparticles for hyperthermia-influence of particle immobilization in tumour tissue on magnetic properties. *Nanotechnology* **2011**, *22*, 265102. [CrossRef] [PubMed]
16. Kratz, H.; Taupitz, M.; de Schellenberger, A.A.; Kosch, O.; Eberbeck, D.; Wagner, S.; Trahms, L.; Hamm, B.; Schnorr, J. Novel magnetic multicore nanoparticles designed for MPI and other biomedical applications: From synthesis to first in vivo studies. *PLoS ONE* **2018**, *13*, e0190214. [CrossRef] [PubMed]
17. Dai, Q.; Lam, M.; Swanson, S.; Yu, R.H.R.; Milliron, D.J.; Topuria, T.; Jubert, P.O.; Nelson, A. Monodisperse Cobalt Ferrite Nanomagnets with Uniform Silica Coatings. *Langmuir* **2010**, *26*, 17546–17551. [CrossRef] [PubMed]
18. Smit, J.; Wijn, H.P.J. *Ferrites*; Philips Technical Library: Eindhoven, The Netherlands, 1959; pp. 69–136.
19. Muller, R.; Schuppel, W. Co spinel ferrite powders prepared by glass crystallization. *J. Magn. Magn. Mater.* **1996**, *155*, 110–112. [CrossRef]
20. Flanders, P.J. Impact-Induced Demagnetization, Magnetostriction and Coercive Force in CO-Doped Iron-Oxide Recording Tapes. *IEEE Trans. Magn.* **1976**, *12*, 770–772. [CrossRef]
21. Dutz, S.; Hergt, R.; Murbe, J.; Topfer, J.; Muller, R.; Zeisberger, M.; Andra, W.; Bellemann, M.E. Magnetic nanoparticles for biomedical heating applications. *Z. Phys. Chem.* **2006**, *220*, 145–151. [CrossRef]
22. Berry, F.J.; Skinner, S.; Thomas, M.F. Fe-57 Mossbauer spectroscopic examination of a single crystal of Fe_3O_4. *J. Phys. Condens. Matter* **1998**, *10*, 215–220. [CrossRef]
23. Landers, J.; Stromberg, F.; Darbandi, M.; Schoppner, C.; Keune, W.; Wende, H. Correlation of superparamagnetic relaxation with magnetic dipole interaction in capped iron-oxide nanoparticles. *J. Phys. Condens. Matter* **2015**, *27*, 026002. [CrossRef] [PubMed]
24. Sickafus, K.E.; Wills, J.M.; Grimes, N.W. Structure of spinel. *J. Am. Ceram. Soc.* **1999**, *82*, 3279–3292. [CrossRef]
25. Hou, Y.H.; Zhao, Y.J.; Liu, Z.W.; Yu, H.Y.; Zhong, X.C.; Qiu, W.Q.; Zeng, D.C.; Wen, L.S. Structural, electronic and magnetic properties of partially inverse spinel $CoFe_2O_4$: A first-principles study. *J. Phys. D Appl. Phys.* **2010**, *43*, 445003. [CrossRef]
26. Na, J.G.; Lee, T.D.; Park, S.J. Effects of cation distribution on the magnetic and electrical-properties of cobalt ferrite. *IEEE Trans. Magn.* **1992**, *28*, 2433–2435. [CrossRef]
27. Sawatzky, G.A.; Vanderwo, F.; Morrish, A.H. Cation Distributions in Octahedral and Tetrahedral Sites of Ferrimagnetic Spinel $CoFe_2O_4$. *J. Appl. Phys.* **1968**, *39*, 1204–1205. [CrossRef]
28. Schatz, G.; Weidinger, A. *Nukleare Festkörperphysik: Kernphysikalische Messmethoden und ihre Anwendungen*; Teubner Verlag: Stuttgart, Germany, 1997.

29. Peddis, D.; Yaacoub, N.; Ferretti, M.; Martinelli, A.; Piccaluga, G.; Musinu, A.; Cannas, C.; Navarra, G.; Greneche, J.M.; Fiorani, D. Cationic distribution and spin canting in $CoFe_2O_4$ nanoparticles. *J. Phys. Condens. Matter* **2011**, *23*, 426004. [CrossRef]
30. Yafet, Y.; Kittel, C. Antiferromagnetic arrangements in ferrites. *Phys. Rev.* **1952**, *87*, 290–294. [CrossRef]
31. Tronc, E.; Prene, P.; Jolivet, J.P.; Dormann, J.L.; Greneche, J.M. Spin canting in gamma-Fe_2O_3 nanoparticles. *Hyperfine Interact.* **1998**, *112*, 97–100. [CrossRef]
32. Lucht, N.; Friedrich, R.P.; Draack, S.; Alexiou, C.; Viereck, T.; Ludwig, F.; Hankiewicz, B. Biophysical Characterization of (Silica-coated) Cobalt Ferrite Nanoparticles for Hyperthermia Treatment. *Nanomaterials* **2019**, *9*, 1713. [CrossRef]
33. Karaagac, O.; Yildiz, B.B.; Kockar, H. The influence of synthesis parameters on one-step synthesized superparamagnetic cobalt ferrite nanoparticles with high saturation magnetization. *J. Magn. Magn. Mater.* **2019**, *473*, 262–267. [CrossRef]
34. Brown, W.F. Thermal Fluctuations of a Single-Domain Particle. *Phys. Rev.* **1963**, *130*, 1677–1686. [CrossRef]
35. Dutz, S.; Hergt, R. Magnetic nanoparticle heating and heat transfer on a microscale: Basic principles, realities and physical limitations of hyperthermia for tumour therapy. *Int. J. Hyperth.* **2013**, *29*, 790–800. [CrossRef] [PubMed]
36. Hergt, R.; Dutz, S.; Mueller, R.; Zeisberger, M. Magnetic particle hyperthermia: Nanoparticle magnetism and materials development for cancer therapy. *J. Phys. Condens. Matter* **2006**, *18*, S2919–S2934. [CrossRef]
37. Kallumadil, M.; Tada, M.; Nakagawa, T.; Abe, M.; Southern, P.; Pankhurst, Q.A. Suitability of commercial colloids for magnetic hyperthermia. *J. Magn. Magn. Mater.* **2009**, *321*, 1509–1513. [CrossRef]
38. Stins, M.F.; Badger, J.; Kim, K.S. Bacterial invasion and transcytosis in transfected human brain microvascular endothelial cells. *Microb. Pathog.* **2001**, *30*, 19–28. [CrossRef] [PubMed]
39. Baehring, F.; Schlenk, F.; Wotschadlo, J.; Buske, N.; Liebert, T.; Bergemann, C.; Heinze, T.; Hochhaus, A.; Fischer, D.; Clement, J.H. Suitability of Viability Assays for Testing Biological Effects of Coated Superparamagnetic Nanoparticles. *IEEE Trans. Magn.* **2013**, *49*, 383–388. [CrossRef]
40. Theumer, A.; Graefe, C.; Baehring, F.; Bergemann, C.; Hochhaus, A.; Clement, J.H. Superparamagnetic iron oxide nanoparticles exert different cytotoxic effects on cells grown in monolayer cell culture versus as multicellular spheroids. *J. Magn. Magn. Mater.* **2015**, *380*, 27–33. [CrossRef]
41. Xu, M.L.; McCanna, D.J.; Sivak, J.G. Use of the viability reagent PrestoBlue in comparison with alamarBlue and MTT to assess the viability of human corneal epithelial cells. *J. Pharmacol. Toxicol. Methods* **2015**, *71*, 1–7. [CrossRef]
42. Braun, K.; Sturzel, C.M.; Biskupek, J.; Kaiser, U.; Kirchhoff, F.; Linden, M. Comparison of different cytotoxicity assays for in vitro evaluation of mesoporous silica nanoparticles. *Toxicol. In Vitro* **2018**, *52*, 214–221. [CrossRef]
43. De Simone, U.; Spinillo, A.; Caloni, F.; Avanzini, M.A.; Coccini, T. In vitro evaluation of magnetite nanoparticles in human mesenchymal stem cells: Comparison of different cytotoxicity assays. *Toxicol. Mech. Methods* **2020**, *30*, 48–59. [CrossRef] [PubMed]
44. Kroll, A.; Pillukat, M.H.; Hahn, D.; Schnekenburger, J. Interference of engineered nanoparticles with in vitro toxicity assays. *Arch. Toxicol.* **2012**, *86*, 1123–1136. [CrossRef] [PubMed]
45. Fisichella, M.; Dabboue, H.; Bhattacharyya, S.; Saboungi, M.L.; Salvetat, J.P.; Hevor, T.; Guerin, M. Mesoporous silica nanoparticles enhance MTT formazan exocytosis in HeLa cells and astrocytes. *Toxicol. In Vitro* **2009**, *23*, 697–703. [CrossRef] [PubMed]
46. Aam, B.B.; Fonnum, F. Carbon black particles increase reactive oxygen species formation in rat alveolar macrophages in vitro. *Arch. Toxicol.* **2007**, *81*, 441–446. [CrossRef] [PubMed]
47. Ahmad, F.; Zhou, Y. Pitfalls and Challenges in Nanotoxicology: A Case of Cobalt Ferrite ($CoFe_2O_4$) Nanocomposites. *Chem. Res. Toxicol.* **2017**, *30*, 492–507. [CrossRef]
48. Xing, J.Z.; Zhu, L.J.; Jackson, J.A.; Gabos, S.; Sun, X.J.; Wang, X.B.; Xu, X. Dynamic monitoring of cytotoxicity on microelectronic sensors. *Chem. Res. Toxicol.* **2005**, *18*, 154–161. [CrossRef]
49. Durr, S.; Lyer, S.; Mann, J.; Janko, C.; Tietze, R.; Schreiber, E.; Herrmann, M.; Alexiou, C. Real-time Cell Analysis of Human Cancer Cell Lines after Chemotherapy with Functionalized Magnetic Nanoparticles. *Anticancer Res.* **2012**, *32*, 1983–1989.
50. Otero-Gonzalez, L.; Sierra-Alvarez, R.; Boitano, S.; Field, J.A. Application and Validation of an Impedance-Based Real Time Cell Analyzer to Measure the Toxicity of Nanoparticles Impacting Human Bronchial Epithelial Cells. *Environ. Sci. Technol.* **2012**, *46*, 10271–10278. [CrossRef]

51. Horev-Azaria, L.; Baldi, G.; Beno, D.; Bonacchi, D.; Golla-Schindler, U.; Kirkpatrick, J.C.; Kolle, S.; Landsiedel, R.; Maimon, O.; Marche, P.N.; et al. Predictive Toxicology of cobalt ferrite nanoparticles: Comparative in-vitro study of different cellular models using methods of knowledge discovery from data. *Part. Fibre Toxicol.* **2013**, *10*, 32. [CrossRef] [PubMed]
52. Ansari, S.M.; Bhor, R.D.; Pai, K.R.; Mazumder, S.; Sen, D.; Kolekar, Y.D.; Ramana, C.V. Size and Chemistry Controlled Cobalt-Ferrite Nanoparticles and Their Anti-proliferative Effect against the MCF-7 Breast Cancer Cells. *ACS Biomater. Sci. Eng.* **2016**, *2*, 2139–2152. [CrossRef]
53. Khalafalla, S.E.; Reimers, G.W. Preparation of dilution-stable aqueous magnetic fluids. *IEEE Trans. Magn.* **1980**, *16*, 178–183. [CrossRef]
54. Hörsten, U. Pi Program Package. Available online: www.unidue.de/hm236ap/hoersten/home.html (accessed on 15 March 2020).
55. Dutz, S.; Muller, R.; Eberbeck, D.; Hilger, I.; Zeisberger, M. Magnetic nanoparticles adapted for specific biomedical applications. *Biomed. Eng. Biomed. Tech.* **2015**, *60*, 405–416. [CrossRef] [PubMed]

© 2020 by the authors. Licensee MDPI, Basel, Switzerland. This article is an open access article distributed under the terms and conditions of the Creative Commons Attribution (CC BY) license (http://creativecommons.org/licenses/by/4.0/).

Article

Influence of Experimental Parameters of a Continuous Flow Process on the Properties of Very Small Iron Oxide Nanoparticles (VSION) Designed for T_1-Weighted Magnetic Resonance Imaging (MRI)

Thomas Vangijzegem [1,*], Dimitri Stanicki [1], Adriano Panepinto [2], Vlad Socoliuc [3], Ladislau Vekas [3,4], Robert N. Muller [5] and Sophie Laurent [1,5,*]

1. Department of General, Organic and Biomedical Chemistry, NMR and Molecular Imaging Laboratory, University of Mons, B-7000 Mons, Belgium; dimitri.stanicki@umons.ac.be
2. Chimie des Interactions Plasma-Surface (ChIPS), University of Mons, 23 Place du Parc, B-7000 Mons, Belgium; Adriano.panepinto@umons.ac.be
3. Laboratory of Magnetic Fluids, Center for Fundamental and Advanced Technical Research, Romanian Academy—Timisoara Branch, 300223 Timisoara, Romania; vsocoliuc@gmail.com (V.S.); vekas.ladislau@gmail.com (L.V.)
4. Research Center for Complex Fluids Systems Engineering, Politehnica University of Timisoara, M. Viteazu Ave. #1, 300222 Timisoara, Romania
5. Center for Microscopy and Molecular Imaging, Rue Adrienne Bolland, 8, B-6041 Gosselies, Belgium; robert.muller@umons.ac.be
* Correspondence: thomas.vangijzegem@umons.ac.be (T.V.); sophie.laurent@umons.ac.be (S.L.); Tel.: +32-(0)65-373-525 (T.V.)

Received: 12 March 2020; Accepted: 30 March 2020; Published: 15 April 2020

Abstract: This study reports the development of a continuous flow process enabling the synthesis of very small iron oxide nanoparticles (VSION) intended for T_1-weighted magnetic resonance imaging (MRI). The influence of parameters, such as the concentration/nature of surfactants, temperature, pressure and the residence time on the thermal decomposition of iron(III) acetylacetonate in organic media was evaluated. As observed by transmission electron microscopy (TEM), the diameter of the resulting nanoparticle remains constant when modifying the residence time. However, significant differences were observed in the magnetic and relaxometric studies. This continuous flow experimental setup allowed the production of VSION with high flow rates (up to 2 mL·min^{-1}), demonstrating the efficacy of such process compared to conventional batch procedure for the scale-up production of VSION.

Keywords: iron oxide nanoparticles; flow synthesis; contrast agents; magnetic resonance imaging

1. Introduction

Due to their remarkable superparamagnetic properties and their relatively harmless nature, iron oxide nanoparticles have been at the center of several research in the biomedical field for various applications including magnetic resonance imaging (MRI) [1,2], drug delivery [3–5], induced magnetic hyperthermia [6,7] and cell labeling [8,9]. MRI constitutes the main application field for such material, where their use as contrast agents drastically enhances the sensitivity of the technique. These agents are generally divided in two categories depending on whether they can induce a positive (bright contrast) or a negative contrast (dark contrast) [10]. Previously used commercial contrast agents, based on superparamagnetic iron oxide contrast agents (Endorem, Resovist) [11,12], belong to the second class of contrast agents (the so-called T_2 agents). However, the clinical uses of such agents have been limited due to disadvantages stemming from both, the nature of the magnetic nanoparticles (inducing dark contrast and prone to magnetic susceptibility artefact) [13] and the competitiveness of gadolinated contrast agent acting as T_1 agents, which are preferred for the ease of diagnosis in clinical trials. On the other hand, concerns about the potential toxicity of gadolinium-based contrast agents have been pointed out for several years with the well-known nephrogenic systemic fibrosis (NSF) [14,15], and more recently, with cases of brain gadolinium deposition in rats [16,17]. For these reasons, the development of nanoparticulate T_1 contrast agents, based on iron oxide nanoparticles, has become a forward-looking area of research for MRI. On the contrary to the classical clinically used T_2 contrast agents (Endorem), very small (<5 nm) iron oxide nanoparticles (VSION) are potential candidates for T_1-weighted imaging, due to their low magnetization and the presence of a surface spin-canted layer [18–20]. However, one must stress that control over the size and size distribution is an essential prerequisite to observe VSION-induced contrast enhancement on T_1-weighted images.

In recent years, researchers have developed a wide range of synthetic procedures enabling the synthesis of superparamagnetic particles with fine control over their physicochemical properties (shape, size and size distribution) [21]. Among the existing synthesis protocols, the thermal decomposition of organometallic precursors (iron complex) has emerged, over time, as the most efficient method for the preparation of highly calibrated iron oxide nanoparticles. The typical thermal decomposition method relies on the decomposition of an iron complex (iron oleate, iron acetylacetonate) in high boiling point organic solvent (i.e., dibenzylether, oleylalcohol) in the presence of organic surfactants (i.e., oleylamine, oleic acid). Although the method has proven its efficacy, the high level of control required for synthesis, and the harsh experimental conditions, could be limiting factors, especially for the exploitation of the process on larger scales [22].

Bearing in mind these issues, researchers have focused on developing new alternatives, in order to avoid these limitations. For some years, we have witnessed the emergence of continuous flow processes for the preparation of inorganic nanoparticles [23]. Due to a very high surface to volume ratio, flow reactors provide substantial advantages over the conventional batch procedures, very rapid heat transfer enables fast cooling or heating of a solution, as well as precise temperature control [24]. These features are particularly interesting for the implementation of the thermal decomposition synthesis of iron oxide nanoparticles.

In this study, we developed a continuous flow system preparation of VSION by thermal decomposition. A parametric study is proposed, focusing on the influence of different experimental parameters (concentration and nature of surfactants, temperature, pressure, residence time, and capillary inner diameter) on nanoparticles characteristics, such as the size, the magnetic and relaxometric properties by means of the study of their nuclear magnetic resonance dispersion (NMRD) profiles.

2. Experimental Section

2.1. Materials

Oleylamine (98%), oleic acid (90%), dibenzylether (99%), iron(III) acetylacetonate (99.9%), nitric acid and hydrogen peroxide were purchased from Sigma-Aldrich (Overijse, Belgium). Oleyl alcohol (>60%) and 1,2-hexadecanediol (98%) were purchased from TCI Chemicals (Zwijndrecht, Belgium). Ethanol (EtOH; 100%) and tetrahydrofuran (THF; 99.9%) were purchased from Chemlab (Zedelgem, Belgium). All the materials mentioned above were used without further purification.

2.2. Batch Synthesis of VSION

Typically, a mixture of oleic acid (2 mmol; 635 µL), oleylamine (2 mmol; 658 µL) and 1,2-hexadecanediol (10 mmol; 2.58 g) was solubilized in oleyl alcohol (10 mL) and heated at 300 °C for 10 min under nitrogen and magnetic stirring. Then, a solution of iron (III) acetylacetonate (2 mmol; 706 mg) in oleyl alcohol (10 mL) was rapidly injected into the flask. The mixture was heated at 300 °C for a further 30 min, then rapidly cooled. The particles were isolated after pouring an excess of ethanol

(40 mL) into the cooled solution, followed by magnetic decantation. Finally, the as-obtained precipitate was redispersed in tetrahydrofuran (10 mL) and centrifuged to remove any undissolved materials (16.000 g; 10 min).

2.3. Flow Synthesis of VSION

A flow-adapted thermal decomposition process was used for the synthesis of very small iron oxide nanoparticles. In typical experiment, the following stock solution was prepared: iron acetylacetonate (10 mmol; 3.53 g), oleylamine (40 mmol; 10.7 g; 13.2 mL) and oleic acid (40 mmol; 11.3 g; 12.6 mL) were mixed in a 1:1 mixture of dibenzylether and oleylalcohol to reach a final volume of 100 mL (iron precursor concentration of 100 mM). This stock solution was preheated at 70 °C to form a homogenous solution. The stock solution was then pumped using a High Performance Liquid Chromatography (HPLC) pump through a 1 m polytetrafluoroethylene (PTFE) tube reactor immerged in a heating device where the temperature was maintained at elevated temperatures (temperatures of 200, 225, 250, 275 and 300 °C). The system pressure was controlled by a back-pressure regulator assembly (BPRA). Five BPRA inducing counter-pressure values of 5, 20, 40, 75 and 100 psi were evaluated. The effect of surfactant concentration and surfactant nature was studied by varying the composition of the stock solution. The total surfactant concentration was varied between 40 and 120 mM using three different conditions: (i) Oleic acid alone, (ii) oleylamine and oleic acid in equimolar concentration, and (iii) Oleylamine alone. The effect of residence time/flow rate and capillary internal diameter were investigated using tubing with internal diameters (ID) of 0.5, 0.75, 1 and 2.4 mm and flow rates of 0.05, 0.1, 0.5, and 1 mL/min and 2 mL/min* (* Only for the 2.4 mm (ID) PTFE reactor). After being cooled to room temperature, the particles were isolated by pouring an excess of a 1:1 mixture of ethanol and acetone (40 mL) into the suspension, followed by magnetic decantation. Finally, the as-obtained precipitate was resuspended in tetrahydrofuran and centrifuged to remove any undissolved materials (16.000 g; 10 min).

2.4. Characterizations

Transmission electron microscopy (TEM) was used to obtain detailed morphological information and was carried out using a Microscope Fei Tecnai 10 operating at an accelerating voltage of 80 kV (Oregon, OR, USA). The samples were prepared by placing a drop of diluted suspension on carbon-coated copper-grid (300 mesh; NP in organic media), allowing the liquid to dry in air at room temperature. The statistical treatment of the TEM images was performed by iTEM (Münster, Germany) on multiple images for each sample. The mean diameter, the standard deviation and the polydispersity index (PDI) were calculated by measuring the particle diameter. The number of nanoparticles counted ranged from 300 to 500.

The measurements of the size distribution of the nanoparticles suspended in organic medium were performed on a Zetasizer nano zs (Malvern Instruments, Malvern, UK) using laser He-Ne (633 nm). For each measurement, the suspensions were diluted in THF to approximately 1 mM in iron concentration.

Nuclear Magnetic Relaxation Dispersion (NMRD) profiles were recorded on samples in organic media with a field cycling relaxometer (STELAR, Mede, Italy) measuring the longitudinal relaxation rates (R_1) in a magnetic field range extending from 0.24 mT to 1 T. The temperature of the samples was adjusted to 37 °C with a precision of 0.1 °C. The theoretical adjustment of the NMRD profiles was performed using classical relaxation models [25,26] assuming a diffusion coefficient of THF of 3.27×10^{-9} cm$^2 \cdot$s^{-1} [27].

Longitudinal (R_1) and transverse (R_2) relaxation rate measurements at 0.47 and 1.41 T were obtained on Minispec mq 20 and mq 60 spin analysers (Bruker, Mannheim, Germany) respectively. The relaxation rates were measured as a function of the iron molar concentration at 0.47 and 1.41 T, in order to calculate the r_1 and r_2 relaxivities (defined as the enhancement of the relaxation rate of water protons in 1 mmol·L^{-1} solution of contrast agents). The relaxivities were calculated as the slope of relaxation rate (R_i obs) versus iron concentration according to the equation,

$$R_i^{obs} = \frac{1}{T_i^{obs}} = r_i[Fe] + \frac{1}{T_i^{dia}} \quad (1)$$

with r_i being the relaxivities and T_i^{dia} being the proton relaxation times of the solvent without nanoparticle.

The total iron concentration was determined by the measurement of the longitudinal relaxation rate R_1 according to the method previously described [28]. Briefly, the samples were mineralized by microwave digestion (MLS-1200 Mega, Milestone, Analis, Belgium) and the R_1 value of the resulting solutions was recorded at 0.47 T and 37 °C, which determined the iron concentration using equation,

$$[Fe] = (R_1^{sample} - R_1^{dia}) \times 0.0915 \quad (2)$$

where R_1^{dia} (s^{-1}) is the diamagnetic relaxation rate of acidified water (0.36 s^{-1}) and 0.0915 (s·mM) is the slope of the calibration curve.

The yield was determined after measuring the iron content of purified suspensions. The yield is defined as:

$$\text{Reaction yield} = \frac{n_{Fe}}{n_{Fe \text{ Stock solution}}} \times 100\% \quad (3)$$

where n_{Fe} is iron molar content in the purified suspension, $n_{Fe \text{ Stock solution}}$ is the iron molar content in the stock solution.

The magnetization of the samples was measured by means of Vibrating Sample Magnetometry (VSM) at room temperature using an ADE Technologies VSM880 magnetometer (Massachusetts, MA, USA). The samples saturation magnetization was determined as the intercept of the linear fit of M = M(1/H) at H large (i.e., 1/H → 0) [29]. The magnetic diameter distribution of the nanoparticles was obtained by means of magnetogranulometry, i.e., the nonlinear regression of the magnetization curve with Ivanov and coworkers second order modified mean field theory for highly concentrated polydisperse samples [30]. The log-normal distribution was assumed for nanoparticles magnetic diameter [31],

$$f(D) = \frac{1}{D\sigma\sqrt{2\pi}} \cdot e^{-\frac{1}{2\sigma^2} \cdot (ln\frac{D}{D_0})^2} \quad (4)$$

where D_0 and σ are the fit parameters. Using D_0 and σ, the calculation of the mean diameter (<D>) and standard deviation (StDev) is straight forward.

The phase constitution of the powders was evaluated by X-ray diffraction (XRD) using a PANalytical Empyrean diffractometer working with Cu K$_{\alpha 1}$ radiation (λ = 0.1546 nm) at an incidence angle, Ω, of 3°. The powders were continuously rotated during the measurements to ensure a homogeneous exposure to the X-Rays. The resulting patterns were compared to the JCPDS card of magnetite Fe$_3$O$_4$ (JCPDS card, No. -080-6407). The X-ray source voltage was fixed at 45 kV and the current at 40 mA. The grain size (G_s) was calculated from the most intense diffraction peak, i.e., [311] using the following Scherrer equation [32],

$$G_s = \frac{K \lambda}{\beta \cos \theta} \quad (5)$$

where K is a dimensionless shape factor, λ is the X-ray wavelength, β is the line broadening at half the maximum (FWHM) and θ is the Bragg angle.

3. Results and Discussion

3.1. VSION Formation by Thermal Decomposition Using Batch Process

In standard batch conditions, two experimental procedures are commonly used for the thermal decomposition process: the "heating-up" [33] and the "hot-injection" methods [34]. Both procedures allow the formation of nearly monodisperse iron oxide nanoparticles, with tunable sizes, by simply varying the experimental parameters based on the precise separation between the nucleation and the growth steps involved in the formation mechanism of the nanocrystals [35]. For the preparation of VSION, we decided to apply a previously established process [20] which consisted in the direct injection of a solution containing the iron organometallic complex solubilized in oleylalcohol in a solution of hydrophobic surfactants (e.g., oleic acid, oleylamine and 1,2-hexadecanediol in a 1:1:5 molar ratio) in a dibenzylether solution preheated at 300 °C. As a result, VSION with well-defined properties were readily obtained. Figure 1A presents a TEM image of the resulting nanoparticles which show spherical morphology and exhibiting an average diameter of 3.5 ± 0.6 nm (PDI: 1.1). The monodispersity of the sample was also confirmed by photon correlation spectroscopy (PCS), for which a narrow size distribution has been observed (Figure 1B), as well as a hydrodynamic diameter centered at around 7 nm (PDI: 0.090).

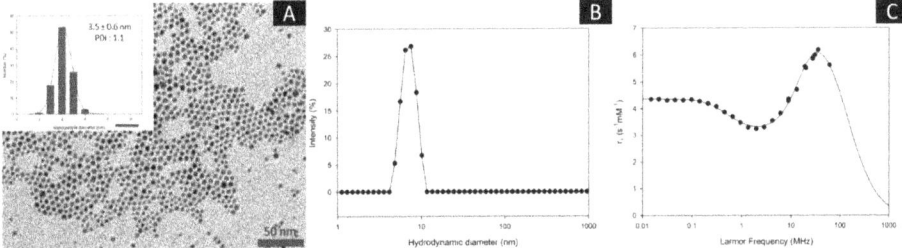

Figure 1. TEM image of the VSION in THF. The scale bar corresponds to 50 nm. The inset shows the size distribution: 3.5 ± 0.6 nm (PDI: 1.1) (**A**); Size distribution in intensity as measured by PCS (**B**); ^{1}H NMRD profile, recorded at 37 °C in THF (**C**).

The relaxation properties of magnetic nanoparticles were obtained by the study of their nuclear magnetic resonance dispersion profiles (NMRD profiles; Figure 1C). These curves show the evolution of the proton longitudinal relaxation (r_1) as a function of the applied magnetic field (Larmor frequency). The shape of this profile corresponds to the typical curve expected for superparamagnetic colloidal suspensions. At low field, the small dispersion is a main feature of small-sized individual iron oxide nanoparticles characterized by a low anisotropy energy [25]. After this dispersion, the hump present at stronger magnetic field is due to the evolution of the Curie magnetization. The fit of the NMRD curve using the classical relaxation models allowed to estimate a saturation magnetization (MSAT of 39.6 $A \cdot m^2 \cdot kg^{-1}$) and a NMRD diameter ($D_{NMRD}$ of 6.24 nm). One could underline the good correlation between D_{NMRD} and D_{PCS}. Relaxometric properties of batch-prepared VSION, Dotarem®and Resovist®are shown in Table S1.

3.2. Influence of Experimental Parameters on Nanoparticles Formation by Thermal Decomposition Using Flow Process

The flow synthesis was performed using PTFE tube reactors with a fixed length (1 m) and internal diameters of 0.5, 0.75, 1, and 2.4 mm. Figure 2 depicts the experimental setup designed for the flow synthesis of the nanoparticles. It consists of a HPLC pump operating at flow rates between 10 $\mu L \cdot min^{-1}$ and 10 $mL \cdot min^{-1}$ which inject the reaction mixture in a PTFE tubing heated at constant temperature. The system pressure is controlled by a back-pressure regulator assembly (BPRA) located at the output of the capillary. The as-synthesized nanoparticles were cooled at room temperature at the exit of the capillary and then collected. Any unreacted iron complexes, as well as surfactant excess were removed by means of several washing steps using ethanol/acetone mixture. After redispersion in THF, the as-obtained suspensions were finally centrifuged to remove any undissolved materials.

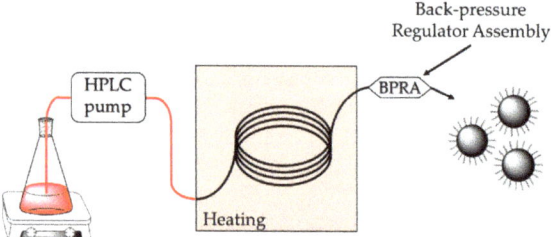

Figure 2. Schematic representation of the experimental setup used for the flow synthesis of VSION.

Due to the high surface-to-volume ratios characteristic of flow reactors [22], fast heat transfer can be achieved in order to emulate the hot-injection batch process. This sudden increase of temperature induces the decomposition of the iron organometallic precursor into iron monomers (iron oxo-clusters) [36,37] leading to a burst nucleation phenomenon. Subsequent growth of these nuclei results in the formation of iron oxide nanoparticles in the capillary.

In order to adapt and optimize the synthesis to the continuous flow process, slight modifications were made to the batch recipe. The major one implied the withdrawing of 1,2-hexadecanediol owing to its low solubility in the reaction mixture at room temperature (to avoid clogging of the HPLC pump). As a result, some adjustments in the oleic acid/oleylamine ratio had to be done.

3.3. Influence of the Ligand/Precursor Molar Ratio and Ligand Nature

Among the numerous studies treating about the thermal decomposition for the synthesis of iron oxide nanoparticles, many reports show the strong impact of the surfactant/precursor molar ratio on the properties of the obtained nanoparticles with different tendencies reported by several research groups [38,39]. In order to evaluate the influence of the surfactant concentration and the surfactant nature, the flow synthesis has been conducted with different ratios of (i) oleic acid only, (ii) oleic acid and oleylamine and (iii) oleylamine only. All the experiments were carried out in the same solvent mixture (i.e., dibenzylether and oleylalcohol 1/1) and three surfactants/precursor molar ratios were tested (i.e. 4, 8 and 12 equivalents of surfactants). Experiments were carried out using a temperature of 250 °C and BPRA of 40 psi.

Iron oxide nanoparticles synthesized with the three different concentrations of pure oleic acid exhibited very poor stability in THF, leading to diluted suspensions with very low yields (less than 10%). Moreover, the TEM images (and the corresponding histograms) (Figure S1; SI) show nanoparticles with no well-defined shape and very broad size distributions (PDI > 1.3). These results correlate well with other report [40] in which the synthesis of iron oxide nanoparticles using oleic acid as surfactant resulted in low yields. The lack of stabilisation using oleic acid was recently investigated by Harris et al. [41] using molecular mechanics modelling to evidence lowest binding energy between oleic acid and the nanoparticle's surface compared to systems combining oleic acid and oleylamine. In their study, they demonstrated that oleic acid only cannot yield to stable suspensions. It was suggested the binding of deprotonated oleic acid molecules onto the nanoparticles leads to a neutral average surface charge and at the same time to an increase in free proton concentration in the media. As a consequence of these two phenomena, the zeta potential will overtime tend toward zero and the formed nanoparticles tend to agglomerate.

Interestingly, when oleylamine was used as surfactant. A linear decrease of the mean particle size when increasing oleylamine amount was observed (Figure 3A). Qi et al. [42] reported a similar trend using three different aliphatic amines (hexylamine, oleylamine and octadecylamine). However, substantial differences are observed in the NMRD profiles recorded for these samples (Figure 3B). Indeed, smaller nanoparticles exhibit higher longitudinal relaxivity on the overall range of Larmor frequency.

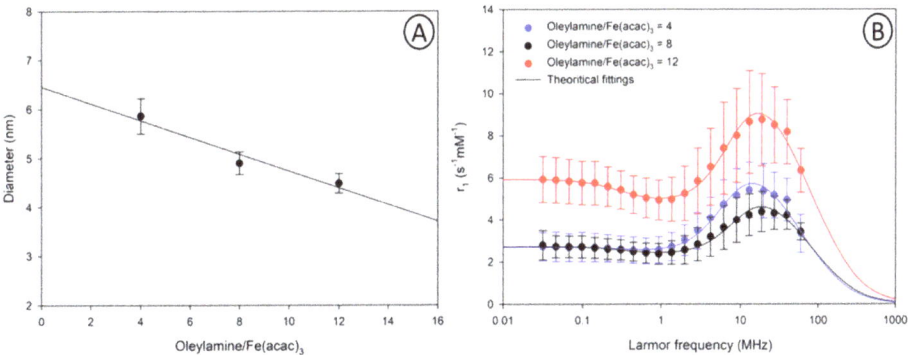

Figure 3. (**A**) Evolution of TEM diameter as a function of the oleylamine amount; (**B**) ^{1}H NMRD profiles of samples obtained by flow synthesis with various amount of oleylamine.

Although these results seem contradictory with the superparamagnetic relaxation theory, differences in the magneto-crystalline state of the nanoparticles could explain this phenomenon. Oleylamine, can act as a reducing agent besides its role as a capping agent. The presence of a large excess of oleylamine can therefore provide a strong reductive environment affecting the crystallinity of the obtained nanoparticles [43]. The magnetic properties of nanoparticles were evaluated, using VSM measurements, and compared to the parameters extracted from the fitting of the NMRD profiles (Table S3; SI). Again, a good correlation between the M_S values, measured by VSM and the M_S values, determined from the superparamagnetic theory, was obtained. Therefore, these results demonstrate that tuning the oleylamine to iron precursor ratio is an experimental parameter, which can be easily implemented to produce iron oxide nanoparticles with variable sizes and variable degrees of crystallinity, using our continuous flow process.

On the other hand, conducting the synthesis with various amounts of oleylamine and oleic acid (in equimolar quantities), yielded to stable and concentrated colloidal suspensions. Size changes were measured with TEM (Figure 4). These results show that a ratio of surfactants to iron precursor of 4:1 leads to smaller nanoparticles with narrower size distribution compared to synthesis performed with larger (6:1) and lesser (2:1) amounts of surfactants. This suggests that there is an optimum ratio at which the organic surfactants provide a better stability to the nuclei formed in situ, and is therefore, favourable to the formation of smaller nanoparticles having appropriate relaxometric properties (r_2/r_1 ratios) (Table S2; SI). Therefore, the surfactants to iron precursor of 4:1 was used for the study of the other experimental parameters, described in the following sections.

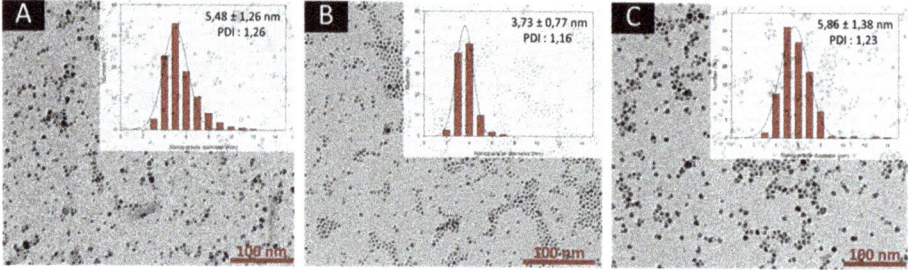

Figure 4. TEM images of the nanoparticles obtained through flow synthesis with various equivalents of oleic acid and oleylamine: Two equivalents (**A**), four equivalents (**B**) and six equivalents (**C**) in the 1 mm capillary reactor. The scale bar corresponds to 100 nm. Insets show the size distributions determined by statistical analysis.

3.4. Influence of Temperature

To study the influence of the temperature, synthesis experiments were carried out using a PTFE tube reactor with an internal diameter of 1 mm and at a flow rate of 1 mL·min^{-1}. In a typical synthesis, a BPRA inducing a back pressure of 40 psi was used. Synthesis were carried out using oleic acid and oleylamine as surfactants (with a surfactant to precursor ratio of 4/1) considering the results abovementioned (i.e., flow synthesis using four equivalents of oleic acid and oleylamine in equimolar concentration). Five batches were prepared by working at temperatures of 200, 225, 250, 275 and 300 °C. Surprisingly, no significant differences in the mean diameter has been observed neither by PCS nor by TEM. The resulting suspensions were all characterized by mean hydrodynamic diameters below 10 nm (Figure 5B; Table S4; SI) and a mean inorganic diameter below 4 nm (Figure 5C) accompanied with polydispersity indexes between 1.10 and 1.16. The relaxometric properties of these samples were investigated by the study of their NMRD profiles (Figure 5A).

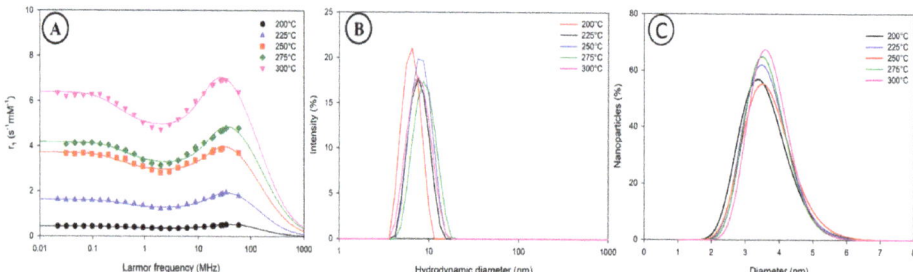

Figure 5. ^{1}H NMRD profiles of iron oxide nanoparticles obtained at various temperatures recorded at 37 °C in THF (**A**). The continuous lines correspond to the fitting by the superparamagnetic relaxation theory; Size distribution in intensity as measured by PCS of VSION obtained at various temperatures (**B**); size distributions as determined by TEM (**C**).

The results indicate a marked increase of the relaxivity values on the overall range of Larmor frequency when increasing the temperature in the capillary reactor. The samples obtained at lower temperature (i.e., 200 and 225 °C) exhibit very low values of longitudinal relaxivities. Despite the significant differences observed in their relaxometric properties, the small dispersion observed for each sample still indicates that their anisotropy energy remains low and characteristic of VSION. Theoretical parameters (M_{SAT} and D_{NMRD}) extracted from the fitting of these profiles are shown in Table S4 along with the nanoparticle sizes determined by PCS and TEM. The estimated D_{NMRD} values are in good correlation with the hydrodynamic sizes measured by PCS. Taken together, these theoretical and experimental measurements indicate no influence of the temperature on the size of the synthesized nanoparticles. However, the extracted values of M_{SAT} are markedly increasing with the reaction temperature (Table S4; SI). At this stage, the formulated hypothesis relies on differences in the magneto-crystalline state of the VSION. Previous report from Belaid et al. [44] showed the same tendency when synthesizing iron oxide nanoparticles in classical thermal decomposition synthesis at different temperatures. XRD and VSM measurements showed the increase of the particles crystallinity degree with the temperature, giving rise to increased relaxometric properties.

Since the iron oxide nanoparticles obtained at a temperature of 250 °C display relaxometric properties similar to those of obtained for batch samples, this temperature was chosen as the working temperature for the following flow syntheses. Moreover, this temperature was chosen due to the nature of the capillary reactor i.e., PTFE which is known to be thermally resistant up to 250 °C (for long-term continuous service) [45].

3.5. Influence of Pressure

The influence of pressure on the particles properties was studied using back-pressure regulator assemblies inducing counter-pressure varying between 5 and 100 psi (BPR of 5, 20, 40, 75 and 100 psi were tested). With the 5 and 20 psi BPR, flow cavitation was observed in the capillary reactor and the resulting nanoparticles were characterized by very low relaxometric properties (Table 1). On the other hand, BPR of 40, 75 and 100 psi delivered good flow stability and a steady flow was straightforwardly reached in the capillary reactor. The TEM images of VSION obtained with these three BPR are presented in Figure 6. Size histograms were fitted using a log-normal function and standard deviation and polydispersity index were calculated.

Table 1. Relaxometric properties of VSION obtained at different pressures.

BPRA (psi)	20 MHz		60 MHz	
	r_1 (s^{-1}mM^{-1})	r_2 (s^{-1}mM^{-1})	r_1 (s^{-1}mM^{-1})	r_2 (s^{-1}mM^{-1})
5	0.5	0.7	0.6	1.2
20	1.4	1.9	1.5	3.3
40	3.7	5.3	3.7	8.8
75	3.6	5	3.7	8.6
100	3.6	5	3.7	8.6

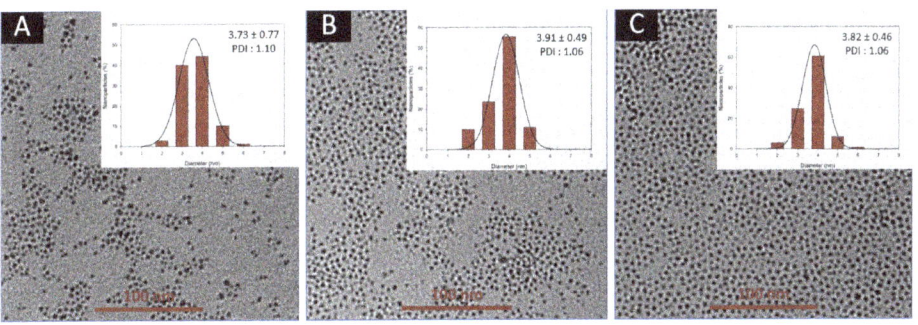

Figure 6. TEM images of the nanoparticles obtained through flow synthesis with BPR of 40 psi (A), 75 psi (B) and 100 psi (C). The scale bar corresponds to 100 nm. Insets show the size distributions determined by statistical analysis.

For each sample, VSION with a mean diameter smaller than 4 nm are observed. NMRD profiles of these three samples of VSION (Figure S2; SI) are stackable, showing no incidence of the pressure on both, the size and relaxometric properties of the obtained VSION (Table S5). Therefore, we concluded that pressure only influences the stability of the flow in the capillary reactor and we investigated the influence of other parameters such as flow rate and residence with the 40 psi BPR.

3.6. Influence of Residence Time in the Capillary Reactor

The residence time in the capillary reactor was modified by varying two experimental parameters, i.e., the capillary inner diameter and the flow rate. The influence of these flow parameters was initially investigated by using three capillary reactors with internal diameters of 0.5; 0.75 and 1 mm. For each capillary, experiments have been conducted with four flow rates (0.05; 0.1; 0.5 and 1 mL·min^{-1}). Photon correlation spectroscopy analysis confirmed for each sample the formation of nanoparticles characterized by a narrow monomodal distribution with a mean diameter below 10 nm. Determination of nanoparticles size and polydispersity index was performed using TEM (Figure 7).

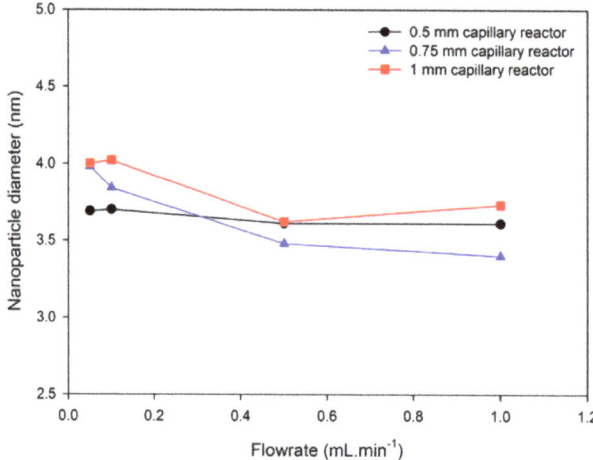

Figure 7. Average diameter of VSION as a function of flowrate in capillary reactors with different internal diameters.

No matter the experimental conditions, it was observed that the nanoparticle diameter remained roughly constant when working at different flow rates. All samples are characterized by mean diameter between 3.4 and 4 nm and no significant tendency of size increase is observed when decreasing the flow rate. Such observations are quite different from other report from Jiao et al. [46], whereby a significant impact of the residence time on the size properties of iron oxide nanoparticles was reported using a similar flow process. This suggests a different mechanism of nanoparticle formation using our flow process. The decomposition of the iron precursor depends on the solvents and the amounts of surfactants present in the media. In our experimental setup, the reaction conditions seem to lead to a burst nucleation phenomenon yielding VSION with constant size. However, if no clear influence was observed by TEM, significant differences were observed when analysing the NMRD profiles (Figure 8).

Within each capillary reactor, we observe a global increase of the relaxometric properties of the VSION when using slower flow rate. The increase of the residence time induces an increase of proton longitudinal relaxivity in the overall studied range of Larmor frequency. Notably, is the shift of the hump (maximum proton longitudinal relaxivity) towards lower frequency values when slower flow rates are used, this indicates that an increased residence time is favourable to slightly larger nanoparticles, which is not observed from the TEM analysis. This could also be provoked by a structural change with different heating time. This observation can be confirmed by the NMRD diameters extracted by the theoretical fittings of the NMRD profiles (Table 1). The NMRD diameters are much higher than the sizes determined by TEM, given that the superparamagnetic relaxation theory was developed for crystalline superparamagnetic compounds, the hypothesis explaining this observation would rely on differences in the crystalline state of the nanoparticles. Nanoparticles synthesized at higher flow rates have a residence time of less than 2 min in the capillary reactor. Such short residence time might not be sufficient to yield well-crystallized iron oxide nanoparticles. These differences in the nanoparticles relaxometric properties can, therefore, be attributed to differences in the nanoparticle's crystallinity degree and consequently to their magnetic and relaxometric properties. Nonetheless, when compared to their batch-prepared derivatives, VSION prepared at slower flow rates exhibit similar or even better relaxometric properties (high longitudinal relaxivity values and low r_2/r_1 ratios) for applications as T_1 contrast agents (Table 2).

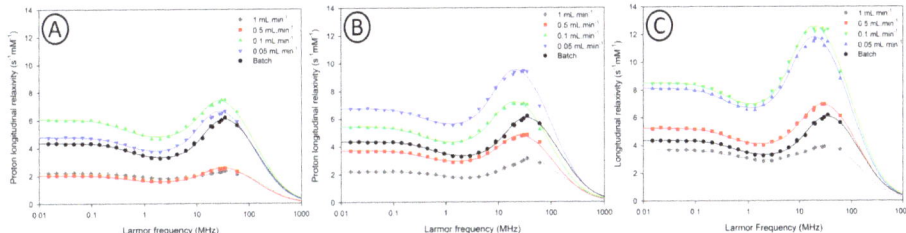

Figure 8. ^{1}H NMRD profiles of VSION obtained in the 0.5 mm capillary reactor (**A**), the 0.75 mm capillary reactor (**B**) and 1 mm capillary reactor (**C**). The black curve is the NMRD profile of the batch-prepared VSION. The continuous lines correspond to the fitting by the superparamagnetic relaxation theory.

Table 2. Summary of the experimental parameters and results obtained by PCS, TEM and relaxometry for the study of flow parameters influences.

Capillary Inner Diameter (mm)	Flow Rate (mL·min^{-1})	Residence Time (min)	Diameter			NMRD Data		Relaxivity Measurements	
			D_{PCS} (nm)	D_{TEM} (nm)	PDI	D_{NMRD} (nm)	M_s (A·m^2·kg^{-1})	r_2/r_1 (20 MHz)	r_2/r_1 (60 MHz)
0.5	0.05	3.93	6.1	3.69 ± 0.81	1.16	6.82	38.4	1.58	2.42
0.5	0.1	1.96	6.2	3.70 ± 0.84	1.14	6.50	41.2	1.54	2.30
0.5	0.5	0.39	8.8	3.70 ± 0.92	1.15	7.18	25.4	1.45	2.26
0.5	1	0.20	6.6	3.61 ± 0.72	1.15	5.82	27	1.40	2.19
0.75	0.05	8.84	9.1	3.98 ± 1.00	1.23	7.34	41.6	1.57	2.63
0.75	0.1	4.42	8.8	3.84 ± 0.76	1.16	7.76	34.8	1.56	2.60
0.75	0.5	0.88	8.1	3.48 ± 0.76	1.20	7.28	31	1.56	2.53
0.75	1	0.44	6.8	3.40 ± 0.82	1.23	5.68	30.5	1.50	2.49
1	0.05	15.71	7.1	4.00 ± 0.94	1.19	7.52	44.8	1.63	2.55
1	0.1	7.85	9.5	4.02 ± 0.80	1.14	7.52	46.6	1.73	2.74
1	0.5	1.57	8.1	3.62 ± 0.74	1.17	6.70	39.8	1.58	2.48
1	1	0.79	7.7	3.73 ± 0.77	1.10	6.96	33.9	1.53	2.43
2.4	0.5	9	7.3	4.09 ± 0.81	1.15	7.44	47.3	1.84	2.69
2.4	1	4.5	9.3	3.98 ± 0.77	1.21	7.04	44.2	1.63	2.59
2.4	2	2.25	8.1	3.74 ± 0.64	1.15	6.62	38.6	1.51	2.46

To confirm these differences in the magnetic properties of the iron oxide nanoparticles, magnetometry measurements were conducted on samples obtained in the 1 mm capillary reactor. Those results are presented in Figure 9.

Magnetization curves obtained for VSION synthesized at all flow rates show the superparamagnetic nature of the nanoparticles for each sample. As demonstrated in Figure 9, all samples have a superparamagnetic behavior characterized by the absence of hysteresis (absence of coercivity and absence of remanent magnetization). Fitting the magnetometry data using Ivanov 2nd order modified field theory [30] allowed the the magnetic diameter distribution to be determined. Saturation magnetization (MS) was measured using Chantrell method [29] (Table S6; SI). The results are well fitted with the relaxometry measurements (M_S and D_{NMRD} obtained by the fitting of the NMRD profiles). From these results, a same trend towards the increase of the magnetic properties when working at slower flow rates was observed.

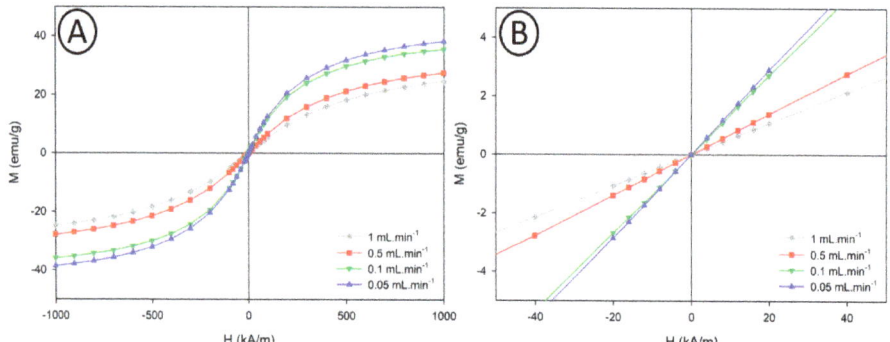

Figure 9. Magnetization curves of samples obtained in the 1 mm capillary reactor as a function of the applied magnetic field (**A**); Zoomed region around origin (**B**).

To go further in the parametric study and the production objective intended for the synthesis of VSION, the influence of flow parameters was investigated using a capillary with a larger internal diameter (2.4 mm). Considering the dead volume of this capillary, flow rates of 0.5, 1 and 2 mL·min^{-1} were used. TEM images (and corresponding histograms fitted using a log-normal function) of samples obtained at flow rates of respectively 0.5, 1 and 2 mL·min^{-1} are presented in Figure 10 with the corresponding NMRD profiles recorded for each sample.

TEM and relaxometry results lead to the same conclusions: flow rate and consequently the residence time do not have a marked influence on the size properties of the as-synthesized nanoparticles. For all three samples, nanoparticles were characterized by a mean size around 4 nm and relatively low polydispersity index. However, these three samples exhibit interesting relaxometric properties compared to their batch-prepared derivatives. As seen on the NMRD profiles, samples obtained at 0.5 mL·min^{-1} and 1 mL·min^{-1} exhibit proton longitudinal relaxivity higher than the batch sample over the whole range of proton Larmor frequencies. In addition, sample obtained at 2 mL·min^{-1} and the batch sample display analogous values of longitudinal relaxivities, confirming that a residence time of approximately 2 min is sufficient to yield VSION with properties similar to batch-prepared ones.

Furthermore, by comparing the properties of all samples obtained in tubings with different internal diameters, no clear proportionality between the residence time and the nanoparticles properties can be pointed out. Indeed, even if a global increase of the relaxometric properties with the residence time is observed for each capillary. Samples obtained with similar residence time in capillaries with different internal diameters do not exhibit properties (Table 2). An additional parameter which could potentially explain these discrepancies concern the thickness of the PTFE tubings. All tubings are made of 1/8 PTFE with varying internal diameters. As a result, the wall thickness is not the same for each tubing. This thickness difference might affect the heat transfer in the tubing and therefore affect the properties of the synthesized nanoparticles.

As mentioned above, these differences in relaxometric properties may be attributed to differences in the nanoparticle's magneto-crystalline properties. To confirm these differences in the crystalline state potentially affecting the magnetic and therefore the relaxometric properties of the VSION, VSM measurements were performed on the three samples, but also X-ray diffraction experiments, to obtain a qualitative estimation of the crystallinity of the three samples (Figure 11).

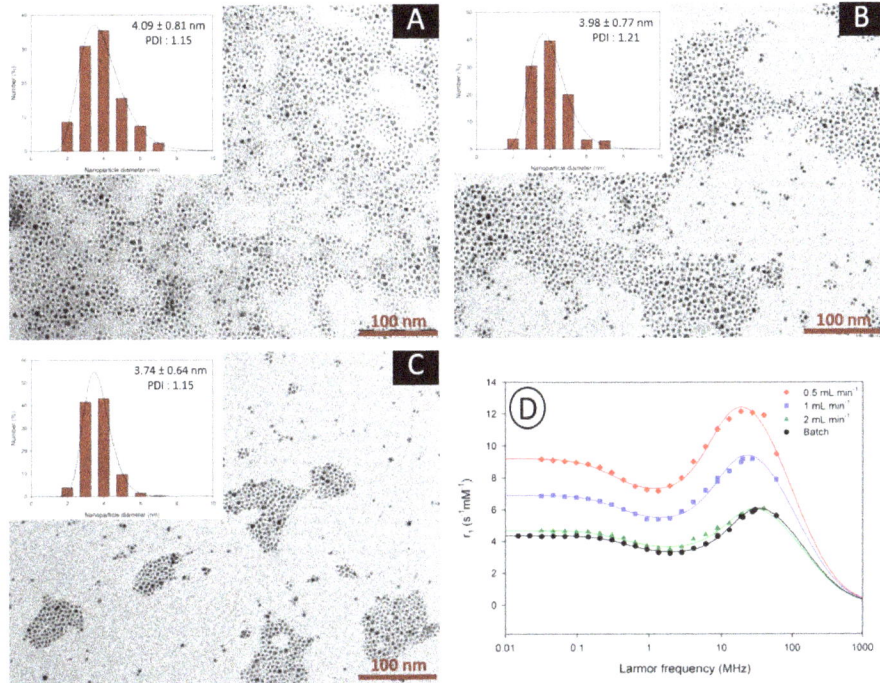

Figure 10. TEM images of the nanoparticles obtained through flow synthesis with flow rates of 0.5 mL·min^{-1} (**A**), 1 mL·min^{-1} (**B**) and 2 mL·min^{-1} (**C**) in the 2.4 mm capillary reactor. The scale bar corresponds to 100 nm. Insets show the size distributions determined by statistical analysis; (**D**) ^{1}H NMRD profiles of the three samples (colored curves) and the batch-prepared sample (black curve). The continuous lines correspond to the fitting by the superparamagnetic relaxation theory.

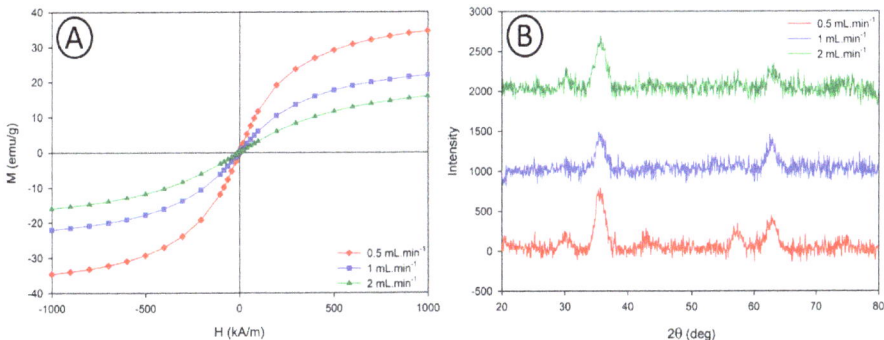

Figure 11. Magnetization curves (**A**) and XRD diffraction patterns (**B**) of samples obtained in the 2.4 mm capillary reactor.

Similarly to the VSION synthesized in the 1 mm PTFE capillary, the VSION are all characterized by a superparamagnetic behaviour and the magnetization saturations obtained by the Chantrell method follow the same trend as previously shown with the 1 mm capillary reactor. Differences in the crystallinity state of these 3 samples was confirmed using X-ray diffraction experiments. The characteristics diffraction peaks observed at 30.2°, 35.5°, 42.8°, 57.2° and 62.7° and their relative intensities match well with the standard XRD data of magnetite Fe$_3$O$_4$ (JCPDS card No. 080-6407). The XRD peak intensity increases with the residence time in the capillary reactor, combined with the line broadening for samples obtained at higher flow rate (indicating the presence of smaller crystallites when using shorter residence time), this demonstrates the presence of nanoparticles with increased crystallinity when working at slower flow rate. An estimation of the magnetite nanoparticles size was made with both, characterization techniques, using magnetogranulometry and the Scherrer formula (taking the highest intensity peak corresponding to the [311] plane at $2\theta = 35.5°$) for VSM, and XRD, respectively (Table 3).

Table 3. Summary of the results obtained with TEM, VSM and XRD for samples obtained in the 2.4 mm capillary reactor.

Flow Rate (mL·min^{-1})	Diameter			M_{SAT} (emu/g)
	D^{TEM} (nm)	D^{VSM} (nm)	D^{XRD} (nm)	
0.5	3.71 ± 0.76	4.16 ± 0.91	5.02	40.1
1	3.98 ± 0.77	3.90 ± 0.76	4.59	26.4
2	3.74 ± 0.64	3.51 ± 0.58	4.16	20.4

Whereas, the estimated size of the nanoparticles, determined by magnetogranulometry, is in relatively good agreement with the TEM diameter, the estimated particle diameters measured from the XRD measurements, using the Scherrer formula are slightly higher than the measured size determined from the TEM analysis, this overestimation of the nanoparticles size might be due to the very small size of the nanoparticles limiting the use of the Scherrer equation.

4. Conclusions

We have developed a continuous flow process for the thermal decomposition synthesis of very small iron oxide nanoparticles. The influence of experimental parameters (ligand concentration and nature, temperature, pressure capillary inner diameter and residence time) was investigated to determine their inflluence on the nanoparticles magneto-crystalline and relaxometric properties. This study showed that experimental parameters have almost no incidence on the nanoparticles size properties. However, a clear impact was evidenced based on the relaxometric and magnetic properties of the particles. As a consequence, nanoparticles obtained with increased residence times exhibit higher relaxometric and magnetic properties. With respect to the application of these VSION for MRI, an important feature defining their efficacy as T_1 contrast agents is their r_2/r_1 ratios. The ideal T_1 MRI contrast agent is characterized by high r_1 values and low r_2/r_1 ratios in order to reach their highest imaging efficacy [47,48]. In our case, VSION synthesized using this flow process exhibit higher r_1 values than commercially used T_1 contrast agents [49] such as Dotarem®or other paramagnetic agents as well as r_2/r_1 ratios below three at standard clinical fields (1.5 T). This flow process allowed the production of VSION with superior throughput than classical batch methods.

Supplementary Materials: The following are available online at http://www.mdpi.com/2079-4991/10/4/757/s1. Table S1: Relaxometric properties of Dotarem®, Resovist®and batch-prepared VSION; Figure S1: TEM images of the nanoparticles obtained through flow synthesis with various equivalents of oleic acid: 4 equivalents (left), 8 equivalents (middle) and 12 equivalents (right) in the 1 mm capillary reactor. The scale bar corresponds to 100 nm. Insets show the size distributions determined by statistical analysis; Table S2: Size and relaxometric properties of VSION synthesized using various surfactant concentrations (oleic acid and oleylamine); Table S3: Size and magnetic properties of VSION synthesized using various oleylamine concentrations; Table S4: Size and relaxometric properties of VSION synthesized at different temperatures using continuous flow process; Table S5: size and magnetic properties of VSION synthesized using different BPR; Figure S2: ^{1}H NMRD profiles of iron oxide nanoparticles obtained at various pressure recorded at 37 °C in THF; Table S6: Magnetic properties extracted from the fitting of the magnetization curves of samples obtained in the 1 mm capillary reactor.

Author Contributions: T.V. and D.S. performed the synthesis of the samples, analyzed the data and wrote the manuscript. A.P. performed the XRD analysis. V.S. performed the VSM analysis. L.V. supervised and analyzed the VSM data. S.L. and R.N.M. supervised and analyzed the NMR data. All authors have read and agreed to the published version of the manuscript.

Funding: This research received no external funding.

Acknowledgments: This work was supported by the Fond National de la Recherche Scientifique (FNRS), UIAP VII, ARC Programs of the French Community of Belgium, COST actions and the Walloon region (ProtherWal and Interreg projects). V. Socoliuc and L. Vekas acknowledge the financial support from Romanian Academy—Timisoara Branch 2016-2020 research programme. The authors thank the laboratories of D. Nonclercq for the use of TEM.

Conflicts of Interest: The authors declare no conflict of interest.

References

1. Shen, S.; Wu, A.; Chen, X. Iron oxide nanoparticles-based contrast agents for magnetic resonance imaging. *Mol. Pharm.* **2016**, *14*, 1352–1364. [CrossRef] [PubMed]
2. Qiao, R.; Yang, C.; Gao, M. Superparamagnetic iron oxide nanoparticles: From preparations to in vivo MRI applications. *J. Mater. Chem.* **2009**, *19*, 6274–6293. [CrossRef]
3. Vangijzegem, T.; Stanicki, D.; Laurent, S. Magnetic iron oxide nanoparticles for drug delivery: Applications and characteristics. *Expert Opin. Drug Deliv.* **2019**, *16*, 69–78. [CrossRef] [PubMed]
4. Mou, X.; Ali, Z.; Li, S.; He, N. Applications of magnetic nanoparticles in targeted drug delivery system. *J. Nanosci. Nanotechnol.* **2015**, *15*, 54–62. [CrossRef] [PubMed]
5. Wahajuddin, S.A. Superparamagnetic iron oxide nanoparticles: Magnetic nanoplatforms as drug carriers. *Int. J. Nanomed.* **2012**, *7*, 3445–3471. [CrossRef] [PubMed]
6. Hedayatnasab, Z.; Abnisa, F.; Daud, W.M.A.W. Review on magnetic nanoparticles for magnetic nanofluid hyperthermia application. *Mater. Design* **2014**, *123*, 174–196. [CrossRef]
7. Blanco-Andujar, C.; Walter, A.; Cotin, G.; Bordeianu, C.; Mertz, D.; Felder-Flesch, D.; Begin-Colin, S. Design of iron oxide-based nanoparticles for MRI and magnetic hyperthermia. *Nanomedicine* **2016**, *11*, 1889–1910. [CrossRef]
8. Li, L.; Jiang, W.; Luo, K.; Song, H.; Lan, F.; Wu, Y.; Gu, Z. Superparamagnetic iron oxide nanoparticles as MRI contrast agents for non-invasive stem cell labeling and tracking. *Theranostics* **2013**, *3*, 595–615. [CrossRef]
9. Tefft, B.J.; Uthamaraj, S.; Harburn, J.J.; Klabusay, M.; Dragomir-Daescu, D.; Sandhu, G.S. Cell labeling and targeting with superparamagnetic iron oxide nanoparticles. *J. Vis. Exp.* **2015**, *104*, 1–9. [CrossRef]
10. Merbach, A.; Helm, L.; Tóth, É. *The Chemistry of Contrast Agents in Medical Magnetic Resonance Imaging*; John Wiley & Sons: Hoboken, NJ, USA, 2013.
11. Wang, Y.J. Superparamagnetic iron oxide based MRI contrast agents: Current status of clinical application. *Quant. Imaging Med. Surg.* **2011**, *1*, 35–40.
12. Reimer, P.; Balzer, T. Ferucarbotran (Resovist): A new clinically approved RES-specific contrast agent for contrast-enhanced MRI of the liver: Properties, clinical development, and applications. *Eur. Radiol.* **2003**, *13*, 1266–1276. [CrossRef] [PubMed]
13. Kim, B.H.; Lee, N.; Kim, H.; An, K.; Park, Y.I.; Choi, Y.; Shin, K.; Lee, Y.; Kwon, S.G.; Na, H.B.; et al. Large-scale synthesis of uniform and extremely small-sized iron oxide nanoparticles for high-resolution T_1 magnetic resonance imaging contrast agents. *J. Am. Chem. Soc.* **2011**, *133*, 12624–12631. [CrossRef] [PubMed]
14. Ramalho, J.; Semelka, R.C.; Ramalho, M.; Nunes, R.H.; Alobaidy, M.; Castillo, M. Gadolinium-based contrast agent accumulation and toxicity: An update. *Am. J. Neuroradiol.* **2016**, *37*, 1192–1198. [CrossRef] [PubMed]
15. Wertman, R.; Altun, E.; Martin, D.R.; Mitchell, D.G.; Leyendecker, J.R.; O'Malley, R.B.; Parsons, D.J.; Fuller, E.R.; Semelka, R.C. Risk of nephrogenic systemic fibrosis: Evaluation of gadolinium chelate contrast agents at four American universities. *Radiology* **2008**, *248*, 799–806. [CrossRef]
16. Robert, P.; Violas, X.; Grand, S.; Lehericy, S.; Idée, J.; Ballet, S.; Corot, C. Linear gadolinium-based contrast agents are associated with brain gadolinium retention in healthy rats. *Investig. Radiol.* **2016**, *51*, 73–82. [CrossRef]
17. Arena, F.; Bardini, P.; Blasi, F.; Gianolio, E.; Marini, G.M.; La Cava, F.; Valbusa, G.; Aime, S. Gadolinium presence, MRI hyperintensities, and glucose uptake in the hypoperfused rat brain after repeated administrations of gadodiamide. *Neuroradiology* **2019**, *61*, 163–173. [CrossRef]
18. Zhou, Z.; Wang, L.; Chi, X.; Bao, J.; Yang, L.; Zhao, W.; Chen, Z.; Wang, X.; Chen, X.; Gao, J. Engineered iron-oxide-based nanoparticles as enhanced T_1 contrast agents for efficient tumor imaging. *ACS Nano* **2013**, *7*, 3287–3296. [CrossRef]
19. Shin, T.; Choi, Y.; Kim, S.; Cheon, J. Recent advances in magnetic nanoparticle-based multi-modal imaging. *Chem. Soc. Rev.* **2015**, *44*, 4501–4516. [CrossRef]

20. Vangijzegem, T.; Stanicki, D.; Boutry, S.; Paternoster, Q.; Vander Elst, L.; Muller, R.N.; Laurent, S. VSION as high field MRI T_1 contrast agent: Evidence of their potential as positive contrast agent for magnetic resonance angiography. *Nanotechnology* **2018**, *29*, 1–14. [CrossRef]
21. Laurent, S.; Forge, D.; Port, M.; Roch, A.; Robic, C.; Vander Elst, L.; Muller, R.N. Magnetic iron oxide nanoparticles: Synthesis, stabilization, vectorization, physicochemical characterizations, and biological applications. *Chem. Rev.* **2008**, *108*, 2064–2110. [CrossRef]
22. Laird, T. How to minimise scale up difficulties. *Chem. Ind. Dig.* **2010**, *23*, 51–56.
23. Makgwane, P.R.; Ray, S.S. Synthesis of nanomaterials by continuous-flow microfluidics: A review. *J. Nanosci. Nanotechnol.* **2014**, *14*, 1338–1363. [CrossRef] [PubMed]
24. Fanelli, F.; Parisi, G.; Degennaro, L.; Luisi, R. Contribution of microreactor technology and flow chemistry to the development of green and sustainable synthesis. *Bellstein J. Org. Chem.* **2017**, *13*, 520–542. [CrossRef] [PubMed]
25. Roch, A.; Muller, R.N.; Gillis, P. Theory of proton relaxation induced by superparamagnetic particles. *J. Chem. Phys.* **1999**, *110*, 5403–5411. [CrossRef]
26. Roch, A.; Gossuin, Y.; Muller, R.N.; Gillis, P. Superparamagnetic colloid suspensions: Water magnetic relaxation and clustering. *J. Magn. Mag. Mater.* **2005**, *293*, 532–539. [CrossRef]
27. Hannecart, A.; Stanicki, D.; Vander Elst, L.; Muller, R.N.; Brûlet, A.; Sandre, O.; Schatz, C.; Lecommandoux, S.; Laurent, S. Embedding of superparamagnetic iron oxide nanoparticles into membranes of well-defined poly(ethylene oxide)-block-poly(ε-caprolactone) nanoscale magnetovesicles as ultrasensitive MRI probes of membrane bio-degradation. *J. Mat. Chem. B* **2019**, *7*, 4692–4705. [CrossRef]
28. Boutry, S.; Forge, D.; Burtea, C.; Mahieu, I.; Murariu, O.; Laurent, S.; Vander Elst, L.; Muller, R.N. How to quantify iron in an aqueous or biological matric: A technical note. *Contrast Media Mol. Imaging* **2009**, *4*, 299–304. [CrossRef]
29. Chantrell, R.W.; Popplewell, J.; Charles, S.W. Measurements of particle size distribution parameters in ferrofluids. *IEEE Trans. Magn.* **1978**, *14*, 975–977. [CrossRef]
30. Ivanov, A.O.; Kantorovich, S.S.; Reznikov, E.N.; Holm, C.; Pshenichnikov, A.F.; Lebedev, A.V.; Chremos, A.; Camp, P.J. Magnetic properties of polydisperse ferrofluids: A critical comparison between experiment, theory, and computer simulation. *Phys. Rev.* **2007**, *75*, 1–12. [CrossRef]
31. Cogoni, G.; Grosso, M.; Baratti, R.; Romagnoli, J.A. Time evolution of the PSD in crystallisation operations: An analytical solution based on Ornstein-Ulhenbeck process. *AIChE* **2012**, *58*, 3731–3739. [CrossRef]
32. Scherrer, P. Bestimmung der Größe und der inneren Struktur von Kolloidteilchen mittels Röntgenstrahlen. *Gottingen. Math. Phys. Kl.* **1918**, *2*, 98–100.
33. Kwon, S.G.; Piao, Y.; Park, J.; Angappane, S.; Jo, Y.; Hwang, N.; Park, J.; Hyeon, T. Kinetics of monodisperse iron oxide nanocrystal formation by "heating-up" process. *J. Am. Chem. Soc.* **2007**, *129*, 12571–12584. [CrossRef]
34. Donegá, C.; Liljeroth, P.; Vanmaekelbergh, D. Physicochemical evaluation of the hot-injection method, a synthesis route for monodisperse nanocrystals. *Nanocrystals* **2005**, *12*, 1152–1162.
35. Van Embden, J.; Chesman, A.S.R.; Jasienak, J.J. The heat-up synthesis of colloidal nanocrystals. *Chem. Mater.* **2015**, *27*, 2246–2285. [CrossRef]
36. Lamer, V.K.; Dinegar, R.H. Theory, production and mechanism of formation of monodispersed hydrosols. *J. Am. Chem. Soc.* **1950**, *72*, 4847–4854. [CrossRef]
37. Stanicki, D.; Vander Elst, L.; Muller, R.N.; Laurent, S. Synthesis and processing of magnetic nanoparticles. *Curr. Op. Chem. Eng.* **2015**, *8*, 7–14. [CrossRef]
38. Park, J.; An, K.; Hwang, Y.; Park, J.; Noh, H.; Kim, J.; Park, J.; Hwang, N.; Hyeon, T. Ultra-large-scale syntheses of monodisperse nanocrystals. *Nat. Mater.* **2004**, *3*, 891–895. [CrossRef]
39. Baaziz, W.; Pichon, B.P.; Fleutot, S.; Liu, Y.; Lefevre, C.; Greneche, J.; Toumi, M.; Mhiri, T.; Begin-Colin, S. Magnetic iron oxide nanoparticles: Reproducible tuning of the size and nanosized-dependent composition, defects, and spin canting. *J. Phys. Chem. C* **2014**, *118*, 3795–3810. [CrossRef]
40. Hou, Y.; Xu, Z.; Sun, S. Controlled synthesis and chemical conversions of FeO nanoparticles. *Angew. Chem.* **2007**, *46*, 6329–6332. [CrossRef]
41. Harris, R.A.; Shumbula, P.M.; Van der Walt, H. Analysis of the interaction of surfactants oleic acid and oleylamine with iron oxide nanoparticles through molecular mechanics modeling. *Langmuir* **2015**, *31*, 3934–3943. [CrossRef]

42. Qi, B.; Ye, L.; Stone, R.; Dennis, C.; Crawford, T.M.; Mefford, O.T. Influence of ligand-precursor molar ratio on the size evolution of modifiable iron oxide nanoparticles. *J. Phys. Chem. C* **2013**, *117*, 5429–5435. [CrossRef]
43. Xu, Z.; Shen, C.; Hou, Y.; Gao, H.; Sun, S. Oleylamine as both reducing agent and stabilizer in a facile synthesis of magnetite nanoparticles. *Chem. Mater.* **2009**, *21*, 1778–1780. [CrossRef]
44. Belaïd, S.; Stanicki, D.; Vander Elst, L.; Muller, R.; Laurent, S. Influence of experimental parameters on iron oxide nanoparticles properties synthesized by thermal decomposition: Size and nuclear magnetic resonance studies. *Nanotechnology* **2018**, *29*, 1–30. [CrossRef] [PubMed]
45. Nightingale, A.M.; Krishnadasan, S.H.; Berhanu, D.; Niu, X.; Drury, C.; McIntyre, R.; Valsami-Jones, E.; DeMello, J.C. A stable droplet reactor for high temperature nanocrystal synthesis. *Lab. Chip.* **2011**, *11*, 1221–1227. [CrossRef] [PubMed]
46. Jiao, M.; Zeng, J.; Jing, L.; Liu, C.; Gao, M. Flow synthesis of biocompatible Fe$_3$O$_4$ nanoparticles: Insight into the effects of residence time, fluid velocity, and tube reactor dimension on particle size distribution. *Chem. Mater.* **2015**, *27*, 1299–1305. [CrossRef]
47. Estelrich, J.; Sánchez-Martin, M.J.; Busquets, M.A. Nanoparticles in magnetic resonance imaging: From simple to dual contrast agents. *Int. J. Nanomed.* **2015**, *10*, 1727–1741.
48. Hashemi, R.H.; Bradley, W.G.; Lisanti, C.J. *MRI: The Basics*; Lippincott Williams & Wilkins: Philadelphia, PA, USA, 2004.
49. Laurent, S.; Vander Elst, L.; Muller, R.N. Comparative study of the physicochemical properties of six clinical low molecular weight gadolinium contrast agents. *Contrast Media Mol. Imaging* **2006**, *1*, 128–137. [CrossRef]

© 2020 by the authors. Licensee MDPI, Basel, Switzerland. This article is an open access article distributed under the terms and conditions of the Creative Commons Attribution (CC BY) license (http://creativecommons.org/licenses/by/4.0/).

Article

Synthesis of Magnetic Ferrite Nanoparticles with High Hyperthermia Performance via a Controlled Co-Precipitation Method

Mohamed S. A. Darwish [1,2], Hohyeon Kim [1], Hwangjae Lee [3], Chiseon Ryu [3], Jae Young Lee [3,*] and Jungwon Yoon [1,4,*]

1. School of Integrated Technology, Gwangju Institute of Science and Technology, Gwangju 61005, Korea
2. Petrochemicals Department, Egyptian Petroleum Research Institute, 1 Ahmed El-Zomor Street, El Zohour Region, Nasr City, Cairo 11727, Egypt
3. School of Materials Science and Engineering, Gwangju Institute of Science and Technology, Gwangju 500-712, Korea
4. Research Center for Nanorobotics in Brain, Gwangju Institute of Science and Technology, Gwangju 500-712, Korea
* Correspondence: jaeyounglee@gist.ac.kr (J.Y.L.); jyoon@gist.ac.kr (J.Y.)

Received: 16 July 2019; Accepted: 13 August 2019; Published: 16 August 2019

Abstract: Magnetic nanoparticles (MNPs) that exhibit high specific loss power (SLP) at lower metal content are highly desirable for hyperthermia applications. The conventional co-precipitation process has been widely employed for the synthesis of magnetic nanoparticles. However, their hyperthermia performance is often insufficient, which is considered as the main challenge to the development of practicable cancer treatments. In particular, ferrite MNPs have unique properties, such as a strong magnetocrystalline anisotropy, high coercivity, and moderate saturation magnetization, however their hyperthermia performance needs to be further improved. In this study, cobalt ferrite ($CoFe_2O_4$) and zinc cobalt ferrite nanoparticles ($ZnCoFe_2O_4$) were prepared to achieve high SLP values by modifying the conventional co-precipitation method. Our modified method, which allows for precursor material compositions (molar ratio of $Fe^{+3}:Fe^{+2}:Co^{+2}/Zn^{+2}$ of 3:2:1), is a simple, environmentally friendly, and low temperature process carried out in air at a maximum temperature of 60 °C, without the need for oxidizing or coating agents. The particles produced were characterized using multiple techniques, such as X-ray diffraction (XRD), dynamic light scattering (DLS), transmission electron microscopy (TEM), ultraviolet-visible spectroscopy (UV–Vis spectroscopy), and a vibrating sample magnetometer (VSM). SLP values of the prepared nanoparticles were carefully evaluated as a function of time, magnetic field strength (30, 40, and 50 kA m^{-1}), and the viscosity of the medium (water and glycerol), and compared to commercial magnetic nanoparticle materials under the same conditions. The cytotoxicity of the prepared nanoparticles by in vitro culture with NIH-3T3 fibroblasts exhibited good cytocompatibility up to 0.5 mg/mL. The safety limit of magnetic field parameters for SLP was tested. It did not exceed the 5×10^9 Am^{-1} s^{-1} threshold. A saturation temperature of 45 °C could be achieved. These nanoparticles, with minimal metal content, can ideally be used for in vivo hyperthermia applications, such as cancer treatments.

Keywords: specific loss power; magnetic ferrite; modified co-precipitation; hyperthermia

1. Introduction

Recently, magnetic nanoparticles have shown great potential for application in various biomedical fields such as drug delivery, magnetic separation, imaging, and hyperthermia cancer treatments [1–4]. In particular, the hyperthermia capability of magnetic nanoparticles, by which they convert dissipated magnetic energy into thermal energy, enables cancer treatment. Such hyperthermia treatment depends on heating of the region affected by cancer, where the temperatures between 43 and 45 °C can be reached using magnetic nanoparticles under an alternating current (AC) magnetic field [5–7]. Hyperthermia can destroy the cancer cells with minimal influence on the healthy tissues, so it could potentially be used for localized, scarless, and economical treatments with few side effects. The efficacy of the hyperthermia process depends on many factors, such as properties of the magnetic nanoparticles, the magnitude of the applied AC magnetic field (H), and its frequency (f) and duration (t) of actuation [8,9]. The magnitude and frequency of the applied AC magnetic field must remain within safe limits to prevent unwanted side effects. A commonly prescribed safety limit is that the product of the frequency and field amplitude ($H \times f$) should be no greater than 5×10^9 Am^{-1} s^{-1} to protect the healthy tissues against excessive heating [10]. In addition, particle size is very important in determining the magnetic properties. By controlling the particle size in the transition range (about 20 nm), the particle properties can be moved from superparamagnetic to ferromagnetic, meaning that higher SLP values can be obtained [10]. Thus, it is necessary to enhance the heating ability of nanoparticles by controlling their size to obtain the required temperature rise during hyperthermia. The heating power generated per particle, the specific loss power (SLP), should be as high as possible in the injected material, and ensuring bio-safety is considered the most critical challenge to achieving desirable tumor destruction [7]. In superparamagnetic nanoparticles, the two mechanisms primarily responsible for magnetic relaxation in nanoparticles involve the physical rotation of the individual particles in the fluid (Brownian relaxation) and the collective rotation of the atomic magnetic moments within each particle (Néel relaxation) [1]. Based on these mechanisms, controlling the morphology and composition of the magnetic nanoparticles is an effective method for increasing the SLP. However, due to the problem of toxicity, the number of usable elements is severely limited [3,8]. Cobalt ferrite nanoparticles ($CoFe_2O_4$) have some advantageous unique properties, as compared to ferrous ferrite, including strong magnetocrystalline anisotropy, high coercivity (Hc), and moderate saturation magnetization (Ms) [4,5]. These properties, along with their high oxidative and thermal stability, make these nanoparticles attractive for hyperthermia applications. $CoFe_2O_4$ also have a variety of medical and technological applications due to their high moments at low magnetic fields, together with superparamagnetic properties [11,12]. Furthermore, they are non-toxic, biocompatible, and can be heated remotely by alternating magnetic fields. Control of their morphology and size can be achieved by varying the pH, ionic strength, coating agent, and temperature of the reaction [13–16]. Cobalt ferrite nanoparticles have been prepared by several methods, including sol-gel [11], hydrothermal [12,13], co-precipitation [14], and thermal decomposition methods [15,16]. Using the co-precipitation process, many researchers have made efforts to achieve the smallest possible particles and to improve the magnetic properties and SLP of the cobalt ferrite nanoparticles and cobalt zinc ferrite nanoparticles [14,17–35]. The ferromagnetic-superparamagnetic size threshold for cobalt ferrite nanoparticles has been reported by Pereira et al., who prepared superparamagnetic cobalt ferrite nanoparticles with the tuning of particle size (4.2–4.8 nm) and magnetic properties (Ms 30.6–48.8 emu/g) using a co-precipitation method. Saturation magnetization increases with the increase in particle size until it reaches a threshold size beyond which magnetization is constant and is close to the bulk value [23]. Recently, a high SLP of 237–272 W/g$_{metal}$ was obtained with $CoFe_2O_4$ nanoparticles via the co-precipitation method at 90 °C in the presence of chitosan as a coating agent [17]. In another report, a high SLP of 91.84 W/g$_{metal}$ was obtained using $CoFe_2O_4$ nanoparticles via a co-precipitation method, at room temperature in the presence of a coating agent (polyethylene glycol and oleic acid) [18]. Under safe AC field conditions, a high SLP of 2131 W/g$_{metal}$ was obtained with $CoFe_2O_4$ nanoparticles fabricated by co-precipitation (100 °C in the presence of Lauric acid) [33]. It is necessary to overcome the drawbacks of the reported synthesis processes, such as being environmentally unfavorable, expensive, dependent on high energy, and time-consuming, as well as yielding low SLP values. The solution-based synthesis method is sensitive to the composition and temperature of the precursor materials, which in turn influences the formation of the cobalt ferrite nanoparticles, particle size, and degree of crystallinity. In the current investigation, we modified the conventional co-precipitation process to make it simple, environmentally friendly, and amenable to operation under relatively low temperatures. The properties of $CoFe_2O_4$ and $ZnCoFe_2O_4$, prepared by varying the compositions of the precursor material (molar ratio of Fe^{+3}:Fe^{+2}:Co^{+2}/Zn^{+2} of 3:2:1) at relatively low temperature (60 °C) and without oxidizing or coating agents, were investigated. The magnetic properties, optical activity, and toxicity were evaluated in an in vitro culture with NIH-3T3 fibroblasts. Ensuring the bio-safety of the nanoparticles is considered to be the greatest challenge to the development of desirable treatments. The heat generation performance of the MNPs was studied while varying the magnetic field, timing, concentration, and viscosity of the medium.

2. Experimental Work

2.1. Materials

Iron(III) chloride hexahydrate, iron(II) chloride tetrahydrate, cobalt(II) chloride hexahydrate, zinc(II) chloride, sodium citrate, and ammonium hydroxide were purchased from Sigma Aldrich (St. Louis, MO, USA). We also tested commercial iron oxides, as follows: BNF, the particles of which are magnetite with a shell of dextran dispersed in water, from micromod Partikeltechnologie GmbH (Rostock, Germany). SHA30 and SHA15, which are amine iron oxide nanoparticles dispersed in PBS with sizes of 30 and 15 nm, respectively, from Ocean Nanotech, LLC, Manufacturing Facility and R&D (San Diego, CA, USA). HyperMAG A, HyperMAG B, and HyperMAG C, which are magnetic iron oxide nanoparticles with sizes of 10.3, 11.7, and 15.2 nm, respectively, dispersed in water, from nanoTherics Ltd. (Newcastle, UK). Resovist, which are superparamagnetic iron oxide nanoparticles coated with carboxydextran, from Meito Sangyo Co., Ltd. (Nagoya, Japan).

2.2. Preparation of Cobalt Ferrite and Zinc Cobalt Ferrite Nanoparticles by Co-Precipitation

The synthesis process used herein is a modified version of a previously reported co-precipitation method [14,17–35]. The molar ratio of iron(III) chloride hexahydrate(8.1 g): iron(II) chloride tetrahydrate (3.97 g): cobalt(II) chloride hexahydrate (2.37 g) was precisely set to be 3:2:1 and mixed in 50 mL of distilled water for 15 min to obtain a homogeneous solution at room temperature. The temperature was then increased to 60 °C, and maintained for 5 min, to ensure complete homogenous mixing. With vigorous stirring, 20 ml of ammonium hydroxide (30%) was added in a dropwise manner to induce the particle growth, followed by additional stirring for 30 min at 60 °C to evaporate any excess ammonia. The black precipitate thus formed was washed several times using distilled water to remove possible ammonium salts and then dried for 24 h to obtain a cobalt ferrite nanoparticle powder.

For preparation of zinc cobalt ferrite nanoparticles, the same procedure as described above was used with the addition of zinc chloride (1.36 g).

For coating of nanoparticles with sodium citrate, 0.5 g of sodium citrate was dispersed in 25 mL water by sonication using an ultrasonic bath for 30 min to form a homogenous solution. A total of 0.1 g of nanoparticles was added to the solution and sonicated for 4 h using the ultrasonic bath.

2.3. Characterization

Dynamic light scattering (DLS) and zeta potential measurements were performed using a Zeta-potential and Particle Size Analyzer (ELSZ-2000; Photal Otsuka Electronics, Osaka, Japan). For DLS and zeta potential tests, suspensions of 50 mg nanoparticles in 6 mL deionized water were subjected to ultrasound (5 min) before the analyses. The magnetic properties of the samples were measured using a vibrating sample magnetometer (VSM; Lake Shore 7400 series; Lake Shore Cryotronics, Westerville, OH, USA). The X-ray diffraction (XRD) of the nanoparticles was analyzed using an X-ray diffractometer (Rigaku, Japan). The diffractometer used a copper X-ray tube and Cu Kr radiation. A scan speed of 4°/min and 2θ ranging from 5° to 65° were selected as the measurement parameters. The metal contents of the precursors were analyzed using inductively coupled plasma-optical emission spectroscopy (ICP-OES; Optima 8300, PerkinElmer, Waltham, MA, USA). A total of 0.1 g of the nanoparticles dispersed in 25 mL D.I. water was subjected to ultrasound before the analyses. The quantitation range for cost elements was 50 ppm for ICP-OES. The samples were made using an aqueous nitric acid solution. Additional dilutions were performed to make the sample concentrations according to the specified range. The morphology and structure of the materials were characterized by transmission electron microscopy (TEM) and selected area electron diffraction (SAED) (Tecnai G2S Twin; Philips, USA) at 300 keV. The optical properties were evaluated using ultraviolet-visible spectroscopy (UV–Vis spectroscopy). The clear colloid obtained after sonicating the nanoparticles dispersed in deionized water was used for the measurement and pure deionized water was used as a reference. A band gap energy (E_g) is an intrinsic property of a material and it estimated from the absorption curve. Generally, electrons can jump from one band to another as long as they have the specific minimum amount of energy required for the transition. To investigate direct and indirect transitions, we plotted $(\alpha h\nu)^2$ against 'hν' and $(\alpha h\nu)^{1/2}$ against 'hν'. The band gap energy of the material is related to the absorption coefficient 'α' by the Tauc relation, as follows:

$$(\alpha h\nu)^n = (\text{absorption coefficient} \times \text{energy})^n = (2.303 A h\nu)^n,$$

where A is a constant, hν is the photon energy, and n is a number ($n = 2$ and $\frac{1}{2}$ for direct and indirect transitions, respectively).

$$\text{Band gap energy } (E_g) = hc/\lambda,$$

where h is Planck's constant (6.626×10^{-34} Joules/s), c is the speed of light (3.0×10^8 m/s), and λ is the cut off wavelength ($E_g = 1240$ eV nm/λ) (energy in eV).

Heating efficiency and SLP were measured using our Lab-made system. In this setup, a function generator is used to generate a sinusoidal voltage signal which is amplified to the desired power through an AC Power amplifier, AE Techron 7224. This amplified signal is fed to a litz wire coil wound around a ferrite core to induce the alternating magnetic field (30–50 kA/m at 97 kHz). High voltage capacitors, CSM 150 capacitor, are used to set the resonant frequency and the particle temperature is measured using an Osensa optic temperature measurement cable. The SLP is calculated using the following equation; we take into account only the first few seconds, in which a quasi-adiabatic regime is assumed [29–35]:

$$SLP = (C_p/m) \times (dT/dt),$$

where dT/dt is the initial gradient of the time-dependent temperature curve, C_p is the volumetric specific heat capacity of the sample solution, J/(g °C) (4.184 [water], 2.43 [glycerol]), and m is the mass of elements in the particles.

2.4. In Vitro Cytocompatibility Test

The toxicity of the prepared nanoparticles was studied by culturing NIH-3T3 cells using a WST assay [36]. The cells were seeded at a seeding density of 5×10^4 cells/well in a 24-well plate and incubated for 1 day in Dulbecco's modified Eagle's medium (DMEM) supplemented with 10% fetal bovine serum (FBS) and 1% antibiotics. Then, the culture medium (1 mL) containing the prepared nanoparticles (0.125, 0.25, or 0.5 mg/mL) was added to the cells and incubated for an additional day. Cells without any treatment of nanoparticles were used as the control. The sample solution was removed and washed with Dulbecco's phosphate-buffered saline (DPBS). Then, fresh culture medium (0.5 mL) and WST assay solution (0.05 mL) were added to each well and incubated for 2 h. Finally, 0.1 mL of the solution from each well was transferred to a new plate and the absorbance (A) at 450 nm was measured using a microplate reader. The cell viability was normalized to the control using the following formula:

$$\text{Cell Viability (\%)} = (A_c - A_s)/A_c \times 100,$$

where A_c is the absorbance of the control sample and A_s is the absorbance of the sample solution.

2.5. MNP Uptake (Prussian Blue Staining)

NIH-3T3 cells were seeded at a seeding density of 5×10^4 cells/well in a 24-well plate and incubated for 1 day in Dulbecco's modified Eagle's medium (DMEM) supplemented with 10% fetal bovine serum (FBS) and 1% antibiotics. Culture media were replaced with fresh media containing 4 types of MNPs with a concentration of 12.5 µg/mL and the cells were further incubated for 24 h. Cells without any treatment of nanoparticles were used as the control. After incubation, the cells were washed twice with DPBS and fixed with 4% paraformaldehyde at room temperature for 15 min. Then, Prussian blue solution prepared with equal volumes of 4% potassium ferrocyanide (II) trihydrate and 4% HCl (in PBS) was added to each well and incubated at room temperature for 20 min. Then, the cells were washed with DPBS and imaged using a microscope.

3. Results and Discussion

3.1. Synthesis and Characterization of Magnetic Nanoparticles

Controlling the chemical composition and size of magnetic materials is important to enhance magnetization that effectively leads to an increase in magnetic hyperthermia heating efficiency of magnetic NPs. Cobalt ferrite nanoparticles (CF-MNPs) and zinc cobalt ferrite nanoparticles (ZCF-MNPs) were prepared by a controlled co-precipitation method, which is a simple, environmentally friendly, and low-temperature method. By shaking, the prepared nanoparticles can be dispersed in water, but after a while they tend to aggregate and sediment. As shown in Figure 1, CF-MNPs had a wide size distribution with an average particle size of 8 ± 2 nm. In the case of ZCF-MNPs, the average particle size was 25 ± 5 nm, with a wide size distribution and varied agglomeration behavior. The incorporation of Zn in the NP structure lead to an increase in the particle size and agglomeration. Magnetic particles agglomerate as a result of high surface energy between the nanoparticles and magnetic dipole–dipole interactions [20,21]. The crystalline nature of the prepared nanoparticles was observed using HRTEM. The clear lattice boundary in the HRTEM image illustrates the higher crystallinity of CF-MNPs, as compared to ZCF-MNPs, which is confirmed by XRD (shown later). The corresponding selected area electron diffraction (SAED) image of nanoparticles displays the ring characteristics consistent with a structure composed of small domains with their crystallographic axes randomly oriented with respect to one another. The SAED pattern shows diffuse rings with less intensity that can be indexed to the nanoparticle plane reflections. Results indicate that our method produced smaller nanoparticles with fewer aggregations for CF-MNPs, as compared to ZCF-MNPs, due to a progressive increase in the solubility product constant of the corresponding divalent metal hydroxides [19–21].

As shown in Figure 2, the mean hydrodynamic size obtained by DLS was 50.9 and 575 nm for CF-MNPs and ZCF-MNPs, respectively (which was higher than that obtained using TEM analysis). Wide size distribution may result from the hydrophobic nature of the prepared nanoparticles [37]. The measurements of the zeta potential were used to assess the effects of nanoparticles in the colloidal phase and their aggregates. Higher zeta potentials indicate stable nanoparticle systems [18]. The zeta potential values were +30.59 and +14.69 mV for CF-MNPs and ZCF-MNPs, respectively (Figure 2). Thus, the synthesized cobalt ferrite nanoparticles dispersed in water due to the large electrostatic repulsive forces between the particles. In contrast, zinc cobalt ferrite nanoparticles appeared less stable due to the low electrostatic repulsive forces between them. Obtaining stable colloidal systems is particularly important from the perspective of using synthesized nanoparticles in nano-medicine and biomedical applications [38].

The crystalline properties, such as average crystallite size (nm) and degree of crystallinity, are important for the hyperthermic performance of magnetic nanoparticles. The ferrite crystalline properties of the prepared cobalt ferrite nanoparticles and zinc cobalt ferrite nanoparticles were investigated together with JCPDS data (#221086) using XRD (Figure 3). The diffraction peaks, indexed to (111), (220), (311), (222), (400), (422), (511), and (440), for the prepared nanoparticles are shown in Table 1. Shifting in peak position towards lower 2θ with decreasing intensities of peaks is caused by the presence of Zn in the structure of ZCF-MNPs.

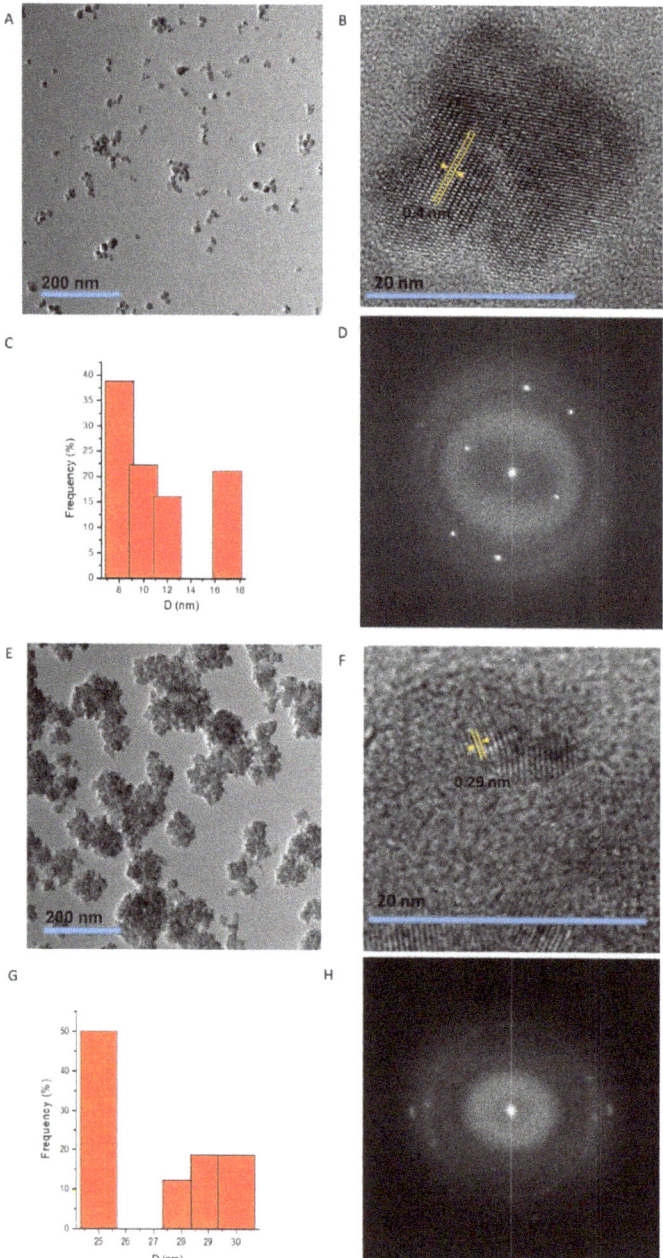

Figure 1. Transmission electron microscopy (TEM), selective area electron diffraction (SAED) and particle size distribution histograms for CF-MNPs (**A–D**) and ZCF-MNPs (**E–H**).

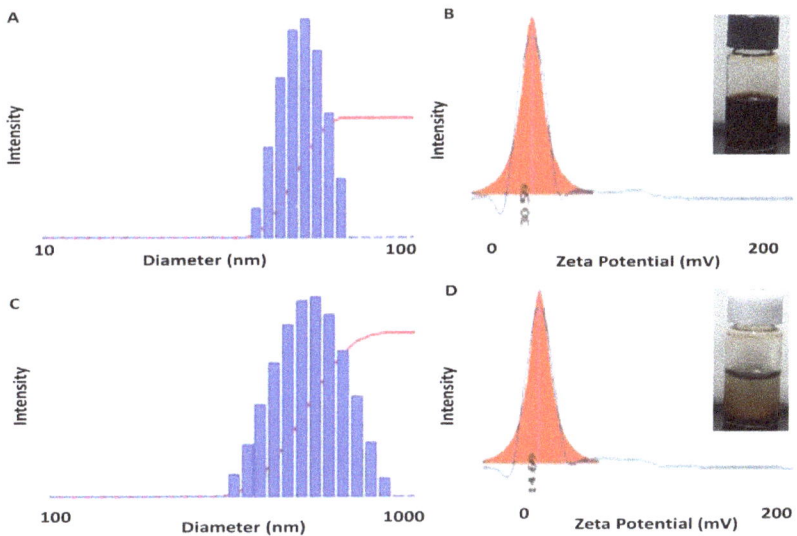

Figure 2. Dynamic light scattering (DLS) and zeta potential results for CF-MNPs (**A**,**B**) and ZCF-MNPs (**C**,**D**).

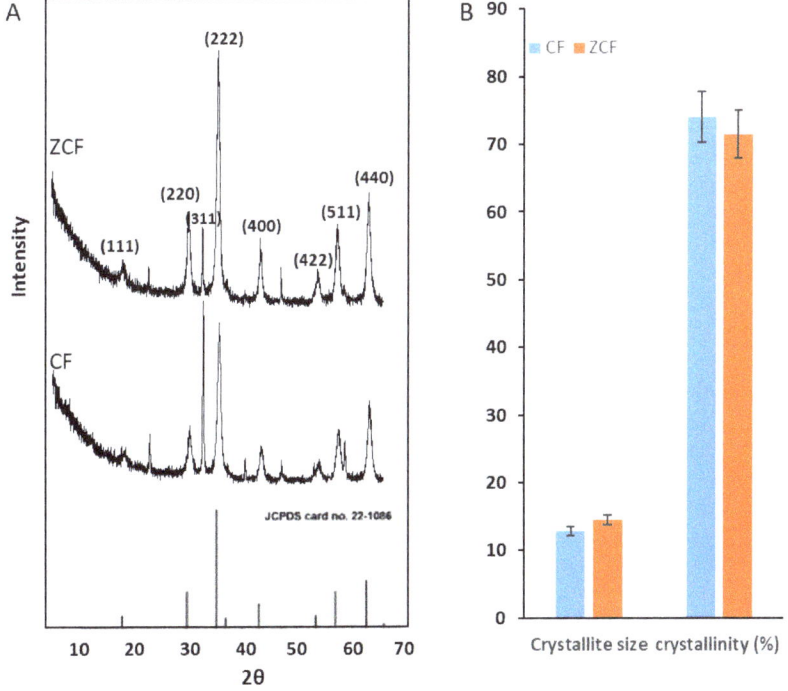

Figure 3. X-ray diffraction (XRD) results for CF-MNPs and ZCF-MNPs, (**A**) average crystallite size (nm) and (**B**) degree of crystallinity (%).

Table 1. The indexed diffraction peaks from XRD.

Indexed	2θ	
	CF	ZCF
(111)	18.22	18.06
(220)	30	29.86
(311)	32.46	32.40
(222)	35.42	35.22
(400)	42.96	42.88
(422)	53.52	53.32
(511)	56.96	56.78
(440)	62.58	62.51

We calculated the crystallite sizes of the nanoparticles based on the Scherrer formula [17], as follows:

$$\text{Crystallite size } (D_p) = K\lambda/(B\cos\theta),$$

where D_p is the average crystallite size (nm), B is the full width at half maximum (FWHM) of XRD peak, λ is the X-ray wavelength (1.5406 Å, K (Scherrer constant [shape parameter]): 0.89), and θ is the XRD peak position.

% Crystallinity = total area of crystalline peaks/total area of all peaks.

The crystallite sizes for the higher intensity peaks of the CF-MNPs were 29.1, 10.5, 8.7, 9.7, 8.9, and 10.4 nm, and the average crystallite size was 12.9 nm. In the case of ZCF-MNPs, the crystallite sizes of the highest intensity peaks were 31.3, 11.5, 9.7, 13.5, 10.3, and 10.8 nm, and the average crystallite size was 14.5 nm. The presence of zinc significantly affected the particle size and degree of crystallinity of the nanostructure, as shown in Figure 3. The crystallite size increased from 12.9 to 14.5 nm, while the degree of crystallinity [39] decreased from 74.03% to 71.5%, as shown in Figure 3, which is also confirmed by TEM.

Ferrites nanoparticles can be prepared from ferrous ions [25,26], ferric ions [17–20], or a mix of ferrous and ferric ions [21,31,34]. Chinnasamy et al. reported that the particle size of the ferrite powders decreased with an increase in ferric ion concentration [21]. The increase in the size of our particles caused an increase in the anisotropy energy, which in turn resulted in an increase in Hc and Ms.

Photocatalysis combined with magnet heating constitutes a typical example of the so-called theranostic agents. The optical absorbance and band energy of the prepared nanoparticles were investigated by UV–Vis spectroscopy. Figure 4 shows the measured optical absorbance spectra of cobalt ferrite nanoparticles and zinc cobalt ferrite nanoparticles at ambient temperatures. The UV–visible absorption spectra of the prepared nanoparticles showed a broad absorption range (300–600 nm) in the visible wavelength range, which we attributed to d-orbital transitions of Fe^{3+}. Especially, the absorption peak was close to 490 nm, which corresponds to the d-d transitions of Fe^{3+} in a tetrahedral coordination environment [40–42].

The calculated direct band gap energies of cobalt ferrite and zinc cobalt ferrite nanoparticles were 3.15 and 2.9 eV, respectively, while the respective calculated indirect band gap energies were 2.6 and 2.3 eV, as shown in Figure 5. It is clear that, as the crystallite size increased from 12.92 to 14.57 nm, the band gap energy decreased from 3.15 to 2.9 eV for cobalt ferrite and zinc cobalt ferrite nanoparticles. The band gap energy of the prepared nanoparticles varied with an inverse relationship with their sizes. This is similar to what has been reported in previous studies [43].

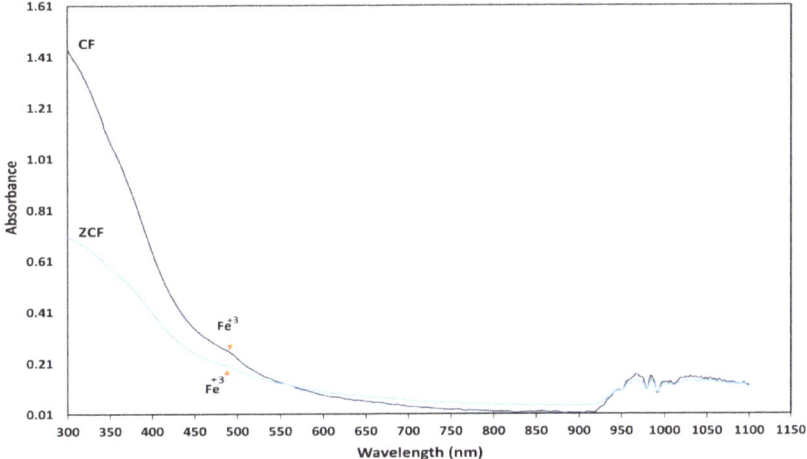

Figure 4. Optical absorbance spectra of CF-MNPs and ZCF-MNPs.

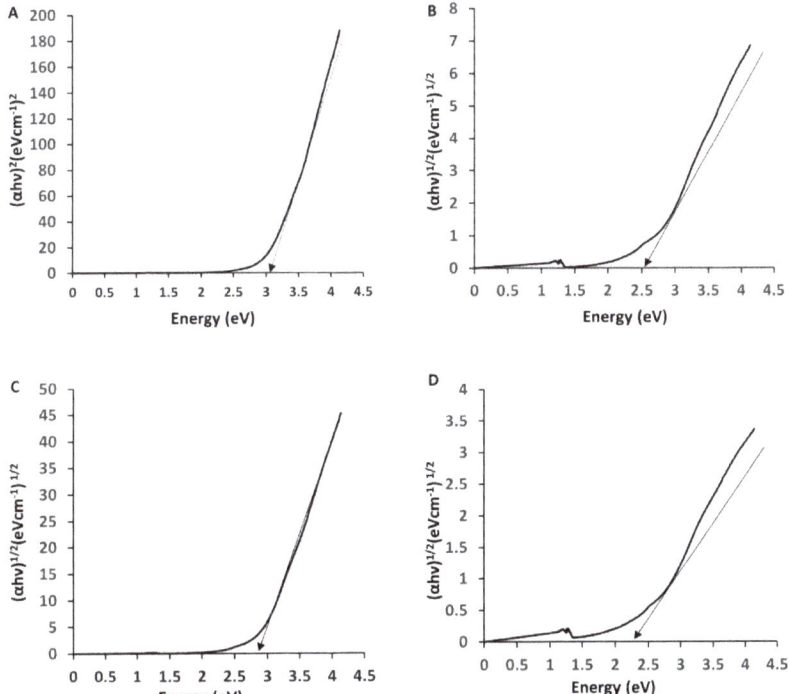

Figure 5. Band gap energy for CF-MNPs (direct (**A**), indirect (**B**)) and for ZCF-MNPs (direct (**C**), indirect (**D**)).

The magnetic properties of the prepared nanoparticles were measured by VSM. The M–H results exhibit a clear hysteresis loop (Figure 6), which indicates that the magnetic nanoparticles were ferromagnetic. The cobalt ferrite nanoparticles had the following values: M_s: 50.61 emu/g, M_r: 10.75 emu/g and H_c: 159.8 Oe. The corresponding values for the zinc cobalt ferrite nanoparticles were as follows: M_s: 50.71 emu/g, M_r: 10.71 emu/g, and H_c: 225 Oe. Hc and Ms of nanoparticles increase with particle size. In addition, as in our case, the Hc and Ms are also increased by the shape anisotropy contribution. Thus, the increase in particle size will increase the anisotropy energy, which in turn results in an increase in the coercivity and magnetization saturation [44]. The squareness (SQ) or reduced remanence value is equal to M_r/M_s. When SQ is greater than or equal to 0.5, the material has a single magnetic domain structure, whereas it has a multi-domain structure when it is below 0.5. In our study, the SQ values were 0.212 and 0.211 for cobalt ferrite and zinc cobalt ferrite nanoparticles, respectively. These values less than 0.5 indicate the formation of a multi-domain structure, as has been observed previously [45].

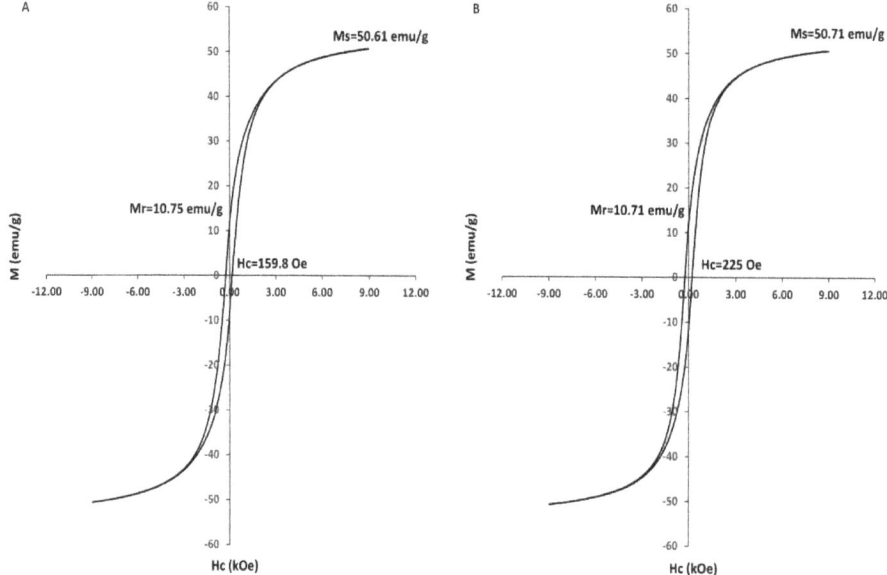

Figure 6. Hysteresis loops of CF-MNPs (**A**) and ZCF-MNPs (**B**).

3.2. Heat Generation Performance of the MNPs

The heating properties of the prepared nanoparticles upon exposure to an AC magnetic field with frequency (97 kHz) at various magnetic field strengths (30, 40, or 50 kA/m) and concentrations (8 or 25 mg/mL) were investigated (Figure 7). The initial rise in temperature over time was approximately linear, then it slowed down gradually until saturation. The rate of heating increased with the concentration and magnetic field strength, as shown in the temperature curves in Figure 7. As the heating rate increased, the samples became heated faster according to magnetic field strength (50 > 40 > 30 kA/m) and concentration (25 > 8 mg/mL). At the highest magnetic field strength (50 kA/m), and with the higher concentration (25 mg/mL) of cobalt ferrite and zinc cobalt ferrite nanoparticles, the temperature rose to higher than 70 and 60 °C, respectively, within 1 min. At the highest magnetic field strength (50 kA/m) and lower concentration (8 mg/mL) of cobalt ferrite and zinc cobalt ferrite nanoparticles, the temperature rose higher than 45 °C within 9 and 3 min, respectively. At a moderate magnetic field strength (40 KA/m) and frequency of 97 kHz with the higher concentration (25 mg/mL) of cobalt ferrite and zinc cobalt ferrite nanoparticles, the temperature rose to over 45 °C within 2 and 3 min, respectively. At a moderate magnetic field strength (40 kA/m) and frequency of 97 kHz for both concentrations (25 and 8 mg/mL) of cobalt ferrite and zinc cobalt ferrite nanoparticles, the temperature did not reach 45 °C, even over a longer period. At the lowest magnetic field strength (30 kA/m) and frequency of 97 kHz with both concentrations (25 and 8 mg/mL) of cobalt ferrite and zinc cobalt ferrite nanoparticles, the temperature did not reach 45 °C, even over longer periods.

Nanoparticles that can increase temperature to 45 °C are suitable for cancer treatment. Faster treatment with a low metal content is highly desirable for hyperthermia applications. Furthermore, for effective therapy, the temperature of cancerous tissue needs to reach 42–45 °C, while temperatures greater than 50 °C cause damage to cancer cells via thermoablation. As shown in Figure 8, the comparison of the saturated temperatures revealed that an increase in the saturation temperature of the nanoparticles was accompanied by the higher concentration and strength of the magnetic field.

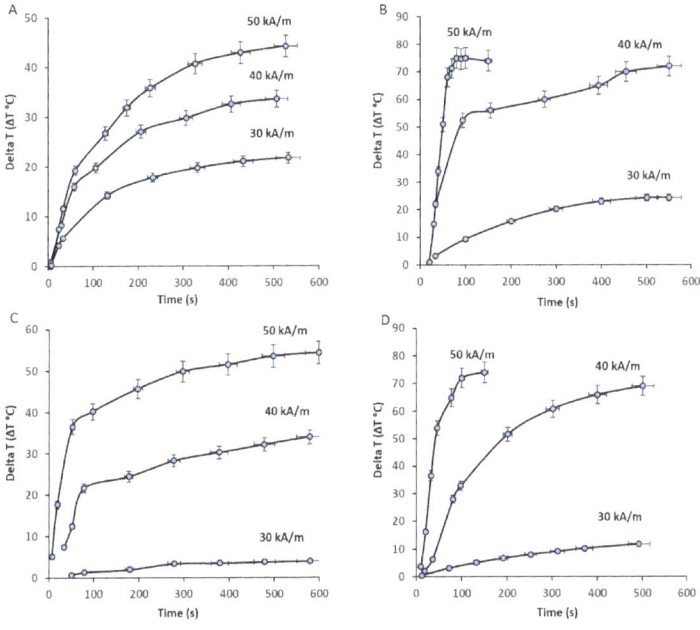

Figure 7. Heating profiles of CF-MNPs ((8 mg/mL) (**A**), (25 mg/mL) (**B**)) and ZCF-MNPs ((8 mg/mL) (**C**), (25 mg/mL) (**D**)) with magnetic field strengths (30, 40, or 50 kA/m) at a frequency of 97 kHz.

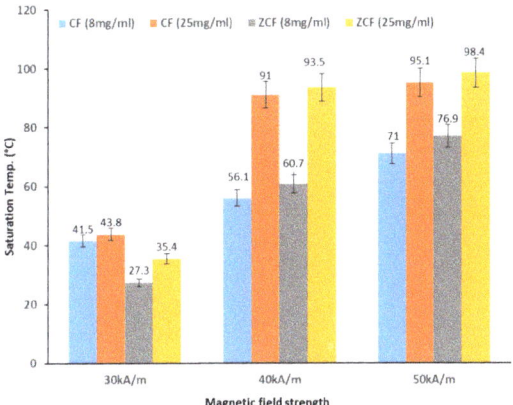

Figure 8. Saturation temperatures of CF-MNPs and ZCF-MNPs by concentration (8 or 25 mg/mL) and magnetic field strengths (30, 40, or 50 kA/m).

The SLP is used as an indicator to measure how much energy is absorbed per mass of magnetic nanoparticles when exposed to an alternating magnetic field.

The metal content in cobalt ferrite was Fe at 56.5% and Co at 7.81% (the total metal content was 64.31%). The metal content in zinc cobalt ferrite was Fe at 53.0%, Co at 6.31%, and Zn at 3.41% (the total metal content was 62.72%). We calculated the SLP (W/g_{metals}) values based on the total metal contents. The SLP of nanoparticles in an external AC magnetic field, as mentioned earlier, can be attributed to two power loss mechanisms, as follows: Néel relaxation and Brownian relaxation. The variation in SLP values is due to several reasons, including sample size, concentrations and the magnitude and frequency of the applied field [29–35,45–47]. With respect to human exposure, it is very important to maintain the product of the magnetic field strength (H), and its frequency (f), below a threshold value. A safety limit that has been commonly prescribed is that the product of the frequency and the field amplitude should remain below $C = H \times f = 5 \times 10^9$ Am^{-1}s^{-1} to minimize any collateral effects of alternating magnetic fields on the human body [10]. The values of C in our experiments were calculated to be 2.9×10^9, 3.8×10^9, and 4.8×10^9 Am^{-1} s^{-1} for 30, 40, and 50 kA/m, respectively. Thus, our experimental conditions for alternative magnetic field did not exceed the safety limit. In the current study, the highest SLP value obtained was 552 W/g_{metal} for zinc cobalt ferrite nanoparticles with a concentration of 8 mg/mL at 50 kA/m and 97 kHz. The lowest SLP obtained was 11.57 W/g_{metal} for zinc cobalt ferrite nanoparticles with a concentration of 25 mg/mL at 30 kA/m and 97 kHz. For ZCF, the magnetic coupling between soft and hard ferrite tune the magnetic anisotropy and, therefore, enhance the SLP values, improving the efficiency of energy conversion for hyperthermia applications. These results show the importance of changing the chemical composition of magnetic materials to enhance magnetization, which effectively leads to an increase in magnetic hyperthermia heating efficiency of magnetic NPs. SLP values are not related to concentration in this study. The observed reduction in SLP is probably caused by an extended aggregation of ferromagnetic nanoparticles, due to the application of the alternating magnetic field, and the power dissipation in the medium decreased. Thus, the heating performance was reduced. When comparing the SLP values of our nanoparticles to those of commercial iron oxide materials under the same magnetic field strength (40 kA/m), our zinc cobalt ferrite nanoparticles were found to show lower SLP values than SHA25, but higher values than the others (BNF, MagA, MagC, Resovist, MagB, and SHA30). Cobalt ferrite nanoparticles showed lower SLP values than SHA25, BNF, and MagA, but higher values than MagC, Resovist, MagB, and SHA30, as shown in Figure 9. Our results are of particular interest for heating therapy applications due to the high SLP obtained under the magnetic field parameters tested and the achievement of a saturation temperature of 45 °C. These characteristics are ideal for cancer treatment applications with minimal metal content.

It is important to differentiate between the contributions of the Néel and Brownian mechanisms to heat generation. For this purpose, we dispersed zinc cobalt ferrite nanoparticles in a high viscosity solvent. When glycerol was used as the dispersing solvent, the heating rate became very low as the exposure time increased (Figure 10). For example, in this study, the highest SLP (552 W/g_{metal}) was obtained for zinc cobalt ferrite nanoparticles with a concentration of 8 mg/mL at 50 kA/m, dispersed in water. The effect of the viscosity of the solvent on the SLP value was surprising; it caused an obvious drop in SLP from 552 W/g_{metal} (water as a dispersed medium) to 35 W/g_{metal} (glycerol as a dispersed medium). Glycerol molecules contain three OH groups per molecule, which allows for extensive hydrogen bonding between the many oxygen and hydrogen atoms, which makes the substance viscous. This indicates that glycerol hinders convective heat transfer under alternating magnetic fields. The charges on the surfaces of these nanoparticles were neutralized by the surrounding solvent ions (OH$^-$), providing stability. Brownian relaxation, which is due to the physical rotation of the particles within the medium, is hindered by the hydrodynamic volume and viscosity of the particles, which in turn tends to inhibit the rotation of the particles in the medium.

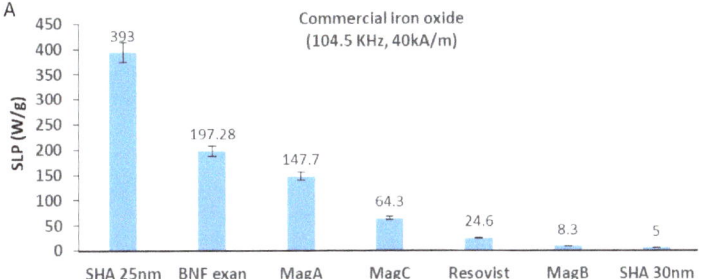

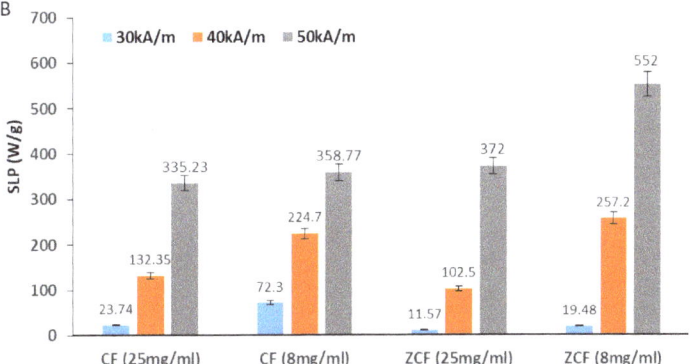

Figure 9. Specific loss power (SLP) values of commercial iron oxides (**A**) and cobalt ferrite nanoparticles (CF-NNPs) and zinc cobalt ferrite nanoparticles (ZCF-NNPs) (**B**).

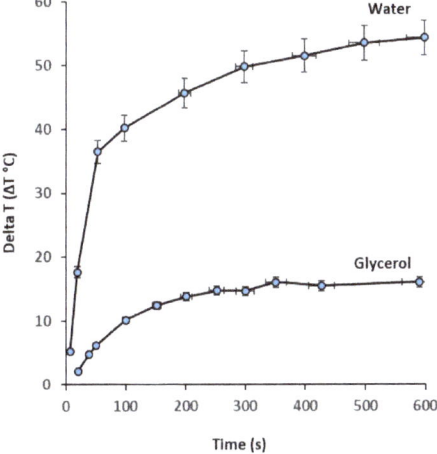

Figure 10. Heating profile of ZCF dispersed in water and glycerol at a concentration of 8 mg/mL, a magnetic field strength of 50 KA/m, and a frequency of 97 kHz.

The Brownian relaxation time (τ_B) is given by

$$\tau_B = 3\eta V_H/(k_B T),$$

where η is the viscosity of the fluid, V_H is the hydrodynamic volume of the particles, k_B is the Boltzmann constant, and T is the temperature [3].

The viscosity affects the heating properties via a Brownian mechanism and, thus, causes a significantly lower SLP, as reported previously [31]. This reduction is primarily due to the high viscosity of glycerol, which is 60 times that of water at room temperature [48]. It is important to note that the large amount of water molecules associated with the nanoparticles may affect their chemical stability and heat-generation ability under an AC magnetic field, which is critical for hyperthermia applications.

The cytotoxicity of the cobalt ferrite and zinc cobalt ferrite nanoparticles was investigated by in vitro culture with NIH-3T3 fibroblasts (Figure 11). The cell viability remained relatively high up to concentrations of 0.5 mg/mL, for both types of nanoparticles compared to the control sample, indicating that the prepared nanoparticles have a low toxicity. The results of our cytocompatibility analysis indicate that the cell viability decreased as the concentration of zinc cobalt ferrite nanoparticles increased to 0.25 and 0.5 mg/mL. This reduction in cell viability may be attributed to the loss of colloidal stability. The incorporation of Zn in the NP structure lead to an increase in the particle size and agglomeration.

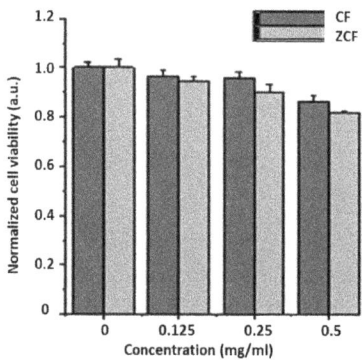

Figure 11. Cytocompatibility tests for CF-MNPs and ZCF-MNPs in in vitro culture with NIH-3T3.

3.3. Characterization of Coated Magnetic Nanoparticles with Sodium Citrate

Modifying the chemical composition of the prepared nanoparticles by the addition of a hydrophilic surface layer (e.g., surfactant) may enhance their dispersion and heating performance. Enhancement of the heating performance of nanoparticles with size controlling and their enhanced SLP value can lead to the improvement of their effectiveness in hyperthermia. To this end, sodium citrate was selected as a surfactant for nanoparticle coating to enhance the surface charge of the MNPs to form a stable colloid with biocompatible behaviors. The effects of surfactant introduction in the colloidal phase on the possible aggregation of the nanoparticles were investigated with zeta potential measurement. The zeta potential values were −49.38 and −48.55 mV for sodium citrate@ CF-MNPs and sodium citrate@ ZCF-MNPs, respectively (Figure 12). In a neutral aqueous solution, the citrate provides negatively charged ions, which become absorbed onto the nanoparticles. The negative surface charges introduced on the surface cause repulsion among the particles, thus preventing particle aggregation and encouraging a stable dispersion. As shown in Figure 12, the mean hydrodynamic size obtained by DLS was 109.2 and 111.1 nm for sodium citrate@ CF-MNPs and sodium citrate@ ZCF-MNPs, respectively. Decrease in the amount of sodium citrate resulted in an increase in the average size of the nanoparticles. In other words, the reduction in the amount of sodium citrate decreases the citrate ions available for stabilizing the particles, which causes small particles to aggregate.

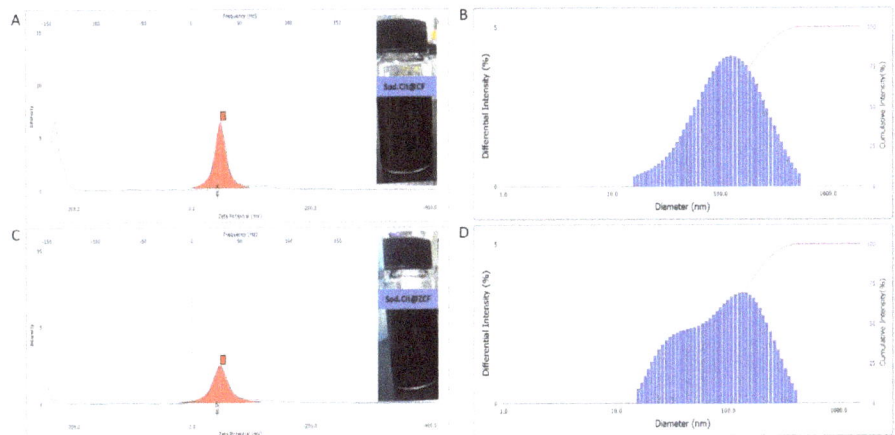

Figure 12. Dynamic light scattering (DLS) and zeta potential results for sodium citrate@ CF-MNPs (**A,B**) and sodium citrate@ ZCF-MNPs (**C,D**).

Particle size distribution and the average size of cobalt ferrite nanoparticles coated with sodium citrate (sodium citrate@ CF-MNPs) and zinc cobalt ferrite nanoparticles coated with sodium citrate (sodium citrate@ ZCF-MNPs) were investigated by TEM. As shown in Figure 13, sodium citrate decreases the aggregation between the prepared nanoparticles and the SAED pattern shows diffuse rings with high intensity that can be indexed to the nanoparticle reflections. The average particle sizes of sodium citrate@ CF-MNPs and sodium citrate@ ZCF-MNPs were 10 ± 4 nm and 30 ± 6 nm, respectively, with a wide size distribution.

The cytotoxicity of the coated nanoparticles was investigated by in vitro culture with NIH-3T3 fibroblasts (Figure 14). The cell viability remained relatively high up to concentrations of 0.5 mg/mL, for both types of nanoparticles compared to the control sample, indicating that the prepared nanoparticles have a low toxicity. However, these results do not prove that these nanoparticles are completely safe for in vivo applications. There may be a dissolution of the nanoparticles, which would lead to high Co^{2+} concentrations in the organism [49]. So, in-depth studies to evaluate the possible interactions between cells and our nanoparticles are currently in progress.

Evaluation of the interaction between cells and nanoparticles was also investigated. MNPs were stained by Prussian blue solution showing blue color (Figure 15). MNPs were located in the cell, indicating that MNPs were internalized into the cell. Citrate coated MNPs were more internalized, likely due to the lesser aggregation of MNPs.

The heating efficiency and SLP values of the prepared magnetic nanoparticles coated with sodium citrate upon exposure to an AC magnetic field with a frequency of 97 kHz at various magnetic field strengths (30, 40, and 50 kA/m) and a at concentration of 4 mg/mL were investigated (Figure 16). The total metal content in cobalt ferrite nanoparticles coated with sodium citrate (sodium citrate@ CF-MNPs) and zinc cobalt ferrite nanoparticles coated with sodium citrate (sodium citrate@ ZCF-MNPs) was 43.6% and 60.87%, respectively. The highest SLP value obtained for coated nanoparticles was 523.45 W/g_{metal} for zinc cobalt ferrite nanoparticles coated with sodium citrate with a concentration of 4 mg/mL at 50 kA/m and 97 kHz. The lowest SLP obtained for coated nanoparticles was 235 W/g_{metal} for cobalt ferrite nanoparticles coated with sodium citrate with a concentration of 4 mg/mL at 30 kA/m and 97 kHz. When comparing the SLP values of our magnetic nanoparticles coated with sodium citrate to those of commercial iron oxide materials under the same magnetic field strength (40 kA/m), our nanoparticles were found to show higher values than the commercial iron oxide materials.

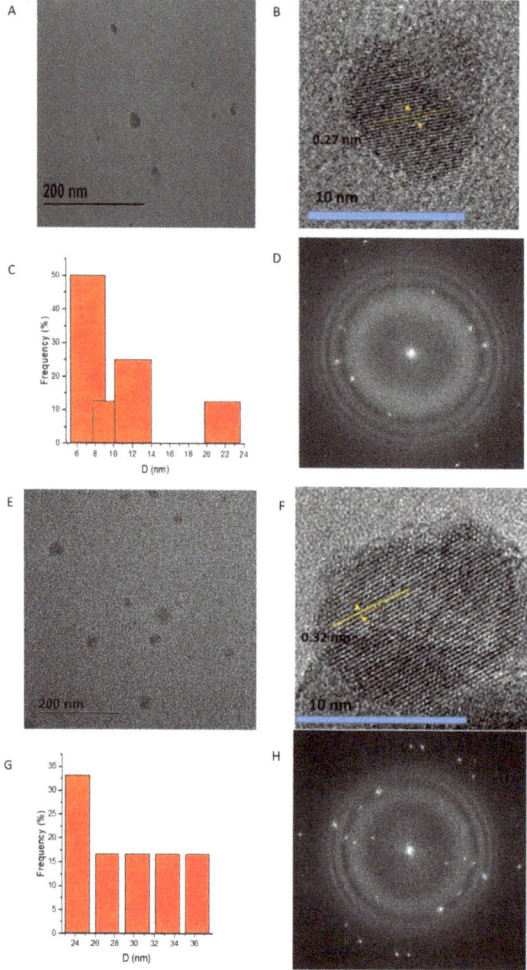

Figure 13. Transmission electron microscopy (TEM), selective area electron diffraction (SAED), and particle size distribution histograms for sodium citrate@ CF-MNPs (**A–D**) and sodium citrate@ ZCF-MNPs (**E–H**).

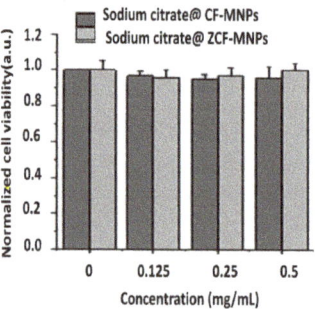

Figure 14. Cytocompatibility tests for sodium citrate@ CF-MNPs and sodium citrate@ ZCF-MNPs in in vitro culture with NIH-3T3.

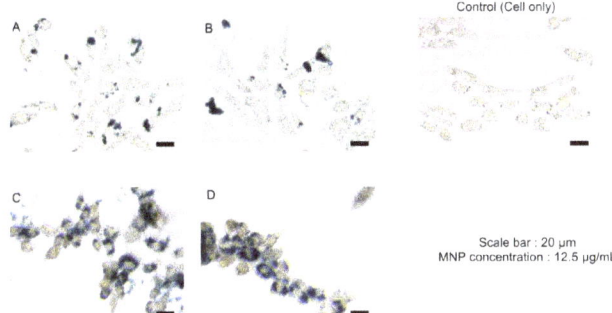

Figure 15. Optical micrographs of cells after the nanoparticle treatment. MNP uptake was shown by Prussian blue staining (**A**) CF-MNPs, (**B**) ZCF-MNPs (**C**) sodium citrate@ CF-MNPs, (**D**) sodium citrate@ ZCF-MNPs.

We compared the results of this study to those obtained using conventional co-precipitation methods, as summarized in Table 2. The comparison was based on the synthesis conditions, magnetic properties, and the maximum SLP obtained under AC field conditions with a note on whether the safety limit was exceeded. Researchers have made many efforts to achieve the smallest nanoparticles possible at high temperatures using a coating agent and to improve the magnetic properties and SLP. Salunkhe et al. and Hoquea et al. prepared $CoFe_2O_4$ nanoparticles at room temperature via a co-precipitation method in the presence of a coating agent [18,29]. Surendra et al. prepared $CoFe_2O_4$ nanoparticles and achieved a high SLP (2131 W/g_{metal}) under safe AC field conditions (18.3 kA/m, 275 kHz) [33]. In our study, the molar ratio of Fe^{+3}:Fe^{+2}:Co^{+2}/Zn^{+2} could be simply controlled to be 3:2:1 and the maximum temperature for the synthesis could be lowered to be 60 °C in air, without the need for oxidizing or coating agents. Our nanoparticles are of particular interest for hyperthermia therapy applications due to their high SLP (552 W/g_{metal}), which was achieved with minimal metal content and magnetic field conditions within the physiologically tolerable limit of 5×10^9 Am^{-1} s^{-1}. A temperature of 45 °C was also achieved, which is ideal for cancer treatment applications.

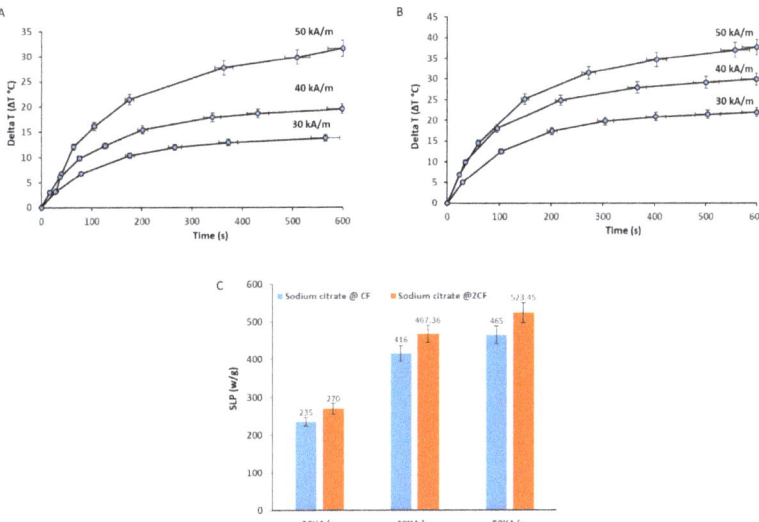

Figure 16. Heating profiles of sodium citrate@ CF-MNPs (**A**) and sodium citrate@ ZCF-MNPs (**B**) with magnetic field strengths (30, 40, and 50 kA/m) at a frequency of 97 kHz and SLP values (**C**).

Table 2. Comparison between conventional co-precipitation methods and the current study.

Sample	Temperature (°C)/Coating Agent	Size (nm)	M_s (emu/g)	SLP (W/g_{metal})	Alternating Current (AC) Field Condition (Safety Limit)	Ref.
$CoFe_2O_4$	70 (*)	14	26	–	–	[19]
$CoFe_2O_4$	90 (*)	6.5–9.7	25–42	–	–	[14]
$CoFe_2O_4$	90 (oleic acid)	14.8	52	–	–	[20]
$CoFe_2O_4$	93 ± 2 (*)	15–20	61	–	–	[21]
$CoFe_2O_4$	100 (*)	5–24	22–74	–	–	[22]
$CoFe_2O_4$	100 (*)	4.2–18.6	30–48	–	–	[23]
$CoFe_2O_4$	100 (*)	22	38	–	–	[24]
$CoFe_2O_4$	100 (*)	46–77	82–91	–	–	[25]
$CoFe_2O_4$	60 (*)	–	13	–	–	[26]
$CoFe_2O_4$	60 (*)	8	36	–	–	[27]
$ZnCoFe_2O_4$	85 (*)	6–10	14–49	–	–	[28]
$CoFe_2O_4$	90 (chitosan)	14	46.1	237–272	30 mT, 342 kHz (not safe)	[17]
$CoFe_2O_4$	Room Temp. (polyethylene glycol, oleic acid)	9.9	60.42	91.84	30 kA/m, 260 kHz (not safe)	[18]
$CoFe_2O_4$	Room Temp. (polyethylene glycol, chitosan)	7	32.3–73.1	11–289	76 mT, 400 kHz (not safe)	[29]
$CoFe_2O_4$	90 (sodium citrate)	13.56	–	82.6	9.4 kA/m, 198 kHz	[30]
$CoFe_2O_4$	80 (sodium citrate)	9.1	–	360	24.8 kA/m, 700 kHz (not safe)	[31]
$CoFe_2O_4$	90 (trisodium citrate dehydrate)	16.2	68	90.2	769 A/m, 400 kHz (safe)	[32]
$CoFe_2O_4$	100 (oleic acid) Heat treatment 100–600 °C	12–20	49–56	114–2131	18.3 kA/m, 275 kHz (safe)	[33]
$CoFe_2O_4$	90–95 (lauric acid)	9–10	59.56	51.8	15 kA/m, 300 kHz (safe)	[34]
$ZnCoFe_2O_4$	80 (*)	13	70.23	114.98	335.2 Oe, 265 kHz (not safe)	[35]
$CoFe_2O_4$	60 (*)	8	50.61	358.77	50 kA/m, 97 kHz (safe)	Current study
$ZnCoFe_2O_4$	60 (*)	25	50.71	552	50 kA/m, 97 kHz (safe)	Current study
$CoFe_2O_4$	60 (sodium citrate)	10	–	465	50 kA/m, 97 kHz (safe)	Current study
$ZnCoFe_2O_4$	60 (sodium citrate)	30	–	523.45	50 kA/m, 97 kHz (safe)	Current study

*: without coating agent; –: not measured.

4. Conclusions

In this study, ferrite nanoparticles, namely cobalt ferrite (8 ± 2 nm) and zinc cobalt ferrite (25 ± 5 nm), were prepared using a controlled co-precipitation process. The synthesis process was a modified version of the conventional co-precipitation methods, with changes in the composition of the precursor materials (molar ratio of Fe^{+3}:Fe^{+2}:Co^{+2}/Zn^{+2} of 3:2:1) and a simple, environmentally friendly, and low-temperature process carried out in air (the maximum temperature was 60 °C and neither oxidizing nor coating agents were required). The prepared nanoparticles exhibited optical activity with moderate magnetic saturation (50 emu/g). SLP was enhanced with respect to time, magnetic field strength, concentration, and medium viscosity. The highest SLP obtained was 552 W/g_{metal} for zinc cobalt ferrite nanoparticles at a concentration of 8 mg/mL with a magnetic field of 50 kA/m and frequency of 97 kHz, while the lowest SLP was 11.57 W/g_{metal}, obtained at a concentration of 25 mg/mL with magnetic field of 30 kA/m and frequency of 97 kHz. The SLP values of our magnetic nanoparticles coated with sodium citrate were found to be higher than those for the commercial iron oxide materials. The nanoparticles exhibited high cell viability and none of the applied AC magnetic fields exceeded the physiologically tolerable limit of 5×10^9 Am^{-1} s^{-1}. The nanoparticles prepared using the presented method achieve high SLP and are promising for biomedical applications, such as hyperthermia cancer treatment.

Author Contributions: Conceptualization, M.S.A.D.; Methodology, M.S.A.D., H.K., H.L. and C.R.; Supervision, J.Y.L. and J.Y.; Writing—original draft, M.S.A.D.; Writing—review & editing, J.Y.L. and J.Y.

Funding: This research was supported by the National Research Foundation (NRF) of Korea (2019M3C1B8090798), the Korea Evaluation Institute of Industrial Technology (KEIT) grant (No. 20003822), and the Korea Health Industry Development Institute (KHIDI) grant (number: HI19C1234).

Conflicts of Interest: The authors declare no conflict of interest.

References

1. Sánchez-Cabezas, S.; Montes-Robles, R.; Gallo, J.; Sancenón, F.; Martínez-Máñez, R. Combining magnetic hyperthermia and dual T1/T2 MR imaging using highly versatile iron oxide nanoparticles. *Dalton Trans.* **2019**, *48*, 3883–3892. [CrossRef]
2. Carvalho, A.; Gallo, J.; Pereira, D.; Valentão, P.; Andrade, P.; Hilliou, L.; Ferreira, P.; Bañobre-López, M.; Martins, J. Magnetic Dehydrodipeptide-Based Self-Assembled Hydrogels for Theragnostic Applications. *Nanomaterials* **2019**, *9*, 541. [CrossRef]
3. Le, T.; Bui, M.P.; Yoon, J. Theoretical Analysis for Wireless Magnetothermal Deep Brain Stimulation Using Commercial Nanoparticles. *Int. J. Mol. Sci.* **2019**, *20*, 2873. [CrossRef] [PubMed]
4. Appa Rao, P.; Srinivasa Rao, K.; Pydi Raju, T.R.K.; Kapusetti, G.; Choppadandi, M.; Chaitanya Varma, M.; Rao, K.H. A systematic study of cobalt-zinc ferrite nanoparticles for self-regulated magnetic hyperthermia. *J. Alloys Compd.* **2019**, *794*, 60–67. [CrossRef]
5. Apostolov, A.; Apostolova, I.; Wesselinowa, J. Specific absorption rate in Zn-doted ferrites for self-controlled magnetic hyperthermia. *Eur. Phys. J. B* **2019**, *92*, 3. [CrossRef]
6. Mai, B.T.; Balakrishnan, P.B.; Barthel, M.J.; Piccardi, F.; Niculaes, D.; Marinaro, F.; Fernandes, S.; Curcio, A.; Kakwere, H.; Autret, G.; et al. Thermoresponsive Iron Oxide Nanocubes for an Effective Clinical Translation of Magnetic Hyperthermia and Heat-Mediated Chemotherapy. *ACS Appl. Mater. Interfaces* **2019**, *11*, 5727–5739. [CrossRef] [PubMed]
7. Gupta, R.; Sharma, D. Evolution of Magnetic Hyperthermia for Glioblastoma Multiforme Therapy. *ACS Chem. Neurosci.* **2019**, *10*, 1157–1172. [CrossRef]
8. Makridis, A.; Curto, S.; van Rhoon, G.C.; Samaras, T.; Angelakeris, M. A standardisation protocol for accurate evaluation of specific loss power in magnetic hyperthermia. *J. Phys. D Appl. Phys.* **2019**, *52*, 255001. [CrossRef]
9. Shaw, S.K.; Biswas, A.; Gangwar, A.; Maiti, P.; Prajapat, C.L.; Meena, S.S.; Prasad, N.K. Synthesis of exchange coupled nanoflowers for efficient magnetic hyperthermia. *J. Magn. Magn. Mater.* **2019**, *484*, 437–444. [CrossRef]
10. Hergt, R.; Dutz, S. Magnetic particle hyperthermia—Biophysical limitations of a visionary tumour therapy. *J. Magn. Magn. Mater.* **2007**, *311*, 187–192. [CrossRef]
11. Niu, Z.P.; Wang, Y.; Li, F.S. Magnetic properties of nanocrystalline Co–Ni ferrite. *J. Mater. Sci.* **2006**, *41*, 5726–5730. [CrossRef]
12. Wang, X.; Zhuang, J.; Peng, Q.; Li, Y. A general strategy for nanocrystal synthesis. *Nature* **2005**, *437*, 121–124. [CrossRef]
13. Sorescu, M.; Grabias, A.; Tarabasanu-Mihaila, D.; Diamandescu, L. Influence of cobalt and nickel substitutions on populations, hyperfine fields, and hysteresis phenomenon in magnetite. *J. Appl. Phys.* **2002**, *91*, 8135–8137. [CrossRef]
14. Rani, S.; Varma, G.D. Superparamagnetism and metamagnetic transition in Fe_3O_4 nanoparticles synthesized via co-precipitation method at different pH. *Phys. B Condens. Mater.* **2015**, *472*, 66–77. [CrossRef]
15. Sun, S.; Zeng, H.; Robinson, D.B.; Raoux, S.; Rice, P.M.; Wang, S.X.; Li, G. Monodisperse MFe_2O_4 (M = Fe Co, Mn) nanoparticles. *J. Am. Chem. Soc.* **2004**, *126*, 273–279. [CrossRef]
16. Jana, N.R.; Chen, Y.; Peng, X. Size-and shape-controlled magnetic (Cr, Mn, Fe Co, Ni) oxide nanocrystals via a simple and general approach. *Chem. Mater.* **2004**, *16*, 3931–3935. [CrossRef]
17. Nasrin, S.; Chowdhury, F.-U.-Z.; Hoque, S.M. Study of hyperthermia temperature of manganese-substituted cobalt nano ferrites prepared by chemical co-precipitation method for biomedical application. *J. Magn. Magn. Mater.* **2019**, *479*, 126–134. [CrossRef]
18. Salunkhe, A.B.; Khot, V.M.; Ruso, J.M.; Patil, S.I. Water dispersible superparamagnetic Cobalt iron oxide nanoparticles for magnetic fluid hyperthermia. *J. Magn. Magn. Mater.* **2016**, *419*, 533–542. [CrossRef]
19. Chinnasamy, C.N.; Jeyadevan, B.; Perales-Perez, O.; Shinoda, K.; Tohji, K.; Kasuya, A. Growth Dominant Co-Precipitation Process to Achieve High Coercivity at Room Temperature in $CoFe_2O_4$ Nanoparticles. *IEEE Trans. Magn.* **2002**, *38*, 5. [CrossRef]

20. Ayyappan, S.; Mahadevan, S.; Chandramohan, P.; Srinivasan, M.P.; Philip, J.; Raj, B. Influence of Co^{2+} Ion Concentration on the Size, Magnetic Properties, and Purity of $CoFe_2O_4$ Spinel Ferrite Nanoparticles. *J. Phys. Chem. C* **2010**, *114*, 6334–6341. [CrossRef]
21. Chinnasamy, C.N.; Senoue, M.; Jeyadevan, B.; Perales-Perez, O.; Shinoda, K.; Tohji, K. Synthesis of size-controlled cobalt ferrite particles with high coercivity and squareness ratio. *J. Colloid Interface Sci.* **2003**, *263*, 80–83. [CrossRef]
22. Safi, R.; Ghasemi, A.; Shoja-Razavi, R.; Tavousi, M. The role of pH on the particle size and magnetic consequence of cobalt ferrite. *J. Magn. Magn. Mater.* **2015**, *396*, 288–294. [CrossRef]
23. Pereira, C.; Pereira, A.M.; Fernandes, C.; Rocha, M.; Mendes, R.; Fernández-García, M.P.; Guedes, A.; Tavares, P.B.; Grenèche, J.M.; Araújo, J.P.; et al. Superparamagnetic MFe_2O_4 (M = Fe, Co, Mn) Nanoparticles: Tuning the Particle Size and Magnetic Properties through a Novel One-Step Coprecipitation Route. *Chem. Mater.* **2012**, *24*, 1496–1504. [CrossRef]
24. Van Berkum, S.; Dee, J.T.; Philipse, A.P.; Erné, B.H. Frequency-Dependent Magnetic Susceptibility of Magnetite and Cobalt Ferrite Nanoparticles Embedded in PAA Hydrogel. *Int. J. Mol. Sci.* **2013**, *14*, 10162–10177. [CrossRef] [PubMed]
25. Tatarchuk, T.; Bououdina, M.; Macyk, W.; Shyichuk, O.; Paliychuk, N.; Yaremiy, I.; Al-Najar, B.; Pacia, M. Structural, Optical, and Magnetic Properties of Zn-Doped $CoFe_2O_4$ Nanoparticles. *Nanoscale Res. Lett.* **2017**, *12*, 141. [CrossRef]
26. Yadavalli, T.; Jain, H.; Chandrasekharan, G.; Chennakesavulu, R. Magnetic hyperthermia heating of cobalt ferrite nanoparticles prepared by low temperature ferrous sulfate based method. *AIP Adv.* **2016**, *6*, 055904. [CrossRef]
27. Kim, Y.I.; Kim, D.; Lee, C.S. Synthesis and characterization of $CoFe_2O_4$ magnetic nanoparticles prepared by temperature-controlled coprecipitation method. *Phys. B Condens. Matter* **2003**, *337*, 42–51. [CrossRef]
28. Sharifi, I.; Shokrollahi, H. Nanostructural, magnetic and Mössbauer studies of nanosized $Co_{1-x}Zn_xFe_2O_4$ synthesized by co-precipitation. *J. Magn. Magn. Mater.* **2012**, *324*, 2397–2403. [CrossRef]
29. Manjura Hoquea, S.; Huanga, Y.; Coccod, E.; Maritimc, S.; Santind, A.D.; Shapiroe, E.M.; Comana, D.; Hyder, F. Improved specific loss power on cancer cells by hyperthermia and MRI contrast of hydrophilic $Fe_xCo_{1-x}Fe_2O_4$ nanoensembles. *Contrast Media Mol. Imaging.* **2016**, *11*, 514–526. [CrossRef] [PubMed]
30. Kahil, H.; El sayed, H.M.; Elsayed, E.M.; Sallam, A.M.; Talaat, M.; Sattar, A.A. Effect of in vitro magnetic fluid hyperthermia using citrate coated cobalt ferrite nanoparticles on tumor cell death. *Rom. J. Biophys.* **2015**, *25*, 209–224.
31. Fortin, J.P.; Wilhelm, C.; Servais, J.; Ménager, C.; Bacri, J.C.; Gazeau, F. Size-Sorted Anionic Iron Oxide Nanomagnets as Colloidal Mediators for Magnetic Hyperthermia. *J. Am. Chem. Soc.* **2007**, *129*, 2628–2635. [CrossRef]
32. Durneata, D.; Hempelmann, R.; Caltun, O.; Dumitru, I. High-Frequency Specific Absorption Rate of $Co_xFe_{1-x}Fe_2O_4$ Ferrite Nanoparticles for Hipertermia Applications. *IEEE Trans. Magn.* **2014**, *50*, 5201104. [CrossRef]
33. Surendra, M.K.; Dutta, R.; Ramachandra Rao, M.S. Realization of highest specific absorption rate near superparamagnetic limit of $CoFe_2O_4$ colloids for magnetic hyperthermia applications. *Mater. Res. Express* **2014**, *1*, 026107. [CrossRef]
34. Pradhan, P.; Giri, J.; Samanta, G.; Sarma, H.D.; Mishra, K.P.; Bellare, J.; Banerjee, R.; Bahadur, D. Comparative evaluation of heating ability and biocompatibility of different ferrite-based magnetic fluids for hyperthermia application. *J. Biomed. Mater. Res. Part B Appl. Biomater.* **2007**, *81*, 12–22. [CrossRef]
35. Nikam, D.S.; Jadhav, S.V.; Khot, V.M.; Phadatare, M.R.; Pawar, S.H. Study of AC magnetic heating characteristics of $Co_{0.5}Zn_{0.5}Fe_2O_4$ nanoparticles for magnetic hyperthermia therapy. *J. Magn. Magn. Mater.* **2014**, *349*, 208–213. [CrossRef]
36. Thirunavukkarasu, G.K.; Cherukula, K.; Lee, H.; Jeong, Y.Y.; Park, I.-K.; Young Lee, J.Y. Magnetic field-inducible drug-eluting nanoparticles for image-guided thermo-chemotherapy. *Biomaterials* **2018**, *180*, 240–252. [CrossRef] [PubMed]
37. Takahashi, K.; Kato, H.; Saito, T.; Matsuyama, S.; Kinugasa, S. Precise measurement of the size of nanoparticles by dynamic light scattering with uncertainty analysis. *Part Part Syst. Charact.* **2008**, *8*, 31–38. [CrossRef]
38. Darwish, M.S.A.; Stibor, I. Pentenoic Acid-Stabilized Magnetic Nanoparticles for Nanomedicine Applications. *J. Dispers. Sci. Technol.* **2016**, *37*, 1793–1798. [CrossRef]

39. Park, S.; Baker, J.; Himmel, M.; Parill, P.; Johnson, D. Cellulose crystallinity index: Measurement techniques and their impact on interpreting cellulase performance. *Biotechnol. Biofuels* **2010**, *3*, 10. [CrossRef]
40. O'Leary, S.K.; Lim, P.K. On determining the optical gap associated with an amorphous semiconductor: A generalization of the Tauc model. *Solid State Commun.* **1997**, *104*, 17–21. [CrossRef]
41. Mallick, P.; Dash, B.N. X-ray diffraction and UV–Visible characterizations of γ–Fe_2O_3 nanoparticles annealed at different temperature. *Nanosci. Nanotechnol.* **2013**, *3*, 130–134.
42. El Ghandoor, H.; Zidan, H.M.; Khalil, M.M.; Ismail, M.I.M. Synthesis and some physical properties of magnetite (Fe_3O_4) nanoparticles. *Int. J. Electrochem. Sci.* **2012**, *7*, 5734–5745.
43. Anjum, S.; Tufail, R.; Rashid, K.; Zia, R.; Riaz, S. Effect of cobalt doping on crystallinity, stability, magnetic and optical properties of magnetic iron oxide nano-particles. *J. Magn. Magn. Mater.* **2017**, *432*, 198–207. [CrossRef]
44. El-Okr, M.M.; Salem, M.A.; Salim, M.S.; El-Okr, R.M.; Ashoush, M.; Talaat, H.M. Synthesis of cobalt ferrite nano-particles and their magnetic characterization. *J. Magn. Magn. Mater.* **2011**, *323*, 920–926. [CrossRef]
45. Prabhakaran, T.; Hemalatha, J. Combustion synthesis and characterization of cobalt ferrite nanoparticles. *Ceram. Int.* **2016**, *42*, 14113–14120. [CrossRef]
46. Darwish, M.S.A. Effect of carriers on heating efficiency of oleic acid-stabilized magnetite nanoparticles. *J. Mol. Liq.* **2017**, *231*, 80–85. [CrossRef]
47. Darwish, M.S.A.; El-Sabbagh, A.; Stibor, I. Hyperthermia properties of magnetic polyethylenimine core/shell nanoparticles: Influence of carrier and magnetic field strength. *J. Polym. Res.* **2015**, *22*, 239. [CrossRef]
48. Segur, J.B.; Oberstar, H.E. Viscosity of glycerol and its aqueous solutions. *Ind. Eng. Chem.* **1951**, *43*, 2117. [CrossRef]
49. Leyssens, L.; Vinck, B.; Van Der Straeten, C.; Wuyts, F.; Maes, L. Cobalt toxicity in humans-A review of the potential sources and systemic health effects. *Toxicology* **2017**, *387*, 43–56. [CrossRef]

© 2019 by the authors. Licensee MDPI, Basel, Switzerland. This article is an open access article distributed under the terms and conditions of the Creative Commons Attribution (CC BY) license (http://creativecommons.org/licenses/by/4.0/).

Article

Magnetic Nanoparticles Functionalized Few-Mode-Fiber-Based Plasmonic Vector Magnetometer

Yaofei Chen [1], Weiting Sun [2], Yaxin Zhang [2], Guishi Liu [1], Yunhan Luo [1,*], Jiangli Dong [1], Yongchun Zhong [1], Wenguo Zhu [1], Jianhui Yu [1] and Zhe Chen [1]

[1] Key Laboratory of Optoelectronic Information and Sensing Technologies of Guangdong Higher Education Institutes, Jinan University, Guangzhou 510632, China; chenyaofei@jnu.edu.cn (Y.C.); guishiliu@163.com (G.L.); jldong@jnu.edu.cn (J.D.); ychzhong@163.com (Y.Z.); zhuwg88@163.com (W.Z.); kensomyu@gmail.com (J.Y.); thzhechen@163.com (Z.C.)

[2] Department of Optoelectronic Engineering, College of Science and Engineering, Jinan University, Guangzhou 510632, China; sunweiting130@126.com (W.S.); yaxinzhang98@163.com (Y.Z.)

* Correspondence: luoyunhan@jnu.edu.cn

Received: 6 May 2019; Accepted: 17 May 2019; Published: 22 May 2019

Abstract: In this work, we demonstrate a highly-sensitive vector magnetometer based on a few-mode-fiber-based surface plasmon resonance (SPR) sensor functionalized by magnetic nanoparticles (MNPs) in liquid. To fabricate the sensor, a few-mode fiber is side-polished and coated with a gold film, forming an SPR sensor that is highly sensitive to the surrounding refractive index. The vector magnetometer operates based on the mechanism whereby the intensity and orientation of an external magnetic field alters the anisotropic aggregation of the MNPs and thus the refractive index around the fiber SPR device. This, in turn, shifts the resonance wavelength of the surface plasmon. Experimental results show the proposed sensor is very sensitive to magnetic-field intensity and orientation (0.692 nm/Oe and −11.917 nm/°, respectively). These remarkable sensitivities to both magnetic-field intensity and orientation mean that the proposed sensor can be used in applications to detect weak magnetic-field vectors.

Keywords: magnetic nanoparticles; few-mode fiber; surface plasmonic resonance; magnetic vector

1. Introduction

Magnetic sensing is important in the industry, power transmission, military, etc. In recent years, optical-fiber magnetic-field sensors, integrated with magnetic nanoparticles (MNPs) in liquid that is called magnetic fluids (MFs) or ferrofluids, have attracted considerable attention because of their ease of fabrication, high sensitivity and low cost [1]. MFs, a type of paramagnetic material, consist of MNPs uniformly dispersed in a base liquid with the aid of surfactant [2]. When exposed to an applied magnetic field, the MNPs realign and form a chain- or cluster-like structure [3], endowing the MF with various outstanding magneto-optical properties, including a magneto-controllable refractive index (RI), absorption coefficient, birefringence, and so on [4]. Taking advantage of these properties, researchers have proposed diverse magnetic-field sensors based on MFs and various fiber-optical structures or schemes. Some examples include tapered two-mode fiber interferometers [5], Sagnac interferometers [6], multimode interferometers [7], whispering gallery mode resonators [8,9] or interferometers [10,11], microfiber-based interferometers [12,13], two-core fiber-based interferometers [14], photonic crystal fibers [15–17] and asymmetric-tapered fiber [18]. Furthermore, based on the mechanism whereby protein binding can tailor the response to a magnetic field, these devices have recently been used to measure protein concentrations [19,20], further extending the range of applications to biosensing. However, most of these sensors were designed to detect the magnetic intensity while ignoring the magnetic orientation, due to the lack of exploration on the microstructure of MF around the optical fiber.

In 2016, Zhang et al. [21] demonstrated the phenomenon that a magnetic field could induce a non-uniform distribution of MNPs, and thus a non-uniform distribution of RI, around an optical fiber. This RI distribution can be tuned by modulating either the intensity or the orientation of the magnetic field. This work revealed a new way to detect both the magnitude and orientation of a magnetic field. By creating a non-circularly-symmetric light-field in a fiber, Yin et al. (2017) [22,23] developed two MF-based magnetic-vector sensors by sandwiching a piece of double-clad photonic-crystal-fiber or a thin core fiber between two single-mode fibers with a lateral offset. In 2018, based on a similar mechanism, magnetic-vector sensing was achieved by Layeghi et al., who used a tapered Hi-Bi fiber inserted in a fiber loop mirror [24]. In 2019, Cui et al. [25] and Lu et al. [26] respectively proposed using a single-mode-fiber fused with capillary structure and an excessively tilted fiber grating to realize the sensing to the magnetic vector. Note that in the above cases, the used fibers still remained in a cylindrically symmetric structure. In 2018, Violakis et al. demonstrated that an MF-encapsulated, D-shaped fiber (a non-circularly-symmetric structure) can also respond to the changes of an external magnetic field's direction [27], which further paved the ways to use an asymmetric fiber structure. Although a number of fiber-based magnetic-vector sensors have been developed, the simultaneous high sensitivity to both magnetic intensity and orientation still remains as a challenging issue.

Surface plasmon resonance (SPR) sensors are typically constructed by coating a metal film over a dielectric substrate, and they are powerful tools for detecting tiny changes in the RI over a metal surface thanks to their high RI sensitivities, which even can exceed 10^4 nm per refractive index unit (RIU) around the RI of 1.33. Therefore, combining MFs with SPR technology could provide a scheme for highly-sensitive magnetic-field sensing, and several works in that direction have already been published [21,28–32]. Ying et al. made a numerical study of the magnetic-field response based on a Kretschmann prism SPR configuration [28] which, unfortunately, is limited in practical use because of its bulky size. Weng et al. [29] and Liu et al. [30] theoretically investigated magnetic-field sensors based on side-hole fiber SPRs and D-shaped photonic-crystal-fiber SPRs, respectively. Schwendtner et al. experimentally boosted the sensitivity up to ~1 nm/Oe by using a tapered fiber as an SPR substrate [31]. Unfortunately, these SPR-based MF sensors were only proposed to detect magnetic field intensity, but not orientation. Later, Zhang et al. demonstrated a plasmonic fiber-optic vector magnetometer based on directional scattering between polarized plasmon waves and ferro-magnetic nanoparticles, and achieved the sensitivities of 0.18 nm/Oe and 2 nm/° to intensity and orientation, respectively [21]. Recently, we proposed a fiber SPR sensor, based on side-polished multimode fiber, for the sensing to magnetic vector, and further improved the sensitivities to 0.599 nm/Oe and −5.63 nm/° [33]. However, to realize the sensing to a weaker magnetic field, the magnetic fields' sensitivities to orientation and intensity need further development.

In this paper, aiming to develop a higher sensitivity magnetometer with the ability of simultaneous sensing to the magnetic intensity and orientation, we propose and investigate a plasmonic vector magnetometer based on a side-polished few-mode-fiber (FMF) functionalized by MNPs. This customized SPR device's high sensitivity to RI means this sensor is highly-sensitive to magnetic-field intensity (up to 0.692 nm/Oe). In addition, the non-circularly-symmetric geometry of the side-polished fiber and the non-uniform distribution of the MF around the fiber together allow the sensor to sense the magnetic field's orientation with the high sensitivity of −11.917 nm/°. The excellent characteristics of this sensor and its operating mechanism are comprehensively analyzed herein, offering an effective solution for high-sensitivity magnetic-vector sensing.

2. Principles and Structure

A schematic diagram of the proposed sensor is shown in Figure 1. In this work, the side-polished fiber was made from an FMF. Its core diameter is less than that of a multimode fiber but greater than that of a single-mode fiber, which alleviates the problems of low sensitivity and low signal-to-noise ratio that plagues SPR sensors based on side-polished multi-mode and single-mode fibers [34,35]. The sensor consists of a side-polished FMF (OFS Fitel, LLC) coated with a gold film and immersed in

MF (EMG 605, Ferrotec. Inc., Santa Clara, USA) which is sealed in a capillary tube. The MNPs with the material of Fe_3O_4 that evenly dispersed in the MF have the average diameter of ~10 nm and the volume concentration of 3.9%, i.e., the ratio of the total volume of MNPs to the volume of MF is 3.9%. When the light in the FMF propagates over the gold-coated region, the p-polarized component of the resonance wavelength will couple with a surface plasmon wave in the gold film, if their wave vectors match. As a result, the transmittance spectrum will feature a narrow-band absorption at the resonance wavelength, which is extremely sensitive to the surrounding RI. Conversely, the RI of the surrounding MF depends on the applied magnetic field [36], so the resonance wavelength depends strongly on the external magnetic field. Therefore, magnetic-field sensing can be implemented by monitoring the resonance wavelength.

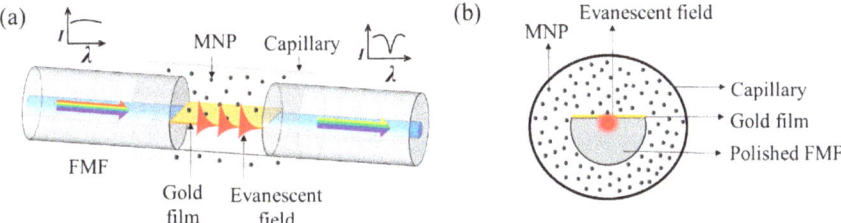

Figure 1. Schematic diagrams of (**a**) the proposed sensor and (**b**) a cross section of the sensor. A broadband incoming light from the left is partially absorbed due to the SPR, producing the notched spectrum of the outgoing spectrum on the right end.

A mode solver (Mode Solution, Lumerical Solutions, Inc.) was used to simulate the mode field distribution. For these simulations, the FMF core and cladding diameters are 19 and 125 μm, respectively; the RIs of the cladding and the FMF core are 1.444 and 1.449, respectively; the RI of the surrounding MF, under zero magnetic field, is 1.385, which is calibrated by a homemade, prism-based SPR testing system; and the thickness of the side-polished FMF is 71 μm, which means that the fiber is polished into the core section. Figure 2 shows the field distributions and the propagation losses of modes LP_{01}, LP_{11a}, LP_{11b}, and LP_{21a} at the wavelength of 650 nm. A strong evanescent field is seen over the gold film surface, indicating SPR excitation.

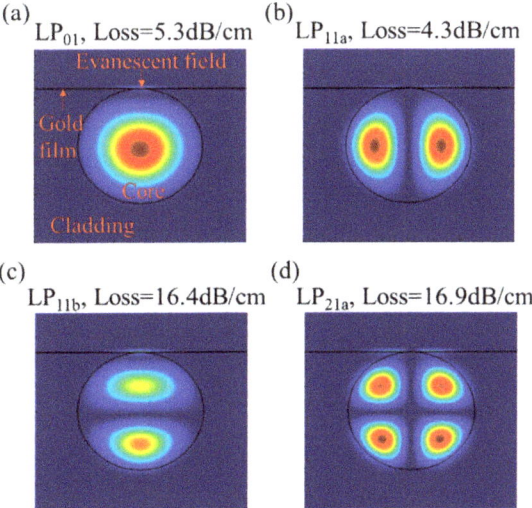

Figure 2. Field distributions in the fiber for modes (**a**) LP_{01}, (**b**) LP_{11a}, (**c**) LP_{11b}, and (**d**) LP_{21a} at 650 nm.

As the evanescent field, which interacts with the surrounding MF, only exists near the gold film surface, the MF microstructure, which is related to the RI over the gold surface, determines the sensor performance. When exposed to an applied magnetic field, the MNPs in the MF become magnetic dipoles and aggregate, forming stable chain-like structures oriented along the applied magnetic field direction [3]. However, the optical fiber in the MF destroys the original equilibrium state, forcing the MNPs to realign and form a new distribution around the fiber [23]. Moreover, the MNPs aggregate in an anisotropic manner: when the fiber surface is parallel (perpendicular) to the magnetic-field direction, they tend to gather to (depart from) the fiber surface [21,22]. For our situation, where the fiber has a D-shaped cross section, Figure 3 shows the state of MNP aggregation around the fiber, which is predicted depending on the published works [21–23]. The comparison of Figure 3a,b shows that the aggregation state of the MNFs over the gold film depends on the magnetic field orientation. First, the chain-like structures realign along the magnetic field direction. Second, when the magnetic field direction changes from parallel to perpendicular (relative to the side-polished surface), some of the MNPs depart from the surface of the gold film, leaving fewer MNPs to interact with the evanescent field. The orientation-dependent distribution of MNPs around the fiber is the basis for sensing the orientation of the magnetic-field vector.

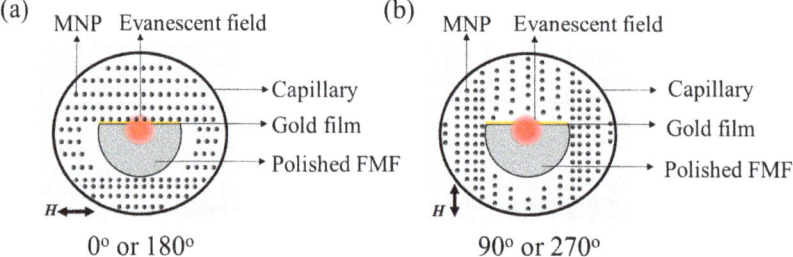

Figure 3. Schematic diagrams of the magnetic nanoparticles (MNPs) distribution around the side-polished fiber, when the external magnetic field is (**a**) parallel and (**b**) perpendicular to the flat surface of the side-polished fiber.

3. Fabrication and Characterization

A piece of FMF was first coarsely polished by a homemade wheel-polishing system for ~4 min, and then finely polished for ~150 min to obtain the desired residual thickness of the polished fiber. The polished fiber profile was characterized with a microscope (Zeiss Axio Scope A1); the results show that the polished region is ~6 mm long, and the residual thickness of the polished region is ~71 μm.

To fabricate an FMF-based SPR sensor, a chromium adhesion layer (~5 nm) and a gold film (~50 nm) were successively vacuum evaporated onto the surface of the polished fiber. The fabricated SPR sensor's sensitivity to the RI was characterized by immersing the sensing section into solutions with different RIs. Figure 4a shows the spectral transmittance of the sensor immersed in various RI solutions. With increasing RI, the resonance spectrum shifts to longer wavelengths because the wave-vector-matching condition between the light and the surface plasmon wave depends on the RI [37]. The resonance wavelength varies nonlinearly with RI over the entire range of 1.33~1.39, as shown in Figure 4b. The blue curve in Figure 4b is a fit to a quadratic polynomial. At the RI of MF (~1.385), the SPR sensor's sensitivity is estimated to be 3693 nm/RIU. The RI sensitivity of the FMF-based SPR sensor exceeds those of optical-fiber RI sensors based on mode interference [38], long-period gratings [39], Fabry–Perot cavities [40], etc., laying a solid foundation for high-sensitivity sensing of magnetic fields.

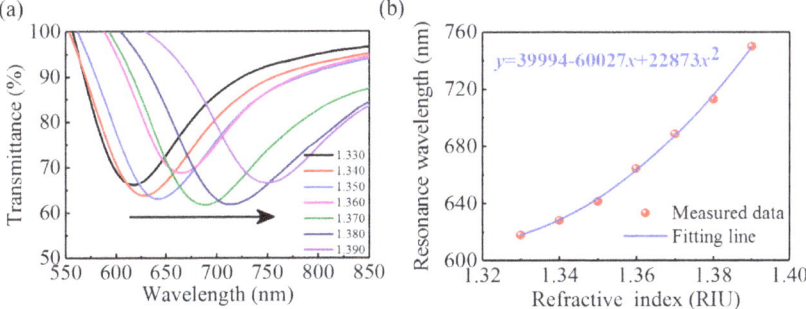

Figure 4. (**a**) Spectral response of the surface plasmon resonance (SPR) sensor to refractive index (RI). (**b**) Measured resonance wavelength as a function of RI.

After characterizing the RI sensing performance, the SPR-based magnetic-field sensor was fabricated by packaging the MF around the side-polished fiber. The side-polished fiber section was first fed into the center of a 30-mm-long capillary tube with an inner (outer) diameter of 0.5 (1.0) mm. The capillary containing the fiber was embedded into a holder consisting of a rectangular groove. Next, the MF was inserted into the capillary using capillary force. Finally, the capillary was sealed by applying UV glue at both ends, completing sensor fabrication. Figure 5a shows a photograph of the fabricated sensor, where the black region is the MF-encapsulated in the capillary. The X-ray diffraction (XRD) characterization for the Fe_3O_4 MNPs was conducted and the XRD pattern is shown in Figure 5b. As we can see, the peaks shown in the pattern agrees well with the typical peaks of Fe_3O_4 standard diffraction at 30.1°, 35.5°, 43.1°, 53.4°, 57.0° and 62.6°, corresponding to the (220), (311), (400), (422), (511) and (440) crystal planes, respectively.

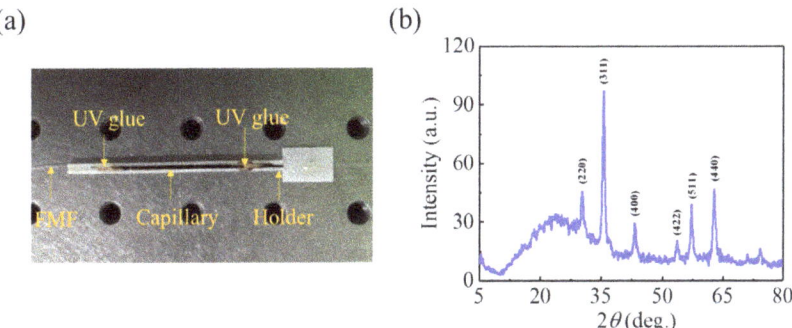

Figure 5. (**a**) Photograph of the fabricated sensor on an optical table. (**b**) XRD pattern for the Fe_3O_4 MNPs.

4. Experiments and Results

4.1. Experimental Setup

Figure 6 shows a schematic diagram of the experimental setup. A tungsten-halogen lamp broadband light source (AvaLight-HAL-(S)-Mini, Apeldoorn, Netherlands) is coupled into the fiber sensor, and a spectrometer (AvaSpec-ULS2048XL, China) is used to record the transmission spectra. An electromagnet generates the applied magnetic field, which is perpendicular to the fiber axis. The intensity is monitored in real time by a gauss meter. The magnetic-field intensity can be tuned by adjusting the applied voltage. The sensor is fixed onto a rotation stage to allow tuning its orientation with the applied magnetic field.

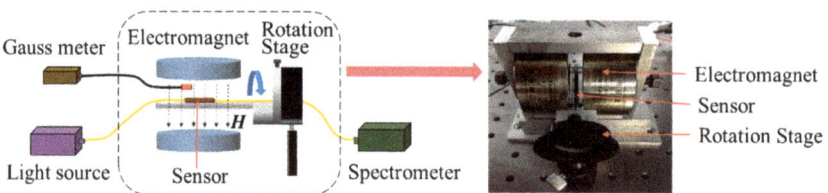

Figure 6. Schematic diagram of experimental setup and photograph of main components, including electromagnet, sensor and rotation stage.

4.2. Response to Magnetic-Field Orientation

To characterize the sensor's response to the magnetic-field direction, it was rotated from 0° to 360° in 4° increment while the magnetic field's intensity was fixed. Here the orientation angle is the one between the magnetic-field direction and the polished surface. Therefore, 0° and 180° (90° and 270°) indicate the magnetic field is parallel (perpendicular) to the flat surface of the side-polished fiber, as shown in Figure 3a,b.

First, the magnetic intensity was fixed at 300 Oe. Upon rotating the sensor counterclockwise (i.e., changing the relative orientation between magnetic field and sensor), the transmission spectrum blueshifts and redshifts repeatedly, indicating the sensor is sensitive to the applied magnetic field direction. Figure 7 shows the spectral response to the relative orientation of the magnetic field: the transmission dip, due to the SPR, blueshifts as the orientation angle goes from 0° to 90° and from 180° to 270°; On the contrary, the dip in transmission redshifts when the orientation angle goes from 90° to 180° and from 270° to 360°.

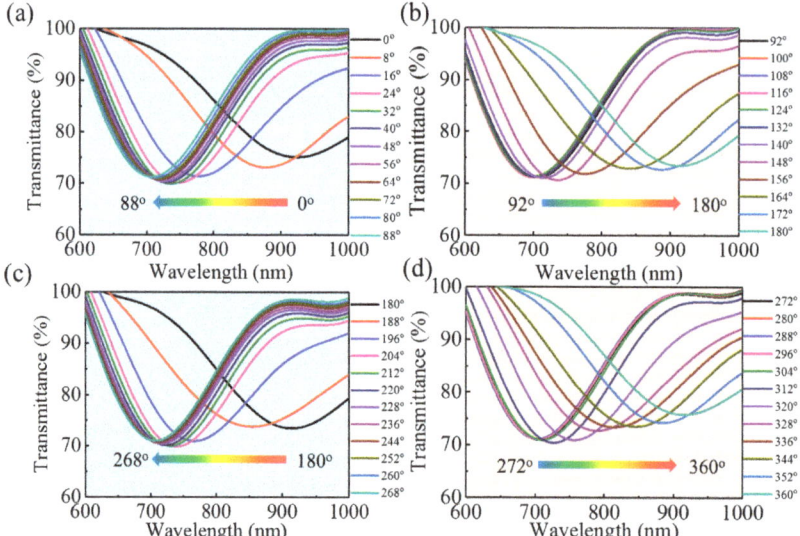

Figure 7. Spectral response of sensor to orientation of magnetic field. The magnetic-field intensity is fixed at 300 Oe, while the orientation is changed (**a**) from 0° to 88°, (**b**) from 92° to 180°, (**c**) from 180° to 268°, and (**d**) from 272° to 360°.

Figure 8a shows the resonance wavelengths at different orientations in a polar coordinate system. The noncircular curve indicates the orientation-dependent sensor response. The resonance wavelength depends on the orientation angle because the MNPs anisotropic aggregation around the side-polished fiber [22,23]: when the magnetic field goes from parallel to perpendicular with respect to the gold film, the MNPs concentration over the gold film gradually decreases, as illustrated in Figure 3. As the RI of the MF correlates positively with the concentration of MNPs, the RI of the MF over the gold film will reach its maximum (minimum) at 0° and 90° (180° and 270°). Therefore, four extreme points appear in the resonance wavelength curve versus orientation angle, as shown in Figure 8a, and the resonance wavelength changes monotonically from one maximum (minimum) to the adjacent minimum (maximum) upon monotonically varying the orientation. In addition, Figure 8a shows that the change around the maxima (0° and 180°) is sharper than that around the minima (90° and 270°), indicating a deviation from parallel orientation induces a larger change in the RI over the gold film than it does from a perpendicular orientation.

Another series of tests to characterize the sensor's orientation response were conducted by the same way, but with a magnetic-field intensity of 60 Oe. Figure 8b shows the resonance wavelength as a function of orientation angle, which presents a similar trend to a magnetic-field intensity of 300 Oe (cf. Figure 8a). However, the curve is more circular, indicating that the lower magnetic-field intensity weakens the orientation dependence, which arises from the smaller change in the RI of the MF at the lower magnetic-field intensity.

To quantitatively assess the orientation response, we plot, in Figure 8c, the resonance wavelength versus orientation angle in Cartesian coordinates and show the corresponding linear fits in the linear regions. Within the range of 144°~180°, the sensitivities to orientation are 5.889 nm/° and 1.562 nm/° for 300 and 60 Oe respectively, indicating that the sensor is more sensitive to orientation at a higher magnetic-field-intensity. Within the range of 184°~200° at 300 Oe, we can achieve the maximal sensitivity of −11.917 nm/°, which is much higher than the 2 nm/° and −5.63 nm/° obtained by using the SPR scheme based on a tilted fiber-Bragg grating [21] and a side-polished multimode fiber [33], respectively. This advantage can be attributed to the non-circularly-symmetric cross section of the side-polished fiber and the smaller core diameter of FMF than that of multimode fiber.

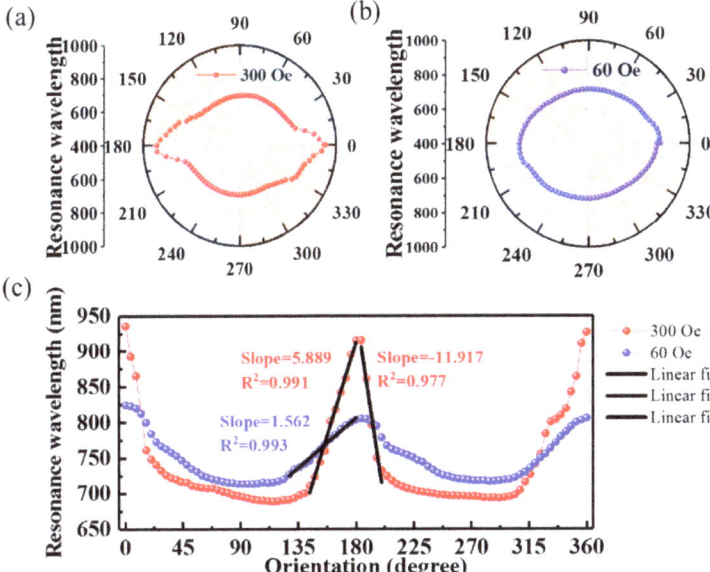

Figure 8. Resonance wavelength as a function of magnetic-field orientation plotted in (**a**,**b**) polar coordinates and (**c**) Cartesian coordinates. The magnetic field intensity is fixed at 300 Oe (red dots) and 60 Oe (blue dots).

4.3. Response to Magnetic-Field Intensity

The sensor's response to the magnetic-field intensity was characterized by gradually changing the intensity from 0 to 400 Oe while keeping the orientation fixed. During the measurements, the orientation was fixed at two specific angles (i.e., 0° and 90°), corresponding to the magnetic field being parallel and perpendicular, respectively, to the flat surface of the polished fiber. Figure 9 shows the transmission spectra for several magnetic-field intensities and the resonance wavelength as a function of magnetic-field intensity.

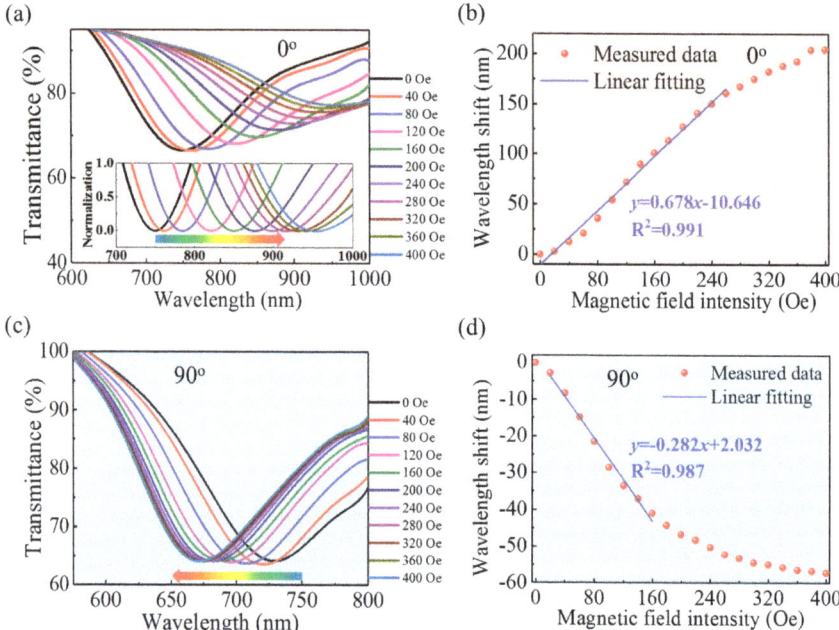

Figure 9. (a,c) Spectra for different magnetic-field intensities and (b,d) corresponding resonance wavelength as a function of magnetic-field intensity. Panels (a,b) correspond to the orientation angle of 0°, and panels (c,d) correspond to the orientation angle of 90°. The inset in (a) shows the normalized transmittance spectra near the resonance wavelengths.

As shown in Figure 9, the transmission dip, due to the SPR, redshifts (blueshifts) with the increase of magnetic-field intensity when the orientation angle is 0° (90°). This indicates the change in the RI over the gold film for the parallel orientation is opposite that for the perpendicular orientation, which is attributed to the MNPs anisotropic aggregation around the side-polished fiber: as illustrated in Figure 3, the MNPs approach (or distance themselves from) the gold film when an external magnetic field is applied parallel (or perpendicular) to the surface of the gold film. Moreover, increasing the magnetic-field intensity further enhances this gathering or evacuation of MNPs, resulting in a further increase or decrease of the RI over the gold film. Figure 9b,d show the resonance wavelength as a function of magnetic-field intensity for parallel and perpendicular orientation, respectively. Overall, the wavelength is nonlinear in magnetic-field intensity, which is mainly attributed to (i) the nonlinear dependence of the RI of the MF on magnetic field, which usually follows a Langevin function [41]; and (ii) the nonlinear sensitivity to RI of the SPR sensor, as shown in Figure 4. Nevertheless, the linear regions were fitted to characterize the sensitivity of the device to magnetic-field intensity. For 0° and 90°, the sensitivity to magnetic-field intensity is 0.692 nm/Oe (0~220 Oe) and −0.282 nm/Oe (20~160 Oe), respectively.

To make a clear comparison, we summarized the relevant-published works that use SPR and MF for magnetic field sensing in Table 1. As we can see, the sensitivity to magnetic-field-intensity achieved in this paper is competitive with the most reported results. More importantly, the employing of side-polished FMF, which has a non-circularly-symmetric geometry and a relatively-smaller core diameter compared to multimode fibers, makes the proposed senor possess the highest sensitivity to magnetic-field-orientation.

Table 1. Comparisons between the smartphone-based and the traditional platforms.

SPR Scheme	Simulation/Experiment	Vector	Sensitivity	Reference
Prism	Simulation	No	0.061°/Oe	[28]
Side-hole fiber	Simulation	No	1.063 nm/Oe	[29]
D-shaped PCF	Simulation	No	0.087 nm/Oe	[30]
Metal-dielectric-Metal	Simulation	No	0.027 nm/Oe	[42]
Tapered fiber	Experiment	No	~1.0 nm/Oe	[31]
No-core fiber	Experiment	No	0.303 nm/Oe	[32]
Tilted FBG	Experiment	Yes	0.18 nm/Oe 2 nm/°	[21]
Side-polished MMF	Experiment	Yes	0.599 nm/Oe −5.63 nm/°	[33]
Side-polished FMF	Experiment	Yes	0.692 nm/Oe −11.917 nm/°	This work

4.4. Control Experiment

In the experiments, the applied magnetic-field's relative orientation is modified by rotating the sensor, which inevitably twists the optical fiber. To nullify any effect from a twisted fiber, which may cause a further change in the fiber's state of polarization and birefringence [43,44], we did a control experiment in which the MF was replaced with distilled water and the sensor was rotated over several full rotations. Figure 10a shows the transmission spectra for several rotation angles. Although the spectrum fluctuates slightly, no clear shift in the transmission dip appears. The corresponding resonance wavelengths are presented in Figure 10b. The largest shift in the resonance wavelength induced by rotation is only ~2 nm, which is much less than what occurs when the SPR sensor is immersed in the MF, as shown in Figures 7 and 8. Therefore, we conclude that, in these experiments, the shift in resonance wavelength is induced by the magnetic-field intensity and the MF's orientation-dependent optical properties instead of by any twisting of the fiber that may be induced by rotating the sensor.

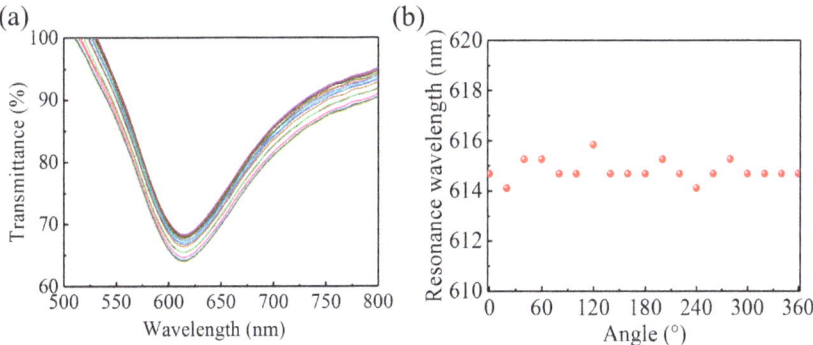

Figure 10. (a) Transmission spectra for different orientations and with no magnetic field, and (b) resonance wavelength as a function of rotation angle.

4.5. Discussion

We have demonstrated a highly sensitive device to detect magnetic-field vectors. The device is based on a side-polished-FMF SPR immersed in an MF. By exploiting the intrinsic high sensitivity of the SPR to the RI and the strong magneto-optical effect in the MF, the proposed sensor is highly sensitive to magnetic-field intensity. Additionally, the nonsymmetrical configuration of the side-polished fiber and the anisotropic aggregation of MNPs around the fiber make the sensor highly-sensitive to the magnetic field's relative orientation. At a fixed magnetic-field intensity, the RI of the MF in the vicinity of the side-polished surface is maximized when the magnetic field is parallel to the surface. Moreover, in the parallel orientation, the RI of the MF increases with increasing magnetic-field intensity, causing the resonance wavelength to redshift. On the contrary, in the perpendicular orientation, the resonance wavelength blueshifts with increasing of magnetic-field intensity because the RI of the MF is smaller in this orientation.

The measurement resolution is calculated using $R = \sigma/S$ [45], where σ is the standard deviation of the output noise, measured multiple times under the same conditions, and S is the sensor's sensitivity. In the present work, $\sigma = 0.156$ nm, which is determined mainly by the spectrometer's performance. As a result, the sensor resolution is calculated to be 0.225 Oe and 0.013° for magnetic-field intensity and orientation, respectively.

A common problem with such MF-based sensors is the temperature's effect on their performance, which is due to the thermo-optical effect (about -10^{-4} RIU/°C depending on the concentration of MNPs) [41]. In our case, the RI sensitivity of the SPR sensor was determined to be 3693 nm/RIU for a RI ~1.385 (i.e., the RI of the MF under zero magnetic field). Therefore, we calculate the sensor's temperature sensitivity to be -0.369 nm/°C. Then, considering the measured sensitivities of 0.692 nm/Oe (sensitivity to magnetic-field intensity) and -11.917 nm/° (sensitivity to magnetic-field orientation), the error induced by the temperature fluctuation will be 0.533 Oe/°C and 0.031°/°C for magnetic-field intensity and orientation, respectively. The temperature sensitivity is less than the magnetic-field sensitivity, indicating that temperature is a minor factor in determining the proposed sensor's performance. Nevertheless, a method to eliminate or alleviate the temperature effect is desired and will be the subject of a future presentation.

5. Conclusions

In summary, we proposed and investigated an optical-fiber-based magnetic-vector sensor that offers high sensitivity to both the magnetic-field intensity and orientation. The proposed sensor consists of an SPR sensor based on a side-polished FMF functionalized by immersion in an MF. Experiments show that the maximal sensitivities of the proposed sensor can reach as high as 0.692 nm/Oe (to magnetic-field-intensity) and -11.917 nm/° (to magnetic-field-orientation). This high sensitivity, compact size, and online detection scheme give the magnetic-vector sensor significant potential for applications involving the detection of weak magnetic-field vectors.

Author Contributions: Conceptualization, Y.C. and Y.L.; methodology, W.S. and Y.C.; formal analysis, Y.Z., G.L. and Y.L.; investigation, Y.Z.; data curation, W.S. and Y.Z.; writing—original draft preparation, Y.C.; writing—review and editing Y.L.; supervision, J.D., Y.Z., W.Z., J.Y. and Z.C.; project administration, Y.C.; funding acquisition, Y.L.

Funding: This research was funded by National Natural Science Foundation of China (NSFC) (61575084, 61805108, 61705087); Special Research Fund for Central Universities (21618404, 21617332); Science and Technology Projects of Guangdong Province (2017A010101013, 2017A030313359); Science & Technology Project of Guangzhou (201707010500, 201807010077, 201704030105, 201605030002); Joint Fund of Pre-research for Equipment, Ministry of Education of China (6141A02022124).

Conflicts of Interest: The authors declare no conflict of interest.

References

1. Zhao, Y.; Liu, X.; Lv, R.-Q.; Zhang, Y.-N.; Wang, Q. Review on Optical Fiber Sensors Based on the Refractive Index Tunability of Ferrofluid. *J. Lightwave Technol.* **2017**, *35*, 3406–3412. [CrossRef]

2. Clark, N.A. Soft-matter physics: Ferromagnetic ferrofluids. *Nature* **2013**, *504*, 229. [CrossRef]
3. Yusuf, N.A. Field and concentration dependence of chain formation in magnetic fluids. *J. Phys. D Appl. Phys.* **1989**, *22*, 1916. [CrossRef]
4. Torres-Díaz, I.; Rinaldi, C. Recent progress in ferrofluids research: Novel applications of magnetically controllable and tunable fluids. *Soft Matter* **2014**, *10*, 8584. [CrossRef] [PubMed]
5. Sun, B.; Fang, F.; Zhang, Z.; Xu, J.; Zhang, L. High-sensitivity and low-temperature magnetic field sensor based on tapered two-mode fiber interference. *Opt. Lett.* **2018**, *43*, 1311–1314. [CrossRef]
6. Zu, P.; Chan, C.C.; Lew, W.S.; Jin, Y.; Zhang, Y.; Liew, H.F.; Chen, L.H.; Wong, W.C.; Dong, X. Magneto-optical fiber sensor based on magnetic fluid. *Opt. Lett.* **2012**, *37*, 398–400. [CrossRef] [PubMed]
7. Chen, Y.; Han, Q.; Liu, T.; Lan, X.; Xiao, H. Optical fiber magnetic field sensor based on single-mode-multimode-single-mode structure and magnetic fluid. *Opt. Lett.* **2013**, *38*, 3999–4001. [CrossRef]
8. Mahmood, A.; Kavungal, V.; Ahmed, S.S.; Farrell, G.; Semenova, Y. Magnetic-field sensor based on whispering-gallery modes in a photonic crystal fiber infiltrated with magnetic fluid. *Opt. Lett.* **2015**, *40*, 4983–4986. [CrossRef]
9. Yu, Z.; Jiang, J.; Zhang, X.; Liu, K.; Wang, S.; Chen, W.; Liu, T. Fiber Optic Magnetic Field Sensor Based On Magnetic Nanoparticle Assembly In microcapillary ring resonator. *IEEE Photonics J.* **2017**, *9*, 1–9. [CrossRef]
10. Liu, T.; Chen, Y.; Han, Q.; Lu, X. Magnetic Field Sensor Based on U-Bent Single-Mode Fiber and Magnetic Fluid. *IEEE Photonics J.* **2014**, *6*, 1–7. [CrossRef]
11. Chen, Y.; Liu, T.; Han, Q.; Yan, W.; Yu, L. Fiber loop ring-down cavity integrated U-bent single-mode-fiber for magnetic field sensing. *Photonics Res.* **2016**, *4*, 322–326. [CrossRef]
12. Luo, L.; Pu, S.; Tang, J.; Zeng, X.; Lahoubi, M. Reflective all-fiber magnetic field sensor based on microfiber and magnetic fluid. *Opt. Express* **2015**, *23*, 18133–18142. [CrossRef] [PubMed]
13. Liu, H.; Zhang, H.; Liu, B.; Song, B.; Wu, J.; Lin, L. Ultra-sensitive magnetic field sensor with resolved temperature cross-sensitivity employing microfiber-assisted modal interferometer integrated with magnetic fluids. *Appl. Phys. Lett.* **2016**, *109*, 042402. [CrossRef]
14. Li, Z.; Liao, C.; Song, J.; Wang, Y.; Zhu, F.; Wang, Y.; Dong, X. Ultrasensitive magnetic field sensor based on an in-fiber Mach–Zehnder interferometer with a magnetic fluid component. *Photonics Res.* **2016**, *4*, 197–201. [CrossRef]
15. Liang, H.; Liu, Y.; Li, H.; Han, S.; Zhang, H.; Wu, Y.; Wang, Z. Magnetic-ionic-liquid-functionalized photonic crystal fiber for magnetic field detection. *IEEE Photonics Technol. Lett.* **2018**, *30*, 359–362. [CrossRef]
16. Gao, R.; Jiang, Y.; Abdelaziz, S. All-fiber magnetic field sensors based on magnetic fluid-filled photonic crystal fibers. *Opt. Lett.* **2013**, *38*, 1539–1541. [CrossRef]
17. Chen, Y.; Han, Q.; Liu, T.; Yan, W.; Yao, Y. Magnetic Field Sensor Based on Ferrofluid and Photonic Crystal Fiber With Offset Fusion Splicing. *IEEE Photonics Technol. Lett.* **2016**, *28*, 2043–2046. [CrossRef]
18. Deng, M.; Liu, D.; Li, D. Magnetic Field Sensor based on Asymmetric Optical Fiber Taper and Magnetic Fluid. *Sens. Actuators A* **2014**. [CrossRef]
19. Ma, R.; Kong, R.; Xia, Y.; Li, X.; Wen, X.; Pan, Y.; Dong, X. Microfiber polarization modulation in response to protein induced self-assembly of functionalized magnetic nanoparticles. *Appl. Phys. Lett.* **2018**, *113*, 033702. [CrossRef]
20. Ma, R.; Li, X.; Dong, X.; Xia, Y.; Ma, R.; Li, X.; Dong, X.; Xia, Y.; Ma, R.; Li, X. Magnetic field modulating in-line fiber polarization modulator based on microfiber and magnetic fluid. *Appl. Phys. Lett.* **2017**, *111*, 093503. [CrossRef]
21. Zhang, Z.; Guo, T.; Zhang, X.; Xu, J.; Xie, W.; Nie, M.; Wu, Q.; Guan, B.O.; Albert, J. Plasmonic fiber-optic vector magnetometer. *Appl. Phys. Lett.* **2016**, *108*, 289. [CrossRef]
22. Yin, J.; Ruan, S.; Liu, T.; Jiang, J.; Wang, S.; Wei, H.; Yan, P. All-fiber-optic vector magnetometer based on nano-magnetic fluids filled double-clad photonic crystal fiber. *Sens. Actuators B* **2017**, *238*, 518–524. [CrossRef]
23. Yin, J.; Yan, P.; Chen, H.; Yu, L.; Jiang, J.; Zhang, M.; Ruan, S. All-fiber-optic vector magnetometer based on anisotropic magnetism-manipulation of ferromagnetism nanoparticles. *Appl. Phys. Lett.* **2017**, *110*, 231104. [CrossRef]
24. Layeghi, A.; Latifi, H. Magnetic field vector sensor by a nonadiabatic tapered Hi-Bi fiber and ferrofluid nanoparticles. *Opt. Laser Technol.* **2018**, *102*, 184–190. [CrossRef]

25. Cui, J.; Qi, D.; Tian, H.; Li, H. Vector optical fiber magnetometer based on capillaries filled with magnetic fluid. *Appl. Opt.* **2019**, *58*, 2754–2760. [CrossRef]
26. Lu, T.; Sun, Y.; Moreno, Y.; Sun, Q.; Zhou, K.; Wang, H.; Yan, Z.; Liu, D.; Zhang, L. Excessively tilted fiber grating-based vector magnetometer. *Opt. Lett.* **2019**, *44*, 2494–2497. [CrossRef] [PubMed]
27. Violakis, G.; Korakas, N.; Pissadakis, S. Differential loss magnetic field sensor using a ferrofluid encapsulated D-shaped optical fiber. *Opt. Lett.* **2018**, *43*, 142–145. [CrossRef]
28. Ying, Y.; Zhao, Y.; Lv, R.-Q.; Hu, H.-F. Magnetic field measurement using surface plasmon resonance sensing technology combined with magnetic fluid photonic crystal. *IEEE Trans. Instrum. Meas.* **2016**, *65*, 170–176. [CrossRef]
29. Weng, S.; Pei, L.; Wang, J.; Ning, T.; Li, J. High sensitivity side-hole fiber magnetic field sensor based on surface plasmon resonance. *Chin. Opt. Lett.* **2016**, *14*, 19–22.
30. Liu, H.; Li, H.; Wang, Q.; Wang, M.; Ding, Y.; Zhu, C.; Cheng, D. Temperature-compensated Magnetic Field sensor based on Surface Plasmon Resonance and directional Resonance coupling in a D-shaped photonic crystal fiber. *Optik* **2018**, *158*, 1402–1409. [CrossRef]
31. Rodríguez-Schwendtner, E.; Díaz-Herrera, N.; Navarrete, M.; González-Cano, A.; Esteban, Ó. Plasmonic sensor based on tapered optical fibers and magnetic fluids for measuring magnetic fields. *Sens. Actuators A* **2017**, *264*, 58–62. [CrossRef]
32. Zhou, X.; Li, X.; Li, S.; An, G.-W.; Cheng, T. Magnetic Field Sensing Based on SPR Optical Fiber Sensor Interacting With Magnetic Fluid. *IEEE Trans. Instrum. Meas.* **2018**, *68*, 234–239. [CrossRef]
33. Jiang, Z.; Dong, J.; Hu, S.; Zhang, Y.; Chen, Y.; Luo, Y.; Zhu, W.; Qiu, W.; Lu, H.; Guan, H. High-sensitivity vector magnetic field sensor based on side-polished fiber plasmon and ferrofluid. *Opt. Lett.* **2018**, *43*, 4743–4746. [CrossRef] [PubMed]
34. Wang, Y.; Dong, J.; Luo, Y.; Tang, J.; Lu, H.; Yu, J.; Guan, H.; Zhang, J.; Chen, Z. Indium Tin Oxide Coated Two-Mode Fiber for Enhanced SPR Sensor in Near-Infrared Region. *IEEE Photonics J.* **2017**, *9*, 1–9. [CrossRef]
35. Jang, H.S.; Park, K.N.; Kang, C.D.; Kim, J.P.; Sim, S.J.; Lee, K.S. Optical fiber SPR biosensor with sandwich assay for the detection of prostate specific antigen. *Opt. Commun.* **2009**, *282*, 2827–2830. [CrossRef]
36. Yang, S.; Chieh, J.; Horng, H.; Hong, C.-Y.; Yang, H. Origin and applications of magnetically tunable refractive index of magnetic fluid films. *Appl. Phys. Lett.* **2004**, *84*, 5204–5206. [CrossRef]
37. Tagoudi, E.; Milenko, K.; Pissadakis, S. Intercore Coupling Effects in Multicore Optical Fiber Tapers Using Magnetic Fluid Out-Claddings. *J. Lightwave Technol.* **2016**, *34*, 5561–5565. [CrossRef]
38. Liu, D.; Mallik, A.K.; Yuan, J.; Yu, C.; Farrell, G.; Semenova, Y.; Wu, Q. High sensitivity refractive index sensor based on a tapered small core single-mode fiber structure. *Opt. Lett.* **2015**, *40*, 4166–4169. [CrossRef]
39. Shen, F.; Wang, C.; Sun, Z.; Zhou, K.; Zhang, L.; Shu, X. Small-period long-period fiber grating with improved refractive index sensitivity and dual-parameter sensing ability. *Opt. Lett.* **2017**, *42*, 199–202. [CrossRef] [PubMed]
40. Tian, J.; Lu, Y.; Zhang, Q.; Han, M. Microfluidic refractive index sensor based on an all-silica in-line Fabry–Perot interferometer fabricated with microstructured fibers. *Opt. Express* **2013**, *21*, 6633–6639. [CrossRef]
41. Chen, Y.; Yang, S.; Tse, W.; Horng, H.; Hong, C.-Y.; Yang, H. Thermal effect on the field-dependent refractive index of the magnetic fluid film. *Appl. Phys. Lett.* **2003**, *82*, 3481–3483. [CrossRef]
42. Ren, K.; Ren, X.; He, Y.; Han, Q. Magnetic-field sensor with self-reference characteristic based on a magnetic fluid and independent plasmonic dual resonances. *Beilstein J. Nanotechnol.* **2019**, *10*, 247–255. [CrossRef]
43. Galtarossa, A.; Palmieri, L. Measure of Twist-Induced Circular Birefringence in Long Single-Mode Fibers: Theory and Experiments. *J. Lightwave Technol.* **2002**, *20*, 1149–1159. [CrossRef]
44. El-Khozondar, H.J.; Muller, M.S.; Buck, T.C.; El-Khozondar, R.J. Experimental investigation of polarization rotation in twisted optical fibers. In Proceedings of the International Symposium on Optomechatronic Technologies, Istanbul, Turkey, 21–23 September 2009; pp. 219–222.
45. Piliarik, M.; Homola, J. Surface plasmon resonance (SPR) sensors: Approaching their limits? *Opt. Express* **2009**, *17*, 16505–16517. [CrossRef]

© 2019 by the authors. Licensee MDPI, Basel, Switzerland. This article is an open access article distributed under the terms and conditions of the Creative Commons Attribution (CC BY) license (http://creativecommons.org/licenses/by/4.0/).

www.ingramcontent.com/pod-product-compliance
Lightning Source LLC
LaVergne TN
LVHW070508100526
838202LV00014B/1815

MDPI
St. Alban-Anlage 66
4052 Basel
Switzerland
Tel. +41 61 683 77 34
Fax +41 61 302 89 18
www.mdpi.com

Nanomaterials Editorial Office
E-mail: nanomaterials@mdpi.com
www.mdpi.com/journal/nanomaterials